Five Millennium Canon of Lunar Eclipses

Volume 1: –1999 to 0

2nd Edition

Fred Espenak and Jean Meeus

August, 2021

Five Millennium Canon of Lunar Eclipses, Volume 1: -1999 to 0

Second Edition

Five Millennium Canon of Lunar Eclipses, Volume 1: -1999 to 0 (2nd Edition) is published through Astropixels Publishing. For more information, contact the publisher:

Astropixels Publishing
P.O. Box 16197
Portal, AZ 85632

Astropixels Publishing Website: *astropixels.com/pubs*

This book may be ordered at: *astropixels.com/pubs/5MCLE2.html*

Printed in the United States of America

ISBN 978-1-941983-41-6

Astropixels Publication: AP039 (Version 1.0a)

Front Cover: Woodcut illustrating the geometry of an eclipse of the moon from *Sphaericum opusculum* (1485) by Johannes de Sacro Bosco, printed by Erhard Ratdolt.

Back Cover: Photo of Fred Espenak is Copyright © 1999 by Patricia Espenak

Table of Contents

Preface – 1st Edition

Theodor von Oppolzer's 1887 *Canon der Finsternisse* (Canon of Eclipses) stands as one of the greatest accomplishments in computational astronomy of the nineteenth century. It contains the elements of all 8000 solar eclipses and 5200 lunar eclipses occurring between the years –1207 and +2161, together with maps showing the approximate positions of the central lines of total, annular and hybrid solar eclipses.

To make this remarkable achievement possible, a number of approximations were used in the calculations and maps. Furthermore, the 19th century ephemerides for the Sun and Moon, which are critical to eclipse predictions, cannot compare to the accuracy and precision of modern ephemerides. Finally, the 1887 canon did not take into account the shifts in latitude and timing of ancient eclipses due to Earth's variable rotation rate and the secular acceleration of the Moon.

Nevertheless, von Oppolzer's canon remained the seminal reference on predictions of eclipses until well into the 20th century. With the arrival of the electronic computer, the time was ripe to produce updated solar[1] and lunar eclipse canons. In 1979, Meeus and Mucke published *Canon of Lunar Eclipses: –2002 to +2526* containing the Besselian elements of 10,936 lunar eclipses. It was intended mainly for historical research and served as the modern day successor of von Oppolzer's great canon.

The Meeus-Mucke work also contains data on penumbral eclipses that are not included in von Oppolzer's canon. Neither of these publications offers diagrams or maps to illustrate the geometry or visibility of each eclipse. Espenak's *Fifty Year Canon of Lunar Eclipses* (1989) includes individual Moon-shadow geometry diagrams and eclipse visibility maps of all lunar eclipses, but it covers a relatively short period from +1986 to +2035.

Both of these recent lunar eclipse canons are based on *Newcomb's Tables of the Sun* (1895) and Brown's lunar theory (1905), subject to later modifications in the *Improved Lunar Ephemeris* (1954). These were the best ephemerides of their day but they have since been superseded.

The *Canon of Lunar Eclipses 1500 B.C.–A.D. 3000* (Liu and Fiala, 1992) uses modern theories of the Sun and the Moon prepared by the *Bureau des Longitudes* of Paris. However, it does not contain individual eclipse geometry diagrams or maps. Instead, it offers a series of map templates to approximate the geographic regions of eclipse visibility, and an optional computer program do generate these figures for any eclipse in the canon.

The present publication is the first to offer eclipse geometry diagrams and visibility maps for every lunar eclipse (12,064 eclipses) over a period covering five thousand years from –1999 to +3000. The following points highlight the features and characteristics of this work.

- based on modern theories of the Sun and the Moon constructed at the *Bureau des Longitudes* of Paris rather than the older Newcomb and Brown ephemerides
- ephemerides and eclipse predictions performed in Terrestrial Dynamical Time
- covers historical period of eclipses as well as one millennium into the future
- diagrams for each eclipse depict the Moon's path through Earth's penumbral and umbral shadows
- world maps identify geographic regions of visibility for each phase of every eclipse
- visibility maps are based of the most current determination of the historical values of ΔT
- estimates of eclipse visibility map accuracy based on the uncertainty in the value of ΔT (i.e., standard error in ΔT)

---

[1] Several new solar eclipse canons were published in the second half of the 20th century. Meeus, Grosjean, and Vanderleen published *Canon of Solar Eclipses* (1966) containing the Besselian elements of all solar eclipses from +1898 to +2510, together with central line tables and maps. The aim of this work was principally to provide data on future eclipses. The Mucke and Meeus work *Canon of Solar Eclipses: –2003 to +2526* (1983) was intended mainly for historical research and served as a modern day successor of von Oppolzer's great canon. Espenak's *Fifty Year Canon of Solar Eclipses* (1987) includes individual detailed maps and central path data for all solar eclipses from +1986 to +2035. Finally, the Espenak and Meeus work *Five Millennium Canon of Solar Eclipses: –1999 to +3000* (2006) contains individual maps of every solar eclipse and uses modern ephemerides of the Sun and Moon.

A primary goal of this work is to assist historians and archeologists in the identification and dating of eclipses found in references and records from antiquity. For example, an ancient mechanical calculator known as the Antikythera mechanism was apparently designed to calculate eclipses and other astronomical phenomena (Freeth, et. al., 2008). The decoding of this device was possible in part by comparing its combination of wheel positions with the dates of lunar eclipses.

Correlating historical records with specific eclipses is no easy task since there are usually several possible candidates. Accurate visibility maps using the best available values of ΔT coupled with estimates in the standard error of ΔT are critical in discriminating among potential eclipses candidates. Ultimately, historical eclipse identification can lead to improved chronologies in the time line of a particular culture.

The maps can also be used to quickly estimate the approximate circumstances for any geographic location during each eclipse without any calculations. The position of the moonrise and moonset curves for each eclipse contact show which eclipse phases can be seen from any location.

The *Canon* will also be of value to educators, planetariums, and anyone interested in knowing when and where past and future eclipses occur. The general public is fascinated by eclipses—with each major eclipse, the question always arises as to when a particular location experienced its last and next eclipses. The maps presented here are ideally suited to addressing such queries.

— Fred Espenak and Jean Meeus, 2009 January

Preface – 2nd Edition

The *Five Millennium Catalog of Lunar Eclipses: -1999 to +3000* was first published in October 2009 (NASA TP-2009-214173). As a NASA Technical Publication it had a limited publication run and distribution. The available supply of hard copies was depleted within 12 months of publication although the PDF version continues to be available.

A decade later, the *Five Millennium Canon of Lunar Eclipses* is available again in hard copy in this Second Edition through Astropixels Publishing. Because of the large number of pages in the publication it has been necessary to create a completely new layout for the book, and to divide it into two parts.

Volume 1 covers eclipses for the years -1999 to 0, while volume 2 covers eclipses for the years 1 to 3000. A new section (Section 4: Eclipses and the Moon's Orbit) has also been added to this second edition.

— Fred Espenak and Jean Meeus, 2021 August

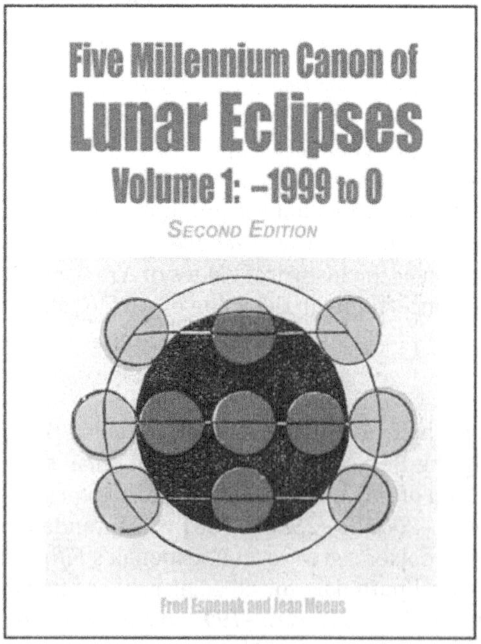

See: *http://astropixels.com/pubs/5MCLE2.html*

Section 1: Eclipse Figures, Maps, and Predictions

1.1 Introduction

Earth will experience 12,064 eclipses of the Moon during the 5000-year period from –1999 to +3000 (2000 BCE to 3000 CE[2]). An individual diagram and visibility map for every lunar eclipse over the five-millennium interval is presented in the Appendix. The Moon's path through Earth's penumbral and umbral shadows illustrates the eclipse geometry, and the accompanying equidistant cylindrical projection[3] map shows the geographic region of visibility during every phase of each eclipse.

1.2 Explanation of Lunar Eclipse Figures

The figure for each eclipse consists of two diagrams. The top one depicts the Moon's path through Earth's penumbral and umbral shadows with Celestial North directed up. The Moon's orbital motion is from west to east (right to left)[4] with respect to the shadows. Each phase of the eclipse is defined by the instant when the Moon's limb is externally or internally tangent to the penumbra or umbra as follows.

Penumbral lunar eclipses have two primary contacts. Neither of these events is observable.

P_1 — Instant of first exterior tangency of the Moon with the Penumbra. (Penumbral Eclipse Begins)
P_4 — Instant of last exterior tangency of the Moon with the Penumbra. (Penumbral Eclipse Ends)

Partial lunar eclipses have two additional contacts as the Moon's limb enters and exits the umbral shadow (U_1 and U_4, respectively). At these two instants, the partial phase of the eclipse begins and ends.

P_1 — Instant of first exterior tangency of the Moon with the Penumbra. (Penumbral Eclipse Begins)
U_1 — Instant of first exterior tangency of the Moon with the Umbra. (Partial Umbral Eclipse Begins)
U_4 — Instant of last exterior tangency of the Moon with the Umbra. (Partial Umbral Eclipse Ends)
P_4 — Instant of last exterior tangency of the Moon with the Penumbra. (Penumbral Eclipse Ends)

Total lunar eclipses have two additional umbral contacts at the instants when the Moon's entire disk is first and last internally tangent to the umbra (U_2 and U_3, respectively). These are the times when the total phase of the eclipse begins and ends.

P_1 — Instant of first exterior tangency of the Moon with the Penumbra. (Penumbral Eclipse Begins)
U_1 — Instant of first exterior tangency of the Moon with the Umbra. (Partial Umbral Eclipse Begins)
U_2 — Instant of first interior tangency of the Moon with the Umbra. (Total Umbral Eclipse Begins)
U_3 — Instant of last interior tangency of the Moon with the Umbra. (Total Umbral Eclipse Ends)
U_4 — Instant of last exterior tangency of the Moon with the Umbra. (Partial Umbral Eclipse Ends)
P_4 — Instant of last exterior tangency of the Moon with the Penumbra. (Penumbral Eclipse Ends)

The instant when the Moon passes closest to the shadow axis is known as the instant of greatest eclipse. This corresponds to the maximum phase of the eclipse, and the Moon's position at this instant is also shown in the eclipse path diagrams.

The bottom diagram in each figure is an equidistant cylindrical projection map of Earth showing the geographic region of visibility at each phase of the eclipse. This is accomplished using a series of curves showing where Moonrise and

---

[2] The terms BCE and CE are abbreviations for "Before the Common Era" and "Common Era," respectively. They are the secular equivalents to the BC and AD dating conventions. A major advantage of the BCE/CE convention is that both terms are suffixes, whereas BC and AD are used as a suffix and prefix, respectively.

[3] The equidistant cylindrical projection (also known as the equirectangular projection) is a simple x-y Cartesian map projection where lines of latitude and longitude are represented by straight, equally spaced horizontal and vertical lines.

[4] Maps are plotted north up, west to the left, and east to the right. This orientation assumes the viewer is looking down on Earth. But if the viewer turns around and looks up at the sky, the directions are reversed with east to the left and west to the right.

Moonset occur at each eclipse contact. The map is also shaded to indicate eclipse visibility. The entire eclipse is visible from the zone with no shading. Conversely, none of the eclipse can be seen from the zone with the darkest shading.

At the instant of greatest eclipse, the Moon is deepest in Earth's shadow. A vertical line running through the middle of the clear zone (complete eclipse visibility) of a map indicates the meridian, or line of longitude, that the Moon is then crossing. An observer positioned on this line would then see the Moon at its highest point in the sky either due south or due north, depending on the observer's latitude and the Moon's declination. The geographic location where the Moon appears in the zenith at greatest eclipse is shown by a black dot on the meridian of greatest eclipse. All salient features of the eclipse figures are identified in Figure 1-1, which serves as the key to the figures in the Appendix.

Figure 1-1. Key to Eclipse Figures

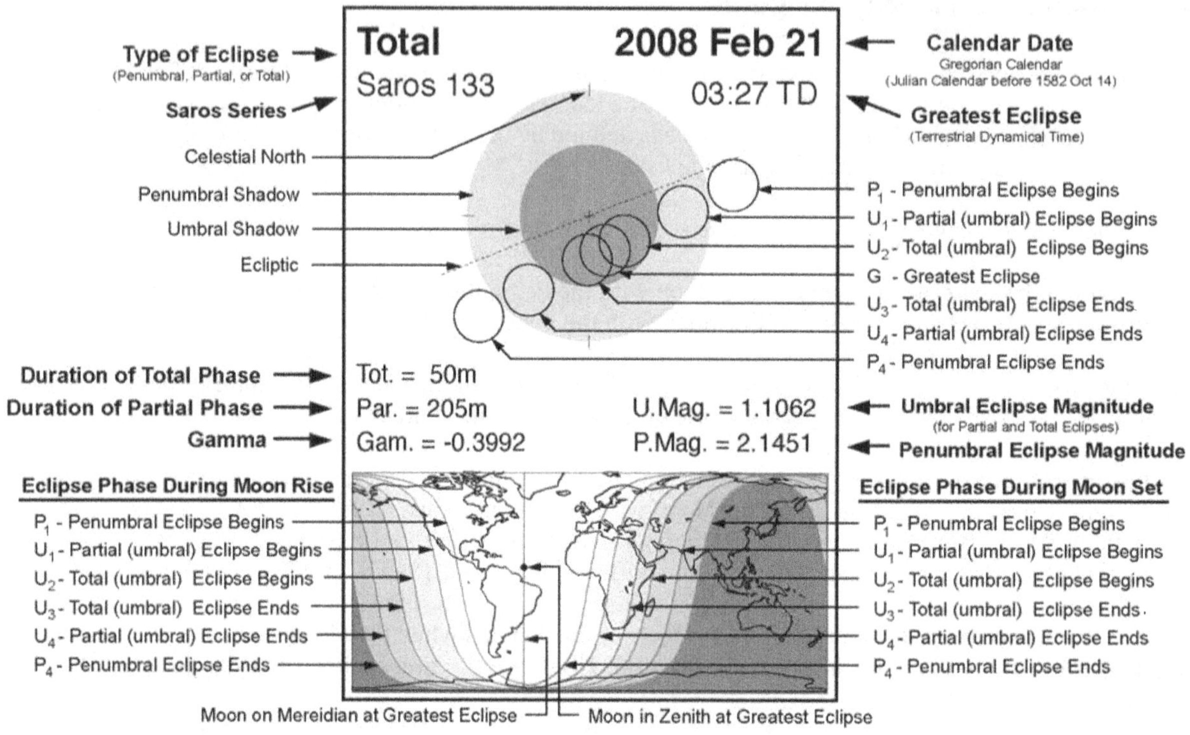

Data relevant to a lunar eclipse appear in the corners of each figure. To the top left, are the eclipse type (penumbral, partial, or total) and the Saros series of the eclipse. To the top right are the Gregorian calendar date (Julian calendar dates are used prior to 1582 Oct 14) and the time of greatest eclipse (Terrestrial Dynamical Time). The lower left corner lists the duration of the major phases of the eclipse in minutes. Depending on the eclipse type, the duration of the penumbral ($P_4 - P_1$), partial ($U_4 - U_1$), or total ($U_3 - U_2$) phases are given. Beneath the eclipse durations is the quantity gamma—the minimum distance of the Moon's center from the axis of Earth's penumbral and umbral shadow cones (in Earth equatorial radii) at the instant of greatest eclipse. The umbral and penumbral eclipse magnitudes[5] are given to the lower right.

A detailed explanation of the data in the lunar eclipse figures appears in the following sections.

[5] The eclipse magnitude is defined as the fraction of the Moon's diameter immersed in either Earth's penumbral or umbral shadows. It is a unit less parameter given at the instant of greatest eclipse.

1.2.1 Lunar Eclipse Type

There are three basic types of lunar eclipses:

1) **Penumbral** — Moon traverses through Earth's penumbral shadow (Moon is partly or completely within the penumbra)
2) **Partial** — Part of the Moon traverses through Earth's umbral shadow (Moon is partly within the umbra)
3) **Total** — Entire Moon traverses through Earth's umbral shadow (Moon is completely within the umbra)

It is also possible for a penumbral eclipse to occur in which the Moon's entire disk is enveloped within the penumbra. Such events are called total penumbral eclipses. They are indicated by "(T)" after the penumbral eclipse type [i.e., Penumbral (T)].

With regard to total (umbral) lunar eclipses, most are non-central in that the Moon's disk misses the central axis of the umbral shadow cone. If a total eclipse is central, it is indicated by a "+" or "–" after the eclipse type, depending on whether the Moon's center passes north or south of the shadow axis (e.g., Total +).

1.2.2 Saros Series Number

Each lunar eclipse belongs to a Saros series (Sect. 5.3) using a numbering system first introduced by van den Bergh (1955). This system has been expanded to include negative values from the past, as well as additional series in the future. The eclipses with an odd Saros number take place at the descending node of the Moon's orbit, while those with an even Saros number take place at the ascending node.

The Saros is a period of 223 synodic months (approximately 18 years, 11 days, and 8 hours). Eclipses separated by this interval belong to the same Saros series and share similar geometry and characteristics.

1.2.3 Calendar Date

All eclipse dates from 1582 Oct 15 onwards use the modern Gregorian calendar currently found throughout most of the world. The older Julian calendar is used for eclipse dates prior to 1582 Oct 04. Because of the Gregorian Calendar Reform, the day following 1582 Oct 04 (Julian calendar) is 1582 Oct 15 (Gregorian calendar).

Pope Gregory XIII decreed the use of the Gregorian calendar in 1582 in order to correct a problem in a drift of the seasons. It adopts the convention of a year containing 365 days. Every fourth year is a leap year of 366 days if it is divisible by 4 (e.g., 2004, 2008, etc.). However, whole century years (e.g., 1700, 1800, 1900) are excluded from the leap year rule unless they are also divisible by 400 (e.g., 2000). This dating scheme was designed to keep the vernal equinox on, or within a day of, March 21.

Prior to the Gregorian Calendar Reform of 1582, the Julian calendar was in wide use. It was less complicated than the Gregorian calendar in that all years divisible by 4 were counted as 366-day leap years, but this simplicity came at a cost. After more than 16 centuries of use, the Julian calendar date of the vernal equinox had drifted 11 days from March 21. It was this failure in the Julian calendar that prompted the Gregorian Calendar Reform.

The Julian calendar does not include the year 0, so the year 1 BCE is followed by the year 1 CE. This is awkward for arithmetic dating calculations. The dates used here are counted using the astronomical numbering system, which recognizes the year 0. Historians should note the numerical difference of one year between astronomical dates and BCE dates. The astronomical year 0 corresponds to 1 BCE, and the astronomical year –100 equals 101 BCE, etc.

There are a number of ways to write the calendar date through variations in the order of day, month, and year. The International Organization for Standardization's (ISO) 8601 advises a numeric date representation, which organizes the elements from the largest to the smallest. The exact format is YYYY-MM-DD where YYYY is the calendar year, MM is the month of the year between 01 (January) and 12 (December), and DD is the day of the month between 01 and 31. For example, the 27th day of April in the year 1943 would then be expressed as 1943-04-27. The authors of this report support the ISO convention, but have replaced the month number with the three-letter English abbreviation of the month name for additional clarity. From the previous example, the date is expressed as 1943 Apr 27.

1.2.4 Greatest Eclipse

The instant of greatest eclipse occurs when the distance between the center of the Moon and the axis or Earth's umbral shadow cone reaches a minimum. Because of Earth's flattening, the instant of greatest eclipse differs slightly from the instant of greatest magnitude. In practice, Earth's atmosphere diffuses the edges of the penumbral and umbral shadows such that the difference between greatest eclipse and greatest magnitude cannot be distinguished observationally.

Lunar eclipses occur when the Moon is near one of the nodes of its orbit and, therefore, moving at an angle of about 5° to the ecliptic. Hence, unless the eclipse is perfectly central, the instant of greatest eclipse does not coincide with that of apparent ecliptic conjunction with Earth's shadow (i.e., Full Moon), nor with the time of conjunction in Right Ascension.

Greatest eclipse is given in Terrestrial Dynamical Time or TD (Sect. 2.3), which is a time system based on International Atomic Time. As such, TD is the atomic time equivalent to its predecessor Ephemeris Time (Sect. 2.2) and is used in the theories of motion for bodies in the solar system. To determine the geographic visibility of an eclipse, TD is converted to Universal Time (Sect. 2.4) using the parameter ΔT (Sects. 2.6 and 2.7).

1.2.5 Duration of Eclipse Phases

The duration of a penumbral eclipse "Pen." is the time between first and last external tangencies of the Moon with the penumbral shadow (i.e., $P_4 - P_1$). Similarly, the duration of a partial eclipse "Par." is the time between first and last external tangencies of the Moon with the umbral shadow (i.e., $U_4 - U_1$). Finally, the duration of a total eclipse "Tot." is the time between first and last internal tangencies of the Moon with the umbral shadow (i.e., $U_3 - U_2$).

The duration given for each eclipse phase is rounded to the nearest minute.

1.2.6 Gamma

The quantity gamma is the minimum distance from the center of the Moon to the axis of Earth's umbral shadow cone, in units of Earth's equatorial radius. This distance is positive or negative, depending on whether the Moon passes north or south of the shadow cone axis.

The change in the value of gamma after one Saros period is larger when Earth is near aphelion (June–July) than when it is near perihelion (December–January). Table 1-1 illustrates this point using eclipses from Saros series 129 (near aphelion— table on left) and Saros series 134 (near perihelion—table on right).

Table 1-1. Variation in Gamma at Aphelion vs. Perihelion

Date	Gamma	Δ Gamma	Date	Gamma	Δ Gamma
1928 Jun 03	-0.3175	-	1928 Nov 27	0.3952	-
1946 Jun 14	-0.2324	0.0851	1946 Dec 08	0.3864	-0.0088
1964 Jun 25	-0.1461	0.0863	1964 Dec 19	0.3801	-0.0063
1982 Jul 06	-0.0579	0.0882	1982 Dec 30	0.3758	-0.0043
2000 Jul 16	0.0302	0.0881	2001 Jan 09	0.3720	-0.0038
2018 Jul 27	0.1168	0.0866	2019 Jan 21	0.3684	-0.0036
2036 Aug 07	0.2004	0.0836	2037 Jan 31	0.3619	-0.0065
2054 Aug 18	0.2806	0.0802	2055 Feb 11	0.3526	-0.0093

A similar situation exists in the case of solar eclipses. The explanation can be found in van den Bergh (1955).

1.2.7 Eclipse Magnitude

The eclipse magnitude is defined as the fraction of the Moon's diameter immersed in Earth's shadows. Because there are two shadows—penumbral and umbral—there are two corresponding eclipse magnitudes. The eclipse magnitudes appearing in the figures are given at the instant of greatest eclipse when the Moon passes closest to the axis of the two shadow cones.

The penumbral eclipse magnitude of penumbral eclipses in this Canon of Lunar Eclipses (hereafter referred to as the Canon) ranges from 0.0004 to 1.0858 (Sect. 3.5). For most penumbral eclipses, the penumbral magnitude is less than 1 meaning only a fraction of the Moon's disk enters the penumbra. When the penumbral magnitude is greater than or equal to 1.0, the Moon's entire disk is immersed in the penumbra and the event is termed a total penumbral eclipse (Sect. 3.11). It should be noted that penumbral eclipses are subtle events (Sect. 1.7.1). In fact, they cannot be detected visually—with or without optical aid—unless the eclipse magnitude is greater than ~0.6. The umbral eclipse magnitude of a penumbral eclipse is always negative. It is a measure of the distance of the Moon's limb to the edge of the umbral shadow in units of the Moon's diameter.

During a partial lunar eclipse, some fraction of the Moon's disk enters the umbral shadow. The umbral magnitude for partial eclipses in the Canon ranges from 0.0001 to 0.9998 (Sect. 3.6). The Moon also passes through the penumbra during a partial eclipse so the penumbral magnitude is usually greater than 1.

In the case of a total lunar eclipse, the Moon's entire disk passes through Earth's umbral shadow. During totality, the Moon can take on a range of colors from bright orange, to deep red, dark brown, or even very dark grey (Sect. 1.7.2). The only light reaching the Moon at this time is heavily filtered and attenuated by Earth's atmosphere. The umbral magnitude for total eclipses in the Canon ranges from 1.0001 to 1.8821 (Sect. 3.7).

1.2.8 Additional Elements

Two additional parameters are listed at the bottom of each page of the Appendix. The first element is ΔT (Sect. 2.6). It is the arithmetic difference between TD (Sect. 2.3) and Universal Time (Sect. 2.4). The value given is specific to the first eclipse on the page. To the right of ΔT is its corresponding standard error. This is the estimated uncertainty in ΔT (Sect. 2.8) and is given both in seconds, and in the equivalent shift in longitude east or west.

1.3 Solar and Lunar Coordinates

The coordinates of the Moon used in the eclipse predictions have been calculated on the basis of the VSOP87 theory constructed by Bretagnon and Francou (1988) at the *Bureau des Longitudes*, Paris. This theory gives the ecliptic longitude and latitude of each planet, and its radius vector, as sums of periodic terms. In the calculations presented here, the complete set of periodic terms of version D[6] of VSOP87 were used.

For the Moon, use has been made of the theory ELP-2000/82 of Chapront-Touzé and Chapront (1983), again of the *Bureau des Longitudes*. This theory contains a total of 37,862 periodic terms, namely 20,560 for the Moon's longitude, 7,684 for the latitude, and 9,618 for the distance to Earth. However, many of these terms are very small: some have an amplitude of only 0.00001 arcsec for the longitude or the latitude, and of just 2 cm for the distance. In the computer program used for the Canon, all periodic terms were neglected with coefficients smaller than 0.0005 arcsec in longitude and latitude, and smaller than 1 m in distance. Because of neglecting the very small periodic terms, the Moon's positions calculated in the program have a mean error (as compared to the full ELP theory) of about 0.0006 seconds of time in right ascension, and about 0.006 arcsec in declination. The corresponding error in the calculated times of the contacts of a lunar eclipse is of the order of 1/40 second, which is much smaller than the uncertainties in predicted values of ΔT, and several orders of magnitude smaller than the uncertainty due to the diffuse edge of Earth's shadows caused by the terrestrial atmosphere.

Improved expressions for the mean arguments L', D, M, M' and F have been taken from Chapront, Chapront-Touzé, and Francou [2002]. A major consequence of this work is to bring the secular acceleration of the Moon's longitude (–

[6] Version D of VSOP87 gives solar (and planetary) positions referred to the mean equinox of the date.

25.858 "/cy^2, where arcsec/cy^2 is arc seconds per Julian century squared[7]) into good agreement with Lunar Laser Ranging (LLR) observations from 1972 to 2001 (Sect. 1.4).

A fundamental parameter used in eclipse predictions is the Moon's radius k expressed in units of Earth's equatorial radius. The Moon's actual radius varies as a function of position angle and libration because of irregularities in the surface topography of the Moon and the projection of these features along the line of sight onto the lunar limb profile. The current work uses a value for the lunar radius of k=0.2724880 representing a mean over the Moon's topographic features. This is the same value used for partial solar eclipses in the *Five Millennium Canon of Solar Eclipses: –1999 to +3000* (Espenak and Meeus, 2006).

The center of figure of the Moon does not coincide exactly with its center of mass. The magnitude of this difference is ~0.5 arcsec and is much smaller than the uncertainty in the edges of the penumbral and umbral shadows because of diffusion by Earth's atmosphere (Sect. 1.5). Thus, the difference between center-of-figure and center-of-mass positions has no practical impact on the present work.

1.4 Secular Acceleration of the Moon

Ocean tides are caused by the gravitational pull of the Moon (and, to a lesser extent, the Sun). The resulting tidal bulge in Earth's oceans is dragged ahead of the Moon in its orbit due to the daily rotation of Earth. As a consequence, the ocean mass offset from the Earth-Moon line exerts a pull on the Moon and accelerates it in its orbit. Conversely, the Moon's gravitational tug on this mass exerts a torque that decelerates the rotation of Earth. The length of the day gradually increases as energy is transferred from Earth to the Moon, causing the lunar orbit and period of revolution about Earth to increase.

This secular acceleration of the Moon is small but it has a cumulative effect on the Moon's position when extrapolated over many centuries. Direct measurements of the acceleration have only been possible since 1969 using the Apollo retro-reflectors left on the Moon. The results from Lunar Laser Ranging show that the Moon's mean distance from Earth is increasing by 3.8 cm per year (Dickey, et al., 1994). The corresponding acceleration in the Moon's ecliptic longitude is –25.858 arcsec/cy^2 (Chapront, Chapront-Touzé, and Francou, 2002). This is the value we have adopted in our lunar ephemeris calculations.

There is a close correlation between the Moon's secular acceleration and changes in the length of the day. However, the relationship is not exact because the lunar orbit is inclined anywhere from about 18.5° to 28.5° to Earth's equator. The parameter ΔT (sections 2.6 and 2.7) is a measure of the accumulated difference in time between an ideal clock and one based on Earth's rotation as it gradually slows down. Published determinations of ΔT from historical eclipse records have assumed a secular acceleration of –26 "/cy^2 (Morrison and Stephenson, 2004). Since we have adopted a slightly different value for the secular acceleration, we must make a small correction "c" to the published values of ΔT as follows:

$$c = -\,0.91072\,(\,-25.858 + 26.0\,)\,u^2, \qquad (1\text{-}1)$$

where: $u = (\text{year}-1955)/100$

Then:

$$\Delta T\,(\text{corrected}) = \Delta T + c \qquad (1\text{-}2)$$

Evaluation of the correction at 1,000 year intervals over the period spanned by the *Canon* is found in Table 1-2.

---

[7] This unit, arcsec/cy^2, is used in discussing secular changes in the Moon's longitude over long time intervals.

Table 1-2. Corrections to ΔT Due to Secular Acceleration

Year	Correction (seconds)
-2000	-202
-1000	-113
0	-49
+1000	-12
+2000	0
+3000	-14

Clearly the correction is only important for negative years although it is significantly smaller than the actual uncertainty in ΔT itself (Sect. 2.8).

The secular acceleration of the Moon is very poorly known and may not be constant. Careful records for its derivation only go back about a century. Before then, spurious and often incomplete eclipse and occultation observations from medieval and ancient manuscripts comprise the data base. In any case, the current value implies an increase in the length of day [LOD] of about 2.3 milliseconds per century. Such a small amount may seem insignificant, but it has very measurable cumulative effects. At this rate, time as measured through Earth's rotation is losing about 84 seconds per century2 when compared to atomic time.

1.5 Lunar Eclipse Contacts

The major stages of a lunar eclipse are defined by the instants when the Moon's limb is externally or internally tangent to the penumbral or umbral shadows.

Penumbral lunar eclipses have two primary contacts. Neither of these events is observable.

> **P_1** — Instant of first exterior tangency of the Moon with the Penumbra. (Penumbral Eclipse Begins)
> **P_4** — Instant of last exterior tangency of the Moon with the Penumbra. (Penumbral Eclipse Ends)

Partial lunar eclipses have two additional contacts as the Moon's limb enters and exits the umbral shadow (U_1 and U_4, respectively). At these two instants, the partial phase of the eclipse begins and ends.

> **P_1** — Instant of first exterior tangency of the Moon with the Penumbra. (Penumbral Eclipse Begins)
> **U_1** — Instant of first exterior tangency of the Moon with the Umbra. (Partial Umbral Eclipse Begins)
> **U_4** — Instant of last exterior tangency of the Moon with the Umbra. (Partial Umbral Eclipse Ends)
> **P_4** — Instant of last exterior tangency of the Moon with the Penumbra. (Penumbral Eclipse Ends)

Total lunar eclipses have two additional umbral contacts at the instants when the Moon's entire disk is first and last internally tangent to the umbra (U_2 and U_3, respectively). These are the times when the total phase of the eclipse begins and ends, i.e., totality.

> **P_1** — Instant of first exterior tangency of the Moon with the Penumbra. (Penumbral Eclipse Begins)
> **U_1** — Instant of first exterior tangency of the Moon with the Umbra. (Partial Umbral Eclipse Begins)
> **U_2** — Instant of first interior tangency of the Moon with the Umbra. (Total Umbral Eclipse Begins)
> **U_3** — Instant of last interior tangency of the Moon with the Umbra. (Total Umbral Eclipse Ends)
> **U_4** — Instant of last exterior tangency of the Moon with the Umbra. (Partial Umbral Eclipse Ends)
> **P_4** — Instant of last exterior tangency of the Moon with the Penumbra. (Penumbral Eclipse Ends)

The instant when the Moon passes closest to the shadow axis is known as the instant of greatest eclipse. This corresponds to the maximum phase of the eclipse.

1.6 Enlargement of Earth's Shadows

In the early 1700s, Philippe de La Hire made a curious observation about Earth's umbra. The predicted radius of the shadow needed to be enlarged by about 1/41 in order to fit timings made during a lunar eclipse (La Hire 1707). Additional observations over the next two centuries revealed that the shadow enlargement was somewhat variable from one eclipse to the next. According to Chauvenet (1891):

"This fractional increase of the breath of the shadow was given by Lambert as 1/40, and by Mayer as 1/60. Beer and Maedler found 1/50 from a number of observations of eclipses of lunar spots in the very favorable eclipse of December 26, 1833."

Chauvenet adopted a value of 1/50, which has become the standard enlargement factor for lunar eclipse predictions published by many national institutes worldwide. The enlargement enters into the definitions of the penumbral and umbral shadow radii as follows.

penumbral radius: $R_p = 1.02 * (0.998340 * P_m + S_s + P_s)$ (1-3)

umbral radius: $R_u = 1.02 * (0.998340 * P_m - S_s + P_s)$ (1-4)

where: P_m = Equatorial horizontal parallax of the Moon,
S_s = Geocentric semi-diameter of the Sun, and
P_s = Equatorial horizontal parallax of the Sun.

The factor 1.02 is the enlargement of the shadows by 1/50. Earth's true figure approximates that of an oblate ellipsoid with a flattening of ~1/300. Furthermore, the axial tilt of the planet towards or away from the Sun throughout the year means the shape of the penumbral and umbral shadows vary although the effect is small. It is sufficient to use a mean radius of Earth at latitude 45° to approximate the departure from perfectly circular shadows. The *Astronomical Almanac*[8] uses a factor of 0.998340 to scale the Moon's equatorial horizontal parallax to account for this (i.e., 0.998340 = 1 - 0.5 * 1/300).

In an analysis of 57 eclipses covering a period of 150 years, Link (1969) found a mean shadow enlargement of 2.3%. Furthermore, timings of crater entrances and exits through the umbra during four lunar eclipses from 1972 to 1982 (Table 1-3) closely support the Chauvenet value of 2%. From a physical point of view, there is no abrupt boundary between the umbra and penumbra. The shadow density actually varies continuously as a function of radial distance from the central axis out to the extreme edge of the penumbra. However, the density variation is most rapid near the theoretical edge of the umbra. Kuhl's (1928) contrast theory demonstrates that the edge of the umbra is perceived at the point of inflexion in the shadow density. This point appears to be equivalent to a layer in Earth's atmosphere at an altitude of about 120 to 150 km. The net enlargement of Earth's radius of 1.9% to 2.4% corresponds to an umbral shadow enlargement of 1.5% to 1.9%, in reasonably good agreement with the conventional value.

Table 1-3. Umbral Shadow Enlargement from Craters Timings

Lunar Eclipse Date	Crater Entrances Percent Enlargement	Crater Exits Percent Enlargement	Sky & Telescope Reference
1972 Jan 30	1.69 [420]	1.68 [295]	Oct 1972, p.264
1975 May 24	1.79 [332]	1.61 [232]	Oct 1975, p.219
1982 Jul 05	2.02 [538]	2.24 [159]	Dec 1982, p.618
1982 Dec 30	1.74 [298]	1.74 [90]	Apr 1983, p.387

Note: Values in "[ ]" are the number of observations included in each shadow enlargement measurement.

Some authorities dispute Chauvenet's shadow enlargement convention. Danjon (1951) notes that the only reasonable way of accounting for a layer of opaque air surrounding Earth is to increase the planet's radius by the altitude of the layer. This can be accomplished by proportionally increasing the parallax of the Moon. The radii of the umbral and penumbral shadows are then subject to the same absolute correction and not the same relative correction employed

[8] The *Astronomical Almanac* is published annually by the Almanac Office of the U.S. Naval Observatory.

in the traditional Chauvenet 1/50 convention. Danjon estimates the thickness of the occulting layer to be 75 km and this results in an enlargement of Earth's radius and the Moon's parallax of about 1/85.

Since 1951, the French almanac *Connaissance des Temps* has adopted Danjon's method for the enlargement Earth's shadows in their eclipse predictions as shown below.

penumbral radius: $R_p = 1.01 * Pm + S_s + P_s$ (1-5)

umbral radius: $R_u = 1.01 * Pm - S_s + P_s$ (1-6)

where: P_m = Equatorial horizontal parallax of the Moon,
S_s = Geocentric semi-diameter of the Sun,
P_s = Equatorial horizontal parallax of the Sun, and
$1.01 \approx 1 + 1/85 - 1/594$.

The factor 1.01 combines the 1/85 shadow enlargement term with a 1/594 term[9] to correct for Earth's oblateness at a latitude of 45°.

Danjon's method correctly models the geometric relationship between an enlargement of Earth's radius and the corresponding increase in the size of its shadows. Meeus and Mucke (1979) and Espenak (2006) both use Danjon's method. However, the resulting umbral and penumbral eclipse magnitudes are smaller by approximately 0.006 and 0.026, respectively, as compared to predictions using the traditional Chauvenet convention of 1/50.

For instance, the umbral magnitude of the partial lunar eclipse of 2008 Aug 16 was 0.813 according to the *Astronomical Almanac* (2008) using Chauvenet's method, but only 0.806 according to Meeus and Mucke (1979) using Danjon's method.

Of course, the small magnitude difference between the two methods is difficult to observe because the edge of the umbral shadow is poorly defined. The choice of shadow enlargement method has the largest effect in certain limiting cases where a small change in magnitude can shift an eclipse from one type to another. For example, an eclipse that is barely total according to Chauvenet's method will be a large magnitude partial eclipse if calculated using Danjon's method. Table 1-4 shows five such instances where a shallow total eclipse calculated with Chauvenet's method becomes a deep partial eclipse with Danjon's.

Table 1-4. Total (Chauvenet) vs. Partial (Danjon) Lunar Eclipses: 1501–3000

Calendar Date	Umbral Magnitude (Chauvenet)	Umbral Magnitude (Danjon)	Magnitude Difference
1540 Sep 16	1.0007	0.9947	0.0060
1856 Oct 13	1.0017	0.9960	0.0057
2196 Jul 10	1.0007	0.9960	0.0047
2413 Nov 08	1.0042	0.9993	0.0049
2669 Feb 08	1.0016	0.9951	0.0065

Similarly, small umbral magnitude partial eclipses using Chauvenet's method must be reclassified as penumbral eclipses of large penumbral magnitude when calculated with Danjon's method. A recent example was the eclipse of 1988 Mar 03, which was partial with an umbral magnitude of 0.0028 according to Chauvenet's method, but was penumbral with an umbral magnitude –0.0017[10] by Danjon's method. A similar case will occur on 2042 Sep 29. For a list of all such cases from 1501 through 3000, see Table 1-5.

[9] *Connaissance des Temps* uses a value of 1/297 for Earth's flattening. At latitude 45°: 1/594 = 0.5 * 1/297.

[10] A negative umbral magnitude means that the Moon lies fully outside the umbral shadow and is, therefore, a penumbral eclipse.

Table 1-5. Partial (Chauvenet) vs. Penumbral (Danjon) Lunar Eclipses: 1501–3000

Calendar Date	Umbral Magnitude (Chauvenet)	Umbral Magnitude (Danjon)	Magnitude Difference
1513 Sep 15	0.0036	-0.0003	0.0039
1900 Jun 13	0.0012	-0.0040	0.0052
1988 Mar 03	0.0028	-0.0017	0.0045
2042 Sep 29	0.0027	-0.0031	0.0058
2429 Dec 11	0.0020	-0.0033	0.0053
2581 Oct 13	0.0017	-0.0054	0.0071
2678 Aug 24	0.0007	-0.0036	0.0043
2733 Aug 17	0.0037	-0.0040	0.0077

Finally, in some cases, the shadow enlargement convention can make the difference between a shallow penumbral eclipse (Chauvenet) or no eclipse at all (Danjon). Table 1–6 lists nine small magnitude penumbral eclipses over a 500-year interval as determined using Chauvenet's method (Liu and Fiala, 1992). When the eclipse predictions are repeated using Danjon's method, no lunar eclipses are found on these dates.

Table 1-6. Penumbral Lunar Eclipses (Chauvenet): 1801–2300

Calendar Date	Penumbral Magnitude (Chauvenet)
1864 Apr 22	0.0237
1872 Jun 21	0.0008
1882 Oct 26	0.0059
1951 Feb 21	0.0068
2016 Aug 18	0.0165
2042 Oct 28	0.0077
2194 Mar 07	0.0085
2219 Apr 30	0.0008
2288 Feb 18	0.0204

Practically speaking, the faint and indistinct edge of the penumbral shadow makes the penumbral eclipse contacts (P_1 and P_4) completely unobservable. So the small magnitude differences discussed here are only of academic interest. Still, it is important to note which shadow enlargement convention is used because it is critical in comparing predictions from different sources.

In the *Canon*, Earth's penumbral and umbral shadow sizes have been calculated by using Danjon's enlargement method. Other sources using Danjon's method include Meeus and Mucke (1979), Espenak (2006) and *Connaissance des Temps*. Several sources using Chauvenet's method are Espenak (1989), Liu and Fiala (1992), and *Astronomical Almanac*.

1.7 Map Accuracy

The accuracy of the eclipse maps depends principally on two factors. The first is the rigorousness of the solar and lunar ephemerides used in the calculations (Sect. 1.3). The Moon's close proximity to Earth coupled with its relatively low mass, results in orbital perturbations that make the Moon's position far more difficult to predict compared to the Sun's position.

Nevertheless, the lunar ephemeris is accurate to better than an arc second within several centuries of the present. Even for eclipses occurring in the year –1999 (2000 BCE), the Moon's position is correct to within a small fraction of a degree. Such positional discrepancies correspond to errors in predicted eclipse paths that are below the resolution threshold of the figures presented in the *Canon*.

A far greater source of error in the geographic zones of eclipse visibility is due to the uncertainty in ΔT (Sect. 2.6). This parameter is the arithmetic difference between TD (Sect. 2.3) and Universal Time or UT (Sect. 2.4). TD can be thought of as time measured with an idealized or perfect clock. In contrast, UT is based on Earth's rotation, which is gradually slowing down. TD is used to calculate solar system ephemerides and eclipse predictions, but UT is used for defining world time and longitudes.

Earth was rotating faster in the past so eclipse predictions generated in TD must first be converted to UT (via ΔT) before the geographic zones of eclipse visibility can be determined. In other words, the physical impact of ΔT on eclipse predictions is to shift the visibility zones east relative to the position calculated from TD. Because 1° in longitude corresponds to 4 min of time, a ΔT value of 240 seconds would shift the visibility zones 1° east of their TD positions. The maps in the Canon already include the ΔT translation of visibility zones from TD to UT; thus, they depict the actual geographic regions of visibility of each eclipse.

The problem with ΔT is that it is an observationally determined quantity. In the distant past or future, the value of ΔT must be estimated from historical trends. The further removed such evaluations are from actual measurements, the greater the uncertainty in the extrapolated value of ΔT. Small deviations can quickly propagate into large uncertainties over the course of 1,000 years.

At the bottom of each appendix page is the value of ΔT and its corresponding standard error. This is an estimate of the uncertainty in the longitude determination of each map. For years in which the standard error is greater than 480 seconds (2.0° in longitude), the maps include an uncertainty zone to graphically depict the longitudinal range of solutions within the standard error. The years prior to –630 and after +2428 have standard errors of this magnitude or greater.

The meridian representing the nominal value of ΔT is plotted as a line of longitude running vertically through each map in the center of the zone of visibility. The uncertainty zones then take the form of two dashed meridians of longitude east and west of the nominal ΔT meridian. They indicate the range of uncertainty in the position of the nominal ΔT meridian, given the standard error (σ) in ΔT (i.e., ΔT ± σ). This means that the entire map beneath the eclipse visibility zones can be shifted east or west by this amount to produce an acceptable solution that falls within the standard error of the estimated value of ΔT.

Figure 1-2a identifies the components of the uncertainty zone for the partial lunar eclipse of –1987 Apr 16 (1988 BCE). The instant of greatest eclipse is 12:21 TD (upper right corner) and the estimated value of ΔT (bottom) is 46,150 seconds or 12 h 49 min. The time expressed in UT is then:

$$UT = TD - \Delta T = 12{:}21 - 12{:}49 = 23{:}32\ UT \qquad (1\text{–}7)$$

In this case, the conversion to UT shifts the instant of greatest eclipse into the previous day (–1987 Apr 15). The time in UT is needed to determine the shift of the geographic zone of eclipse visibility relative to its TD position. The result shows the geographic region of eclipse visibility (Fig. 1-2a), taking into account the fact that Earth rotated faster 4,000 years ago. In Fig. 1-2a, the nominal ΔT meridian is identified as "ΔT", while the eastern and western uncertainty zone longitudes are labeled "ΔT_1" (= ΔT + σ) and "ΔT_2" (= ΔT – σ), respectively. These lines of longitude are located ±15.3° (σ = ±3,676 seconds) with respect to the nominal ΔT meridian and show the range that the underlying map can be rotated to give a solution, given the uncertainty in ΔT.

The map in Figure 1-2b shows the solution when ΔT_1 (= ΔT + σ) is used to calculate UT. In this case, ΔT_1 is equal to ΔT plus the standard error:

$$\Delta T_1 = \Delta T + \sigma = 46{,}150\ s + 3{,}676\ s = 49{,}826\ s = 13h\ 50m \qquad (1\text{–}8)$$

The corresponding time in UT is then

$$UT = TD - \Delta T1 = 12{:}21 - 13{:}50 = 22{:}31\ UT \qquad (1\text{–}9)$$

Once again, the conversion to UT shifts the instant of greatest eclipse into the previous day (–1987 Apr 15). The global map shows that entire eclipse visibility zone is shifted 15.3° east of the nominal solution for ΔT = 46,150 seconds.

Finally, the map in Fig. 1-2c shows the solution when ΔT_2 (= $\Delta T - \sigma$) is used to calculate UT. In this case, ΔT_2 is equal to ΔT minus the standard error:

$$\Delta T_2 = \Delta T - \sigma = 46{,}150 \text{ s} - 3{,}676 \text{ s} = 42{,}474 \text{ s} = 11\text{h } 48\text{m} \qquad (1\text{–}10)$$

The corresponding time in UT (–1987 Apr 16) is then

$$UT = TD - \Delta T_2 = 12{:}21 - 11{:}48 = 00{:}33 \text{ UT} \qquad (1\text{–}11)$$

The global map shows that the entire visibility zone is now shifted 15.3° west of the nominal solution for ΔT = 46,150 seconds.

The uncertainty zones plotted on eclipse maps prior to the year –630 and after the year +2428 can be used to estimate the range of uncertainty in the geographic visibility of an eclipse given the standard error in the corresponding value of ΔT.

Figures 1-2a – 1-2c. Lunar Eclipse Maps and Standard Error (σ) in ΔT

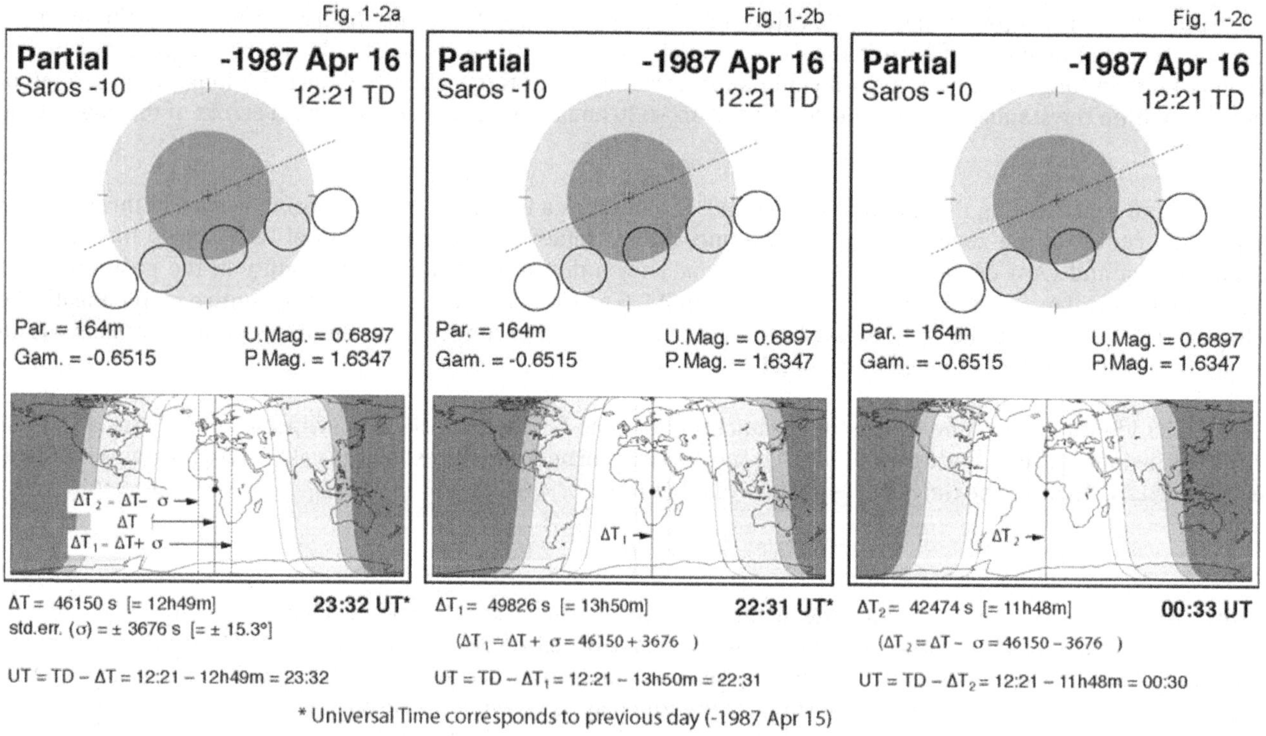

1.8 Visual Appearances of Lunar Eclipses

The visual appearance of penumbral, partial, and total lunar eclipses differs significantly from each other. While penumbral eclipses are pale and difficult to see, partial eclipses are easy naked-eye events, while total eclipses are colorful and dramatic.

1.8.1 Appearance of Penumbral and Partial Lunar Eclipses

Earth's penumbral shadow forms a diverging cone that expands into space in the anti-solar direction. From within this zone, Earth blocks part but not the entire disk of the Sun. Thus, some fraction of the Sun's direct rays continues to reach the most deeply eclipsed parts of the Moon during a penumbral eclipse.

The primary penumbral contacts (P_1 and P_4), as well as the early and late stages of a penumbral eclipse, are completely invisible to the eye with or without optical aid. A penumbral magnitude greater than ~0.6 is needed before skilled observers can detect faint shading across the Moon's disk.

Even when one edge of the Moon is 9/10 of the way into the penumbral shadow, approximately 10% of the Sun's rays still reach the Moon's deepest limb. Under such conditions, the Moon remains relatively bright with only a subtle gradient across its disk. The penumbral eclipse only becomes readily apparent when it is within ~0.05 magnitudes of becoming a partial eclipse.

In comparison, partial eclipses are easy to see with the naked eye. The lunar limb extending into the umbral shadow usually appears very dark or black. This is primarily due to a contrast effect because the remaining portion of the Moon in the penumbra may be brighter by a factor of ~500x. Because the umbral shadow's diameter is typically ~2.7x the Moon's diameter, it appears as though a semi-circular bite has been taken out of the Moon.

Aristotle (384–322 BCE) first proved that Earth was round using the curved umbral shadow seen at partial eclipses. In comparing observations of several eclipses, he noted that Earth's shadow was round no matter where the eclipse took place, whether the Moon was high in the sky or low near the horizon. Aristotle correctly reasoned that only a sphere casts a round shadow from every angle.

1.8.2 Appearance of Total Lunar Eclipses

The total lunar eclipse is the most dramatic and visually compelling type of lunar eclipse. The Moon's appearance can vary enormously throughout the period of totality and from one eclipse to the next. Obviously, the geometry of the Moon's path through the umbra plays an important role. Not as apparent is the effect that Earth's atmosphere has on a total eclipse. Although the physical mass of Earth blocks all direct sunlight from the umbra, the planet's atmosphere filters, attenuates and bends some of the Sun's rays into the shadow.

The molecules in Earth's atmosphere scatter short wavelength light (e.g., yellow, green, blue) more than long wavelength light (e.g., orange, red). This process, which is responsible for making sunsets red, also gives total eclipses their characteristic red-orange color. However, the exact color can vary considerably in both hue and brightness.

Because the lowest layers of the atmosphere are thicker than the upper layers, they absorb more sunlight and refract it through larger angles. About 75% of the atmosphere's mass is concentrated in the bottom 10 km (troposphere) as well as most of the water vapor, which can form massive clouds that block even more light. Just above the troposphere lies the stratosphere (~10 km to 50 km), a rarified zone above most of the planet's weather systems. The stratosphere is subject to important photochemical reactions due to the high levels of solar ultraviolet radiation that penetrates the region. The troposphere and stratosphere act together as a ring-shaped lens that refracts heavily reddened sunlight into Earth's umbral shadow. Because the higher atmospheric layers in the stratosphere contain less gas, they refract sunlight through progressively smaller angles into the outer parts of the umbra. In contrast, lower atmospheric layers containing more gas refract sunlight through larger angles to reach the inner parts of the umbra.

As a consequence of this lensing effect, the amount of light refracted into the umbra tends to increase radially from center to edge. However, inhomogeneities in the form of asymmetric amounts of cloud and dust at differing latitudes can cause significant variations in brightness throughout the umbra.

Besides water (clouds, mist, precipitation), Earth's atmosphere also contains aerosols or tiny particles of organic debris, meteoric dust, volcanic ash, and photochemical droplets. This material can significantly attenuate sunlight before it is refracted into the umbra. For instance, major volcanic eruptions in 1963 (Agung) and 1982 (El Chichón) each dumped huge quantities of gas and ash into the stratosphere and were followed by several years of very dark eclipses (Keen, 1983).

The same thing occurred after the eruption of the Philippine volcano Pinatubo in 1991. While most of the solid ash fell to Earth several days after circulating through the troposphere, a sizable volume of sulphur dioxide (SO_2) and water vapor reached the stratosphere where it produced sulfuric acid (H2SO4). This high-altitude volcanic haze layer can severely dim sunlight that must travel several hundred kilometers horizontally through the layer before being refracted into the umbral shadow. Consequently, the total eclipses following large volcanic eruptions are unusually dark. For instance, the total lunar eclipse of 1992 Dec 09 (1.5 years after Pinatubo) was so dark that the Moon's dull gray disk was difficult to see with the naked eye (Espenak, 2008, personal observation).

1.8.3 Danjon Scale of Lunar Eclipse Brightness

The French astronomer A. Danjon proposed a useful five-point scale for evaluating the visual appearance and brightness of the Moon during total lunar eclipses. The L values for various luminosities are defined as follows:

L=0	Very dark eclipse (Moon is almost invisible, especially at mid-totality)
L=1	Dark eclipse, gray or brownish in coloration (Details are distinguishable only with difficulty)
L=2	Deep red or rust-colored eclipse (Very dark central shadow, while outer umbra is relatively bright)
L=3	Brick-red eclipse (Umbral shadow usually has a bright or yellow rim)
L=4	Very bright copper-red or orange eclipse (Umbral shadow has a bluish, very bright rim)

The Danjon scale illustrates the range of colors and brightness the Moon can take on during a total lunar eclipse. It is also a useful tool to visual observers in characterizing the appearance of an eclipse. The evaluation of an L value is best done with the naked eye, binoculars, or a small telescope near the time of mid-totality. It is also helpful to examine the Moon's appearance just after the beginning and just before the end of totality. The Moon is then near the edge of the shadow, providing an opportunity to assign an L value to the outer umbra. In making such evaluations, the instrumentation and the time must be also recorded.

1.9 Five Millennium Catalog of Lunar Eclipses

A catalog listing concise details about each eclipse in the entire 2-volume *Canon* is available in the associated publication

Five Millennium Catalog of Lunar Eclipses -1999 to +3000

All the information found on the *Canon* figures is included such as calendar date, Dynamical Time, eclipse type, Saros number, magnitude, and duration. In addition, a number of other useful data are listed including ΔT, and the geographic coordinates of greatest eclipse. The eclipse type is augmented to indicate the first/last eclipse in a Saros series.

Details about this publication can be found at: ***http://astropixels.com /pubs/5MKLE2.html***

Section 2: Time

2.1 Greenwich Mean Time

For thousands of years, time has been measured using the length of the solar day. This is the interval between two successive returns of the Sun to an observer's local meridian. Unfortunately, the length of the apparent solar day can vary by tens of seconds over the course of a year. Earth's elliptical orbit around the Sun and the 23.5° inclination of Earth's axis of rotation are responsible for these variations. Apparent solar time was eventually replaced by mean solar time because it provides for a uniform time scale. The key to mean solar time is the mean solar day which has a constant length of 24 hours throughout the year.

Mean solar time on the 0° longitude meridian in Greenwich, England is known as Greenwich Mean Time (GMT). At the International Meridian Conference of 1884, GMT[11] was adopted as the reference time for all clocks around the world. It was also agreed that all longitudes would be measured east or west with respect to the Greenwich meridian. In 1972, Greenwich Mean Time was replaced by Coordinated Universal Time (UTC) as the international time reference. Nevertheless, UTC is colloquially referred to as Greenwich Mean Time although this is technically not correct.

2.2 Ephemeris Time

During the twentieth century, it was found that the rotational period of Earth (length of the day) was gradually slowing down. For the purposes of orbital calculations, time using Earth's rotation was abandoned for a more uniform time scale based on Earth's orbit about the Sun. In 1952, the International Astronomical Union (IAU) introduced Ephemeris Time to address this problem. The ephemeris second was defined as a fraction of the tropical year for 1900 Jan 01 as calculated from Newcomb's Tables of the Sun [1895]. Ephemeris Time was used for Solar System ephemeris calculations until it was replaced by Terrestrial Dynamical Time in 1979.

2.3 Terrestrial Dynamical Time

Terrestrial Dynamical Time (TD) was introduced by the International Astronomical Union (IAU) in 1979 as the coordinate time scale for an observer on the surface of Earth. It takes into account relativistic effects and is based on International Atomic Time (TAI) which is a high-precision standard using several hundred atomic clocks worldwide. As such Terrestrial Dynamical Time is the atomic time equivalent to its predecessor Ephemeris Time (ET) and is used in the theories of motion for bodies in the Solar System. To ensure continuity with Ephemeris Time, TD was defined to match ET for the date 1977 Jan 01. In 1991, the IAU refined the definition of TD to make it more precise. It was also renamed Terrestrial Time (TT) although we prefer and use the older name Terrestrial Dynamical Time.

2.4 Universal Time

For many centuries, the fundamental unit of time was the rotational period of Earth with respect to the Sun. Greenwich Mean Time (GMT) was the standard time reference based on the mean solar time on the 0° longitude meridian in Greenwich, England. Universal Time (UT) is the modern counterpart to GMT and is determined from Very Long Baseline Interferometry (VLBI) observations of the diurnal motion of quasars. Unfortunately, Universal Time is not a uniform time scale because Earth's rotational period is (on average) gradually increasing.

The change is primarily due to tidal friction between Earth's oceans and its rocky mantle through the gravitational attraction of the Moon and, to a lesser extent, the Sun. This secular acceleration (Sect. 1.4) gradually transfers angular momentum from Earth to the Moon. As Earth loses energy and slows down, the Moon gains this energy and its orbital

[11] GMT was originally reckoned from noon to noon. In 1925, some countries shifted GMT by 12 hours so that it would begin at Greenwich midnight. This new definition is the one in common usage for world time and in the navigational publications of English-speaking countries. The designation Greenwich Mean Astronomical Time (GMAT) is reserved for the reckoning of time from noon (and previously called GMT).

period and distance from Earth increase. Shorter period fluctuations in terrestrial rotation also exist which can produce an accumulated clock error of ±20 seconds in one or more decades. These decade variations are attributed to several geophysical mechanisms including fluid interactions between the core and mantle of Earth. Climatological changes and variations in sea-level may also play significant roles since they alter Earth's moment of inertia.

The secular acceleration of the Moon implies an increase in the length of day [LOD] of about 2.3 milliseconds per century. Such a small amount may seem insignificant, but it has very measurable cumulative effects. At this rate, time as measured through Earth's rotation is losing about 84 seconds per century^2 when compared to atomic time.

2.5 Coordinated Universal Time

Coordinated Universal Time (UTC) is the present day basis of all civilian time throughout the world. Derived from International Atomic Time (TAI), the length of the UTC second is defined in terms an atomic transition of the element cesium and is accurate to approximately one nanosecond (billionth of a second) per day. Since most daily life is still organized around the solar day, UTC was defined to closely parallel Universal Time. However, the two time systems are intrinsically incompatible since UTC is uniform while UT is based on Earth's rotation which is gradually slowing. In order to keep the two times within 0.9 seconds of each other, a leap second is added to UTC about once every 12 to 18 months.

2.6 Delta T (ΔT)

The orbital positions of the Sun and Moon required by eclipse predictions, are calculated using Terrestrial Dynamical Time (TD) because it is a uniform time scale. World time zones and daily life, however, are based on Universal Time (UT)[12]. In order to convert eclipse predictions from TD to UT, the difference between these two time scales must be known. The parameter delta-T (ΔT) is the arithmetic difference, in seconds, between the two as:

$$\Delta T = TD - UT \qquad (2\text{-}1)$$

Past values of ΔT can be deduced from the historical records. In particular, hundreds of eclipse observations (both solar and lunar) were recorded in early European, Middle Eastern and Chinese annals, manuscripts, and canons. In spite of their relatively low precision, these data represent the only evidence for the value of ΔT prior to 1600 CE. In the centuries following the introduction of the telescope (circa 1609 CE), thousands of high quality observations have been made of lunar occultations of stars. The number and accuracy of these timings increase from the seventeenth through the twentieth century, affording valuable data in the determination of ΔT. A detailed analysis of these measurements fitted with cubic splines for ΔT from –500 to +1950 is presented in Table 2-1 and includes the standard error for each value (Morrison and Stephenson, 2004).

[12] World time zones are actually based on Coordinated Universal Time (UTC). It is an atomic time synchronized and adjusted to stay within 0.9 seconds of astronomically determined Universal Time (UT). Occasionally, a "leap second" is added to UTC to keep it in sync with UT (which changes due to variations in Earth's rotation rate).

Table 2-1. Values of ΔT Derived from Historical Records

Year	ΔT (seconds)	Standard Error (seconds)
-500	17190	430
-400	15530	390
-300	14080	360
-200	12790	330
-100	11640	290
0	10580	260
100	9600	240
200	8640	210
300	7680	180
400	6700	160
500	5710	140
600	4740	120
700	3810	100
800	2960	80
900	2200	70
1000	1570	55
1100	1090	40
1200	740	30
1300	490	20
1400	320	20
1500	200	20
1600	120	20
1700	9	5
1750	13	2
1800	14	1
1850	7	<1
1900	-3	<1
1950	29	<0.1

In modern times, the determination of ΔT is made using atomic clocks and radio observations of quasars, so it is completely independent of the lunar ephemeris. Table 2-2 gives the value of ΔT every five years from 1955 to 2005 (*Astronomical Almanac* for 2006, page K9).

Table 2-2. Recent Values of ΔT from Direct Observations

Year	ΔT (seconds)	5-Year Change (seconds)	Average 1-Year Change (seconds)
1955.0	+31.1	—	—
1960.0	+33.2	2.1	0.42
1965.0	+35.7	2.5	0.50
1970.0	+40.2	4.5	0.90
1975.0	+45.5	5.3	1.06
1980.0	+50.5	5.0	1.00
1985.0	+54.3	3.8	0.76
1990.0	+56.9	2.6	0.52
1995.0	+60.8	3.9	0.78
2000.0	+63.8	3.0	0.60
2005.0	+64.7	0.9	0.18

The average annual change of ΔT was 0.99 second from 1965 to 1980. However, the average annual increase was just 0.63 second from 1985 to 2000, and only 0.18 second from 2000 to 2005. Future changes and trends in ΔT cannot be predicted with certainty since theoretical models of the physical causes are not of high enough precision. Extrapolations from the table weighted by the long period trend from tidal braking of the Moon offer reasonable estimates of +67 seconds in 2010, +93 seconds in 2050, +203 seconds in 2100, and +442 seconds in the year 2200.

Outside the period of observations (500 BCE to 2005 CE), the value of ΔT can be extrapolated from measured values using the long-term mean parabolic trend:

$$\Delta T = -20 + 32\,u^2 \text{ seconds,} \qquad (2\text{-}2)$$

where: u = (year–1820)/100, and is defined as time measured in centuries.

2.7 Polynomial Expressions for ΔT

Using the ΔT values derived from the historical record and from direct observations (Tables 2-1 and 2-2, respectively), a series of polynomial expressions have been created to simplify the evaluation of ΔT for any time during the interval –1999 to +3000. We define the decimal year "y" as follows:

$$y = \text{year} + (\text{month} - 0.5)/12 \qquad (2\text{-}3)$$

This gives "y" for the middle of the month, which is accurate enough given the precision in the known values of ΔT. The following polynomial expressions can be used to calculate the value of ΔT (in seconds) over the interval of the *Canon*.

Before the year –500, calculate:

$$\Delta T = -20 + 32\,u^2, \qquad (2\text{-}4)$$

where: u = (y–1820)/100.

Between years –500 and +500, we use the data from Table 2-1, except that for the year –500 we changed the value 17190 to 17203.7 in order to avoid a discontinuity with the previous formula at that epoch. The value for ΔT is given by a polynomial of the 6th degree, which reproduces the values in Table 2-1 with an error not larger than 4 seconds:

$$\Delta T = 10583.6 - 1014.41\,u + 33.78311\,u^2 - 5.952053\,u^3 - 0.1798452\,u^4 + 0.022174192\,u^5 + 0.0090316521\,u^6, \qquad (2\text{-}5)$$

where: u = y/100.

Between years +500 and +1600, we again use the data from Table 2-1. Calculate u = (y–1000)/100. The value for ΔT is given by the following polynomial of the 6th degree with a divergence from Table 2-1 not larger than 4 seconds:

$$\Delta T = 1574.2 - 556.01\,u + 71.23472\,u^2 + 0.319781\,u^3 - 0.8503463\,u^4 - 0.005050998\,u^5 + 0.0083572073\,u^6, \qquad (2\text{-}6)$$

where: u = (y–1000)/100.

Between years +1600 and +1700, calculate:

$$\Delta T = 120 - 0.9808\,t - 0.01532\,t^2 + t^3 / 7129, \qquad (2\text{-}7)$$

where: t = y – 1600.

Between years +1700 and +1800, calculate:

$$\Delta T = 8.83 + 0.1603\,t - 0.0059285\,t^2 + 0.00013336\,t^3 - t^4 / 1174000, \qquad (2\text{-}8)$$

where: t = y – 1700.

Between years +1800 and +1860, calculate:

$$\Delta T = 13.72 - 0.332447\,t + 0.0068612\,t^2 + 0.0041116\,t^3 - 0.00037436\,t^4 + 0.0000121272\,t^5 - 0.0000001699\,t^6 + 0.000000000875\,t^7, \qquad (2\text{-}9)$$

where: $t = y - 1800$.

Between years 1860 and 1900, calculate:

$$\Delta T = 7.62 + 0.5737\,t - 0.251754\,t^2 + 0.01680668\,t^3 - 0.0004473624\,t^4 + t^5 / 233174, \quad (2\text{-}10)$$

where: $t = y - 1860$.

Between years 1900 and 1920, calculate:

$$\Delta T = -2.79 + 1.494119\,t - 0.0598939\,t^2 + 0.0061966\,t^3 - 0.000197\,t^4, \quad (2\text{-}11)$$

where: $t = y - 1900$.

Between years 1920 and 1941, calculate:

$$\Delta T = 21.20 + 0.84493\,t - 0.076100\,t^2 + 0.0020936\,t^3, \quad (2\text{-}12)$$

where: $t = y - 1920$.

Between years 1941 and 1961, calculate:

$$\Delta T = 29.07 + 0.407\,t - t^2 / 233 + t^3 / 2547, \quad (2\text{-}13)$$

where: $t = y - 1950$.

Between years 1961 and 1986, calculate:

$$\Delta T = 45.45 + 1.067\,t - t^2 / 260 - t^3 / 718, \quad (2\text{-}14)$$

where: $t = y - 1975$.

Between years 1986 and 2005, calculate:

$$\Delta T = 63.86 + 0.3345\,t - 0.060374\,t^2 + 0.0017275\,t^3 + 0.000651814\,t^4 + 0.00002373599\,t^5, \quad (2\text{-}15)$$

where: $t = y - 2000$.

Between years 2005 and 2050, calculate:

$$\Delta T = 62.92 + 0.32217\,t + 0.005589\,t^2, \quad (2\text{-}16)$$

where: $t = y - 2000$.

This expression is derived from estimated values of ΔT in the years 2010 and 2050. The value for 2010 (66.9 seconds) is based on a linearly extrapolation from 2005 using 0.39 seconds/year (average from 1995 to 2005). The value for 2050 (93 seconds) is linearly extrapolated from 2010 using 0.66 seconds/year (average rate from 1901 to 2000).

Between years 2050 and 2150, calculate:

$$\Delta T = -20 + 32\,((y-1820)/100)^2 - 0.5628\,(2150 - y), \quad (2\text{-}17)$$

The last term is introduced to eliminate the discontinuity at 2050.

After 2150, calculate:

$$\Delta T = -20 + 32\,u^2, \quad (2\text{-}18)$$

where: $u = (y-1820)/100$.

All values of ΔT based on Morrison and Stephenson [2004] assume a value for the Moon's secular acceleration of −26 arcsec/cy^2. However, the ELP-2000/82 lunar ephemeris employed in the *Canon* uses a slightly different value of

25.858 arcsec/cy^2. Thus, a small correction "c" must be added to the values derived from the polynomial expressions for ΔT before they can be used in the *Canon*:

$$c = -0.000012932\ (y - 1955)^2. \qquad (2\text{-}19)$$

Since the values of ΔT for the interval 1955 to 2005 were derived independent of any lunar ephemeris, no correction is needed for this period.

2.8 Uncertainty in ΔT

The uncertainty in the value of ΔT is of particular interest in the calculation of eclipse paths in the distant past and future. Unfortunately, estimating the standard error in ΔT prior to 1600 CE is a difficult problem. It depends on a number of factors which include the accuracy of determining ΔT from historical eclipse records and modeling the physical processes producing changes in Earth's rotation. Morrison and Stephenson [2004] propose a simple parabolic relation to estimate the standard error (σ) which is valid over the period 1000 BCE to 1200 CE:

$$\sigma = 0.8\ t^2 \text{ seconds,} \qquad (2\text{-}20)$$

where: t = (year–1820)/100.

Table 2-3 gives the errors in ΔT along with the corresponding uncertainties in the longitude of an eclipse path.

Table 2-3. Uncertainty of ΔT, Part I

Year	σ (seconds)	Longitude
-1000	636	2.65°
-500	431	1.79°
0	265	1.10°
+500	139	0.58°
+1000	54	0.22°
+1200	31	0.13°

The decade fluctuations in ΔT result in an uncertainty of ~20 seconds (0.08°) for the period 1300 CE to 1600 CE.

During the telescopic era (1600 CE to present), records of astronomical observations pin down the decade fluctuations with increasing reliability. The uncertainties in ΔT are presented in Table 2-4 [Stephenson and Houlden, 1986].

Table 2-4. Uncertainty of ΔT, Part II

Year	σ (seconds)	Longitude
+1700	5	0.021°
+1800	1	0.004°
+1900	0.1	0.0004°

The estimation in the uncertainty of ΔT prior to 1000 BCE must rely on a certain amount of modeling and theoretical arguments since no measurements of ΔT are available for this period. Huber (2000) has proposed a Brownian motion model including drift to estimate the standard error in ΔT for periods outside the epoch of measured values. The intrinsic variability in the LOD (length of day) during the 2500 years of observations (500 BCE to 2000 CE) is 1.780 ms/cy with a standard error of 0.56 ms/cy. This rate is not due entirely to tidal friction but includes a drift in LOD from imperfectly understood effects such as changes in sea level due to variations in polar ice caps. Presumably, the same mechanisms operating during the present era also operated prior to 1000 BCE as well as one millennium into the future.

Huber's derived estimate for the total standard error (fluctuations plus drift) in ΔT is as follows.

$$\sigma = 365.25\ N\ SQRT\,[\,(N\ Q / 3)\ (1 + N / M)\,] / 1000, \qquad (2\text{-}21)$$

where:

N = Difference between target year and calibration year
M = 2500 years (–500 to +2000);
this covers period of observed ΔT measurements
Q = 0.058 ms^2 /yr

The calibration year is taken as –500 for target years before 500 BCE, while the calibration year is 2005 CE for target years in the future. Evaluation of this expression at 1,000 year intervals is found in Table 2-5. It shows estimates in the standard error of ΔT along with the equivalent shift in longitude

Table 2-5. Uncertainty of ΔT, Part III

Year	σ (seconds)	Longitude
-4500	20,717	86.3°
-4000	16,291	67.9°
-3500	12,378	51.6°
-3000	8,978	37.4°
-2500	6,094	25.4°
-2000	3,732	15.6°
-1500	1,900	7.9°
-1000	622	2.6°
—	—	—
+2500	612	2.6°
+3000	1,885	7.9°
+3500	3,711	15.6°
+4000	6,068	25.3°
+4500	8,946	37.3°
+5000	12,341	51.4°

Section 3: Lunar Eclipse Statistics

3.1 Statistical Distribution of lunar Eclipse Types

Eclipses of the Moon can only occur during the Full Moon phase. It is then possible for the Moon to pass through Earth's penumbral and umbral shadows thereby producing an eclipse. There are three types of lunar eclipses:

1) **Penumbral** — Moon passes partially or completely into Earth's penumbral shadow.
2) **Partial** — Moon passes partially into Earth's umbral shadow.
3) **Total** — Moon passes completely into Earth's umbral shadow

During the 5,000-year period from -1999 to +3000 (2000 BCE to 3000 CE), Earth will experience 12,064 eclipses of the Moon. The statistical distribution of the three eclipse types over this interval is shown in Table 3-1.

Table 3-1. Distribution of Basic Lunar Eclipse Types

Eclipse Type	Abbreviation	Number	Percent
All Eclipses	—	12064	100.0%
Penumbral	N	4378	36.3%
Partial	P	4207	34.9%
Total	T	3479	28.8%

During most penumbral eclipses, only part of the Moon passes through Earth's penumbral shadow. Examples of such partial penumbral eclipses include: 2009 Feb 09, 2009 Jul 07, 2009 Aug 06, and 2012 Nov 28. However, it is also possible to have a penumbral eclipse in which the Moon passes completely within Earth's penumbral shadow without entering the inner umbral shadow. Such total penumbral eclipses are quite rare compared to normal (or partial) penumbral eclipses. During the 21st century, there are 87 partial penumbral eclipses, but only 5 total penumbral eclipses: 2006 Mar 14, 2053 Aug 29, 2070 Apr 25, 2082 Aug 08, and 2099 Sep 29. Table 3-2 shows the distribution of the two penumbral eclipse types during the period covered by the Canon. For more on total penumbral eclipses, see Section 3.11.

Table 3-2. Statistics of Penumbral Lunar Eclipses

Eclipse Type	Number	Percent
All Penumbral Eclipses	4378	100.0%
Partial Penumbral	4237	96.8%
Total Penumbral	141	3.2%

Total lunar eclipses through Earth's umbral shadow can be categorized as:

1) **Central** — The Moon passes through the central axis of Earth's umbral shadow.
2) **Non-Central** — The Moon misses the central axis of Earth's umbral shadow.

Using the above categories, the distribution of the 3,479 total eclipses is shown in Table 3-3.

Table 3-3. Statistics of Total Lunar Eclipses

Eclipse Type	Number	Percent
All Total Eclipses	3479	100.0%
Central Total	2074	59.6%
Non-Central Total	1405	40.4%

Examples of central total eclipses include: 2000 Jul 16, 2011 Jun 15, 2018 Jul 27, and 2022 May 16. Several examples of non-central total eclipses are: 2010 Dec 21, 2011 Dec 10, 2014 Apr 15, and 2014 Oct 08.

3.2 Distribution of Lunar Eclipse Types by Century

Table 3-4 summarizes 5,000 years of eclipses by eclipse type over 100-year intervals. The average century contains 241 lunar eclipses of which 88 are penumbral, 84 are partial and 70 are total. Individual centuries deviate from these mean values in interesting ways. For instance, the total number of eclipses in a century varies from a minimum of 225 (century beginning -0899) to a maximum of 259 (centuries beginning 1101 and 2801). The number of penumbral eclipses varies from 76 (century 0301) to 100 (century 2801), while the number of partial eclipses ranges from 54 (century -1599) to 102 (centuries -0699, 1101, and 2801). Finally, the number of total lunar eclipses per century varies from 57 (centuries 0001 and 2801) to 89 (century 0801). Note that century 2801 not only has the most number of eclipses (259), it also has the maximum number of penumbral (100) and partial (102) eclipses, but the minimum number of total eclipses (57).

The last column of Table 3-4 lists the number of total lunar eclipse tetrads per century. A tetrad is a grouping of four consecutive total lunar eclipses each separated by six lunations. Their frequency is quite variable with some centuries having 0 tetrads, while others may have as many as 8. For more on tetrads, see Section 3.16.

When the data in Table 3-4 is displayed graphically (Figure 3-1), other important relationships are revealed. The most apparent feature is a periodic oscillation in the number of all lunar eclipses (Figure 3-1a), as well as in the individual types. By inspection, the period is about 600 years. Using tables from von Oppolzer's Canon der Finsternisse (1887) the Dutch astronomer G. van den Bergh (1954) calculated a period of 586 years. By studying the distributions of tetrad eclipse groups (Section 3.16), Meeus derived an empirical expression showing that this period is slowly decreasing (Meeus 2004b). While its current value is actually 565 years, the period was 618 years in -1999 and it will decrease to 552 years by +3000. A theoretical study by T. Hughes (Hughes 2004) demonstrates that the period is caused by the eccentricity of Earth's orbit, which is gradually decreasing.

There are other interesting features in Figure 3-1. For instance, the number of total lunar eclipses as well as deep central total eclipses is highest when the number of all lunar eclipses is at a minimum. In contrast, numbers of penumbral eclipses, partial eclipses and non-central total eclipses all appear to be synchronized with the overall number of lunar eclipses. Finally, the number of total penumbral eclipses, and total lunar eclipse tetrads are most frequent during epochs when the overall number of lunar eclipses is at a minimum.

Table 3-4a. Number of Lunar Eclipse Types by Century: -1999 to -1000 (2000 BCE to 1001 BCE)

Century Interval	All Lunar Eclipses	All Penum. Eclipses	Partial Penum. Eclipses	Total Penum. Eclipses	All Partial Eclipses	All Total Eclipses	Central Total Eclipses	Non-Central Total Eclipses	Total Eclipse Tetrads
-1999 to -1900	242	88	84	4	92	62	13	49	1
-1899 to -1800	255	98	97	1	97	60	11	49	0
-1799 to -1700	254	94	94	0	99	61	12	49	0
-1699 to -1600	244	90	87	3	87	67	21	46	3
-1599 to -1500	226	85	81	4	54	87	62	25	8
-1499 to -1400	228	86	78	8	60	82	55	27	6
-1399 to -1300	239	83	80	3	90	66	23	43	2
-1299 to -1200	251	92	91	1	99	60	9	51	0
-1199 to -1100	252	96	95	1	95	61	13	48	0
-1099 to -1000	240	88	81	7	86	66	18	48	4

Table 3-4b. Number of Lunar Eclipse Types by Century: -0999 to +3000 (1000 BCE to 3000 CE)

Century Interval	All Lunar Eclipses	All Penum. Eclipses	Partial Penumb. Eclipses	Total Penum. Eclipses	All Partial Eclipses	All Total Eclipses	Central Total Eclipses	Non-Central Total Eclipses	Total Eclipse Tetrads
-0999 to -0900	228	80	72	8	61	87	59	28	7
-0899 to -0800	225	79	76	3	62	84	57	27	7
-0799 to -0700	239	87	85	2	88	64	21	43	1
-0699 to -0600	253	91	91	0	102	60	10	50	0
-0599 to -0500	255	95	95	0	100	60	11	49	0
-0499 to -0400	242	88	86	2	90	64	19	45	1
-0399 to -0300	229	80	79	1	61	88	54	34	7
-0299 to -0200	228	79	76	3	64	85	55	30	6
-0199 to -0100	240	85	84	1	86	69	21	48	3
-0099 to 0000	253	94	93	1	98	61	11	50	0
0001 to 0100	251	93	92	1	101	57	7	50	0
0101 to 0200	239	84	82	2	88	67	23	44	3
0201 to 0300	228	83	75	8	63	82	53	29	6
0301 to 0400	227	76	72	4	69	82	58	24	4
0401 to 0500	244	83	82	1	95	66	18	48	3
0501 to 0600	254	97	96	1	95	62	12	50	0
0601 to 0700	255	97	94	3	100	58	10	48	0
0701 to 0800	240	84	82	2	88	68	23	45	3
0801 to 0900	228	79	74	5	60	89	65	24	8
0901 to 1000	230	82	79	3	70	78	45	33	6
1001 to 1100	246	87	84	3	97	62	14	48	0
1101 to 1200	259	98	98	0	102	59	10	49	0
1201 to 1300	251	95	94	1	96	60	12	48	0
1301 to 1400	231	81	73	8	73	77	41	36	6
1401 to 1500	228	80	77	3	65	83	60	23	4
1501 to 1600	233	82	79	3	74	77	39	38	6
1601 to 1700	249	91	88	3	97	61	11	50	0
1701 to 1800	256	98	98	0	98	60	13	47	0
1801 to 1900	249	90	88	2	97	62	11	51	0
1901 to 2000	229	83	74	9	65	81	48	33	5
2001 to 2100	228	86	81	5	57	85	61	24	8
2101 to 2200	238	81	77	4	88	69	24	45	4
2201 to 2300	252	94	93	1	97	61	11	50	0
2301 to 2400	253	95	95	0	98	60	12	48	0
2401 to 2500	237	83	77	6	85	69	23	46	4
2501 to 2600	227	82	77	5	58	87	64	23	8
2601 to 2700	230	81	80	1	70	79	44	35	7
2701 to 2800	241	86	83	3	91	64	16	48	0
2801 to 2900	259	100	99	1	102	57	9	48	0
2901 to 3000	249	89	89	0	97	63	13	50	1

Figure 3–1. Number of Lunar Eclipses Per Century: -1999 to +3000

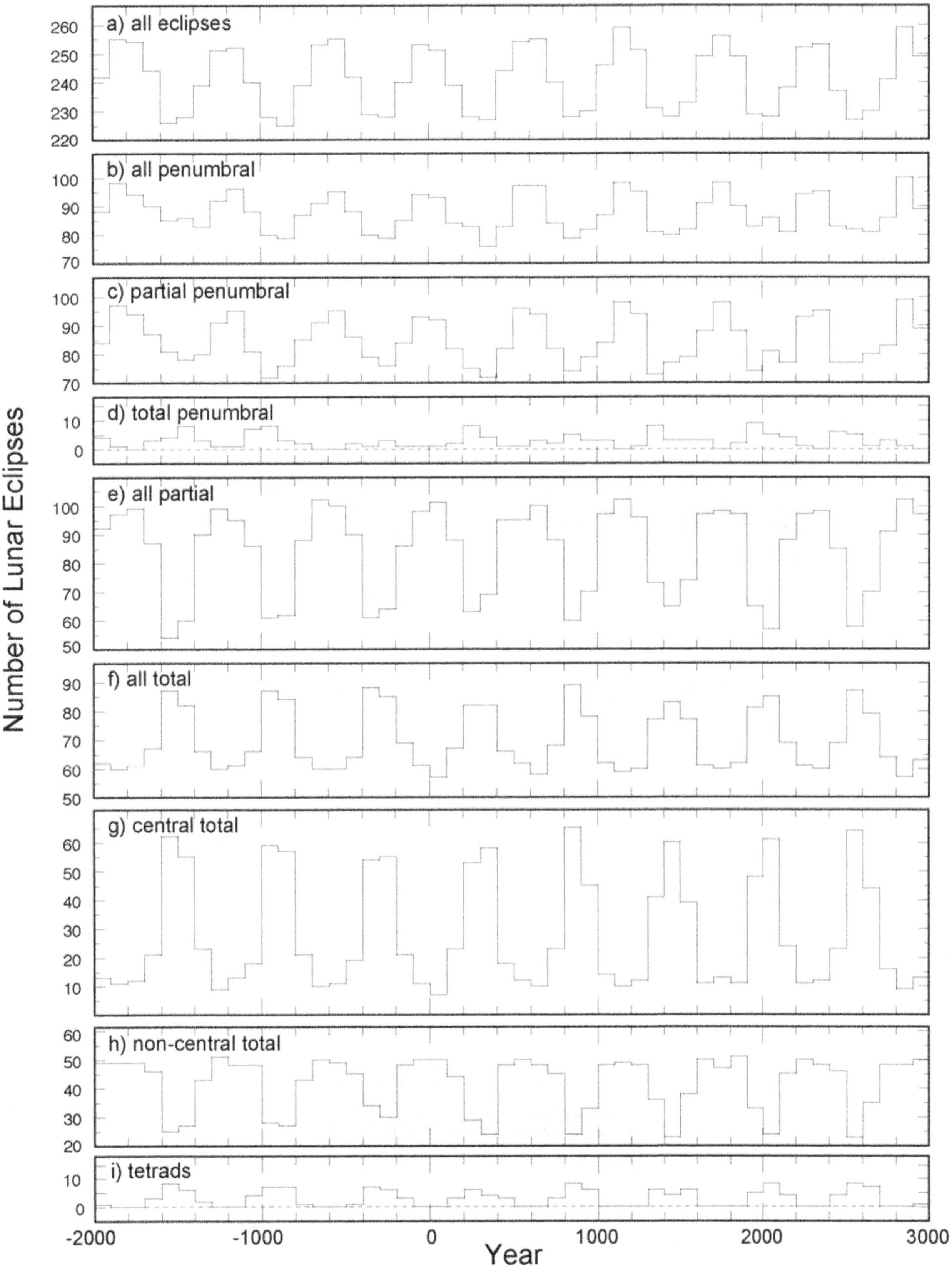

3.3 Distribution of Lunar Eclipse Types by Month

Table 3-5 summarizes 5,000 years of eclipses by eclipse type in each month of the year. The first value in each column is the number of eclipses of a given type for the corresponding month. The second value, in square brackets [ ], is the number of eclipses divided by the number of days in that month. This normalization allows direct comparison of eclipse frequencies in different months.

A brief examination of the normalized values in the column "Number of All Lunar Eclipses" shows that eclipses are equally distributed around the year. The same holds true for partial eclipses; however, the columns for penumbral and total eclipses reveal something interesting. Penumbral eclipses are about 6% more likely during the period of December– January–February compared to the months June–July–August. This effect is attributed to Earth's elliptical orbit. Earth currently reaches perihelion in early January and aphelion in early July. Consequently, the Sun's apparent diameter varies from 1,952 to 1,887 arcsec between perihelion and aphelion. The Sun's larger apparent diameter at perihelion makes Earth's penumbral shadow larger so penumbral eclipses are more frequent at that time.

The opposite argument holds true for total eclipses that are about 4% more likely during the period April to August, compared to the months November to February. In this case, the Sun's smaller apparent size around aphelion increases the diameter of Earth's umbral shadow so the frequency of total eclipses is slightly higher at that time.

Table 3-5. Lunar Eclipse Types by Month: -1999 to +3000 (2000 BCE to 3000 CE)

Month	Number of All Lunar Eclipses	Number of Penumbral Eclipses	Number of Partial Eclipses	Number of Total Eclipses
January	1027 [33.1]	379 [12.2]	352 [11.4]	296 [9.5]
February	936 [33.4]	343 [12.2]	333 [11.9]	260 [9.3]
March	1028 [33.2]	376 [12.1]	361 [11.6]	291 [9.4]
April	986 [32.9]	354 [11.8]	342 [11.4]	290 [9.7]
May_	1025 [33.1]	370 [11.9]	358 [11.5]	297 [9.6]
June	992 [33.1]	351 [11.7]	347 [11.6]	294 [9.8]
July	1025 [33.1]	368 [11.9]	356 [11.5]	301 [9.7]
August	1015 [32.7]	361 [11.6]	355 [11.5]	299 [9.6]
September	990 [33.0]	364 [12.1]	341 [11.4]	285 [9.5]
October	1023 [33.0]	377 [12.2]	356 [11.5]	290 [9.4]
November	993 [33.1]	350 [11.7]	357 [11.9]	286 [9.5]
December	1024 [33.0]	385 [12.4]	349 [11.3]	290 [9.4]

3.4 Lunar Eclipse Frequency and the Calendar Year

There are 2 to 5 lunar eclipses in every calendar year. Table 3-6 shows the distribution in the number of eclipses per year for the 5,000 years covered in the Canon.

Table 3-6. Number of Lunar Eclipses per Year

Number of Eclipses per Year	Number of Years	Percent
2	3,541	70.8%
3	887	17.7%
4	539	10.8%
5	33	0.7%

When two eclipses occur in one calendar year, they can be in any combination of N, P, or T (penumbral, partial, or total, respectively). Table 3-7 lists the frequency of each eclipse combination, along with five recent years when the combination occurs. The table makes no distinction in the order of any two eclipses. For example, the eclipse combination PT includes all years where the order is either PT or TP.

Over 71% of all years containing two eclipses are composed of the combinations PP (34.2%) or TT (37.2%). The NT pair is the rarest combination with only six instances, all within the first two millennia of the Canon, and with five of the six involving a total penumbral eclipse.

Table 3-7. Two Lunar Eclipses in One Year

Eclipse Combinations[a]	Number of Years	Percent	Examples (Years)[b]
NN	116	3.3%	..., 1987, 2016, 2042, 2045, 2053, ...
NP	373	10.5%	..., 1999, 2005, 2006, 2012, 2017, ...
NT	6	0.2%	[-1989, -1971, -1423, -1403, -1385, -0186]
PP	1211	34.2%	..., 1980, 2039, 2041, 2046, 2057, ...
PT	516	14.6%	..., 1997, 2008, 2010, 2019, 2021, ...
TT	1319	37.2%	..., 2003, 2007, 2011, 2014, 2015, ...

a. N = Penumbral, P = Partial, and T = Total.
b. When years are surrounded by square brackets [ ], there are no other examples outside this range.

When three eclipses occur in one calendar year, there are nine possible combinations of N, P, or T. Table 3-8 lists the frequency of each eclipse combination along with five recent years when each combination occurs. The table makes no distinction in the order of eclipses in any combination. For example, the eclipse combination NPT includes all years where the order is NPT, NTP, PNT, PTN, TPN, and TNP. The rarest combinations—NNT and NTT—each occurs six times or less in the five-millennium span of this work. Interestingly, no PPP combinations occur.

Table 3-8. Three Lunar Eclipses in One Year

Eclipse Combinations[a]	Number of Years	Percent	Examples (Years)[b]
NNN	382	43.1%	..., 2002, 2027, 2031, 2049, 2060, ...
NNP	159	17.9%	..., 1958, 2013, 2147, 2168, 2186, ...
NNT	4	0.5%	[-1991, -1405, -0168, 0418]
NPP	158	17.8%	..., 1871, 2075, 2140, 2151, 2169, ...
NPT	27	3.0%	..., 1963, 2001, 2048, 2066, 2504, ...
NTT	6	0.7%	[-1953, -1432, -1367, -0846, 0977, 2466]
PPT	51	5.7%	..., 1898, 2028, 2113, 2178, 2531, ...
PTT	75	8.5%	..., 1852, 2094, 2159, 2224, 2243, ...
TTT	25	2.8%	..., 1917, 1982, 2485, 2550, 2615]

a. N = Penumbral, P = Partial, and T = Total.
b. When a year is bounded by a square bracket "[" or "]", there are no other examples beyond that year.

When four eclipses occur in one calendar year, there are three possible combinations of eclipse types N, P, and T. Table 3-9 lists the frequency of each eclipse combination along with five recent years when each combination occurs. The table makes no distinction in the order of eclipses in the seven combinations. The rarest combination NNPP occurs just 32 times.

Table 3-9. Four Lunar Eclipses in One Year

Eclipse Combinationsa	Number of Years	Percent	Examples (Years)
NNNN	390	72.4%	..., 1944, 2020, 2038, 2056, 2085, ...
NNNP	117	21.7%	..., 1991, 2009, 2150, 2197, 2215, ...
NNPP	32	5.9%	..., 1684, 2205, 2411, 2726, 2819, ...

The maximum number of five lunar eclipses in one calendar year is quite rare. Over the 5,000-year span of the Canon, there are only 33 years containing five lunar eclipses. They occur in two possible combinations of eclipse types where four out of the five eclipses are always of type N. The first eclipse of such a quintet always occurs in the first half of January, while the last eclipse falls in the latter half of December. Table 3-10 lists the frequency of the two eclipse combinations, along with recent years when each combination occurs. The rarest combination NNNPP (actually either PNPNN or NNPNP) occurred just three times. Once again, the table makes no distinction in the order of eclipses in any combination.

Table 3-10. Five Lunar Eclipses in One Year

Eclipse Combinations[a]	Number of Years	Percent	Examples (Years)[b]
NNNNP	30	90.9%	..., 1879, 2132, 2262, 2400, 2653, ...
NNNPP	3	9.1%	[0475, 1694, 1749]

a. N = Penumbral, P = Partial, and T = Total.
b. When years are surrounded by square brackets [ ], there are no other examples outside this range.

3.5 Extremes in Eclipse Magnitude—Penumbral Lunar Eclipses

The penumbral eclipse magnitude is defined as the fraction of the Moon's diameter immersed in Earth's penumbral shadow. It is a unitless parameter that can be expressed numerically by

$$Mp = xp/dm, \qquad (3\text{-}1)$$

where Mp is the penumbral eclipse magnitude, dm is the apparent diameter of the Moon, and xp is the distance measured from the edge of the penumbral shadow to the edge of the Moon deepest in the penumbra.

The penumbral eclipse magnitude reaches its maximum value at the instant of greatest eclipse. This is the value shown in the figures in the Appendix. A search through the 12,064 eclipses in the Canon reveals some interesting cases involving extreme values of the penumbral eclipse magnitude.

Seven penumbral eclipses have a maximum magnitude less than 0.0020 (Table 3-11). These events are all the first or last members in a Saros series. The smallest magnitude was the penumbral eclipse of –0780 Dec 13 with a magnitude of just 0.0004.

Table 3-11. Penumbral Lunar Eclipses with Magnitude ≤ 0.0020

Date (Dynamical Time)	Saros	Gamma	Penumbral Eclipse Magnitude	Penumbral Eclipse Duration
-0780 Dec 13	61	-1.5529	0.0004	5.2m
-0411 Jan 23	27	1.5791	0.0013	11.0m
-0331 May 07	74	1.5502	0.0008	7.8m
-0103 Feb 07	80	1.5540	0.0019	12.0m
0859 May 20	112	1.5700	0.0007	8.1m
2027 Jul 18	110	-1.5758	0.0014	11.8m
2791 Feb 11	175	-1.5670	0.0006	7.4m

Table 3-12 lists the nine penumbral eclipses having a maximum magnitude greater than 1.0800. The greatest penumbral eclipse occurred on 1322 Nov 24 with a maximum magnitude of 1.0951. Because the penumbral magnitudes of these eclipses are all greater than 1, they are classified as total penumbral eclipses, i.e., the Moon is completely immersed within the penumbral shadow.

Table 3-12. Total Penumbral Lunar Eclipses with Magnitude ≥ 1.0800

Date (Dynamical Time)	Saros	Gamma	Penumbral Eclipse Magnitude	Penumbral Eclipse Duration
-1517 Jan 12	0	-0.9901	1.0858	291.2m
-1348 Jan 03	32	0.9948	1.0831	294.4m
-1058 Oct 29	19	0.9917	1.0860	291.6m
0348 Oct 24	90	0.9890	1.0946	296.1m
1322 Nov 24	95	0.9897	1.0951	296.5m
1340 Dec 04	95	0.9970	1.0820	295.1m
1988 Mar 03	113	0.9885	1.0907	293.8m
2429 Dec 11	132	-0.9904	1.0853	290.1m
2732 Apr 02	163	-0.9928	1.0814	293.2m

3.6 Extremes in Eclipse Magnitude—Partial Lunar Eclipses

The umbral eclipse magnitude is defined as the fraction of the Moon's diameter immersed in Earth's umbral shadow. It is a unitless parameter that can be expressed numerically by

$$Mu = xu/dm \qquad (3\text{-}2)$$

where Mu is the umbral eclipse magnitude, dm is the apparent diameter of the Moon, and xu is the distance measured from the edge of the umbral shadow to the edge of the Moon deepest in the umbra.

The umbral eclipse magnitude reaches its maximum value at the instant of greatest eclipse. This is the value shown in the figures in the Appendix. A search through the 12,064 eclipses in the Canon reveals some interesting cases involving extreme values of the umbral eclipse magnitude in partial and total lunar eclipses.

Eleven partial eclipses have a maximum magnitude (at greatest eclipse) less than or equal to 0.0020 (Table 3-13). The partial eclipse with the smallest magnitude (at greatest eclipse) occurred on 1553 Jul 25 and had a magnitude of just 0.0001.

Table 3-13. Partial Lunar Eclipses with Magnitude ≤ 0.0020

Date (Dynamical Time)	Saros	Gamma	Umbral Eclipse Magnitude	Partial Eclipse Duration
-0602 May 03	31	0.9940	0.0015	9.7m
-0519 May 15	32	-1.0261	0.0004	4.3m
-0292 Mar 27	66	1.0016	0.0010	7.5m
1050 Feb 09	108	0.9861	0.0013	8.9m
1416 Nov 05	97	1.0107	0.0019	9.9m
1430 Sep 02	128	1.0229	0.0008	6.3m
1553 Jul 25	131	-1.0253	0.0001	2.5m
1890 Nov 26	114	-0.9994	0.0017	9.8m
2157 Feb 24	145	-0.9868	0.0005	5.6m
2421 Jun 16	156	1.0225	0.0011	7.3m
2627 May 22	160	0.9914	0.0011	8.0m

Nine partial eclipses have a maximum magnitude (at greatest eclipse) greater than or equal to 0.9980 (Table 3-14). The partial eclipse with the largest magnitude (at greatest eclipse) occurred on –1972 Jun 27 with a magnitude of 0.9998.

Table 3-14. Partial Lunar Eclipses with Magnitude ≥ 0.9980

Date (Dynamical Time)	Saros	Gamma	Umbral Eclipse Magnitude	Partial Eclipse Duration
-1972 Jun 27	-2	-0.4505	0.9998	209.0m
-1021 May 16	24	-0.4758	0.9986	192.6m
-0247 Oct 03	52	-0.4532	0.9988	202.7m
0145 Sep 18	78	0.4511	0.9994	205.0m
0274 Apr 08	65	0.4523	0.9993	207.1m
0778 Sep 11	98	0.4670	0.9993	195.0m
1001 Mar 12	87	0.4728	0.9995	191.1m
1165 May 27	107	-0.4727	0.9986	194.5m
2413 Nov 08	152	0.4733	0.9993	189.5m

3.7 Extremes in Eclipse Magnitude—Total Lunar Eclipses

Twelve total eclipses have a maximum magnitude less than or equal to 1.0020 (Table 3-15). The smallest magnitude was the total eclipse of 1529 Oct 17 with a value of just 1.0001 (note the upcoming total eclipse of 2015 Apr 04 and its magnitude of 1.0008). Because the Moon's passage through the umbral shadow is so shallow, these events all have short total phases less than 7 min.

Table 3-15. Total Lunar Eclipses with Magnitude ≤ 1.0020

Date (Dynamical Time)	Saros	Gamma	Umbral Eclipse Magnitude	Total Eclipse Duration
-1573 Jan 31	18	0.4776	1.0008	4.2m
-1338 Jun 09	28	0.4622	1.0012	5.5m
-0318 Aug 10	62	0.4474	1.0009	4.9m
-0200 Mar 30	48	-0.4790	1.0019	6.6m
0767 Apr 19	92	0.4581	1.0002	2.4m
0792 Dec 03	96	0.4413	1.0004	3.1m
1529 Oct 17	109	0.4775	1.0001	1.7m
2015 Apr 04	132	0.4460	1.0008	4.7m
2155 Sep 11	130	-0.4752	1.0003	2.6m
2366 May 25	146	0.4817	1.0007	3.9m
2565 Sep 11	156	0.4700	1.0009	4.6m
2905 Jun 08	164	0.4748	1.0018	6.5m

Eight total eclipses have a maximum magnitude greater than or equal to 1.8700. Their total phase durations all last 98 to 99 min and gamma values are all close to 0.0 indicating that the Moon passes centrally through the middle of the umbral shadow. These eclipses all take place when the Moon is near perigee—the time when the ratio of the Moon's to umbra's diameters is at maximum. The total eclipse with the largest magnitude (1.8821) occurs on 2756 Jun 05.

Table 3-16. Total Lunar Eclipses with Magnitude ≥ 1.8700

Date (Dynamical Time)	Saros	Gamma	Umbral Eclipse Magnitude	Total Eclipse Duration
-0731 Apr 19	39	-0.0054	1.8766	99.4m
-0630 Nov 06	45	-0.0002	1.8703	98.3m
-0051 Apr 01	60	-0.0009	1.8774	99.7m
0564 Sep 06	85	0.0017	1.8764	98.9m
1092 Apr 24	97	0.0080	1.8702	99.3m
1226 Aug 09	106	-0.0021	1.8770	99.6m
1631 May 15	115	0.0052	1.8721	99.8m
2756 Jun 05	152	0.0018	1.8821	99.3m

3.8 Greatest Duration—Penumbral Lunar Eclipses

Nine penumbral eclipses each have durations of 292 min or more. Because the penumbral eclipse magnitude of each of these events is greater than 1, they are all classified as total penumbral eclipses. Each eclipse in Table 3-17 occurs near lunar apogee and within several weeks of aphelion. Earth's penumbra is then largest, while the Moon exhibits a small angular diameter and travels relatively slowly, thereby extending the length of the eclipse.

Table 3-17. Penumbral Lunar Eclipses with a Duration ≥ 292 min

Date (Dynamical Time)	Saros	Gamma	Penumbral Eclipse Magnitude	Penumbral Eclipse Duration
-1367 Dec 23	32	1.0089	1.0588	292.2m
-1348 Jan 03	32	0.9948	1.0831	294.4m
0348 Oct 24	90	0.9890	1.0946	296.1m
1322 Nov 24	95	0.9897	1.0951	296.5m
1340 Dec 04	95	0.9970	1.0820	295.1m
1358 Dec 16	95	1.0039	1.0693	293.6m
1376 Dec 26	95	1.0104	1.0569	292.1m
1988 Mar 03	113	0.9885	1.0907	293.8m
2732 Apr 02	163	-0.9928	1.0814	293.2m

3.9 Greatest Duration—Partial Lunar Eclipses

Seven partial eclipses have durations of 209 min or more. All of these events occur with the Moon close to apogee so its slow trajectory through the umbra tends to prolong the partial phase.

Table 3-18. Partial Lunar Eclipses with a Duration ≥ 209 min

Date (Dynamical Time)	Saros	Gamma	Umbral Eclipse Magnitude	Partial Eclipse Duration
-1274 Jun 21	19	0.4611	0.9768	209.3m
-1133 May 24	22	-0.4592	0.9814	209.6m
-0764 Jan 26	31	0.4540	0.9771	209.1m
-0387 Oct 20	59	-0.4458	0.9908	209.3m
0303 Sep 12	71	0.4510	0.9864	209.6m
1382 Sep 23	117	-0.4458	0.9977	209.6m
2669 Feb 08	144	-0.4424	0.9951	210.0m

3.10 Greatest Duration—Total Lunar Eclipses

Eighteen total eclipses have durations greater than or equal to 106 min. The most recent one was 2000 Jul 16, while the next is 2123 Jun 09. All of these eclipses occur near aphelion and with the Moon close to apogee. Consequently, the Moon has a small angular diameter coupled with a relatively slow orbital motion with respect to the umbra. Furthermore, the umbral shadow is largest when Earth is at aphelion. Such conditions prolong the Moon's passage through the umbra to produce long total lunar eclipses (Meeus, 2002).

Earth's tropical longitude of perihelion is currently increasing by 1.72° per century. This means that the date of aphelion occurred ~17 days earlier for each 1000 years backwards from the present. This too is reflected in the dates seen in Table 3-19.

Table 3-19. Total Lunar Eclipses with a Duration ≥ 106 min

Date (Dynamical Time)	Saros	Gamma	Umbral Eclipse Magnitude	Total Eclipse Duration
-1921 Apr 07	1	0.0302	1.7681	106.2m
-1646 Jun 24	13	0.0164	1.7946	106.3m
-1505 May 27	16	-0.0171	1.7948	106.5m
-1364 Apr 28	19	0.0187	1.7910	106.4m
-0380 Jun 06	53	-0.0308	1.7703	106.2m
-0239 May 09	56	0.0301	1.7713	106.1m
-0098 Apr 11	59	-0.0016	1.8212	106.1m
0054 Aug 07	68	-0.0147	1.7924	106.2m
0177 Jun 28	71	-0.0198	1.7874	106.5m
0318 May 31	74	0.0160	1.7951	106.6m
0459 May 03	77	0.0033	1.8167	106.5m
1443 Jun 12	111	-0.0098	1.8097	106.2m
1584 May 24	114	-0.0065	1.8145	106.1m
1859 Aug 13	126	0.0038	1.8148	106.5m
2000 Jul 16	129	0.0302	1.7684	106.4m
2123 Jun 09	132	0.0406	1.7487	106.1m
2141 Jun 19	132	-0.0446	1.7415	106.1m
2264 May 12	135	0.0121	1.7979	106.2m

3.11 Total Penumbral Lunar Eclipses

During most penumbral eclipses, only part of the Moon passes through Earth's penumbral shadow. However, it is also possible to have a penumbral eclipse in which the entire Moon passes completely within Earth's penumbral shadow without entering the inner umbral shadow. The geocentric apparent diameter of the Moon ranges from an extreme minimum of 1763.0 arcsec (apogee) to an extreme maximum of 2011.8 arcsec (perigee). The penumbral annulus formed by the zone between the outer edges of the penumbra and umbra also undergoes extremes ranging from 1887.7 arcsec (aphelion) to 1951.9 arcsec (perihelion). From these values, it is apparent that the Moon cannot fit entirely within the penumbral annulus when it is near perigee (Meeus, 1997). As a consequence of the restrictive geometry and conditions, total penumbral eclipses are quite rare and account for just 3.2 %, or 141 out of 4378 penumbral eclipses in the Canon. Table 3-20 lists the dates of all 141 total penumbral eclipses.

The frequency of total penumbral eclipse varies considerably with time. If the 5,000-year period covered by the Canon is divided into 100-year intervals, the number of total penumbral eclipses per century varies from 0 to a maximum of 9 (Table 3.4). For instance, the number of total penumbral eclipses in centuries beginning in 1701, 1801, and 1901 are 0, 2, and 9, respectively. The number of total penumbral eclipses appears to be correlated with the number of total eclipses, as well as the number of tetrads per century, and inversely correlated with the overall number of all lunar eclipses per century. When a century has a relatively large number of lunar eclipses, it has fewer total lunar eclipses and few or no total penumbral lunar eclipses or tetrads (Figure 3.1)

Table 3-20. Total Penumbral Lunar Eclipses: –1999 to +3000 (2000 BCE to 3000 CE)

-1989 Dec 01	-1142 Jun 02	-0274 Oct 02	0682 Apr 27	1455 Oct 25	2103 Jan 23
-1981 Jul 07	-1058 Oct 29	-0263 Aug 31	0711 Oct 01	1496 Jul 25	2121 Feb 02
-1971 Dec 12	-1040 Nov 08	-0205 Jan 26	0729 Oct 11	1502 Apr 22	2128 Mar 16
-1953 Dec 23	-1022 Nov 19	-0186 Jun 22	0805 Mar 19	1513 Sep 15	2139 Feb 13
-1840 Jun 08	-1018 Mar 14	-0045 May 25	0816 Aug 11	1542 Aug 25	2222 Aug 23
-1699 May 11	-1007 Aug 07	0096 Apr 26	0822 May 09	1607 Mar 13	2429 Dec 11
-1669 Sep 05	-1004 Nov 30	0154 Sep 09	0823 Sep 24	1637 Jul 07	2447 Dec 22
-1604 Mar 13	-1001 May 05	0197 Nov 12	0862 Sep 12	1665 Jul 27	2458 May 28
-1575 Aug 17	-0986 Dec 11	0215 Nov 24	0945 Mar 31	1806 Jun 30	2466 Jan 01
-1564 Jul 16	-0968 Dec 21	0219 Mar 18	0974 Sep 04	1900 Jun 13	2484 Jan 13
-1558 Apr 13	-0964 Oct 10	0233 Dec 04	0996 Jan 08	1901 May 03	2498 Sep 30
-1517 Jan 12	-0949 Jan 02	0248 Aug 21	1014 Jan 19	1908 Dec 07	2502 Jan 24
-1499 Jan 22	-0946 Oct 21	0251 Dec 15	1032 Jan 30	1926 Dec 19	2520 Feb 04
-1457 Oct 30	-0943 Sep 17	0265 Apr 18	1097 Jul 27	1944 Dec 29	2523 Nov 24
-1452 Jul 08	-0931 Jan 12	0269 Dec 26	1221 May 08	1948 Oct 18	2538 Feb 15
-1439 Nov 10	-0925 Sep 29	0288 Jan 06	1322 Nov 24	1963 Jan 09	2562 Nov 12
-1435 Mar 05	-0860 Apr 06	0312 Oct 02	1340 Dec 04	1981 Jan 20	2656 Oct 25
-1421 Nov 21	-0838 Jul 29	0330 Oct 13	1344 Mar 29	1988 Mar 03	2714 Mar 23
-1406 Aug 09	-0831 Sep 10	0348 Oct 24	1358 Dec 16	1999 Jan 31	2732 Apr 02
-1403 Dec 01	-0755 Feb 15	0388 Mar 09	1373 Sep 02	2006 Mar 14	2761 Sep 06
-1385 Dec 13	-0726 Jul 22	0417 Aug 13	1376 Dec 26	2053 Aug 29	2885 Jun 18
-1367 Dec 23	-0461 Apr 05	0512 Jun 15	1390 Apr 29	2070 Apr 25	
-1348 Jan 03	-0432 Sep 08	0653 May 18	1395 Jan 06	2082 Aug 08	
-1283 Jun 30	-0356 Feb 14	0681 Jun 07	1413 Jan 17	2099 Sep 29	

3.12 Lunar Eclipse Duos

A duo is a pair of eclipses separated by one lunation (synodic month). Of the 12,064 eclipses in the Canon, 3,054 eclipses (25.3%) belong to a duo. One eclipse of a duo always passes north of Earth's shadow axis, while the other eclipse passes to the south. In most cases, both eclipses in a duo are penumbral eclipses; however, there are 51 instances (3.3% of duos) in the Canon where one eclipse is penumbral and the other is partial. In each of these pairs, the penumbral magnitude of the penumbral eclipse is quite small, as is the umbral magnitude of the partial eclipse. The dates and eclipse combinations of all lunar eclipse duos are listed in Table 3-21.

Table 3-21. Lunar Eclipse Duos of Two Types

–1952 May-Jun - NP	–0668 May-Jun - NP	0475 Jun-Jul - NP	1296 Apr-May - NP	2288 Jul-Aug - PN
–1811 Apr-May - NP	–0527 Apr-May - NP	0483 Jul-Aug - PN	1427 Apr-May - PN	2411 Jun-Jul - PN
–1774 May-May - PN	–0519 May-Jun - PN	0606 May-Jun - PN	1430 Aug-Sep - NP	2429 Jun-Jul - PN
–1670 Mar-Apr - NP	–0252 Jul-Jul - NP	0624 Jun-Jul - PN	1553 Jun-Jul - NP	2552 May-Jun - PN
–1492 Mar-Apr - PN	–0111 Jun-Jul - NP	0747 Apr-May - PN	1608 Jul-Aug - PN	2819 Jun-Jul - NP
–1358 Jun-Jul - PN	–0074 Jun-Jul - PN	0888 Mar-Apr - PN	1694 Jun-Jul - NP	2827 Jul-Aug - PN
–1217 May-Jun - PN	0030 May-Jun - NP	0891 Jul-Aug - NP	1749 Jun-Jul - PN	2960 Jun-Jun - NP
–1076 Apr-May - PN	0067 May-Jun - PN	1014 Jun-Jul - NP	1835 May-Jun - NP	
–0935 Mar-Apr - PN	0153 Mar-Apr - NP	1022 Jul-Aug - PN	1958 Apr-May - NP	
–0932 Jul-Aug - NP	0208 Apr-May - PN	1155 May-Jun - NP	2013 Apr-May - PN	
–0809 Jun-Jul - NP	0428 Jun-Jul - NP	1163 Jun-Jul - PN	2147 Aug-Sep - PN	

3.13 Lunar Eclipses Duos in One Calendar Month

There are 57 instances where both members of an eclipse duo occur in one calendar month. In most cases, both eclipses in the duos are penumbral. In three instances, the duo consists of the NP combination (–1774 May, –0252 Jul, and 2960 Jun). The year and month of each duo appears in Table 3-22.

Table 3-22. Two Lunar Eclipses in One Calendar Month

-1883 Sep	-1033 Jan	-0078 Mar	0768 Sep	1694 Dec	2566 Aug
-1861 Jul	-0765 Aug	0019 Dec	0790 Jul	1705 Nov	2577 Jul
-1807 Aug	-0686 May	0074 Jan	0877 May	1716 Oct	2838 Jan
-1796 Jul	-0675 Apr	0269 Jul	1116 Jan	1817 May	2848 Dec
-1785 Jun	-0599 Mar	0291 May	1126 Dec	1904 Mar	2957 Aug
-1785 Jun	-0512 Jan	0356 May	1213 Oct	2172 Oct	2960 Jun
-1706 Mar	-0502 Dec	0367 Apr	1289 Sep	2208 May	2968 Jul
-1286 Aug	-0447 Jan	0443 Mar	1300 Aug	2284 Apr	
-1275 Jul	-0252 Jul	0595 Jan	1311 Jul	2295 Mar	
-1185 Mar	-0165 May	0605 Dec	1376 Jul	2382 Jan	

3.14 January–March Lunar Eclipse Duos

The mean length of one synodic month is 29.5306 days (in year 2000). Because this is longer than the month of February, it is possible to have one member of an eclipse duo in January followed by the second in March. There are six instances of such a rare January/March duo in the Canon: –1109, –0523, –0002, 1915, 2306, and 2371. In all cases, both eclipses in the duos are penumbral.

3.15 Total Lunar Eclipse Multiplets

A total lunar eclipse is usually preceded or succeeded by at least one other total lunar eclipse. Of the 3479 total eclipses in the Canon, 1798 of them (51.7%) are part of a doublet. Another 1023 eclipses (29.4%) belong to a triplet. Finally, 568 total eclipses (16.3%) are part of a quadruplet known as a tetrad. In comparison, only 90 total eclipses (2.6%) occur as solitary singlets.

A key feature of total lunar eclipse multiplets is that the individual members are always separated by six lunations. A summary of the muliplet statistics appears in Table 3-23.

Table 3-23. Total Lunar Eclipse Multiplets

Total Eclipse Multiplet	Number of Eclipses Per Multiplet	Number of Multiplets	Number of Total Eclipses	Percent of Total Eclipses	Recent Examples
All Total Eclipses	-	-	3479	100.0 %	
Singlet	1	90	90	2.6 %	1997 Sep 16; 2021 May 26
Doublet	2	899	1798	51.7 %	1978; 1996; 2022; 2040
Triplet	3	341	1023	29.4 %	2007 to 2008; 2010 to 2011
Tetrad	4	142	568	16.3 %	2003 to 2004; 2014 to 2015

3.16 Lunar Eclipse Tetrads

When four consecutive lunar eclipses are all total, the group is termed a tetrad. As discussed previously, 16.3% (568 out of 3479) of all total eclipses are members of a tetrad. These 142 groupings of total eclipses occur because of the eccentricity of Earth's orbit in conjunction with the timing of eclipse seasons (Section 3.18). During the first 1,000 years of the Canon, the first eclipse of every tetrad occurs sometime from December to May. In later millennia, the first eclipse date occurs later in the year because of precession—during the 3rd Millennium, the period for the first date extends from February to July. For a detailed description of tetrad geometry and an explanation of why tetrads happen, see Meeus (2004b).

Italian astronomer Giovanni Schiaparelli first pointed out that the frequency of tetrads is variable over time. He noticed that tetrads were relatively plentiful during one 300-year interval, while none occurred during the next 300 years. For example, there are no tetrads from 1582 to 1908, but 17 tetrads occur during the following 2 1/2 centuries from 1909 to 2156. This can be seen graphically in Figure 3-1 (i). The 565-year period of the tetrad "seasons" is tied to the slowly decreasing eccentricity of Earth's orbit. Consequently, the tetrad period is gradually decreasing (Hughes, 2004). In the distant future, when Earth's eccentricity is 0, tetrads will no longer be possible.

The umbral magnitudes of the total eclipses making up a tetrad are all relatively small. For the 300-year period 1901 to 2200, the largest umbral magnitude of a tetrad eclipse is 1.4251 on 1949 Apr 13. For comparison, some other total eclipses during this period are much deeper. Two examples are the total eclipses of 2000 Jul 16 and 2029 Jun 26 with umbral magnitudes of 1.7684 and 1.8436, respectively.

Table 3-24 lists the date of the first total lunar eclipse in each of the 142 tetrads in the Canon.

Table 3-24. Date of First Eclipse in Lunar Eclipse Tetrads

-1991 Dec 22	-0999 Mar 14	-0283 Mar 18	0766 Apr 29	1457 Mar 11	2137 Mar 07
-1661 Apr 12	-0981 Mar 25	-0272 Feb 16	0784 May 09	1475 Mar 22	2155 Mar 19
-1643 Apr 22	-0963 Apr 04	-0243 Jan 26	0795 Apr 09	1493 Apr 02	2448 Jun 17
-1625 May 04	-0945 Apr 16	-0225 Feb 06	0802 May 21	1504 Mar 01	2466 Jun 28
-1596 Apr 13	-0923 Feb 12	-0207 Feb 16	0813 Apr 19	1515 Jan 30	2477 May 28
-1585 Mar 13	-0916 Mar 26	-0189 Feb 28	0824 Mar 18	1522 Mar 12	2495 Jun 08
-1578 Apr 24	-0905 Feb 24	-0168 Dec 27	0842 Mar 30	1533 Feb 09	2506 May 08
-1574 Feb 10	-0887 Mar 06	-0149 Jan 07	0860 Apr 09	1562 Jan 20	2524 May 19
-1567 Mar 24	-0876 Feb 03	0162 Apr 17	0878 Apr 20	1580 Jan 31	2542 May 30
-1556 Feb 21	-0869 Mar 17	0180 Apr 27	0889 Mar 21	1909 Jun 04	2564 Mar 29
-1538 Mar 04	-0858 Feb 14	0198 May 08	0900 Feb 18	1927 Jun 15	2571 May 11
-1520 Mar 14	-0847 Jan 14	0227 Apr 19	0918 Feb 28	1949 Apr 13	2582 Apr 09
-1473 Mar 06	-0840 Feb 25	0238 Mar 18	0936 Mar 11	1967 Apr 24	2589 May 21
-1462 Feb 02	-0829 Jan 25	0245 Apr 29	0947 Feb 08	1985 May 04	2600 Apr 20
-1444 Feb 13	-0782 Jan 15	0256 Mar 28	0965 Feb 18	2003 May 16	2611 Mar 20
-1433 Jan 13	-0453 May 06	0267 Feb 26	0976 Jan 19	2014 Apr 15	2618 May 01
-1415 Jan 23	-0377 Apr 06	0285 Mar 08	0994 Jan 30	2032 Apr 25	2629 Mar 31
-1405 Dec 24	-0359 Apr 17	0332 Feb 28	1305 May 09	2043 Mar 25	2640 Feb 29
-1386 Jan 04	-0341 Apr 28	0350 Mar 10	1323 May 21	2050 May 06	2647 Apr 11
-1368 Jan 15	-0330 Mar 28	0361 Feb 06	1341 May 31	2061 Apr 04	2658 Mar 11
-1057 Apr 24	-0319 Feb 24	0390 Jan 17	1352 Apr 30	2072 Mar 04	2676 Mar 22
-1039 May 05	-0312 Apr 07	0408 Jan 29	1370 May 11	2090 Mar 15	2987 Jul 02
-1028 Apr 03	-0301 Mar 07	0437 Jan 08	1399 Apr 20	2101 Feb 14	
-1010 Apr 14	-0290 Feb 04	0455 Jan 19	1428 Mar 31	2119 Feb 25	

3.17 Lunar Eclipses on February 29

There are six instances of a lunar eclipse occurring on February 29. Two eclipses are penumbral, two are partial, and two are total. A list of eclipses on February 29 with physical parameters appears in Table 3-25.

Table 3-25. Lunar Eclipses on February 29

Date (Dynamical Time)	Eclipse Type	Saros	Gamma	Umbral Eclipse Magnitude
-0664 Feb 29	N	22	-1.3596	0.3819[a]
-0588 Feb 29	N	62	1.4929	0.1363[a]
-0124 Feb 29	P	50	-0.5634	0.8114
1420 Feb 29	P	94	-0.9205	0.1614
2268 Feb 29	T-	137	-0.1142	1.6602
2640 Feb 29	T	143	0.3353	1.2478

a. Penumbral Eclipse Magnitude

3.18 Eclipse Seasons

The 5.1° inclination of the lunar orbit around Earth means that the Moon's orbit crosses the ecliptic at two points or nodes. If Full Moon takes place within about 17° of a node[13], then a lunar eclipse will be visible from a portion of Earth.

The Sun makes one complete circuit of the ecliptic in 365.24 days, so its average angular velocity is 0.99° per day. At this rate, it takes 34.5 days for the Sun—and at the opposite node, Earth's umbral and penumbral shadows—to cross the 34° wide eclipse zone centered on each node. Because the Moon's orbit with respect to the Sun has a mean duration of 29.53 days, there will always be one, and possibly two, lunar eclipses during each 34.5-day interval when the Sun (and Earth's shadows) pass through the nodal eclipse zones. These time periods are called eclipse seasons.

The mid-point of each eclipse season is separated by 173.3 days because this is the mean time for the Sun to travel from one node to the next. The period is a little less than half a calendar year because the lunar nodes slowly regress westward by 19.3° per year.

3.19 Quincena

The mean time interval between New Moon and Full Moon is 14.77 days. This is less than half the duration of an eclipse season. As a consequence, the same Sun-node alignment geometry responsible for producing a lunar eclipse always results in a complementary solar eclipse within a fortnight. The solar eclipse may either precede or succeed the lunar eclipse. In either case, the pair of eclipses is referred to here as a quincena[14]. The Quincena Solar Eclipse (QSE) parameter identifies the type of the solar eclipse and whether it precedes or succeeds a particular lunar eclipse. There are four basic types of solar eclipses:

1) p = partial solar eclipse
(Moon's penumbral shadow traverses Earth; umbral/antumbral shadow completely misses Earth)
2) a = annular solar eclipse
(Moon's antumbral[15] shadow traverses Earth; Moon is too far from Earth to completely cover the Sun)
3) t = total solar eclipse
(Moon'sumbral shadow traverses Earth; Moon is close enough to Earth to completely cover the Sun)
4) h = hybrid solar eclipse
(Moon's umbral and antumbral shadows traverse different parts of Earth; eclipse appears either total or annular along different sections of its path—hybrid eclipses are also known as annular-total eclipses)

---

13 The actual value ranges from 15.3° to 17.1° of a node because of the eccentricity of the Moon's (and Earth's) orbit.

14 Quincena is a Spanish word for a period of about 15 days.

15 The cone-shaped umbra gradually narrows to a point. Extending the sides of the umbra beyond this vertex forms an inverted cone. This zone is known as the antumbra. It corresponds to the region in the Moon's shadow where the Moon appears smaller than the Sun. The Moon is then seen in complete silhouette against the Sun's photospheric disk.

The QSE is a two character string consisting of one or more of the above solar eclipse types. The first character in the QSE identifies a solar eclipse preceding a lunar eclipse while the second character identifies a solar eclipse succeeding a lunar eclipse. In most instances, one of the two characters is "–" indicating a single solar eclipse either precedes or succeeds the lunar eclipse. On rare occasions, a double quincena occurs in which a lunar eclipse is both preceded and succeeded by solar eclipses.

3.20 Quincena Combinations with Total Lunar Eclipses

A total lunar eclipse can be preceded or succeeded by a total solar eclipse (7.7%), an annular solar eclipse (10.2%), a hybrid solar eclipse (0.4%), or a partial solar eclipse (42.5%). Double quincenas (a lunar eclipse is both preceded and succeeded by a solar eclipse) occur with a frequency of 39.1% and usually consist of two partial solar eclipses (38.7%). In rare instances, a double quincena consists of a total and a partial solar eclipse (0.4%). A complete list of all QSE combinations with total lunar eclipses appears in Table 3-26. The most recent years when each quincena combination occurs are given in the last column.

Table 3-26 Quincena Combinations with Total Lunar Eclipses

Quincena Solar Eclipse	QSE	Number	Percent	Examples (Years) [a]
– total	-t	127	3.7%	..., 1985, 2003, 2043, 2061, 2072,...
total –	t–	139	4.0%	..., 1957, 1968, 2015, 2033, 2044,...
– annular	-a	178	5.1%	..., 1891, 2003, 2014, 2021, 2032,...
annular –	a–	178	5.1%	..., 1990, 2008, 2026, 2044, 2102,...
– hybrid	-h	12	0.3%	..., 1627, 1645, 1768, 1909, 2050,...
hybrid –	h–	5	0.1%	[-1989, -1848, -1642, 0163, 1986]
– partial	-p	730	21.0%	..., 2000, 2007, 2010, 2014, 2018,...
partial –	p–	749	21.5%	..., 2001, 2004, 2011, 2015, 2019,...
total – partial	tp	8	0.2%	..., -0434, -0159, 1248, 1928, 2912]
partial – total	pt	6	0.2%	[-1310, -1169, -1028, -0026, 2195, 2459]
partial – partial	pp	1347	38.7%	..., 1982, 2000, 2011, 2018, 2029,...

a. When a year is bounded by a square bracket "[" or "]", there are no other examples beyond that year.

3.21 Quincena Combinations with Partial Lunar Eclipses

A partial lunar eclipse can be preceded or succeeded by a total solar eclipse (37.6%), an annular solar eclipse (55.1%), or a hybrid solar eclipse (6.9%). In rare instances, a partial lunar eclipse can be followed by a partial solar eclipse (0.3%). Double quincenas do not occur with partial lunar eclipses. A list of quincena solar eclipse combinations with partial lunar eclipses appears in Table 3-27. The most recent years when each quincena combination occurs are given in the last column.

Table 3-27. Quincena Combinations with Partial Lunar Eclipses

Quincena Solar Eclipse	QSE	Number	Percent	Examples (Years)
– total	-t	782	18.6%	..., 1992, 1999, 2010, 2017, 2021,...
total –	t–	801	19.0%	..., 2001, 2008, 2019, 2026, 2037,...
– annular	-a	1149	27.3%	..., 1995, 2006, 2009, 2013, 2024,...
annular –	a–	1171	27.8%	..., 1994, 2005, 2012, 2023, 2030,...
– hybrid	-h	167	4.0%	..., 1912, 1930, 2209, 2350, 2368,...
hybrid –	h–	124	2.9%	..., 1827, 1845, 2164, 2182, 2323,...
– partial	-p	13	0.3%	..., -0754, -0196, 2086, 2607, 2625,...

3.22 Quincena Combinations with Penumbral Lunar Eclipses

A penumbral lunar eclipse can be preceded or succeeded by a total solar eclipse (39.5%), an annular solar eclipse (51.9%), or a hybrid solar eclipse (8.7%). There are no instances of a quincena involving a partial solar and a penumbral lunar eclipse, nor are there any double quincenas in the Canon. A list of quincena solar eclipse combinations with penumbral lunar eclipses appears in Table 3-28. The most recent years when each quincena combination occurs are given in the last column.

Table 3-28. Quincena Combinations with Penumbral Lunar Eclipses

Quincena Solar Eclipse	QSE	Number	Percent	Examples (Years)
- total	-t	878	20.1%	..., 2002, 2006, 2009, 2020, 2024,...
total -	t-	848	19.4%	..., 1998, 2009, 2012, 2016, 2027,...
- annular	-a	1144	26.1%	..., 1999, 2002, 2017, 2020, 2031,...
annular -	a-	1128	25.8%	..., 2001, 2002, 2009, 2013, 2016,...
- hybrid	-h	176	4.0%	..., 1846, 1908, 2013, 2031, 2049,...
hybrid -	h-	204	4.7%	..., 1987, 2005, 2023, 2172, 2190,...

Section 4: Eclipses and the Moon's Orbit

4.1 Introduction

The Moon revolves around Earth in an elliptical orbit with a mean eccentricity of 0.0549. Thus, the Moon's center-to-center distance from Earth varies with mean values of 363,396 km at perigee to 405,504 km at apogee. The lunar orbital period with respect to the stars (sidereal month) is 27.32166 days (27d 07h 43m 12s). However, there are three other orbital periods or months that are crucial to the understanding and prediction of eclipses. These three cycles and the harmonics between them determine when, where, and how solar (and lunar) eclipses occur.

The mutual gravitational force between the Sun and Moon is over twice as large as between the Moon and Earth. For this reason, the Sun plays a dominant role in perturbing the Moon's motion. The ever changing distances and relative positions between the Sun, Moon, and Earth, the inclination of the Moon's orbit, the oblateness of Earth, and (to a lesser extent) the gravitational attraction of the other planets all act to throw the Moon's orbital parameters into a constant state of change. Although the Moon's position and velocity can be described by the classic Keplerian orbital elements, such osculating elements are only valid for a single instant in time (Chapront-Touze' and Chapront, 1991). Nevertheless, these instantaneous parameters are of value in understanding the Moon's complex motions particularly with respect to the three major orbital cycles that govern eclipses.

4.2 Synodic Month

The most familiar lunar cycle is the synodic month because it governs the well-known cycle of the Moon's phases. The Moon has no light of its own but shines by reflected sunlight. As a consequence, the geometry of its orbital position relative to the Sun and Earth determines the Moon's apparent phase.

The mean length of the synodic month is 29.53059 days (29d 12h 44m 03s). This is nearly 2.21 days longer than the sidereal month. As the Moon revolves around Earth, both objects also progress in orbit around the Sun. After completing one revolution with respect to the stars, the Moon must continue a little farther along its orbit to catch up to the same position it started from relative to the Sun and Earth. This explains why the mean synodic month is longer than the sidereal month.

According to astronomical convention, New Moon is defined as the instant when the geocentric ecliptic longitudes of the Sun and Moon are equal. When the synodic month is measured from New Moon to New Moon, it is sometimes referred to as a lunation, and we will follow that usage here. Historically, the phases of the Moon have been used as the basis of lunar calendars by many cultures around the world. The major problem with such calendars is that the year, based on the solar calendar, is not evenly divisible by a whole number of lunations. Consequently, most lunar calendars are actually lunisolar calendars (e.g., Chinese, Hebrew, and Hindu) that include intercalary months to keep the seasons in step with the year.

The duration of the lunation actually varies from its mean value by up to seven hours. For instance, Table 4-1 contains details for all lunations in 2008. The first column lists the decimal date of every New Moon throughout the year (Terrestrial Dynamical Time), while the second column gives the duration of each lunation. The third column is the difference between the actual and mean lunation. The first lunation of the year (Jan 08) was 03h 23m longer than the mean. Continuing through 2008, the length of each lunation drops and reaches a minimum of 05h 48m shorter than the mean value (Jun 03). The duration now increases with each succeeding lunation until the maximum value of the year is reached of 06h 49m longer than the mean (Dec 27).

Table 4-1 New Moon and Lunation Length in 2008

Date of New Moon (Dynamical Time)	Length of Lunation	Difference From Mean Lunation	Moon's True Anomaly
2008 Jan 08.4849	29d 16h 07m	+03h 23m	242.4°
2008 Feb 07.1567	29d 13h 30m	+00h 46m	280.0°
2008 Mar 07.7190	29d 10h 41m	-02h 03m	310.8°
2008 Apr 06.1642	29d 08h 23m	-04h 21m	332.7°
2008 May 05.5134	29d 07h 04m	-05h 40m	349.4°
2008 Jun 03.8081	29d 06h 56m	-05h 48m	4.4°
2008 Jul 03.0970	29d 07h 54m	-04h 50m	20.1°
2008 Aug 01.4261	29d 09h 45m	-02h 59m	39.2°
2008 Aug 30.8327	29d 12h 14m	-00h 30m	64.9°
2008 Sep 29.3426	29d 15h 02m	+02h 18m	98.7°
2008 Oct 28.9687	29d 17h 41m	+04h 57m	133.4°
2008 Nov 27.7053	29d 19h 28m	+06h 44m	161.9°
2008 Dec 27.5163	29d 19h 33m	+06h 49m	186.6°

What is the cause of this odd behavior? The last column in Table 4-1 gives a clue; it contains the Moon's true anomaly at the instant of New Moon. The true anomaly is the angle between the Moon's position and the point of perigee along its orbit. In other words, it is the orbital longitude of the Moon with respect to perigee. Table 4-1 shows that when New Moon occurs near perigee (true anomaly = 0°), the length of the lunation is at a minimum (e.g., Jun 03). Similarly, when New Moon occurs near apogee (true anomaly = 180°), the length of the lunation reaches a maximum (e.g., Dec 27).

Figure 4–1: Length of Lunation for 2008 – 2010

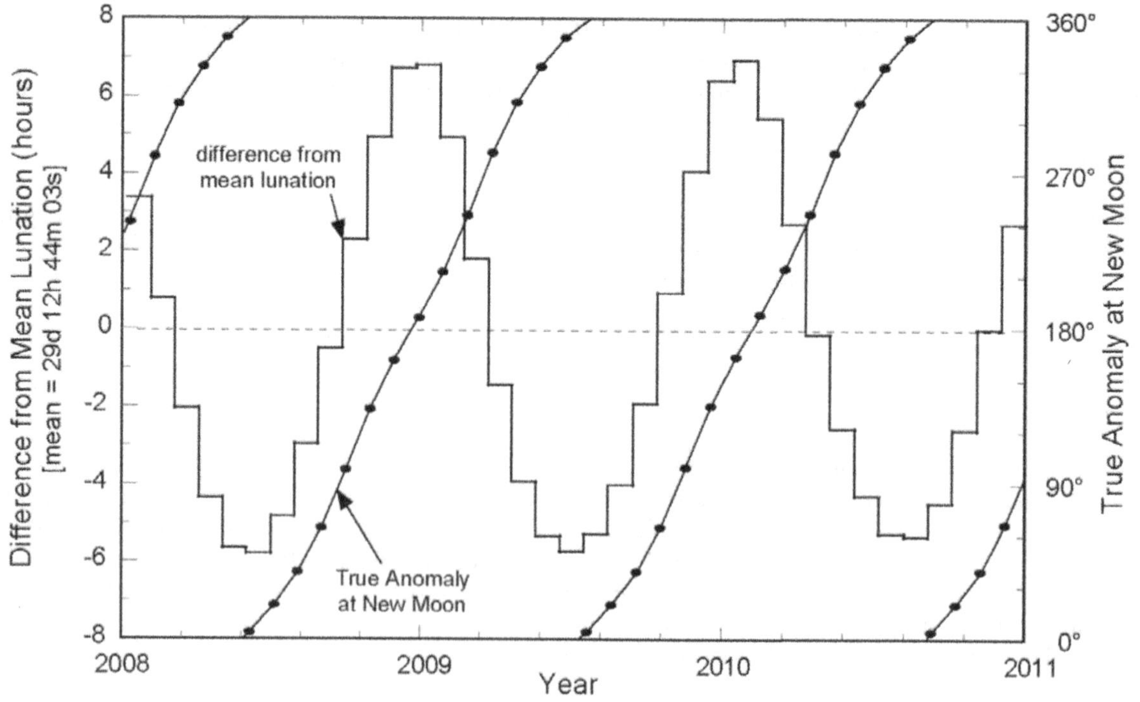

This relationship is quite apparent when viewed graphically. Figure 4-1 plots the difference from mean lunation (histogram) and the Moon's true anomaly (diagonal curves) for every New Moon from 2008 through 2010. The left-hand scale is for the difference from mean lunation, while the right-hand scale is for the true anomaly. The shortest lunations are clearly correlated with New Moon at perigee, while the longest lunations occur at apogee. From the figure, the length of this cycle appears to be about 412 days. The reason why must wait until the next section.

The Moon's orbital period with respect to perigee is the anomalistic month and has a duration of approximately 27.55 days. The lock-step rhythm between the lunation length and true anomaly can be explained with the help of the anomalistic month and Figure 4-2. It illustrates the Moon's orbit around Earth and Earth's orbit around the Sun. The relative sizes and distances of the Sun, Moon, and Earth as well as the eccentricity of the Moon's orbit are all exaggerated for clarity. The major axis of the Moon's orbit marks the positions of perigee and apogee.

Figure 4–2: Moon's Orbit and the Synodic Month

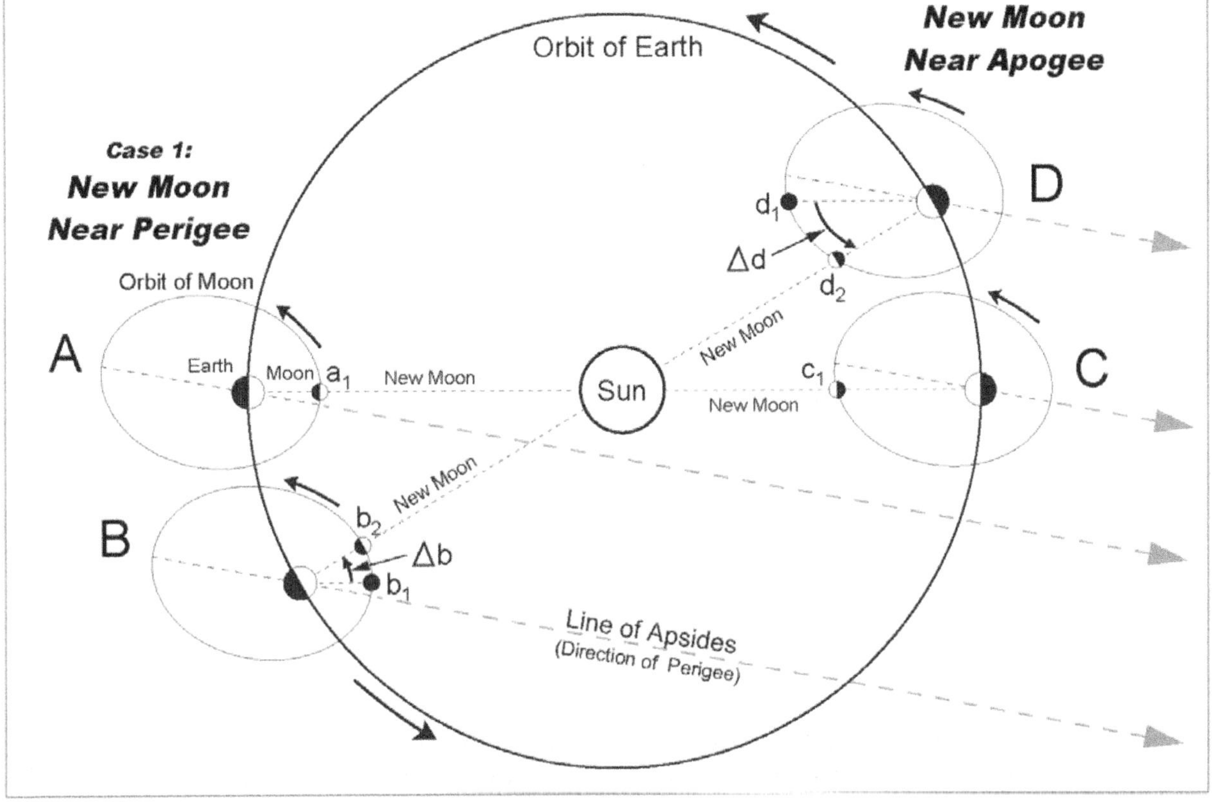

Two distinct cases—each consisting of two revolutions of the Moon around Earth—are depicted in Figure 4-2. The first case covers the New Moon geometry around perigee. The orbit marked A shows New Moon taking place near perigee at position a_1. One anomalistic month later (orbit B), the Moon has returned to the same position relative to perigee (marked b_1). However, Earth has traveled about 30° around its orbit so the Sun's direction relative to the Moon's major axis has shifted. The Moon must travel an additional distance of Δb in its orbit before reaching the New Moon phase at b_2. This graphically demonstrates why the synodic month is longer (~1.98 days) than the anomalistic month.

The second case takes place about half a year later. New Moon then occurs near apogee (orbit C, position c_1). After one anomalistic month, the Moon has returned to the same location with respect to apogee (orbit D, position d_1). Once again, Earth has traveled about 30° around its orbit so the Moon must revolve an additional distance of Δd before reaching the New Moon phase at position d_2.

An inspection of orbits B and D reveals that the orbital arc Δd is longer that Δb. This means that the Moon must cover a greater orbital distance to reach New Moon near apogee as compared to perigee. Furthermore, the Moon's orbital velocity is slower at apogee so it takes longer to travel a given distance. Thus, the length of the lunation is shorter than average when New Moon occurs near perigee and longer than average when New Moon occurs near apogee.

Earth's elliptical orbit around the Sun also factors into the length of the lunation. With an eccentricity of 0.0167, Earth's orbit is about one third as elliptical as the Moon's orbit. Nevertheless, it affects the length of the lunation by producing shorter lunations near aphelion and longer lunations near perihelion.

During the 5000-year period covered in this canon, there are 61841 complete lunations. The shortest lunation began on –1602 Jun 03 and lasted 29.26574 days (29d 06h 22m 40s; 6h 21m 23s shorter than the mean). The longest lunation began on –1868 Nov 27 and lasted 29.84089 days (29d 20h 10m 53s; 7h 26m 50s longer than the mean). Thus, the duration of the lunation varies over a range of 13h 48m 13s during this time interval.

Figure 4–3: Length of Lunation Over 5000 Years

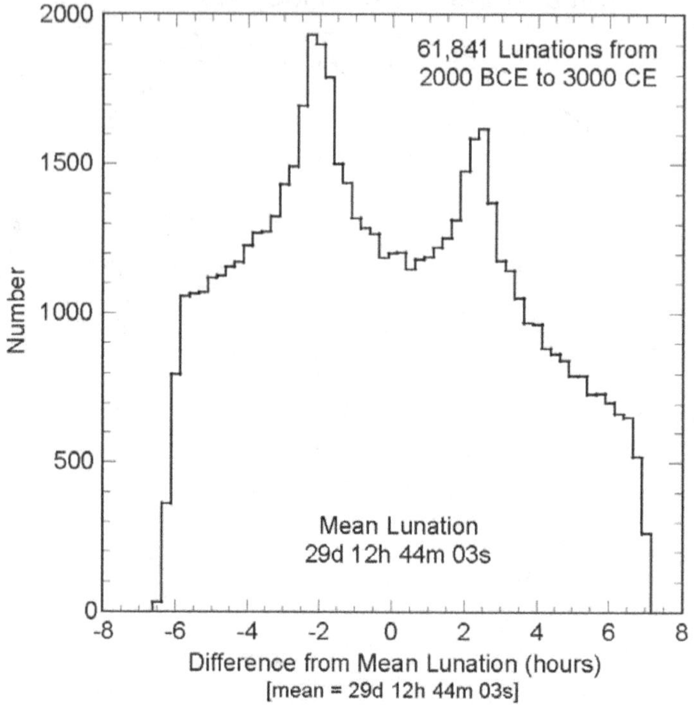

The histogram presented in Figure 4-3 shows the distribution in the length of the lunation over 5000 years. To create the histogram, the durations of individual lunations were binned into 30-minute groups. It might seem reasonable to expect a simple bell-shaped Gaussian curve. However, the results are surprising because the distribution in lunation length has two distinct peaks. This bifurcation can be understood if the lunation length, which depends primarily on the Moon's distance, is considered as a series of sine functions. The extremes of a sine function always occur more frequently than the mean, which is just what is seen in Figure 4-3. For a more detailed discussion, see Meeus (1997).

4.3 Anomalistic Month

The anomalistic month is defined as the revolution of the Moon around its elliptical orbit as measured from perigee to perigee. The length of this period can vary by several days from its mean value of 27.55455 days (27d 13h 18m 33s). Figure 4-4 plots the difference of the anomalistic month from the mean value for the 3-year interval 2008 through 2010. Also plotted is the difference between the mean longitudes of the Sun and perigee. This is just the angle between the Sun and the Moon's major axis in the direction of perigee. The left-hand scale is the length of the anomalistic month minus the mean value, while the right-hand scale is for the difference in longitude (Sun–perigee). For comparison, the lunation length minus its mean value is also plotted (light gray).

Figure 4–4: Length of Anomalistic Month for 2008 – 2010

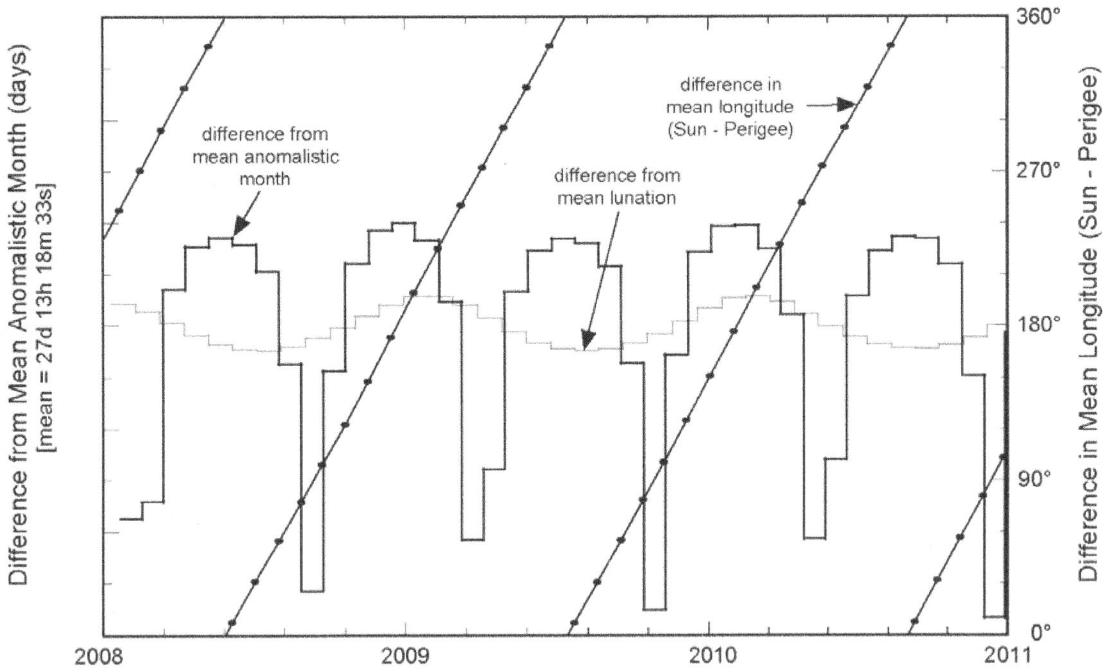

The variation in the length of the anomalistic month is much larger than that of the lunation. Figure 4-4 shows the anomalistic month is typically within 1 day of its mean value. But once or twice every 7 to 8 months, the anomalistic month is significantly shorter than the mean by 2 to nearly 3 days. The difference in longitude of the Sun and perigee show that the shortest anomalistic months are correlated with values of 90° and 270°, when the line of apsides is perpendicular to the Sun's direction.

In comparison, the longest anomalistic months take place when the difference in longitude passes through 0° or 180°. The line of apsides is then directed towards or away from the Sun. The maximum duration of the anomalistic month is then about 28.5 days (1.0 day longer than the mean). The Earth–Sun distance also influences the anomalistic month by causing greater extremes near perihelion. This currently occurs in early January each year.

In an earlier discussion on the synodic month, it was assumed that the lunar orbit's line of apsides has a fixed and permanent direction in space. In fact, the length of the mean anomalistic month (27.55 days) exceeds the mean sidereal month (27.32 days) by 0.23 days. Thus, the Moon's major axis slowly shifts with a mean rate of 0.11140° per day in the direct sense, that is, in the same direction as the Moon's orbital motion. This corresponds to an average of 40.7° per year, so it takes 8.85 years (3231.6 days) for the line of apsides to make one complete revolution with respect to the stars.

What impact do the varying length of the anomalistic month and the direct (eastward) rotation of the Moon's elliptical orbit have on the length of the lunation? To answer this, one must first consider Earth's elliptical orbit around the Sun, which has a mean eccentricity of 0.0167. The center-to-center distance between Earth and the Sun varies with mean values of 147,098,074 km at perihelion to 152,097,701 km at aphelion. The direction of Earth's orbital line of apsides also changes but at a rate far slower than the Moon's. Having a direct (eastward) shift with a mean value of 0.0172° per year, it takes about 20,500 years for Earth's major axis to make one complete revolution. This is only 0.0004 of the lunar rate, so it can be treated as fixed for the purpose of the following discussion.

At certain times, the perigee of the lunar orbit and the perihelion of Earth's orbit can have the same ecliptic longitude. Ignoring the 5.1° tilt of the Moon's obit, the major axes are then essentially parallel to each other and point in the same direction. As time passes, the major axis of the lunar orbit slowly rotates east with respect to Earth's major axis until it becomes perpendicular to it 2.21 years later. In another 2.21 years (4.42 years from the start), the major axes of the orbits are again parallel to each other, but the perigee and the perihelion are 180° apart as they point in opposite directions. After an additional period of 2.21 years, the axes are once more perpendicular. Finally, the Moon's perigee and Earth's perihelion again share the same ecliptic longitude after a total interval of 8.85 years.

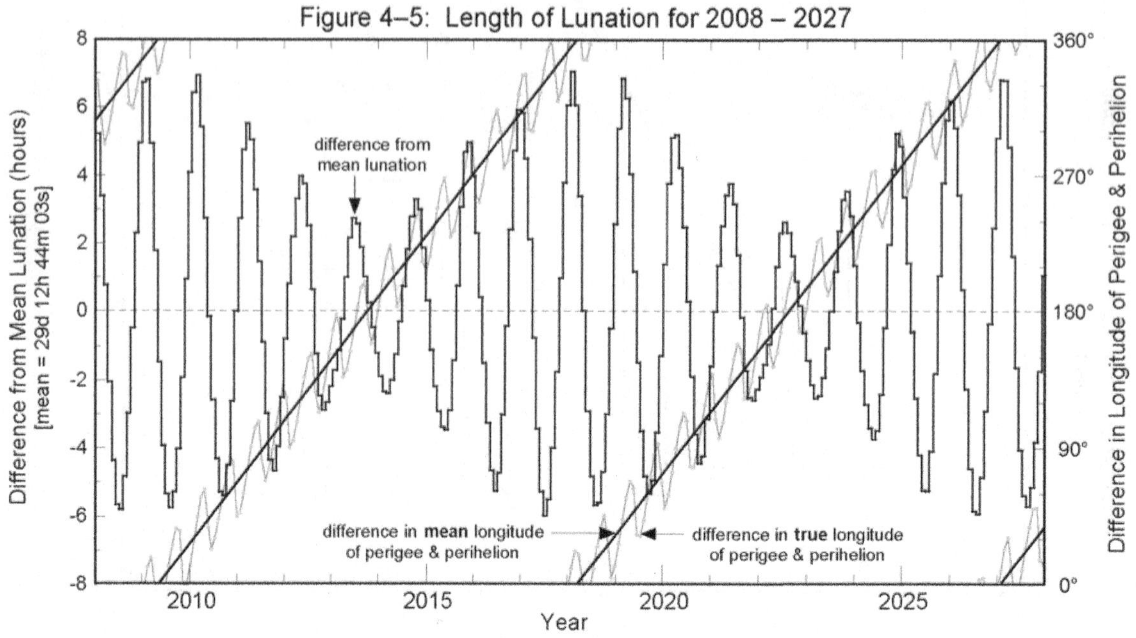

The length of each lunation minus the mean lunation is plotted in Figure 4-5 for the 20-year period from 2008 through 2027. The periodic rhythm between the lunation length and the true anomaly, as described earlier (via Figure 4-1), can now be seen over the course of two decades. The 412-day mean period of this cycle corresponds to the time between two consecutive alignments of the major axis in the direction of the Sun. It is slightly longer than a year because of the slow eastward shift of the Moon's major axis.

An interesting feature revealed in Figure 4-5 is how the extremes in the lunation length slowly vary over a period of nearly 9 years. The envelope defined by the minima and maxima appears to oscillate over a range of values from ±2 h to ±6 h. This behavior is evidence revealing the influence of the 8.85-year cycle in the alignment of the major axes of the orbits of the Moon and Earth.

The amplitude of the envelope is due to the eccentricity of Earth's orbit. When Earth is at perihelion, its orbital velocity is at its maximum value so Earth travels a larger distance around its orbit in a given time as compared to aphelion. Thus, the Moon must travel a greater distance to align with the Sun, which results in a longer lunation. Near aphelion, the opposite conditions produce a shorter lunation.

Using the axis scale on the right, the diagonal lines in Figure 4-5 plot the angle between the Moon's perigee and Earth's perihelion. This is the difference between the Moon's mean longitude of perigee and Earth's true longitude of perihelion. When the angle between the perigee and perihelion is 0°, the length of the lunation varies from a minimum of 29.273 days (-6.17 hours from mean) to a maximum of 29.820 days (+6.93 hours from mean). Similarly, when the angle between the perigee and perihelion is 180°, the length of the lunation varies from a minimum of 29.452 days (-1.88 hours from mean) to a maximum of 29.628 days (+2.33 hours from mean). To summarize, the greatest extremes in the length of the lunation occur when the longitudes of the Moon's perigee and Earth's perihelion are equal. The smallest extremes in the lunation length occur when their longitudes differ by 180°.

Although the Moon's major axis rotates eastward at a mean rate of 0.1114° per day, the true rate varies considerably. Figure 4-5 illustrates the variation by plotting the difference between the true longitudes of the Moon's perigee and Earth's perihelion. This quasi-sinusoidal oscillation about the difference in the mean longitudes shows peak departures of ±30° from average. Indeed, the Moon's major axis can swing both east and west of its mean value, taking on an actual retrograde shift west during some anomalistic months.

This dynamic behavior is due to the gravitational pull of the Sun on the Moon as it orbits Earth. Consequently, a continuous torque is applied to the lunar orbit in an unsuccessful effort to permanently align the major axis towards the Sun. The annual orbit of the Earth–Moon system around the Sun coupled with the Moon's synodic orbit around Earth mean that the conditions for such a permanent alignment are always changing. The overall effect is to twist and distort the shape and orientation of the Moon's elliptical orbit.

Figure 4–6: Moon's Eccentricity for 2008 – 2010

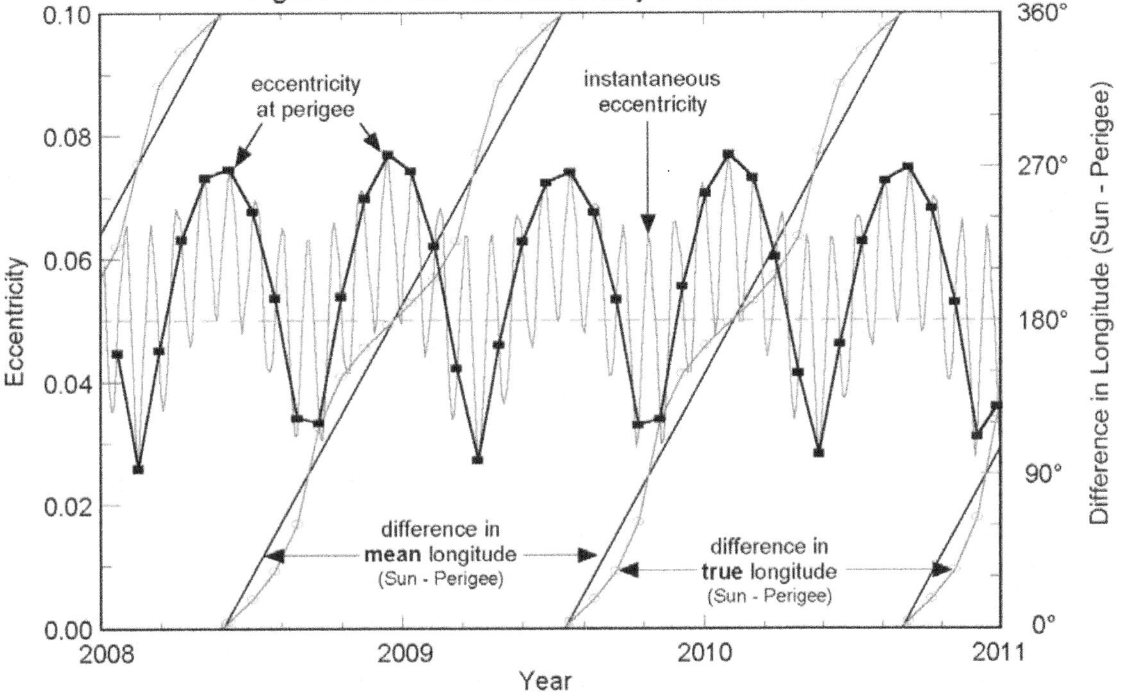

It was stated earlier that the Moon's mean orbital eccentricity is 0.0549, but this too is subject to large changes because of solar perturbations. Figure 4-6 plots the variation in the Moon's orbital eccentricity from 2008 through 2010. The instantaneous eccentricity (light gray curve) oscillates with a period tied to the synodic month and ranges from 0.0266 to 0.0762 over this 3-year interval. Superimposed on the instantaneous eccentricity is the eccentricity at the instant of perigee, which occurs at the beginning of each anomalistic month (heavy black curve). The straight diagonal lines represent the difference between the mean longitudes of the Sun and perigee. In other words, it is the angle between the Moon's perigee-directed major axis and the Sun. Oscillating about this line is the difference between the true longitudes of Sun and perigee. The scale for these angles appears along the right side of Figure 4-6. The extreme range of the Moon's orbital eccentricity at perigee during the 5000 years of the canon is 0.0255 to 0.0775.

Figure 4-6 shows that the eccentricity reaches a maximum when the major axis of the lunar orbit is pointed directly towards or directly away from the Sun (angles of 0° and 180°, respectively). This occurs at a mean interval of 205.9 days, which is somewhat longer than half a year because of the eastward shift of the major axis. The eccentricity reaches a minimum when the major axis of the lunar orbit is perpendicular to the Sun (angles of 90° and 270°).

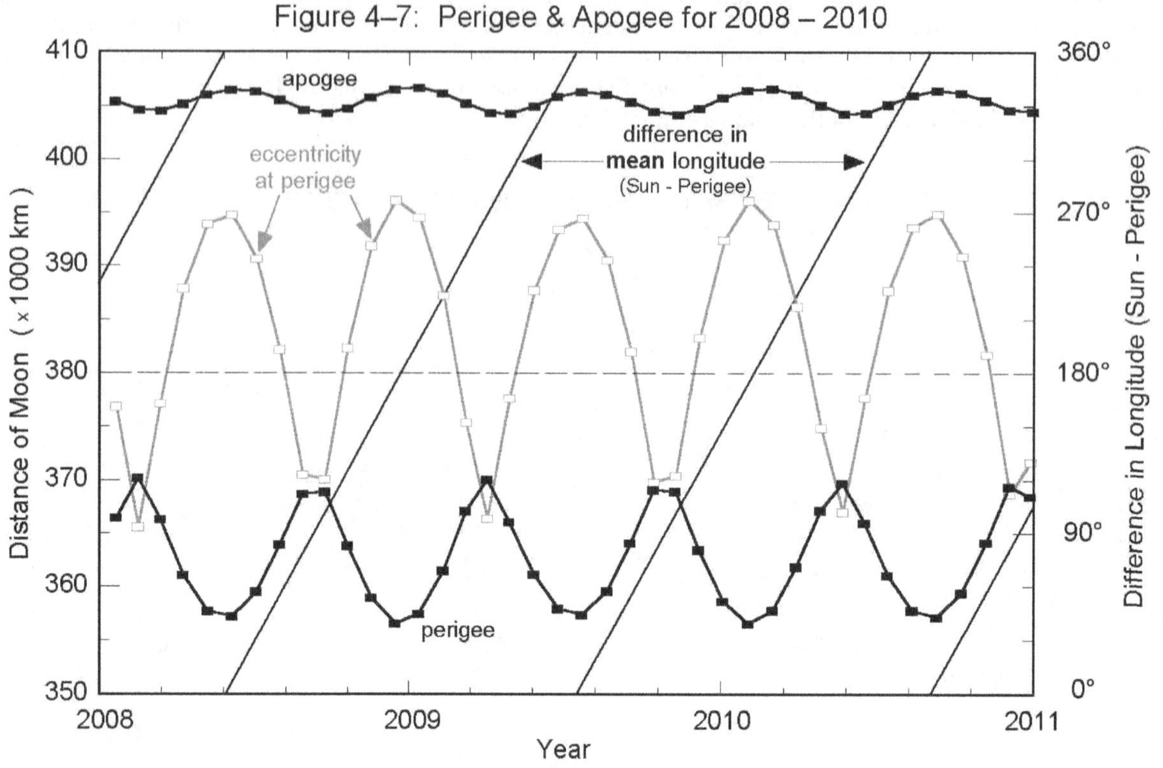

Such changes in orbital eccentricity produce significant variations in the Moon's distance at perigee and apogee. Figure 4-7 plots the Moon's distance for all perigees and apogees from 2008 through 2010. Also shown is the orbital eccentricity at perigee as well as the angle between the perigee directed major axis and the Sun. The closest perigee (minimum perigee distance) and farthest apogee (maximum apogee distance) occur when the eccentricity is at maximum. This corresponds to times when the Moon's major axis points directly towards or directly away from the Sun (angles of 0° and 180°, respectively). The farthest perigee (maximum perigee distance) and closest apogee (minimum apogee distance) occur when the eccentricity is at minimum. At such times, the major axis is oriented perpendicular to the Sun. During the 3-year interval covered in Figure 4-7, the Moon's perigee distance ranges from 356,568 to 370,216 km while the apogee distance ranges from 404,168 to 406,602 km.

Over the 5000-year period of the canon, there are 66,276 perigees and apogees. During this epoch, the distance of the Moon's perigee varies from 356,355 to 370,399 km while the apogee varies from 404,042 to 406,725 km. The minimum and maximum extremes in orbital eccentricity are 0.0255 to 0.0775 and the extremes in the length of the anomalistic month are 24.629 days (2.925 days shorter than the mean) to 28.565 days (1.011 days longer than the mean). A histogram showing the distribution in the length of the anomalistic month is presented in Figure 4-8 where the durations of individual anomalistic months have been binned into 2-hour groups. The sharply asymmetric distribution shows that anomalistic months longer than the mean cluster over a much shorter range of values compared to anomalistic months shorter than the mean.

Figure 4–8: Length of Anomalistic Month Over 5000 Years

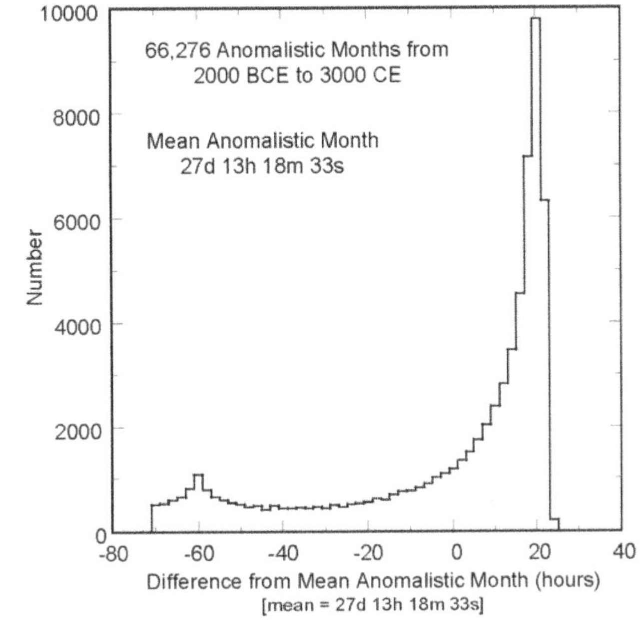

4.4 Draconic Month

The plane of the Moon's orbit is inclined at a mean angle of 5.145° to the plane of Earth's orbit around the Sun. The intersection of these planes defines two points or nodes on the celestial sphere. The node where the Moon's path crosses the ecliptic from south to north is the ascending node, while the node where the Moon's path crosses the ecliptic from north to south is the descending node.

Figure 4–9: Length of Draconic Month for 2008 – 2010

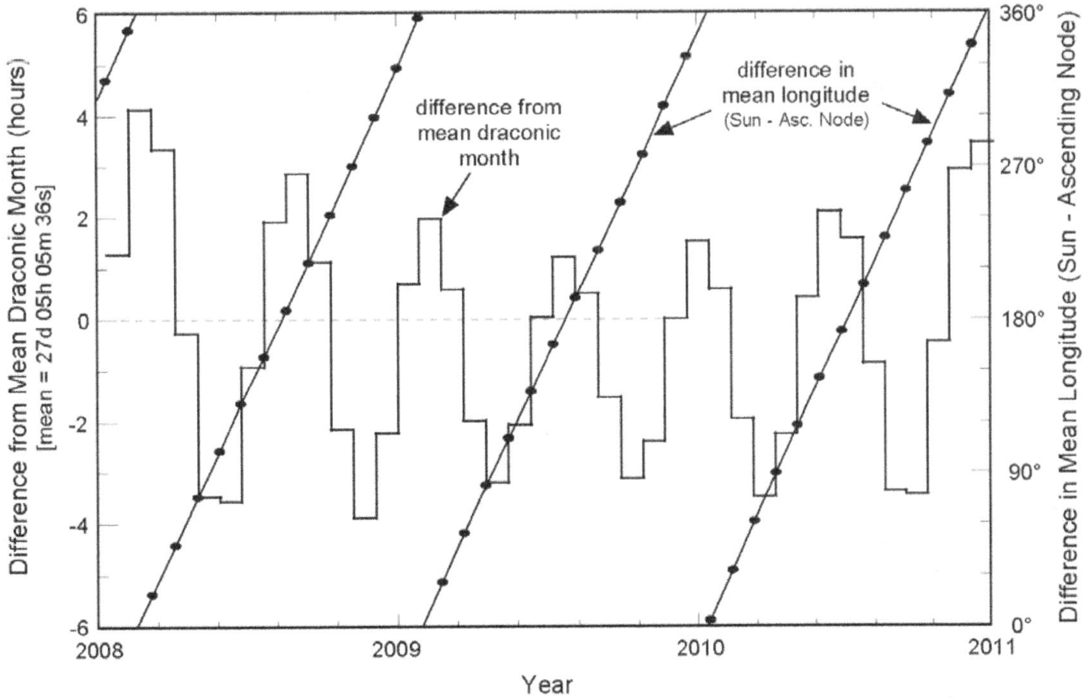

The draconic month is defined as one revolution of the Moon about its orbit with respect to the ascending node. The mean length of this nodical period is 27.21222 days (27d 05h 05m 36s). However, the actual duration can vary by over 6 h from the mean. Figure 4-9 plots the duration of the draconic month minus its mean value for 2008 through 2010. The shortest month over this 3-year period is 27.05115 days (27d 01h 14m), while the longest month is 27.38409 days (27d 09h 13m).

The most significant characteristic of this variation is that it is synchronized with the ascending node relative to the Sun's position along the ecliptic. The mean angle between the Sun and the ascending node (i.e., difference in mean longitude) is also plotted in Figure 4-9 (diagonal lines) to illustrate this relationship. The longitude difference at the start of each draconic month is plotted as a black dot. Longitude values can be read using the scale along the right side of the figure. The longest draconic months occur when the difference in the mean longitudes of the Sun and the ascending node is either 0° or 180°. In contrast, the shortest months occur when the angle between the Sun and the ascending node is either 90° or 270°.

The mean draconic month is 0.10944 day (2h 36m 36s) shorter than the sidereal month. Consequently, the lunar nodes slowly rotate west or retrograde (opposite the Moon's orbital motion) along the ecliptic at a rate of 0.05295° per day. One complete rotation of the ascending node about the ecliptic requires 18.6 years (6793.48 days) with respect to the fixed stars.

Figure 4–10: Lunar Orbit Inclination for 2008 – 2010

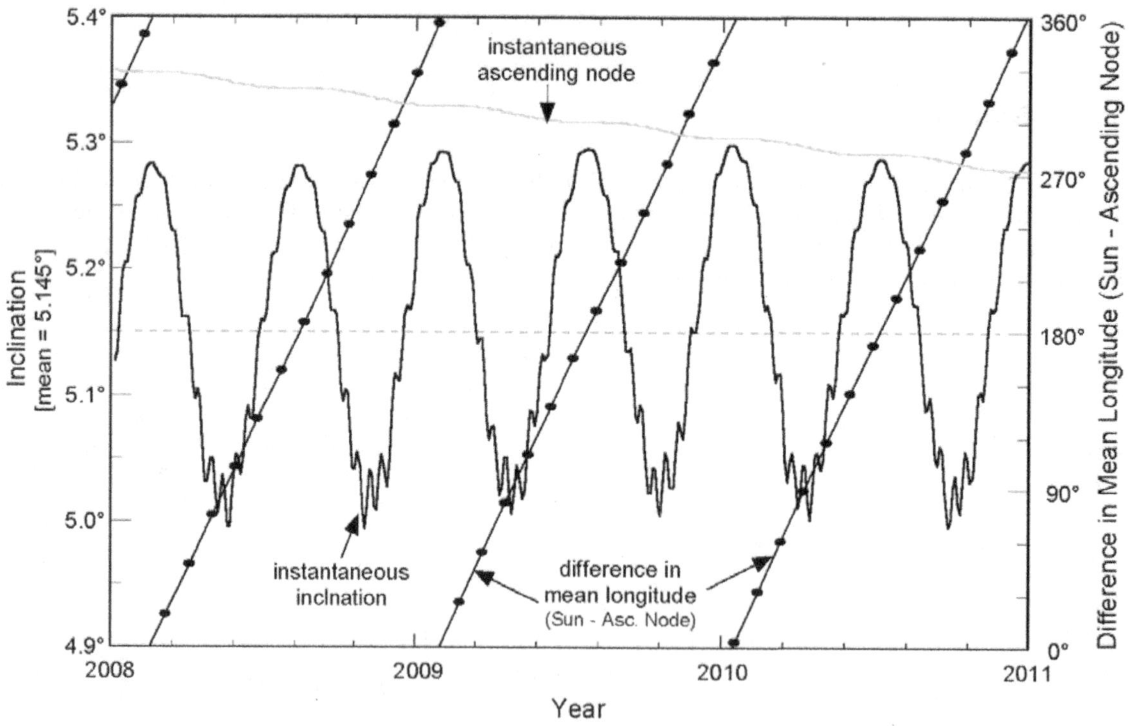

Figure 4-10 plots the instantaneous inclination of the lunar orbit over the 3-year period 2008–2010. The mean angle between the Sun and the ascending node (i.e., difference in mean longitude) is also plotted. The largest inclination of 5.30° occurs when the difference in longitude is either 0° or 180°. In other words, the inclination is always near its maximum value for both solar and lunar eclipses. The smallest inclination of 5.00° occurs when the difference in longitude is either 90° or 270°. Note the small monthly oscillations in the inclination when near its minimum. The figure also plots the longitude of the instantaneous ascending node. Its westward motion draws to a near standstill whenever the Sun aligns with either of the nodes. This corresponds to a difference in longitude of either 0° or 180°.

The mean interval in the periodic variation of both the draconic month and the orbital inclination is 173.3 days. This is the average time it takes for the Sun to travel from one node to the other. It is also equivalent to the interval between the midpoints of two eclipse seasons. The period is slightly less than half a year because of the retrograde motion of the nodes.

Figure 4–11: Draconic Month and Perigee for 2008 – 2017

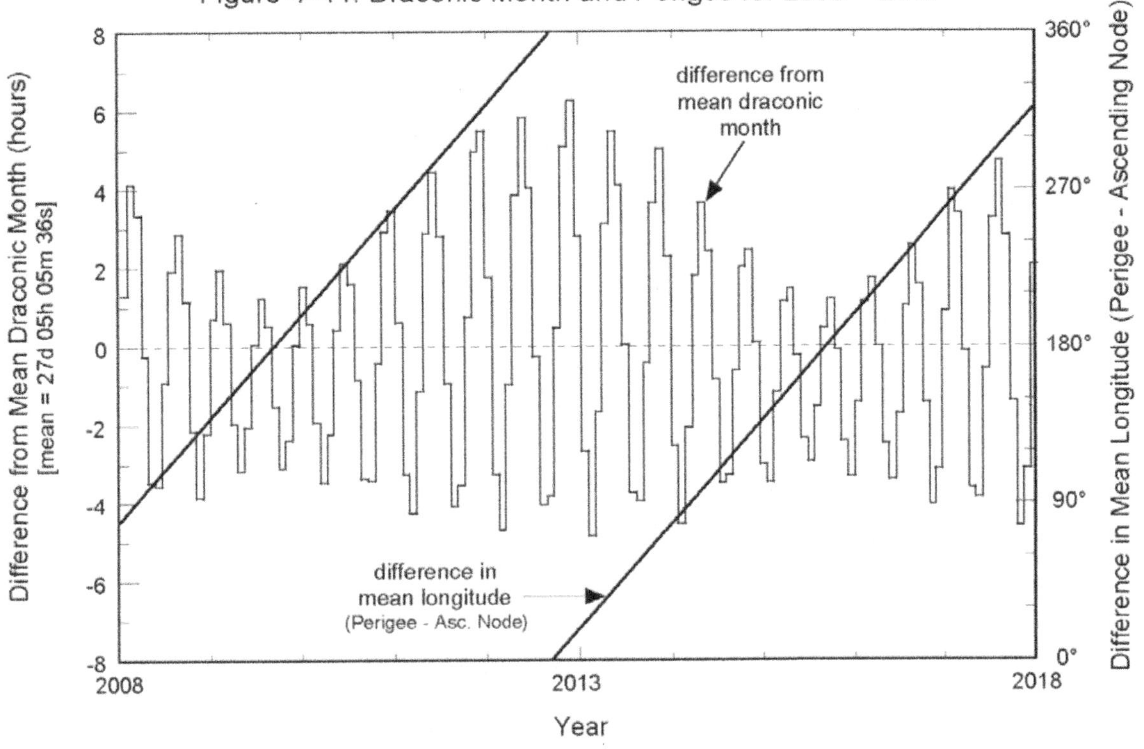

The length of the draconic month is strongly modulated by the position of the nodes with respect to the major axis of the Moon's orbit. The histogram in Figure 4-11 shows how the draconic month changes from 2008 through 2017. The 173-day alignment of the Sun with a node appears as the rapid oscillation in the month length. The quasi-sinusoidal envelopes surrounding the minima and maxima form two longer period oscillations. Over the 10-year period covered in this figure, the minimum month duration varies from 27.089 to 27.011 days (3.0 to 4.8 hours shorter than the mean). The maximum month duration ranges from 27.261 to 27.472 days (1.2 to 6.2 hours longer than the mean).

The difference in the mean longitudes of perigee and the ascending node appear as diagonal lines in Figure 4-11. This is the angle between these orbital parameters measured along the ecliptic. The greatest extremes in the draconic month occur when the angle between perigee and the ascending node is 0°. Likewise, the smallest extremes of the month take place when the difference in longitude is 180°. The mean rates of the major axis and the ascending node are 0.11140° east and 0.05295° west per day, respectively. Therefore, the mean period between alignments of the axis and node is 2190.4 days or 6.0 years. This period is clearly seen in Figure 4-11.

There are 67,111 draconic months during the 5000 years covered in this canon. The shortest and longest months are 27.004 days (0.208 days or 5.0 hours shorter than the mean) and 27.487 days (0.275 days or 6.6 hours longer than the mean), respectively. A histogram of the distribution in the length of the draconic month over the five millennia appears in Figure 4-12 where the duration of individual draconic months have been binned into 30-min groups. The width and bifurcated symmetry of the distribution resemble the distribution for the lunation (synodic month) in Figure 4-4.

Figure 4–12: Length of Draconic Month Over 5000 Years

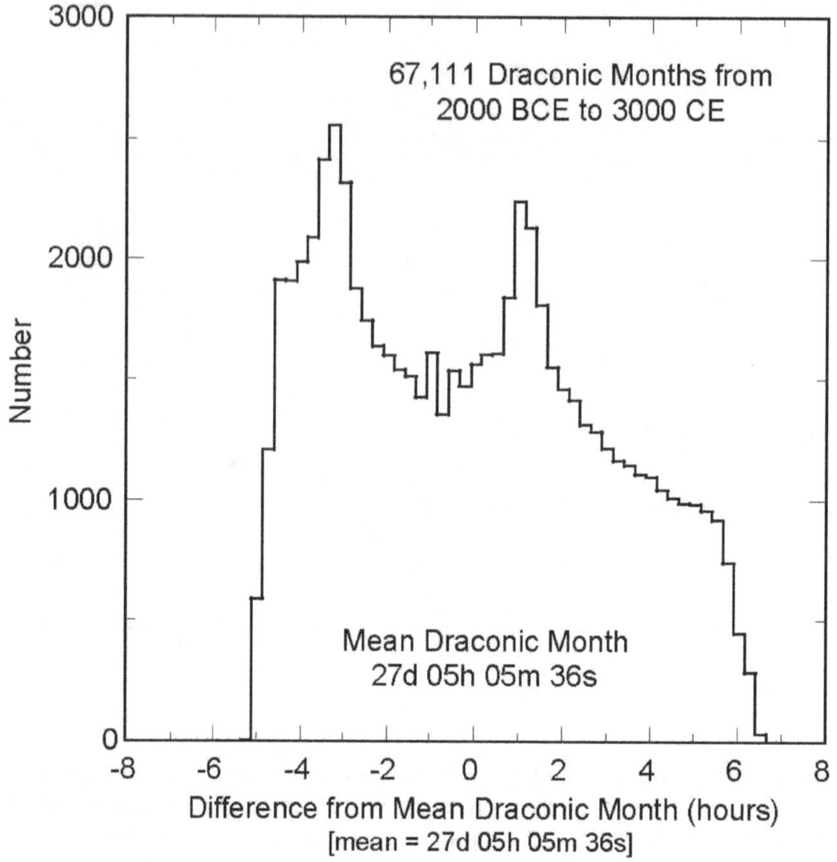

4.5 Eclipse Cycles

The interaction and harmonics of the synodic, anomalistic, and draconic months not only determine how frequently eclipses occur, but they also control the geometric characteristics and classification of each eclipse. The commensurability of these periods over long time scales results in several important eclipse cycles, which will be the subject of the next section.

Section 5: Lunar Eclipse Periodicity

5.1 Interval Between Two Successive Eclipses

The time interval between any two successive lunar eclipses can be either 1, 5 or 6 lunations (synodic months). The distribution of these intervals among the 12,063 in the *Canon* is found in Table 5-1.

Table 5-1. Interval Between Successive Eclipses

Number of Lunations	Number of Eclipses	Percent
1	1527	12.7%
5	2909	24.1%
6	7627	63.2%

5.2 Lunar Eclipse Repetition

Lunar eclipses separated by 1, 5, or 6 lunations are usually quite dissimilar. They are frequently of unlike types (i.e., penumbral, partial, or total) with diverse Sun–Moon–Earth alignment geometries, and with different lunar orbital characteristics (i.e., longitude of perigee and longitude of ascending node). More importantly, these short periods are of no value as predictors of future eclipses because they do not repeat in a recognizable pattern.

A simple lunar eclipse repetition cycle can be found by requiring that certain orbital parameters be repeated. The Moon must be in the full phase with the same longitude of perigee and same longitude of the ascending node. These conditions are met by searching for an integral multiple in the Moon's three major periods—the synodic, anomalistic, and draconic months. A fourth condition might require that an eclipse occur at approximately the same time of year to preserve the axial tilt of Earth and thus, the same season, as well as the distance from the Sun. This last factor controls the apparent diameter of Earth's umbral and penumbral shadows.

5.3 Saros Series

The Saros arises from a harmonic between three of the Moon's orbital cycles. All three periods are subject to slow variations over long time scales, but there current values (2000 CE) are:

Synodic Month (New Moon to New Moon)	= 29.530589 days	= 29d 12h 44m 03s
Anomalistic Month (perigee to perigee)	= 27.554550 days	= 27d 13h 19m 33s
Draconic Month (node to node)	= 27.212221 days	= 27d 05h 06m 36s

One Saros is equal to 223 synodic months, however, 239 anomalistic months and 242 draconic months and are also equal (within a few hours) to this same period:

223 Synodic Months	= 6585.3223 days	= 6585d 07h 43m
239 Anomalistic Months	= 6585.5375 days	= 6585d 12h 54m
242 Draconic Months	= 6585.3575 days	= 6585d 08h 35m

With a period of approximately 6,585.32 days (~18 years 11 days 8 hours), the Saros is a valuable tool in investigating the periodicity and recurrence of eclipses. It was first known to the Chaldeans as an interval when lunar eclipses repeat, but the Saros is applicable to solar eclipses as well.

Any two eclipses separated by one Saros cycle share similar characteristics. They occur at the same node with the Moon at nearly the same distance from Earth and at the same time of year. Because the Saros period is not equal to a whole number of days, its biggest drawback as an eclipse predictor is that subsequent eclipses are visible from different parts of the globe. The extra 1/3 day displacement means that Earth must rotate an additional ~8 hours or

~120° with each cycle. For solar eclipses, this results in a shift of each succeeding eclipse path by ~120° west. Thus, a Saros series returns to approximately the same geographic region every three Saros periods (~54 years and 34 days). This triple Saros cycle is known as the Exeligmos.

Figure 5–1. Lunar Eclipses from Saros 136: 1932 to 2022

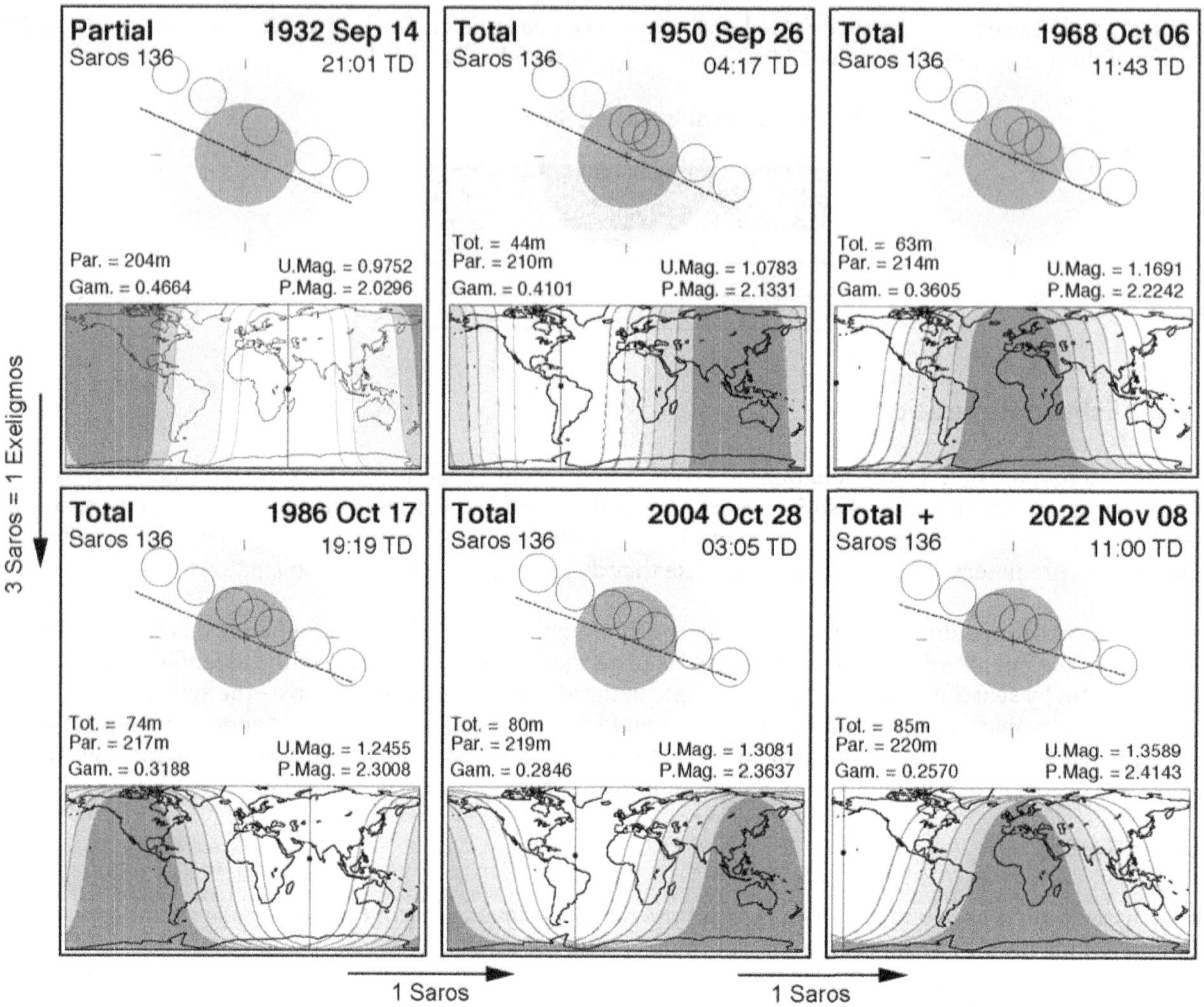

Figure Figure 5-1 shows the path of the Moon through Earth's shadows and the geographic regions of visibility for six lunar eclipses belonging to Saros 136 from 1932 through 2022. The 1932 Sep 14 eclipse was a large magnitude partial eclipse with a visibility zone centered on eastern Europe and Africa. One Saros period later the eclipse was total (1950 Sep 26) with visibility centered on the Americas. After another Saros interval, the eclipse was a larger magnitude event centered on the Pacific Ocean (1968 Oct 06). Finally, after the lapse of one more Saros period (3 Saros = Exeligmos = ~54.1 years), the zone of visibility returned to eastern Europe and Africa with a deeper non-central total lunar eclipse (1986 Oct 17). The same 54.1-year time interval in geographic visibility repeats for the Americas (1950 and 2004) and for the Pacific (1968 and 2022). The westward migration in the zone of eclipse visibility illustrates the effect of the extra 1/3-day in the Saros period. The southward shift of the Moon's path with respect to the shadow axis is due to the progressive decrease in gamma from 0.4664 (1932) to 0.2570 (2022). During this interval, lunar eclipses in the series change from partial (1932) to non-central total (1950) to central total (2022). Although the Moon's path through Earth's shadows is similar from one eclipse to the next, it is not exact. The Moon-shadow geometry changes slowly as do the characteristics of each lunar eclipse in a Saros series.

Saros series do not last indefinitely because the synodic, draconic, and anomalistic months are not perfectly commensurate with one another. In particular, the Moon's node shifts eastward by about 0.48° with each eclipse in a series. The following narrative describes the life cycle of a typical Saros series at the Moon's ascending node. The series begins when the Full Moon occurs approximately 17° east of the node.

A small fraction of the Moon's disk passes through the northern edge of the penumbra and a small penumbral eclipse occurs. One Saros period later, the lunar trajectory shifts a little further south, the Moon passes deeper into the penumbra (gamma decreases) and a penumbral eclipse of slightly larger magnitude results. After ~10 penumbral eclipses, the first partial lunar eclipse occurs as the Moon's southern limb passes through the umbra. Around 20 more partial eclipses occur as the Moon swings increasingly deeper into the umbra after each Saros and the magnitude of each event grows larger. Finally, the Moon passes completely into the umbra and a shallow total eclipse occurs. Over the course of the next 2 centuries, a total lunar eclipse occurs every 18.031 years (= Saros), as the Moon is displaced progressively southward through the umbral shadow with each eclipse. Halfway through this period, the Moon passes through the center of the umbra producing long total eclipses. The last of ~13 total eclipses in the series takes place just inside the southern edge of the umbra. The Saros series winds down now with another string of ~20 partial eclipses following by a final set of ~10 penumbral eclipses. The last eclipse of the series is a small magnitude penumbral event just inside the southern edge of the penumbral shadow. In all, this typical Saros series produces 73 eclipses spanning nearly 13 centuries.

The scenario for a Saros series occurring at the Moon's descending node is similar except that gamma increases as each successive eclipse shifts the Moon's trajectory farther north of the previous one. The magnitude of the shift in gamma is again tied to aphelion and perihelion.

Because of the ellipticity of the orbits of Earth and the Moon, the exact duration and number of eclipses in a complete Saros series is not constant. A series may last 1,226 to 1,587 years and is composed of 69 to 89 eclipses. A series can begin with 6 to 25 penumbral eclipses, followed by 6 to 24 partial eclipses. During the mid-life of a Saros series, there are 11 to 29 total eclipses. Finally, a series ends with 6 to 24 partial eclipses followed by 7 to 25 penumbral eclipses. At present (2008), there are 40 active Saros series numbered 110 to 149. The number of eclipses in each of these series ranges from 70 to 83, however, the majority of them (80%) are composed of 70 to 73 eclipses.

5.4 Gamma And Saros Series

Gamma changes monotonically throughout any single Saros series. As mentioned previously (Sect. 1.2.6), the change in gamma is larger when Earth is near its aphelion (June to July) than when it is near perihelion (December to January). For odd numbered series (descending node), gamma increases, while for even numbered series (ascending node), gamma decreases. This simple rule describes the current behavior of gamma, but this has not always been the case. The eccentricity of Earth's orbit is presently 0.0167, and is slowly decreasing. It was 0.0181 in the year -2000 and will be 0.0163 in +3000. In the past, when the eccentricity was larger, there were Saros series in which the trend in gamma reversed for one or more Saros cycles before resuming its original direction. These instances occur near perihelion when the Sun's apparent motion is highest and may, in fact, overtake the eastward shift of the node. The resulting effect is a relative shift west of the node after one Saros cycle instead of the usual eastward shift. Consequently, gamma reverses direction.

The most unusual case of this occurs in Saros series 13. It began in -2313 with 11 penumbral eclipses, followed by 7 partial eclipses. The series then reverted back to 3 more penumbral eclipse followed by another 11 partial eclipses. The series then produced 13 total, 20 partial, and 8 penumbral eclipses. To understand this odd behavior we must look to gamma. The value of gamma increased positively for the first 15 eclipses of the series. It then reversed direction after the fifth partial and decreased for 5 sequential eclipses while changing from partial back to penumbral. Finally resuming its northward motion, gamma began to increase again and continued to do so for the remainder of the series. Saros 13 produced 73 ending in the year -1015.

Among 200 Saros series examined (-20 to 183), there are many other examples of temporary shifts in the monotonic nature of gamma, although none as unusual as Saros 13. In fact, the first 38 Saros series (series -20 to 17) with members represented in the Canon, experience short reversals in gamma. The reversals are short and rarely last for more than four or five eclipses in a series. Some series even have two separate reversals in gamma (e.g., series -19, -15, -14, -12, -11, -5, -1, 4, 7, and 25). The most recent eclipse with a gamma reversal was on 1648 Jan 10 (Saros 138). The next and last in the Canon will occur on 2353 Jan 20 (Saros 120). In past millennia, the gamma reversals were more frequent because Earth's orbital eccentricity was larger.

5.5 Saros Series Statistics

Lunar eclipses belonging to 204 different Saros series fall within the five millennium span of the Canon. One series (183) has only two eclipses represented, while two series (-20 and 182) have three eclipses each. Another 81 have a larger, but incomplete, subset of their members included (-18 to 19, 12, 25, 139, 142, and 144 to 181). Finally, 120 complete Saros series are contained within the Canon (20 to 23, 26 to 138, 140, 141, and 143).

The number of lunar eclipses in each of these series ranges from 69 to 89. Almost a quarter (22.6%) of the series contain 72 eclipses, another quarter (24.0%) has 73 eclipses, and a sixth (15.7%) consists of 71 eclipses. In other words, nearly 2/3 (62.3%) of all Saros series are composed of 71 to 73 eclipses. If all Saros series with 70 to 74 eclipses are considered, then the percentage jumps to 75.0%. The remaining quarter have either 69 eclipses (Saros 181) or 75 to 89 eclipses.

Table 5-2 presents the statistical distribution of the number of eclipses in each Saros series. The approximate duration (years) as a function of the number of eclipses, is listed along with the first five Saros series containing the corresponding number of eclipses.

All Saros series begin and end with a number of penumbral eclipses. Among the 204 Saros series with members falling within the scope of this Canon, the number of penumbral eclipses in the initial phase ranges from 6 to 25. Similarly, the number of penumbral eclipses in the final phase varies from 7 to 25. The initial penumbral eclipse sequence is followed by a series of 6 to 24 partial eclipses, while the final penumbral eclipses are preceded by a sequence of 6 to 24 partial eclipses. The middle life of a Saros series is composed of total (umbral) eclipses, which range in number from 11 to 29.

Saros 13 is an exception to the normal progression of eclipse types through a series. After beginning with 11 penumbral eclipses followed by 7 partial eclipses, the series reverts back to produce 3 more penumbral eclipses. It then resumes the normal pattern to produce 11 partial eclipses followed by 13 total, 20 partial, and 8 penumbral eclipses. This odd behavior is caused by a reversal in gamma (Sect.5.4).

Figure 5-2 presents data on the number of penumbral, partial, and total eclipses in Saros series 1 through 150. The information is plotted in four separate histograms showing the numbers of all eclipses (Fig. 5-2a), penumbral eclipses (Fig. 5-2b), partial eclipses (Fig. 5-2c), and total eclipses (Fig. 5-2d). Several interesting relationships are revealed in these diagrams. Figure 5-2a shows that Saros series with large numbers of lunar eclipses tend to cluster into tight groups or 3 or 4 series separated by 18 to 19 series. However, most series are composed of 72 to 73 eclipses as already shown in Table 5-2.

Figure 5-2b displays the numbers of both leading (solid) and following (dashed) penumbral eclipses in a series. Note how the two histograms are displaced from each other, but their overlapping sections coincide with Saros series with large numbers of eclipses. Figure 5-2c shows the numbers of leading (solid) and following (dashed) partial eclipses in a series, which are nearly in phase with each other and peak during periods when the number of all lunar eclipses is near the minimum. Finally, Figure 5-2d presents the number of total eclipses in each series, which is inversely correlated with the number of partial lunar eclipses.

Table 5-2. Number of Lunar Eclipses in Saros Series

Number of Eclipses	Duration (Years)	Number of Series	Saros Series
69	1226.0	1	181
70	1244.0	10	126, 147, 148, 163, 166, ...
71	1262.1	32	59, 92, 93, 94, 95, ...
72	1280.1	46	-4, -1, 17, 20, 34, ...
73	1298.1	49	-20, -19, -17, -9, -6, ...
74	1316.2	16	-8, -7, 0, 10, 11, ...
75	1334.2	4	-18, -16, 9, 48
76	1352.2	4	3, 44, 46, 85
77	1370.3	2	5, 140
78	1388.3	2	61, 137
79	1406.3	4	100, 139, 174, 176
81	1442.4	2	156, 158
82	1460.4	5	63, 103, 119, 121, 138
83	1478.4	4	-10, 29, 101, 120
84	1496.5	7	-12, 64, 66, 82, 83, ...
85	1514.5	6	4, 24, 26, 27, 43, ...
86	1532.5	4	6, 8, 47, 65
87	1550.5	3	-15, -13, 25
88	1568.6	2	-14, -11
89	1586.6	1	7

To generalize these relationships, it appears that Saros series that are rich in the number of total eclipses are also rich in the number of penumbral eclipses, but poor in the number of partial eclipses. Conversely, Saros series poor in total eclipses are also poor in penumbral eclipses, but rich in partial eclipses. Finally, Saros series with large numbers of eclipses are centered on series with large numbers of total eclipses. Figure 5-2a also shows that the extremes in the minimum and maximum numbers of lunar eclipses in a Saros series is gradually decreasing.

A concise summary of all 204 Saros series (–20 to 183) is presented in Tables 5-3 to 5-8. The number of lunar eclipses in each series is listed followed by the calendar dates of the first and last eclipses in the Saros. Finally, the chronological sequence of lunar eclipse types in the series is tabulated. The number and type of eclipses varies from one Saros series to the next as reflected in the sequence diversity. Note that the tables make no distinction between central and non-central total lunar eclipses. The following abbreviations are used in the eclipse sequences:

T = Total Lunar Eclipse
P = Partial Lunar Eclipse
N = Partial Penumbral Lunar Eclipse
N* = Total Penumbral Lunar Eclipse

Figure 5–2. Number of Lunar Eclipses vs. Saros Series

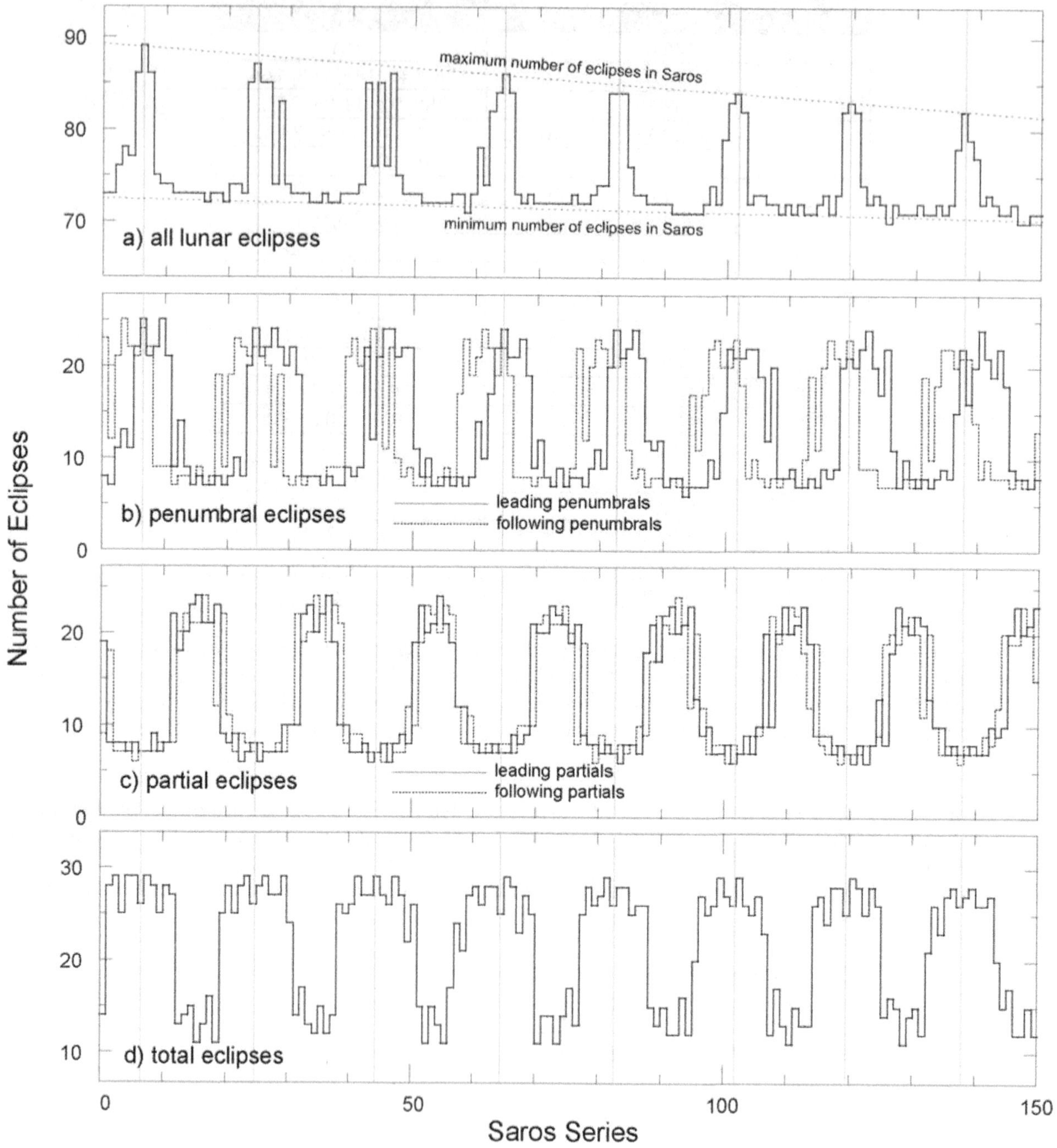

Table 5-3. Summary of Saros Series -20 to 25

Saros Series	Number of Eclipses	First Eclipse	Last Eclipse	Lunar Eclipse Sequence
-20	73	-3250 Apr 01	-1952 May 18	7N 1N* 20P 13T 23P 9N
-19	73	-3221 Mar 11	-1923 Apr 28	7N 9P 27T 22P 8N
-18	75	-3228 Jan 29	-1894 Apr 08	9N 10P 27T 8P 2N* 19N
-17	73	-3109 Mar 03	-1811 Apr 20	7N 8P 26T 8P 24N
-16	75	-3116 Jan 20	-1782 Mar 31	9N 7P 29T 9P 21N
-15	87	-3322 Aug 12	-1771 Feb 27	22N 8P 29T 6P 1N* 21N
-14	88	-3239 Aug 24	-1670 Mar 23	22N 1N* 6P 27T 7P 25N
-13	87	-3192 Aug 14	-1641 Mar 03	21N 7P 29T 8P 22N
-12	84	-3199 Jul 03	-1703 Dec 18	23N 8P 29T 6P 18N
-11	88	-3134 Jul 05	-1565 Feb 01	24N 1N* 6P 26T 8P 23N
-10	83	-3069 Jul 06	-1591 Dec 10	21N 8P 29T 7P 18N
-9	73	-3058 Jun 05	-1760 Jul 23	21N 10P 28T 7P 7N
-8	74	-2993 Jun 07	-1677 Aug 05	23N 1N* 7P 24T 10P 9N
-7	74	-2946 May 28	-1630 Jul 27	9N 2N* 19P 27T 9P 8N
-6	73	-2917 May 08	-1619 Jun 25	8N 23P 18T 17P 7N
-5	73	-2852 May 09	-1554 Jun 26	9N 1N* 21P 11T 23P 8N
-4	72	-2787 May 10	-1507 Jun 17	6N 1N* 22P 14T 22P 7N
-3	73	-2776 Apr 09	-1478 May 28	7N 24P 13T 21P 1N* 7N
-2	73	-2711 Apr 11	-1413 May 29	8N 22P 11T 23P 9N
-1	72	-2646 Apr 12	-1366 May 20	6N 21P 15T 23P 7N
0	74	-2653 Mar 01	-1337 Apr 30	8N 12P 25T 18P 2N* 9N
1	73	-2570 Mar 14	-1272 Apr 30	8N 19P 14T 9P 1N* 22N
2	73	-2523 Mar 03	-1225 Apr 22	7N 8P 28T 18P 12N
3	76	-2567 Dec 30	-1214 Mar 21	11N 8P 29T 7P 1N* 20N
4	78	-2646 Oct 06	-1131 Apr 02	13N 8P 25T 7P 1N* 24N
5	77	-2455 Dec 22	-1084 Mar 24	11N 7P 29T 8P 22N
6	86	-2624 Aug 04	-1091 Feb 10	22N 8P 29T 6P 1N* 20N
7	89	-2595 Jul 16	-1008 Feb 22	25N 7P 26T 7P 24N
8	86	-2494 Aug 08	-0961 Feb 13	21N 7P 29T 7P 22N
9	75	-2501 Jun 26	-1167 Sep 05	22N 9P 28T 7P 9N
10	74	-2454 Jun 17	-1138 Aug 15	24N 1N* 7P 25T 8P 9N
11	74	-2371 Jun 29	-1055 Aug 27	20N 1N* 8P 28T 8P 9N
12	73	-2360 May 28	-1062 Jul 17	9N 22P 27T 8P 7N
13	73	-2313 May 20	-1015 Jul 06	11N 7P 3N* 11P 13T 20P 8N
14	73	-2230 Jun 01	-0932 Jul 19	8N 1N* 20P 14T 22P 8N
15	73	-2219 Apr 30	-0921 Jun 19	7N 23P 15T 21P 7N
16	73	-2172 Apr 21	-0874 Jun 08	8N 24P 11T 21P 1N* 8N
17	72	-2089 May 04	-0809 Jun 11	6N 1N* 21P 13T 24P 7N
18	73	-2078 Apr 02	-0780 May 21	7N 21P 16T 21P 8N
19	73	-2031 Mar 24	-0733 May 11	8N 23P 11T 12P 8N* 11N
20	72	-1948 Apr 05	-0668 May 12	6N 1N* 9P 25T 22P 9N
21	74	-1955 Feb 22	-0639 Apr 23	8N 8P 28T 11P 2N* 17N
22	74	-1926 Feb 02	-0610 Apr 02	10N 9P 25T 7P 1N* 22N
23	73	-1825 Feb 25	-0527 Apr 14	7N 1N* 6P 28T 9P 22N
24	85	-2031 Sep 16	-0516 Mar 14	20N 7P 29T 8P 21N
25	87	-2038 Aug 06	-0487 Feb 21	24N 8P 26T 7P 22N

Table 5-4. Summary of Saros Series 26 to 70

Saros Series	Number of Eclipses	First Eclipse	Last Eclipse	Lunar Eclipse Sequence
26	85	-1919 Sep 09	-0404 Mar 06	20N 1N* 6P 28T 8P 22N
27	85	-1926 Jul 28	-0411 Jan 23	22N 7P 29T 7P 20N
28	74	-1897 Jul 09	-0581 Sep 06	24N 7P 27T 7P 9N
29	83	-1814 Jul 21	-0336 Dec 24	21N 1N* 7P 27T 8P 19N
30	74	-1803 Jun 19	-0487 Aug 18	20N 10P 29T 7P 8N
31	73	-1774 May 30	-0476 Jul 17	22N 10P 24T 10P 7N
32	73	-1673 Jun 23	-0375 Aug 09	12N 7N* 10P 14T 22P 8N
33	73	-1662 May 22	-0364 Jul 10	8N 22P 17T 19P 7N
34	72	-1615 May 13	-0335 Jun 19	8N 23P 13T 20P 1N* 7N
35	72	-1532 May 25	-0252 Jul 01	7N 1N* 20P 12T 24P 8N
36	73	-1521 Apr 24	-0223 Jun 11	7N 22P 15T 22P 7N
37	72	-1492 Apr 03	-0212 May 10	8N 24P 12T 19P 1N* 8N
38	72	-1391 Apr 27	-0111 Jun 03	6N 1N* 19P 14T 23P 9N
39	73	-1380 Mar 26	-0082 May 14	7N 10P 26T 21P 9N
40	73	-1369 Feb 24	-0071 Apr 12	9N 10P 25T 8P 1N* 20N
41	73	-1268 Mar 18	0030 May 06	7N 1N* 7P 26T 9P 23N
42	74	-1275 Feb 04	0041 Apr 05	9N 7P 29T 9P 20N
43	85	-1463 Sep 07	0052 Mar 04	22N 8P 27T 7P 21N
44	76	-1199 Jan 06	0153 Mar 27	11N 1N* 6P 27T 7P 24N
45	85	-1351 Aug 29	0164 Feb 25	21N 7P 29T 7P 21N
46	76	-1358 Jul 19	-0006 Oct 08	24N 8P 27T 6P 1N* 10N
47	86	-1275 Jul 31	0258 Feb 05	23N 1N* 6P 26T 8P 22N
48	75	-1228 Jul 21	0106 Sep 30	21N 8P 29T 7P 10N
49	73	-1217 Jun 21	0081 Aug 08	22N 9P 27T 7P 1N* 7N
50	73	-1134 Jul 03	0164 Aug 20	21N 1N* 8P 22T 12P 9N
51	73	-1105 Jun 13	0193 Jul 31	9N 2N* 19P 26T 10P 7N
52	72	-1076 May 23	0204 Jun 29	8N 23P 15T 19P 1N* 6N
53	72	-0993 Jun 05	0287 Jul 12	9N 1N* 20P 11T 23P 8N
54	72	-0946 May 26	0334 Jul 03	6N 1N* 21P 15T 22P 7N
55	72	-0935 Apr 25	0345 Jun 01	7N 24P 13T 20P 1N* 7N
56	72	-0852 May 07	0428 Jun 13	7N 1N* 21P 11T 23P 9N
57	73	-0823 Apr 16	0475 Jun 05	7N 19P 17T 22P 8N
58	73	-0812 Mar 16	0486 May 04	8N 12P 24T 12P 6N* 11N
59	71	-0711 Apr 09	0551 May 06	7N 12P 21T 8P 1N* 22N
60	73	-0700 Mar 08	0598 Apr 27	8N 8P 27T 11P 19N
61	78	-0780 Dec 13	0609 Mar 26	14N 8P 28T 7P 1N* 20N
62	74	-0624 Feb 08	0692 Apr 06	10N 7P 26T 7P 24N
63	82	-0722 Nov 03	0739 Mar 29	17N 7P 28T 8P 22N
64	84	-0783 Aug 20	0714 Feb 04	22N 8P 28T 7P 19N
65	86	-0736 Aug 11	0797 Feb 16	24N 7P 25T 8P 22N
66	84	-0671 Aug 12	0826 Jan 27	21N 7P 29T 8P 19N
67	73	-0660 Jul 11	0638 Aug 30	21N 9P 28T 7P 1N* 7N
68	72	-0595 Jul 14	0685 Aug 20	23N 8P 23T 10P 8N
69	73	-0530 Jul 15	0768 Sep 01	18N 1N* 9P 27T 10P 8N
70	72	-0519 Jun 13	0761 Jul 21	9N 21P 25T 10P 1N* 6N

Table 5-5. Summary of Saros Series 71 to 115

Saros Series	Number of Eclipses	First Eclipse	Last Eclipse	Lunar Eclipse Sequence
71	72	-0472 Jun 04	0808 Jul 11	11N 1N* 20P 11T 21P 1N* 7N
72	72	-0389 Jun 17	0891 Jul 25	7N 1N* 20P 14T 22P 8N
73	72	-0378 May 16	0902 Jun 23	7N 23P 14T 21P 7N
74	72	-0331 May 07	0949 Jun 13	9N 22P 11T 21P 1N* 8N
75	72	-0266 May 08	1014 Jun 15	7N 21P 14T 23P 7N
76	73	-0255 Apr 07	1043 May 26	8N 19P 17T 20P 9N
77	72	-0190 Apr 09	1090 May 16	8N 21P 13T 8P 2N* 20N
78	72	-0125 Apr 10	1155 May 18	7N 9P 25T 19P 12N
79	73	-0132 Feb 27	1166 Apr 17	9N 8P 28T 8P 1N* 19N
80	74	-0103 Feb 07	1213 Apr 06	11N 8P 26T 6P 1N* 22N
81	74	-0020 Feb 19	1296 Apr 19	9N 7P 27T 9P 22N
82	84	-0208 Sep 21	1289 Mar 08	20N 8P 29T 7P 20N
83	84	-0197 Aug 22	1300 Feb 05	24N 7P 26T 7P 20N
84	84	-0096 Sep 13	1401 Feb 28	20N 1N* 6P 28T 8P 21N
85	76	-0103 Aug 02	1249 Oct 23	22N 8P 28T 7P 11N
86	73	-0074 Jul 13	1224 Aug 30	24N 8P 25T 8P 8N
87	73	0027 Aug 06	1325 Sep 23	20N 1N* 7P 26T 10P 9N
88	72	0038 Jul 05	1318 Aug 12	12N 18P 26T 9P 7N
89	72	0067 Jun 15	1347 Jul 23	11N 21P 15T 17P 1N* 7N
90	72	0150 Jun 27	1430 Aug 04	9N 3N* 17P 13T 22P 8N
91	72	0179 Jun 07	1459 Jul 15	7N 22P 15T 21P 7N
92	71	0208 May 17	1470 Jun 13	8N 23P 12T 20P 1N* 7N
93	71	0291 May 30	1553 Jun 25	7N 1N* 20P 12T 24P 7N
94	71	0320 May 09	1582 Jun 06	6N 21P 16T 21P 7N
95	71	0349 Apr 19	1611 May 26	7N 23P 12T 12P 6N* 11N
96	71	0432 May 01	1694 Jun 07	7N 13P 20T 20P 11N
97	72	0443 Mar 31	1723 May 20	7N 9P 27T 12P 17N
98	74	0436 Feb 18	1752 Apr 28	10N 10P 25T 7P 1N* 21N
99	72	0555 Mar 24	1835 May 12	7N 1N* 7P 26T 8P 23N
100	79	0439 Dec 06	1846 Apr 11	15N 7P 29T 8P 20N
101	83	0360 Sep 11	1839 Feb 28	22N 8P 27T 6P 1N* 19N
102	84	0461 Oct 05	1958 Apr 04	20N 1N* 6P 26T 8P 23N
103	82	0472 Sep 03	1933 Feb 10	21N 7P 29T 7P 18N
104	72	0483 Aug 04	1763 Sep 22	22N 9P 26T 7P 1N* 7N
105	73	0566 Aug 16	1864 Oct 15	21N 1N* 7P 25T 9P 10N
106	73	0595 Jul 27	1893 Sep 25	19N 10P 27T 9P 8N
107	72	0606 Jun 26	1886 Aug 14	12N 20P 23T 10P 7N
108	72	0689 Jul 08	1969 Aug 27	17N 3N* 10P 12T 22P 8N
109	71	0736 Jun 27	1998 Aug 08	7N 1N* 20P 17T 19P 7N
110	72	0747 May 28	2027 Jul 18	8N 23P 13T 20P 1N* 7N
111	71	0830 Jun 10	2092 Jul 19	8N 1N* 20P 11T 23P 8N
112	72	0859 May 20	2139 Jul 12	7N 21P 15T 22P 7N
113	71	0888 Apr 29	2150 Jun 10	7N 23P 13T 18P 2N* 8N
114	71	0971 May 13	2233 Jun 22	7N 1N* 19P 13T 12P 6N* 13N
115	72	1000 Apr 21	2280 Jun 13	7N 9P 26T 19P 11N

Table 5-6. Summary of Saros Series 116 to 160

Saros Series	Number of Eclipses	First Eclipse	Last Eclipse	Lunar Eclipse Sequence
116	73	0993 Mar 11	2291 May 14	9N 9P 27T 8P 1N* 19N
117	71	1094 Apr 03	2356 May 15	8N 9P 24T 7P 23N
118	73	1105 Mar 02	2403 May 07	9N 7P 28T 8P 21N
119	82	0935 Oct 14	2396 Mar 25	20N 8P 28T 6P 1N* 19N
120	83	1000 Oct 16	2479 Apr 07	21N 7P 25T 7P 1N* 22N
121	82	1047 Oct 06	2508 Mar 18	19N 1N* 6P 29T 7P 20N
122	74	1022 Aug 14	2338 Oct 29	22N 8P 28T 7P 9N
123	72	1087 Aug 16	2367 Oct 08	23N 1N* 6P 25T 8P 9N
124	73	1152 Aug 17	2450 Oct 21	20N 8P 28T 8P 9N
125	72	1163 Jul 17	2443 Sep 09	17N 13P 26T 9P 7N
126	70	1228 Jul 18	2472 Aug 19	21N 1N* 8P 14T 19P 7N
127	72	1275 Jul 09	2555 Sep 02	10N 1N* 18P 16T 20P 7N
128	71	1304 Jun 18	2566 Aug 02	7N 23P 15T 19P 1N* 6N
129	71	1351 Jun 10	2613 Jul 24	9N 1N* 21P 11T 21P 8N
130	71	1416 Jun 10	2678 Jul 26	7N 1N* 20P 14T 22P 7N
131	72	1427 May 10	2707 Jul 07	7N 22P 15T 20P 8N
132	71	1492 May 12	2754 Jun 26	8N 21P 12T 11P 7N* 12N
133	71	1557 May 13	2819 Jun 29	6N 1N* 13P 21T 20P 10N
134	72	1550 Apr 01	2830 May 28	8N 10P 26T 10P 1N* 17N
135	71	1615 Apr 13	2877 May 18	9N 10P 23T 7P 1N* 21N
136	72	1680 Apr 13	2960 Jun 01	7N 1N* 7P 27T 8P 22N
137	78	1564 Dec 17	2953 Apr 20	15N 8P 28T 7P 20N
138	82	1521 Oct 15	2982 Mar 30	21N 1N* 7P 26T 6P 21N
139	79	1658 Dec 09	3065 Apr 13	16N 7P 27T 8P 21N
140	77	1597 Sep 25	2968 Jan 06	20N 8P 28T 7P 14N
141	72	1608 Aug 25	2888 Oct 11	24N 7P 26T 7P 8N
142	73	1709 Sep 19	3007 Nov 17	20N 1N* 7P 26T 9P 10N
143	72	1720 Aug 18	3000 Oct 05	19N 10P 27T 8P 8N
144	71	1749 Jul 29	3011 Sep 04	21N 1N* 9P 20T 12P 1N* 7N
145	71	1832 Aug 11	3094 Sep 16	15N 3N* 10P 15T 20P 8N
146	72	1843 Jul 11	3123 Aug 29	9N 20P 17T 19P 7N
147	70	1890 Jul 02	3134 Jul 28	8N 23P 12T 19P 1N* 7N
148	70	1973 Jul 15	3217 Aug 09	7N 1N* 20P 12T 23P 7N
149	71	1984 Jun 13	3246 Jul 20	7N 21P 15T 21P 7N
150	71	2013 May 25	3275 Jun 30	8N 23P 12T 15P 3N* 10N
151	71	2096 Jun 06	3358 Jul 13	7N 1N* 18P 14T 21P 10N
152	72	2107 May 07	3387 Jun 23	8N 10P 25T 15P 1N* 13N
153	71	2136 Apr 16	3398 May 22	9N 10P 24T 8P 1N* 19N
154	71	2237 May 10	3499 Jun 16	7N 8P 25T 8P 23N
155	73	2212 Mar 18	3510 May 17	9N 8P 28T 8P 20N
156	81	2060 Nov 08	3503 Apr 05	20N 8P 27T 6P 1N* 19N
157	73	2306 Mar 01	3604 Apr 27	11N 6P 26T 8P 22N
158	81	2154 Oct 21	3597 Mar 17	20N 7P 28T 8P 18N
159	73	2147 Sep 09	3445 Nov 07	23N 8P 26T 7P 9N
160	72	2248 Oct 03	3528 Nov 19	21N 7P 25T 8P 11N

Table 5-7. Summary of Saros Series 161 to 183

Saros Series	Number of Eclipses	First Eclipse	Last Eclipse	Lunar Eclipse Sequence
161	73	2259 Sep 02	3557 Oct 31	20N 9P 27T 8P 9N
162	71	2288 Aug 12	3550 Sep 19	19N 12P 24T 9P 7N
163	70	2371 Aug 27	3615 Sep 20	19N 2N* 8P 13T 20P 8N
164	71	2400 Aug 05	3662 Sep 11	9N 1N* 18P 18T 18P 7N
165	71	2411 Jul 06	3673 Aug 11	9N 22P 14T 19P 7N
166	70	2494 Jul 18	3738 Aug 13	9N 1N* 19P 11T 22P 8N
167	71	2541 Jul 09	3803 Aug 16	7N 20P 15T 21P 8N
168	71	2552 Jun 08	3814 Jul 15	8N 22P 13T 18P 2N* 8N
169	70	2635 Jun 22	3879 Jul 17	7N 1N* 19P 13T 13P 4N* 13N
170	71	2664 Jun 01	3926 Jul 09	7N 11P 24T 19P 10N
171	71	2675 May 01	3937 Jun 07	8N 10P 26T 8P 1N* 18N
172	70	2758 May 15	4002 Jun 08	8N 9P 23T 8P 22N
173	72	2787 Apr 24	4067 Jun 11	8N 7P 27T 9P 21N
174	79	2635 Dec 16	4042 Apr 18	18N 8P 27T 7P 19N
175	74	2791 Feb 11	4107 Apr 20	14N 7P 25T 7P 21N
176	79	2747 Dec 09	4154 Apr 11	17N 1N* 6P 28T 8P 19N
177	73	2704 Oct 05	4002 Dec 03	21N 8P 28T 6P 1N* 9N
178	70	2769 Oct 07	4013 Nov 01	22N 7P 24T 8P 1N* 8N
179	73	2816 Sep 27	4114 Nov 26	20N 8P 27T 8P 10N
180	71	2827 Aug 28	4089 Oct 03	19N 11P 26T 8P 7N
181	69	2892 Aug 29	4118 Sep 13	21N 1N* 8P 15T 17P 7N
182	70	2957 Aug 31	4201 Sep 26	10N 4N* 14P 15T 20P 7N
183	70	2968 Jul 30	4212 Aug 26	8N 21P 16T 18P 1N* 6N

5.6 Saros and Other Periods

The numbering system used for the Saros series was introduced by van den Bergh in his book *Periodicity and Variation of Solar (and Lunar) Eclipses* (1955). He assigned the number 1 to a pair of solar and lunar eclipse series that were in progress during the second millennium BCE based on an extrapolation from Oppolzer's *Canon der Finsternisse* [1887].

There is an interval of 1, 5 or 6 synodic months between any sequential pair of lunar eclipses. Interestingly, the number of lunations between two eclipses permits the determination of the Saros series number of the second eclipse when the Saros series number of the first eclipse is known. Let the Saros series number of the first eclipse in a pair be "s". The Saros series number of the second eclipse can be found from the relationships in Table 5-8 [Meeus, Grosjean, and Vanderleen, 1966].

Table 5-8. Some Eclipse Periods and their Relationships to the Saros Number

Number of Synodic Months	Length of Time	Saros Series Number	Period Name
1	~1 month	s + 38	Lunation
5	~5 months	s – 33	Short Semester
6	~6 months	s + 5	Semester
135	~11 years - 1 month	s + 1	Tritos
223	~18 years + 11 days	s	Saros
235	~19 years	s + 10	Metonic Cycle
358	~29 years - 20 days	s + 1	Inex
669	~54 years + 33 days	s	Exeligmos (Triple Saros)

5.7 Saros and Inex

A number of different eclipse cycles were investigated by van den Bergh, but the most useful were the Saros and the Inex. The Inex is equal to 358 synodic months (~29 years less 20 days), which is very nearly 388.5 draconic months.

358 Synodic Months = 10,571.9509 days = 10,571d 22h 49m
388.5 Draconic Months = 10,571.9479 days = 10,571d 22h 55m

The extra 0.5 in the number of draconic months means that eclipses separated by one Inex period occur at opposite nodes. Consequently, an eclipse occurring in the northern half of Earth's shadows will be followed one Inex later by an eclipse occurring in the southern half of Earth's shadows, and vice versa.

The mean time difference between 358 synodic months and 388.5 draconic months making up an Inex is only 6 min. In comparison, the mean difference between these two cycles in the Saros is 52 min. This means that after one Inex, the shift of the Moon with respect to the node (+0.04°) is much smaller than for the Saros (−0.48°). While a Saros series lasts 12 to 15 centuries, an Inex series typically lasts 225 centuries and contains about 780 eclipses.

5.8 Saros–Inex Panorama

Van den Bergh placed all 5,200 lunar eclipses in von Oppolzer's Canon der Finsternisse (1887) into a large two-dimensional matrix. (Von Oppolzer did not include penumbral lunar eclipses in his Canon der Finsternisse.) Each Saros series was arranged as a separate column containing every eclipse in chronological order. The individual Saros columns were then staggered so that the horizontal rows each corresponded to a different Inex series. This "Saros–Inex Panorama" proved useful in organizing eclipses. For instance, one step down in the panorama is a change of one Saros period (6585.32 days) later, while one step to the right is a change of one Inex period (10571.95 days) later. The rows and columns were then numbered with the Saros and Inex numbers.

The panorama also made it possible to predict the approximate circumstances of lunar (and solar) eclipses occurring before or after the period spanned by von Oppolzer's Canon. The time interval "t" between any two lunar eclipses can be found through an integer combination of Saros and Inex periods via the following relationship:

$$t = a\,i + b\,s, \qquad (5\text{-}1)$$

where: t = interval in days,
i = Inex period of 10571.95 days (358 synodic months),
s = Saros period of 6585.32 days (223 synodic months), and
a, b = integers (negative, zero, or positive).

From this equation, a number of useful combinations of Inex and Saros periods can be employed to extend von Oppolzer's *Canon* from −1207 back to −1600 using nothing more than simple arithmetic (van den Bergh, 1954). The ultimate goal of the effort was to a produce an eclipse canon for dating historical events prior to −1207. Periods formed by various combinations of Inex and Saros were evaluated in order to satisfy one or more of the following conditions:

1) The deviation from a multiple of 0.5 draconic months should be small
   (i.e., Moon should be nearly the same distance from the node).
2) The deviation from an integral multiple of anomalistic months should be small
   (i.e., Moon should be nearly the same distance from Earth).
3) The deviation from an integral multiple of anomalistic years should be small
   (i.e., eclipse should occur on nearly the same calendar date).

No single Inex–Saros combination meets all three criteria, but there are periods that do a reasonably good job for any one of them. Note that secular changes in the Moon's elements cause a particular period to be of high accuracy for a limited number of centuries. The direct application of the Saros–Inex panorama allows for the determination of eclipse dates in the past (or future); however, the application of the longer Saros–Inex combinations permit the rapid estimation of a number of eclipse characteristics without lengthy calculations. Table 5-9 lists several of the most useful periods.

Table 5-9. Some Useful Eclipse Periods

Period Name	Period (Inex + Saros)	Period (years)	Use
Heliotrope	58i + 6s	1,787	Geographic longitude of eclipse
Accuratissima	58i + 9s	1,841	Geographic latitude of eclipse
Horologia	110i + 7s	3,310	Time of ecliptic conjunction

Modern digital computers using high precision solar and lunar ephemerides can directly predict the dates and circumstances of eclipses. Nevertheless, the Saros and Inex cycles remain useful tools in understanding the periodicity and frequency of eclipses.

5.9 Secular Variations in the Saros and Inex

Because of long secular variations in the average ellipticity of the Moon's and Earth's orbits, the mean lengths of the synodic, draconic, and anomalistic months are slowly changing. The mean synodic and draconic months are increasing by approximately 0.2 and 0.4 s per millennium, respectively. Meanwhile, the anomalistic month is decreasing by about 0.8 s per millennium.

Although small, the cumulative effects of such changes has an impact on both the Saros and Inex. Table 5-10 shows how the number of draconic and anomalistic months change with respect to 223 synodic months (Saros period) over an interval of 7000 years. Of particular interest is the last column, which shows the mean shift of the Moon's node after a period of 1 Saros. It is gradually increasing, which means that the average number of eclipses in a typical Saros series is decreasing. This explains the trend in the number of lunar eclipses seen in Figure 5-2a.

Table 5-10: Number of Anomalistic and Draconic Months in 1 Saros

Year	Anomalistic Months (223 Lunations)	Draconic Months (223 Lunations)	Node Shift (after 1 Saros)
-3000	238.991679	241.998742	0.4529
-2000	238.991763	241.998730	0.4571
-1000	238.991854	241.998717	0.4618
0	238.991950	241.998703	0.4668
1000	238.992051	241.998688	0.4722
2000	238.992157	241.998673	0.4779
3000	238.992267	241.998656	0.4838
4000	238.992379	241.998639	0.4899

Table 5-11 shows how the number of draconic months is changing with respect to 358 synodic months (Inex period) over a 7000-year interval. The mean shift in the lunar node after 1 Inex is much smaller than the Saros and is gradually decreasing. This explains why the lifetime of the Inex is so much longer than the Saros and is still increasing.

Table 5-11: Number of Draconic Months in 1 Inex

Year	Draconic Months (358 Lunations)	Node Shift (after 1 Inex)
-3000	388.500223	-0.0801
-2000	388.500204	-0.0734
-1000	388.500183	-0.0659
0	388.500160	-0.0578
1000	388.500136	-0.0491
2000	388.500111	-0.0400
3000	388.500085	-0.0305
4000	388.500057	-0.0207

Although the Inex possesses a long lifespan, its mean duration is not easily characterized because of the decreasing nodal shift seen in Table 5-11. If the instantaneous mean durations of the synodic and draconic months for the years −2000, +2000, and +4000 are used to calculate the mean duration of the Inex, the resulting lengths are about 14,500, 26,600, and 51,000 years, respectively (Meeus, 2004a).

Abbreviations

arcsec	Arc second
AT	Hybrid eclipse that begins as annular, then changes to total.
ATA	Hybrid eclipse that begins as annular, changes to total, and then reverts back to annular.
BCE	Before the Common Era
CE	Common Era
cm	Centimeter
ET	Ephemeris Time
GMAT	Greenwich Mean Astronomical Time
GMT	Greenwich Mean Time
IAU	International Astronomical Union
ISO	International Standards Organization
LLR	Lunar Laser Ranging
LOD	Length of Day
m	Meter (or minutes in tables)
min	Minutes
ms	Millisecond
s	Second
arcsec/cy2	Arc seconds per Julian century squared
TA	Hybrid eclipse that begins as total and ends as annular.
TAI	International Atomic Time
TD	Terrestrial Dynamical Time
TT	Terrestrial Time
UT	Universal Time
UTC	Coordinated Universal Time
VLBI	Very Long Baseline Interferometry
yr	Year

References

Astronomical Almanac for 2006, Almanac Office, U.S. Naval Observatory, U.S. Government Printing Office, Washington; London: HM Stationery Office (2004).

Astronomical Almanac for 2008, Almanac Office, U.S. Naval Observatory, U.S. Government Printing Office, Washington; London: HM Stationery Office (2006).

Bretagnon P., G. Francou., "Planetary Theories in rectangular and spherical variables: VSOP87 solution", *Astron. Astrophys.*, **202** (309), (1988).

Brown, E.W., "Theory of the Motion of the Moon", *Mem. Royal Astron. Soc.*, Vol. LVII, Part II, pp. 136-141, London (1905).

Chapront-Touzé, M., and J. Chapront, "The Lunar Ephemeris ELP 2000," *Astron. Astrophys.*, **124**(1), pp. 50–62 (1983).

Chapront-Touzé, M., and J. Chapront, *Lunar Tables and Programs from 4000 B.C. to A.D. 8000*, Willmann-Bell, Richmond, Virginia (1991).

Chapront, J., M. Chapront-Touzé, and G. Francou, "A new determination of lunar orbital parameters, precession constant and tidal acceleration from LLR measurements," *Astron. Astrophys.*, **387**, pp. 700–709 (2002).

Chauvenet, W.A., *Manual of Spherical and Practical Astronomy, Vol. 1*, edition of 1891, (Dover reprint, New York, 1960).

Danjon, A., "Les éclipses de Lune par la pénombre en 1951," *L'Astronomie*, **65**, 51–53 (1951).

Dickey, J.O., P.L. Bender, J.E. Faller, X.X. Newhall, R.L. Ricklefs, J.G. Ries,, P.J. Shelus, C. Veillet, A.L. Whipple, J.R. Wiant, J.G. Williams, and C.E. Yoder, "Lunar Laser Ranging: a Continuing Legacy of the Apollo Program," *Science*, **265**, pp. 482–490 (1994).

Espenak, F., *Fifty Year Canon of Solar Eclipses: 1986–2035*, Sky Publishing Corp., Cambridge, Massachusetts (1987).

——, *Fifty Year Canon of Lunar Eclipses: 1986–2035*, Sky Publishing Corp., Cambridge, Massachusetts (1989).

——, Eclipses During 2007, *Observer's Handbook - 2007*, Royal Astronomical Society of Canada (2006).

——, and J. Meeus, *Five Millennium Canon of Solar Eclipses: –1999 to +3000 (2000 BCE to 3000 CE)*, NASA Tech. Pub. 2006–214141, NASA Goddard Space Flight Center, Greenbelt, Maryland (2006).

——, and ——, *Five Millennium Catalog of Lunar Eclipses: –1999 to +3000 (2000 BCE to 3000 CE)*, NASA Tech. Pub. 2009–214173, NASA Goddard Space Flight Center, Greenbelt, Maryland (2009).

Freeth, T., A. Jones, J.M. Steele, and Y. Bitsakis, "Calendars with Olympiad display and eclipse prediction on the Antikythera Mechanism," *Nature*, 07/2008, **454**, pp. 614–617, (2008).

Gingerich, O., (Translator) Canon of Eclipses, Dover Publications, New York (1962) (from the original T.R. von Oppolzer, book, *Canon der Finsternisse*, Wien, [1887]).

Hughes, T., "Addendum," In: *Mathematical Astronomy Morsels III*, Willmann-Bell, Richmond, Virginia, pp. 137–140, (2004b).

Huber, P.J., "Modeling the Length of Day and Extrapolating the Rotation of the Earth," In: *Papers in Honor of Jean Meeus, Astronomical Amusements*, F. Bònoli, S. De Meis, and A. Panaino, Eds., IsIAO, Rome (2000).

Improved Lunar Ephemeris 1952–1959, Nautical Almanac Office, U.S. Naval Observatory, Washington, DC (1954).

Keen, R.A., "Volcanic Aerosols and Lunar Eclipses," *Science*, **222**, pp. 1011–1013 (1983).

Kühl, A., 1928, "Ueber den Einfluss des Grenzkontrastes auf Präzisionsmessungen," *Physikalische Zeitschrift,* **29**, 1- 34.

La Hire, P., *Tabulae Astronomicae* (Paris 1707).

Link, F., *Eclipse Phenomena in Astronomy*, Springer-Verlag, New York (1969).

Liu, B.-L., and A.D. Fiala, *Canon of Lunar Eclipses 1500 B.C.–A.D. 3000*, Willmann-Bell, Richmond, Virginia, p. 215 (1992).

Meeus, J., and H. Mucke, *Canon of Lunar Eclipse: -2002 to +2526*, Astronomisches Büro, Vienna, Austria (1979).

Meeus, J., *Mathematical Astronomy Morsels*, Willmann-Bell, Richmond, Virginia, pp. 108–110, (1997).

——, *More Mathematical Astronomy Morsels*, Willmann-Bell, Richmond, Virginia, pp. 120–126 (2002).

——, *Mathematical Astronomy Morsels III*, Willmann-Bell, Richmond, Virginia, pp. 109–111, (2004a).

——, C.C. Grosjean, and W. Vanderleen, *Canon of Solar Eclipses*, Pergamon Press, Oxford, United Kingdom (1966).

Morrison, L., and Stephenson, F.R., "Historical Values of the Earth's Clock Error DT and the Calculation of Eclipses," J. *Hist. Astron.*, Vol. 35 Part 3, August 2004, No. 120, pp. 327–336 (2004).

Mucke, H., and J. Meeus, *Canon of Solar Eclipses: -2003 to +2526*, Astronomisches Büro, Vienna, Austria (1983).

Newcomb, S., "Tables of the Motion of the Earth on its Axis Around the Sun," *Astron. Papers Amer. Eph.*, Vol. 6, Part I (1895).

Stephenson, F.R., *Historical Eclipses and Earth's Rotation*, Cambridge University Press, Cambridge (1997).

——, and M.A. Houlden, *Atlas of Historical Eclipse Maps, East Asia 1500 BC–AD 1900*, Cambridge University Press, Cambridge/New York (1986).

van den Bergh, *Eclipses in the Second Millennium B.C. -1600 to -1207*, Tjeenk Willink, and Haarlem, Netherlands (1954).

——, *Periodicity and Variation of Solar (and Lunar) Eclipses*, Tjeenk Willink, and Haarlem, Netherlands (1955).

von Oppolzer, T.R., *Canon der Finsternisse*, Vienna, Austria (1887).

Photo 1–1 captures the beginning, middle and end of totality during the total lunar eclipse of 2004 October 27. ©2004 F. Espenak

Photo 1–2 Various phases of the total lunar eclipse of 2000 January 20–21 appear in this composite image. ©2000 F. Espenak

Five Millennium Canon of Lunar Eclipses
Volume 1: –1999 to 0

Appendix

Key to Lunar Eclipse Figures and Maps

Figure 1-1. Key to Eclipse Figures

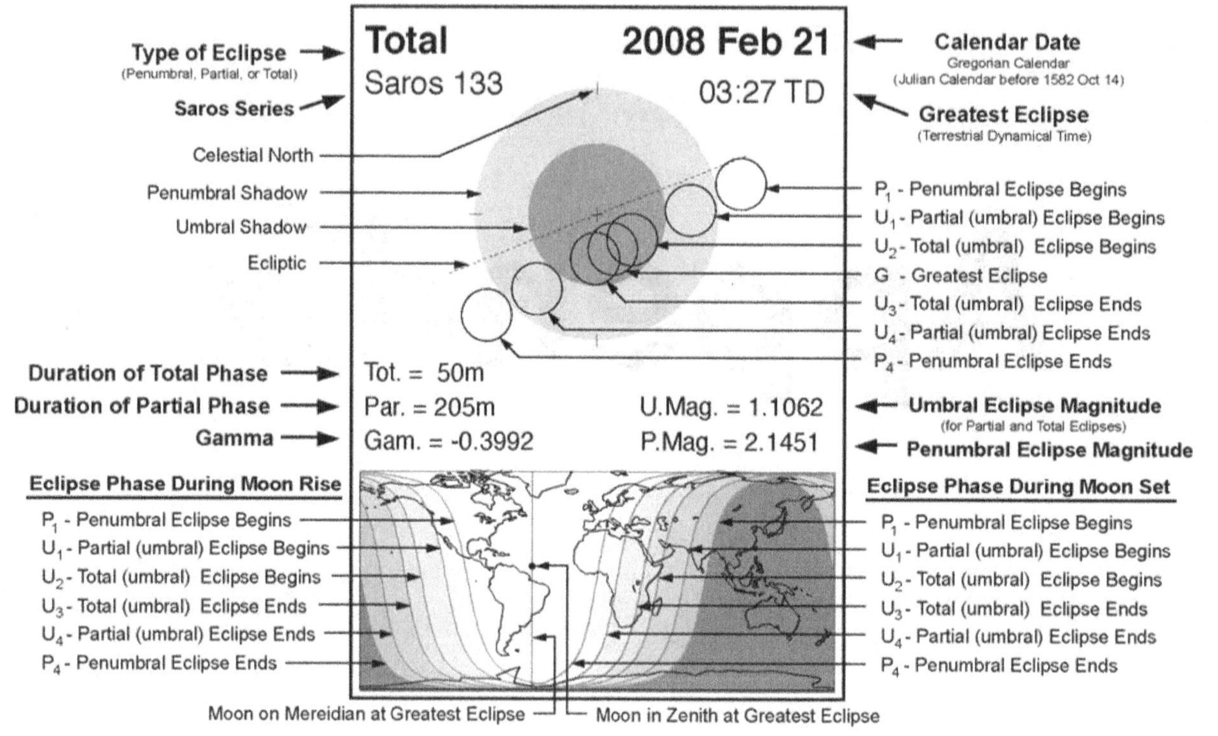

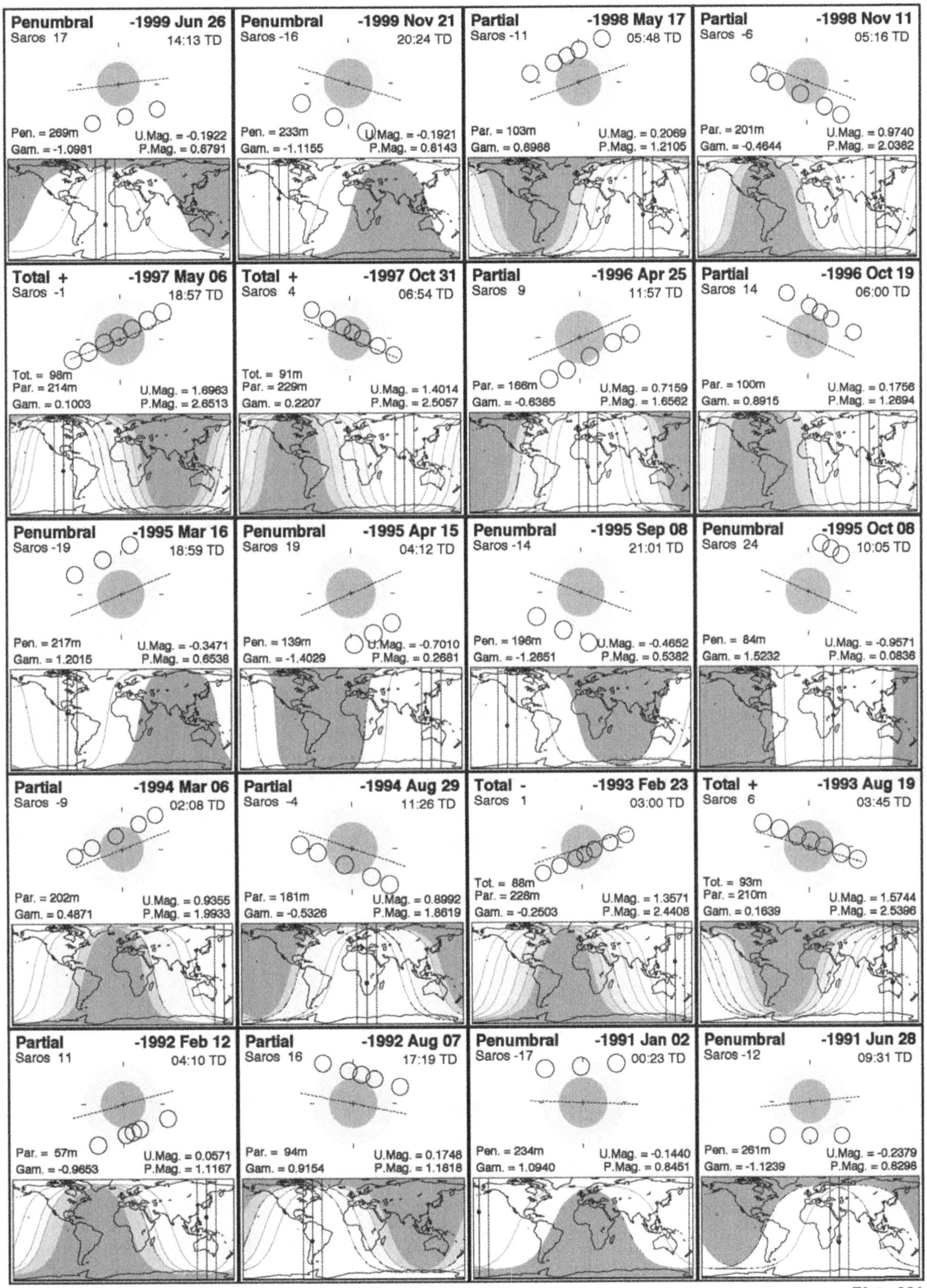

ΔT= 46437 s [= 12h54m] std.err. = ± 3726 s [= ± 15.5°] Plate 001

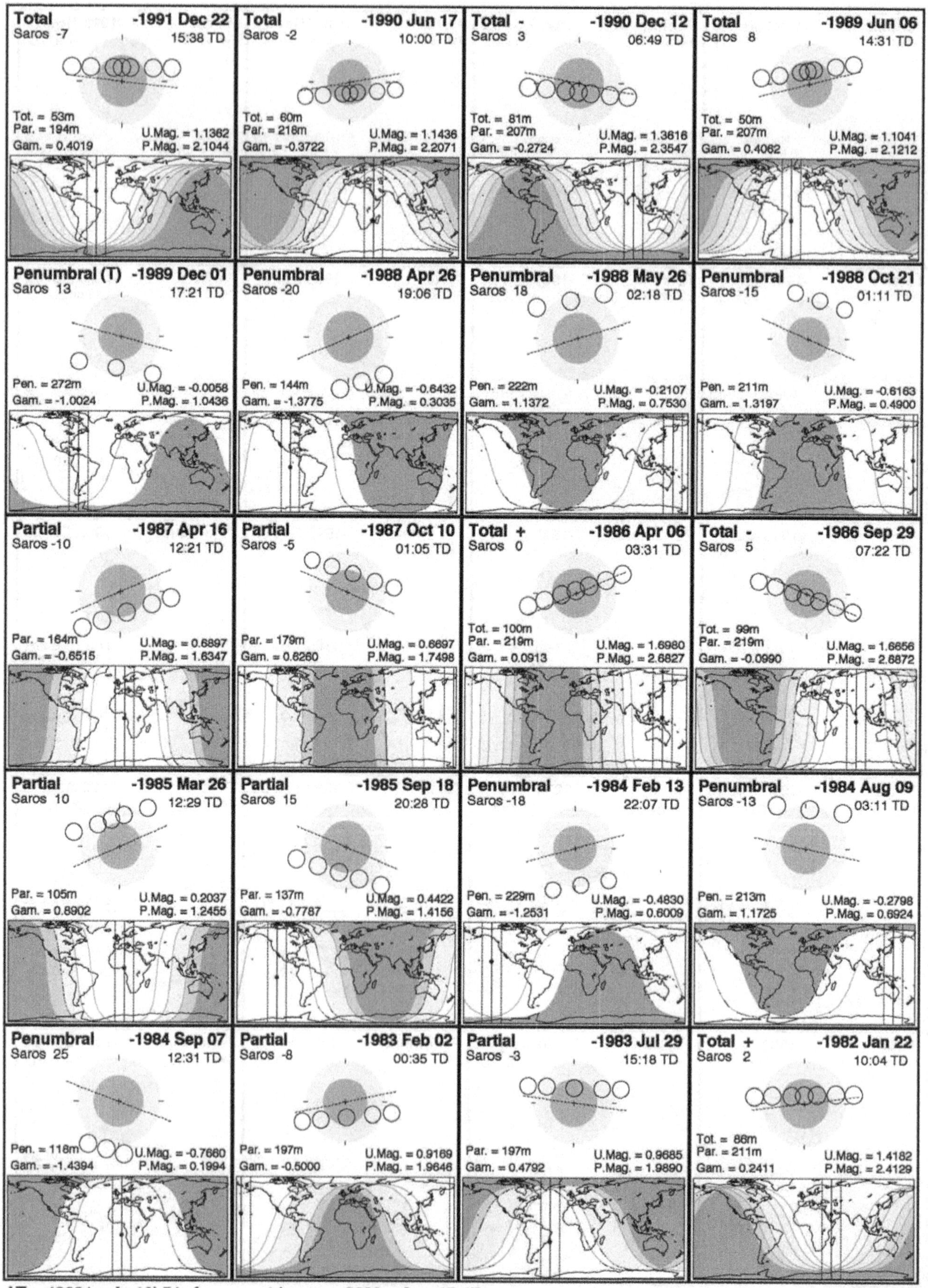

ΔT= 46231 s [= 12h51m] std.err. = ± 3690 s [= ± 15.4°] **Plate 002**

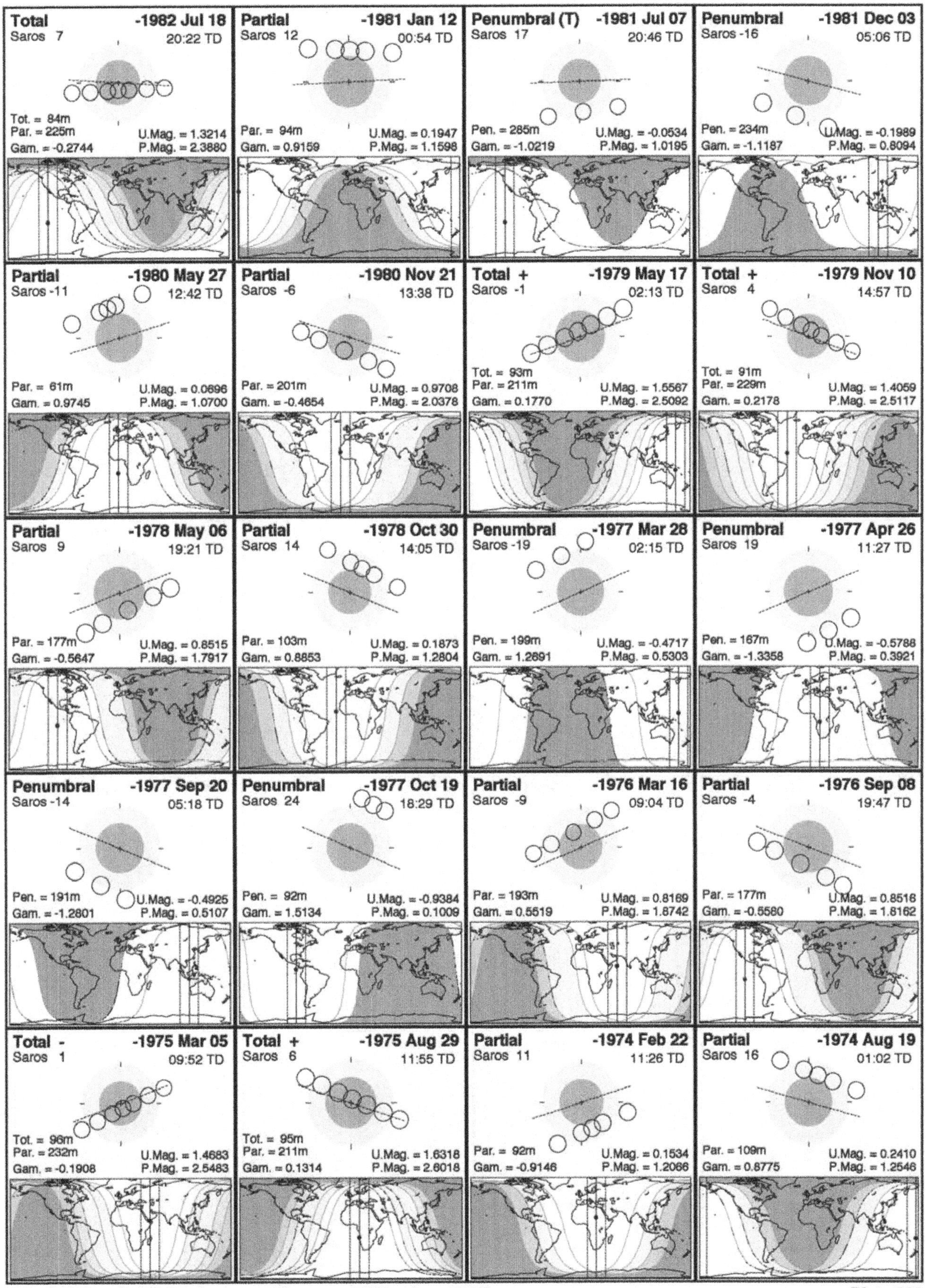

ΔT= 46023 s [= 12h47m] std.err. = ± 3654 s [= ± 15.2°] **Plate 003**

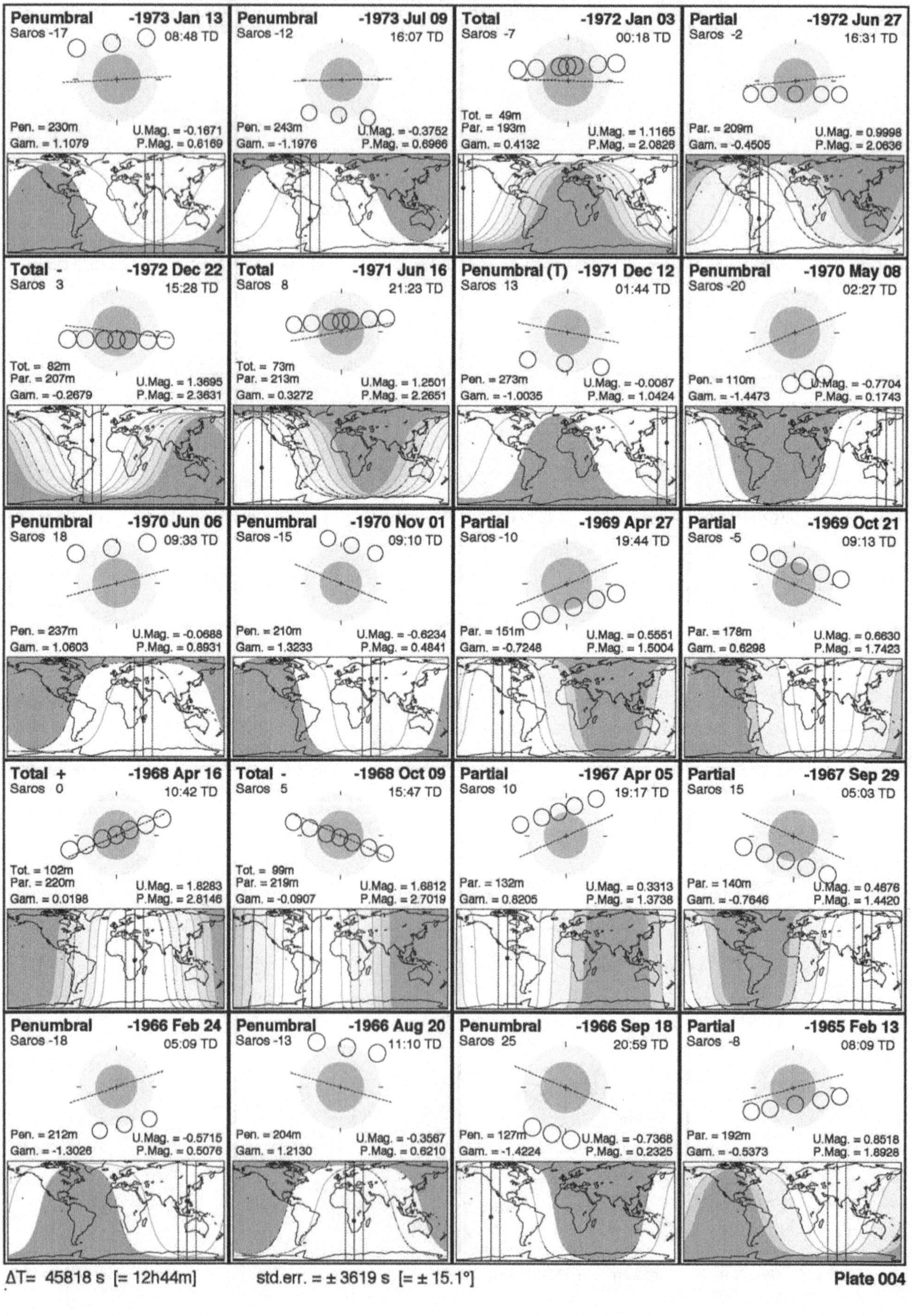

ΔT= 45818 s [= 12h44m] std.err. = ± 3619 s [= ± 15.1°] **Plate 004**

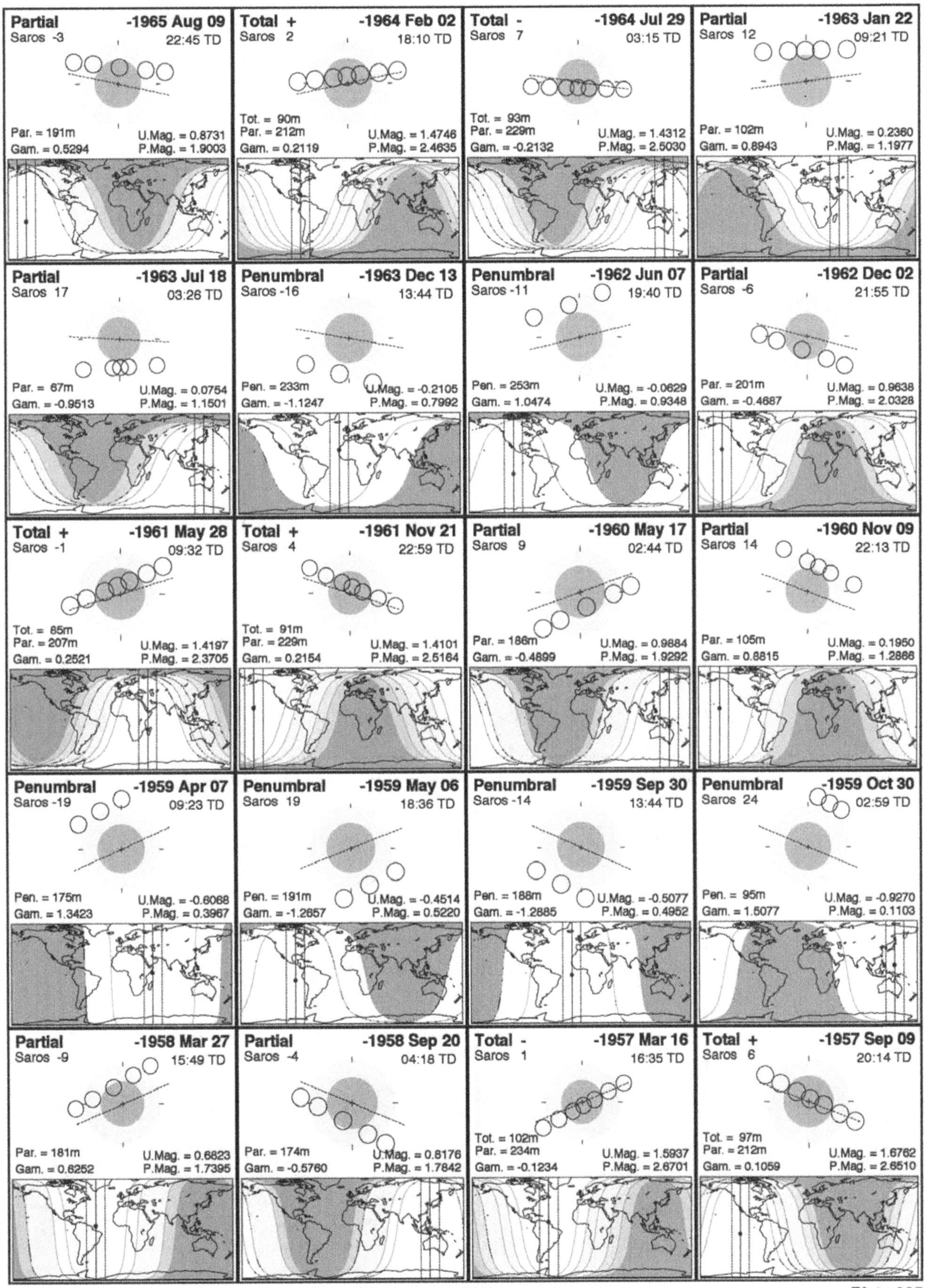

ΔT= 45611 s [= 12h40m] std.err. = ± 3584 s [= ± 14.9°] **Plate 005**

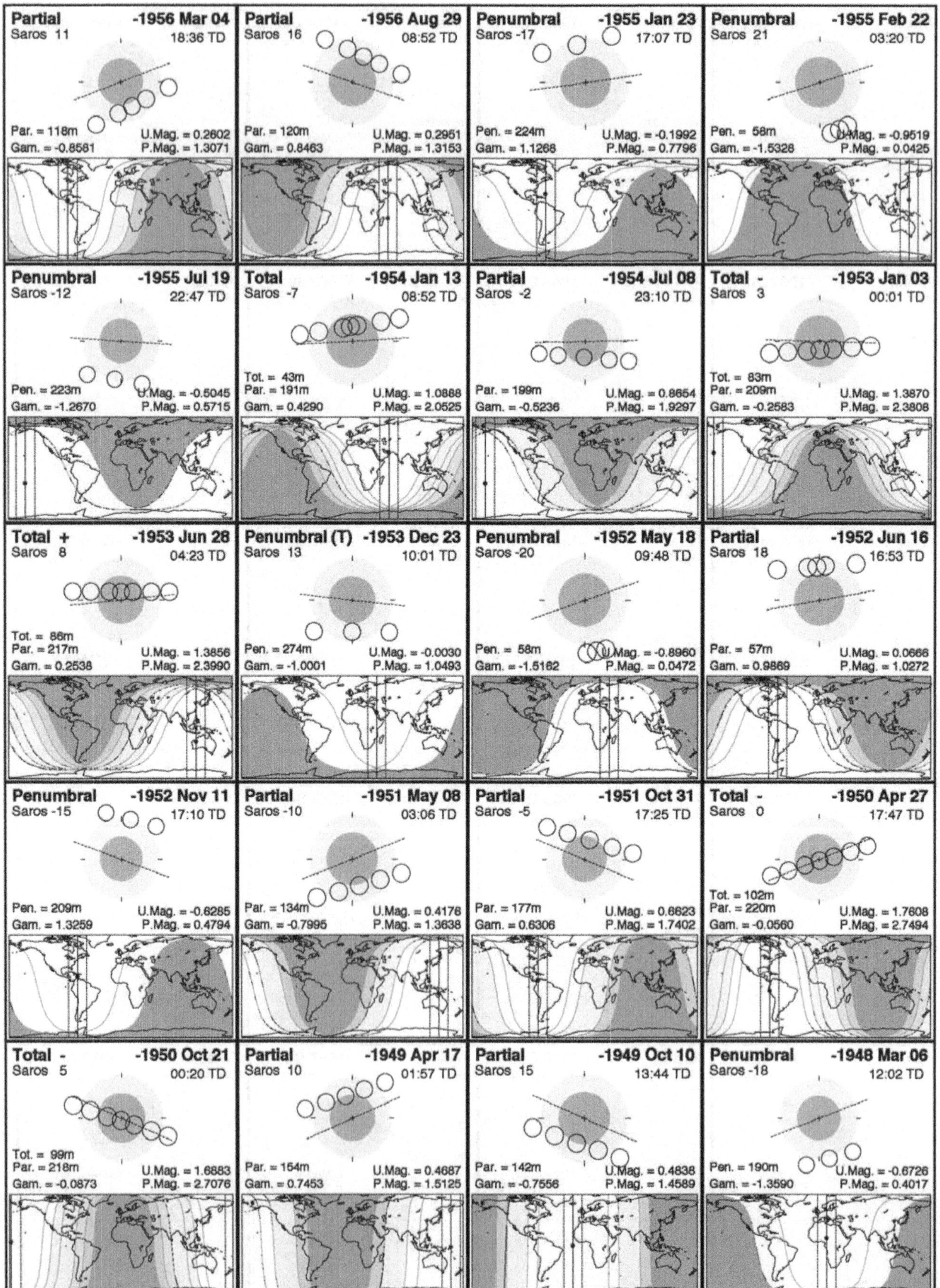

ΔT= 45404 s [= 12h37m] std.err. = ± 3548 s [= ± 14.8°] **Plate 006**

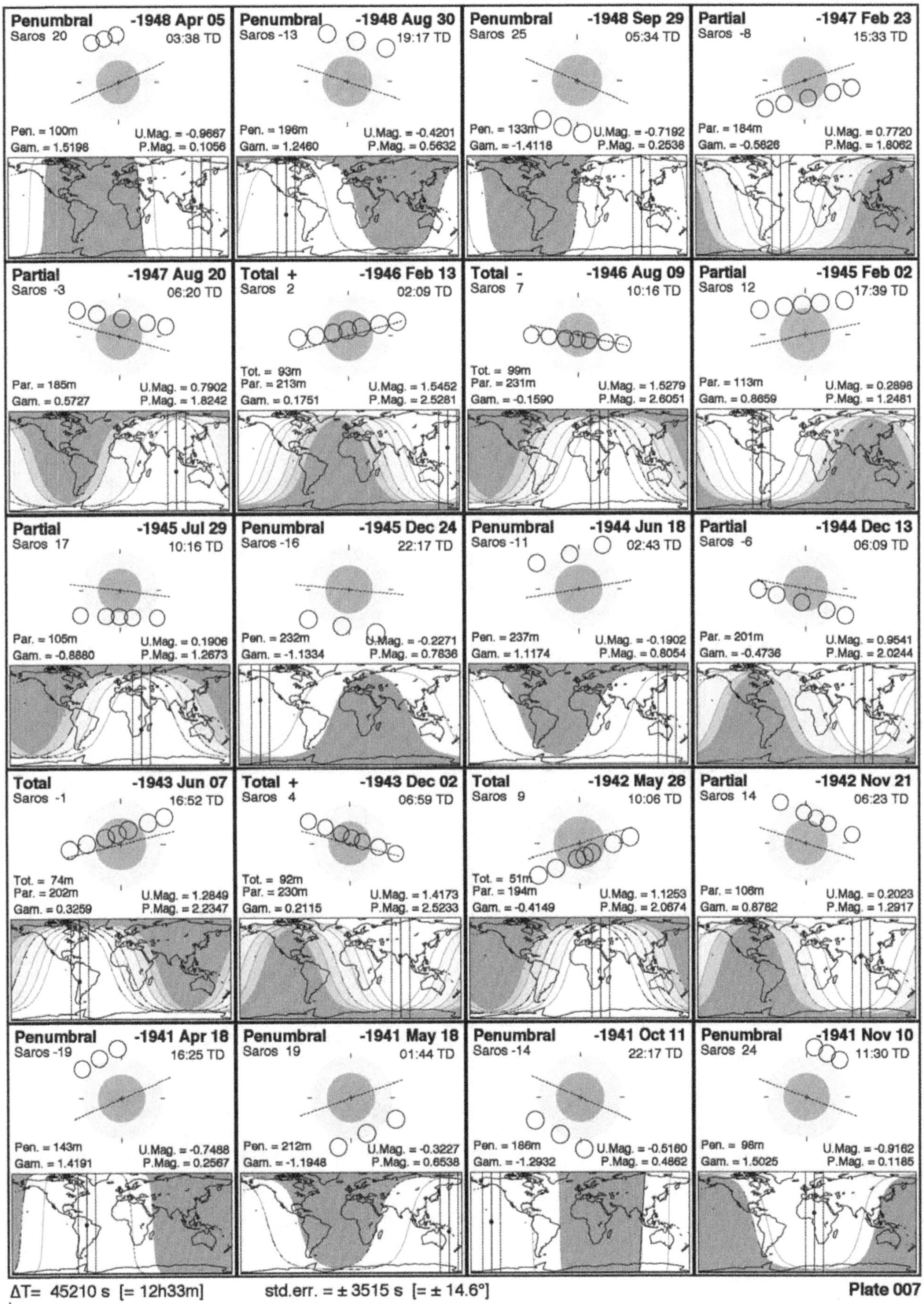

ΔT= 45210 s [= 12h33m] std.err. = ± 3515 s [= ± 14.6°] **Plate 007**

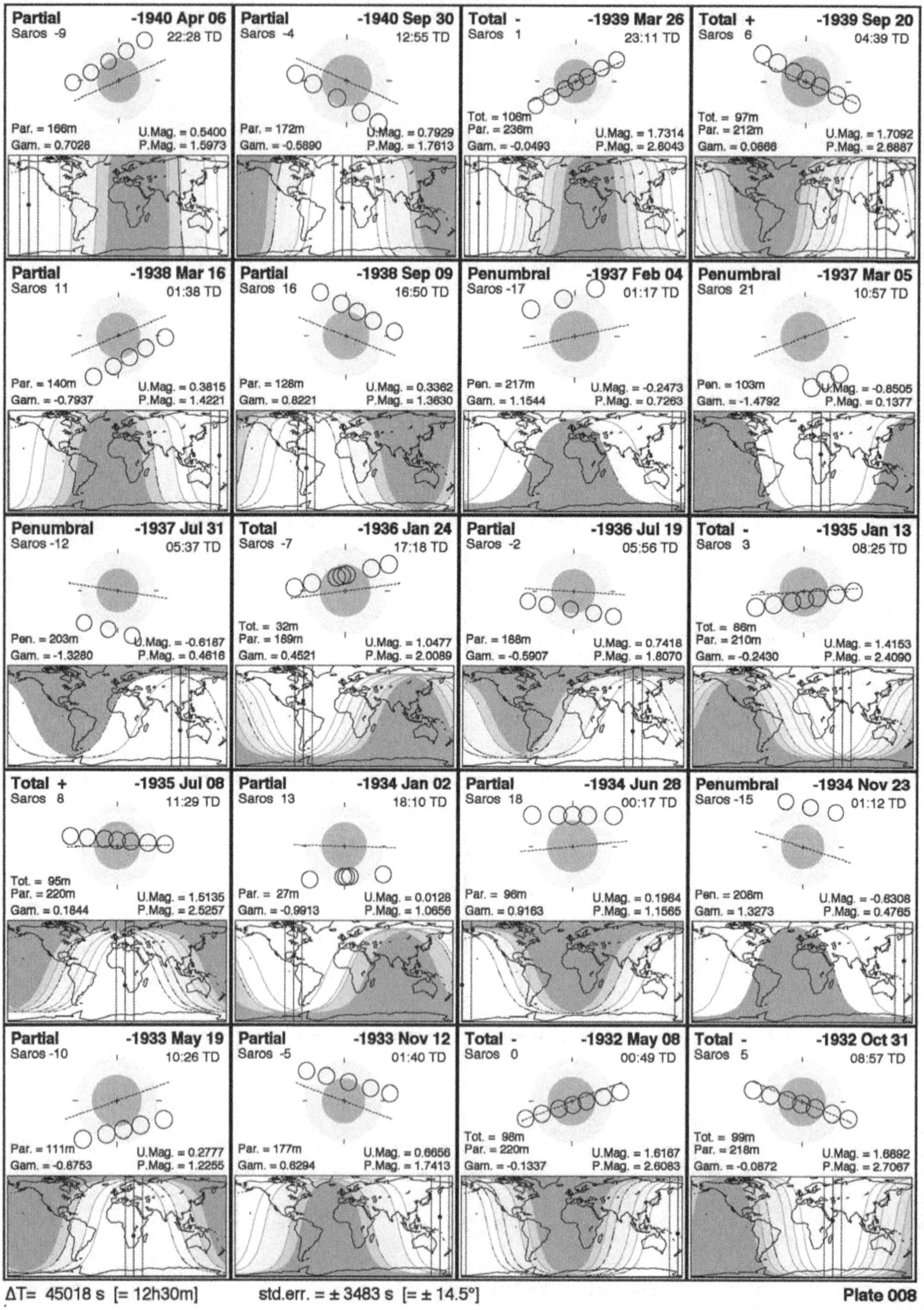

ΔT= 45018 s [= 12h30m] std.err. = ± 3483 s [= ± 14.5°] **Plate 008**

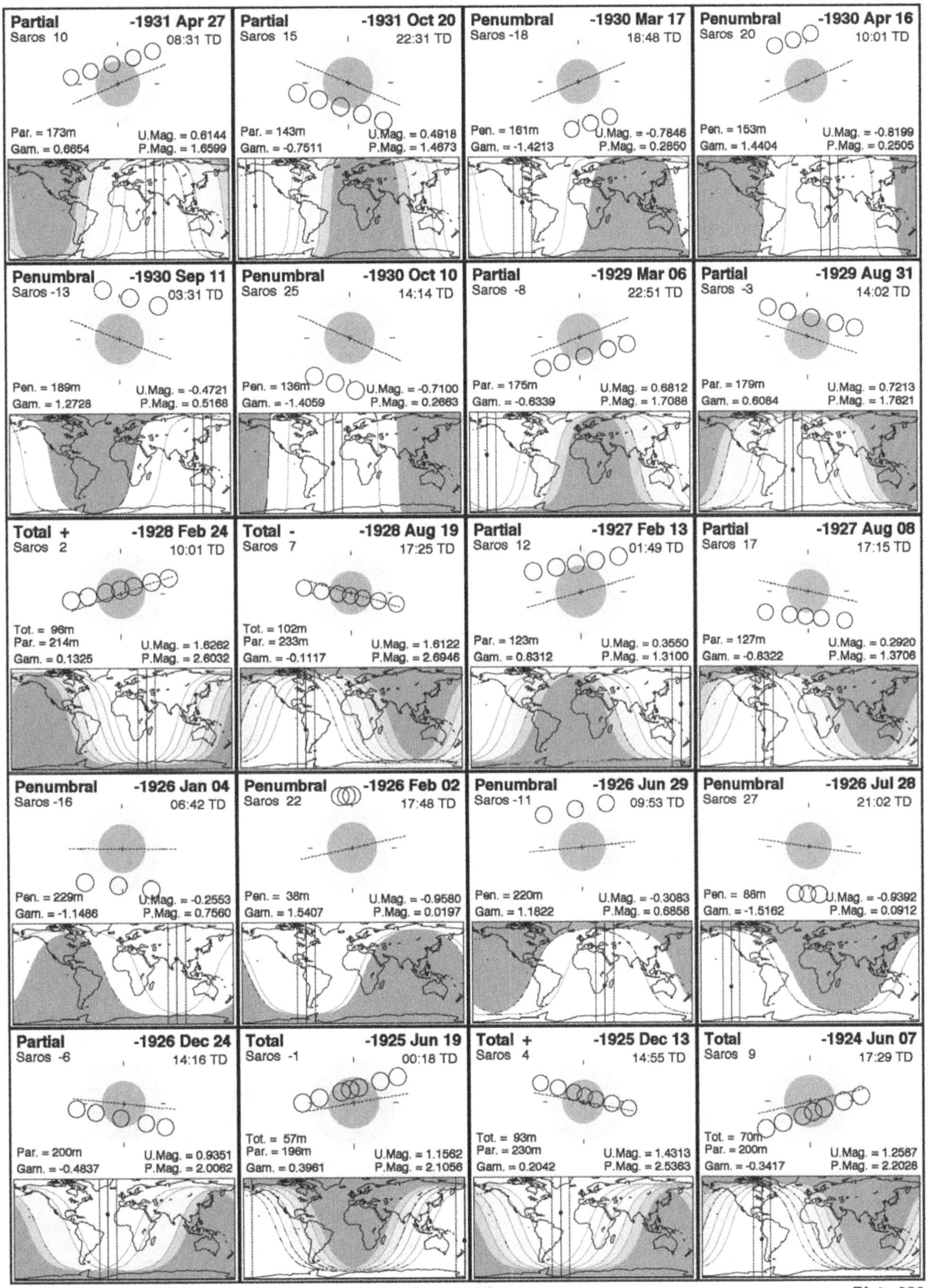

ΔT= 44801 s [= 12h27m] std.err. = ± 3446 s [= ± 14.4°] **Plate 009**

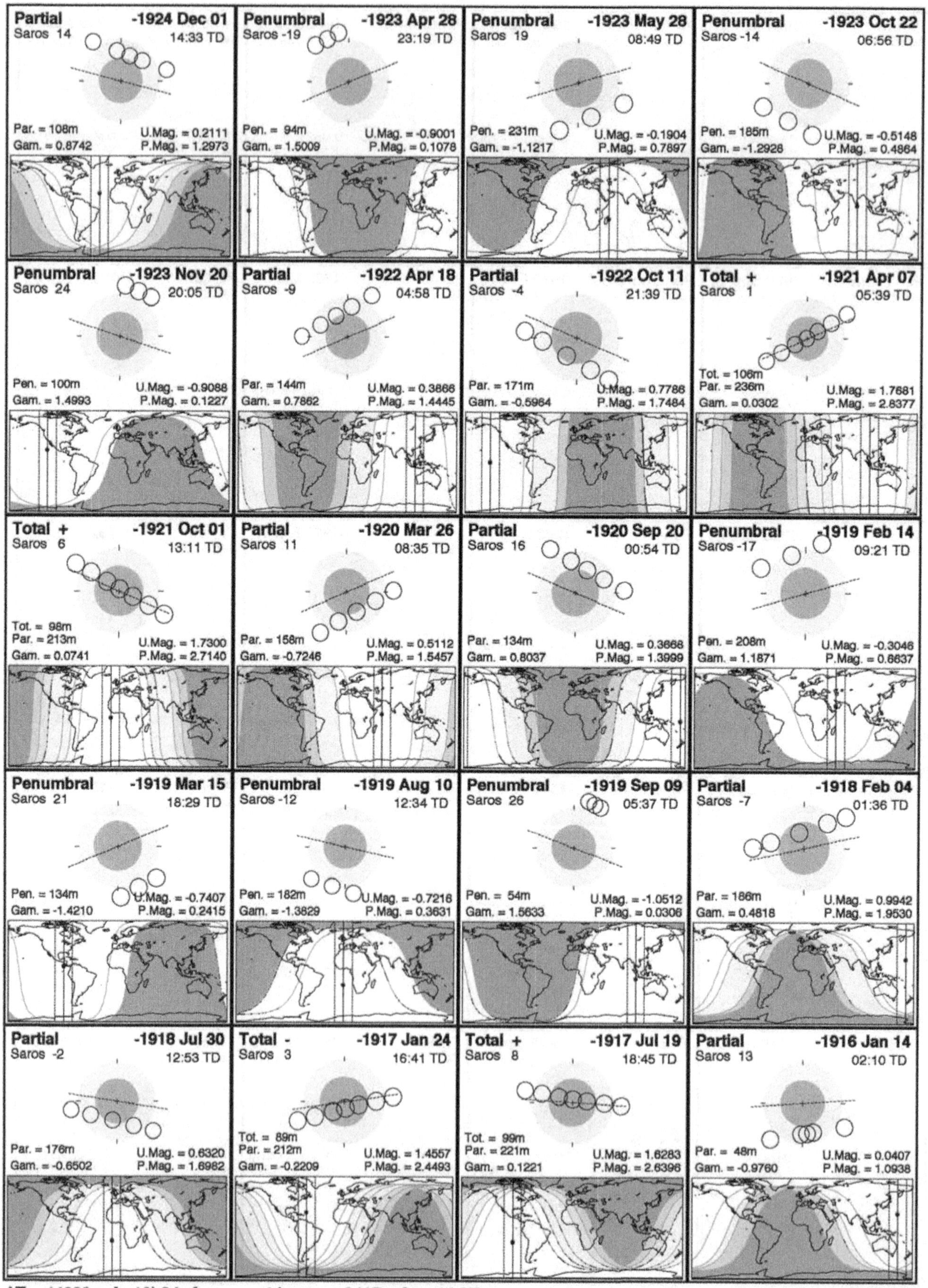

ΔT= 44620 s [= 12h24m] std.err. = ± 3415 s [= ± 14.2°] **Plate 010**

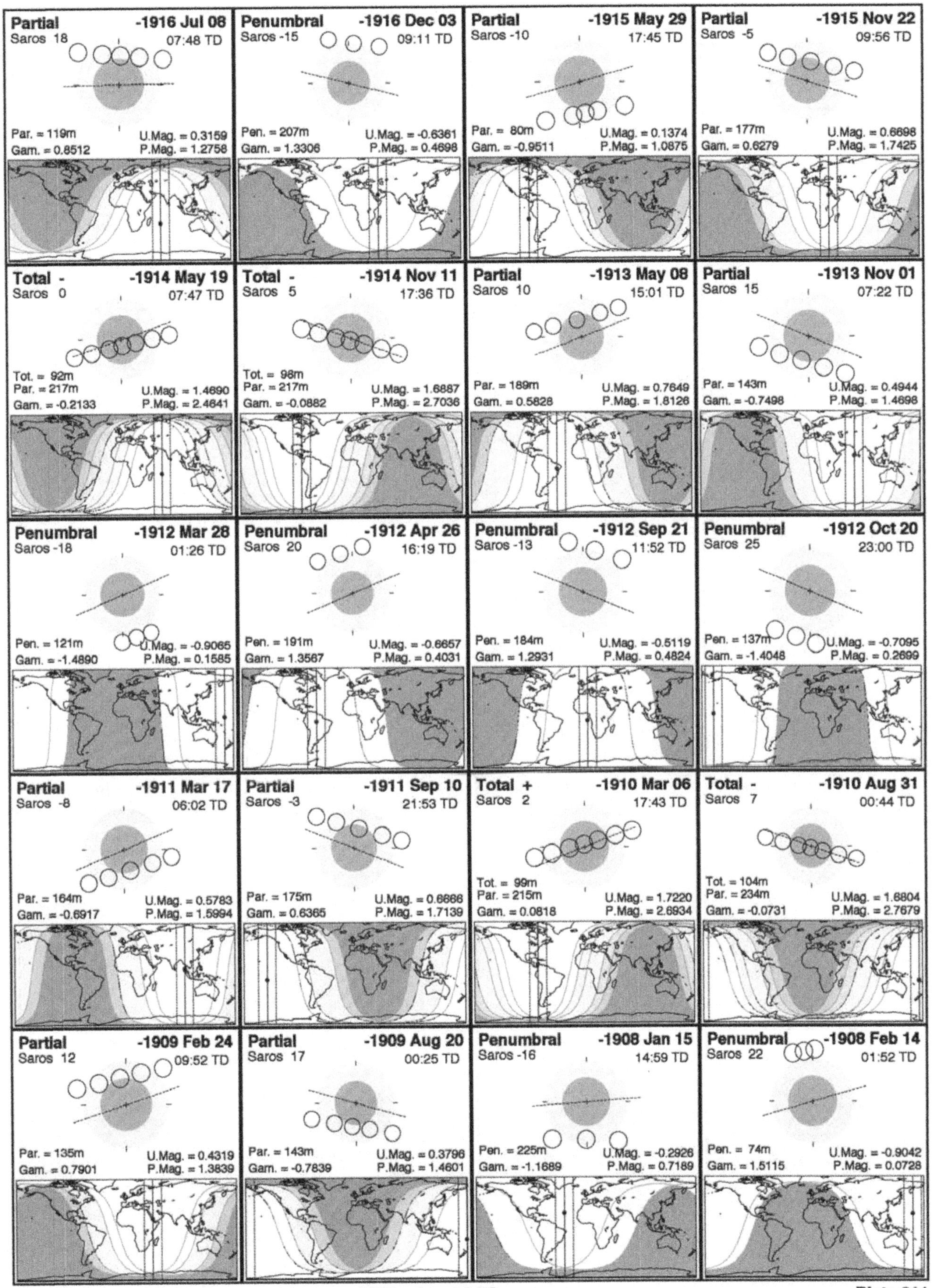

ΔT= 44438 s [= 12h21m] std.err. = ± 3385 s [= ± 14.1°] **Plate 011**

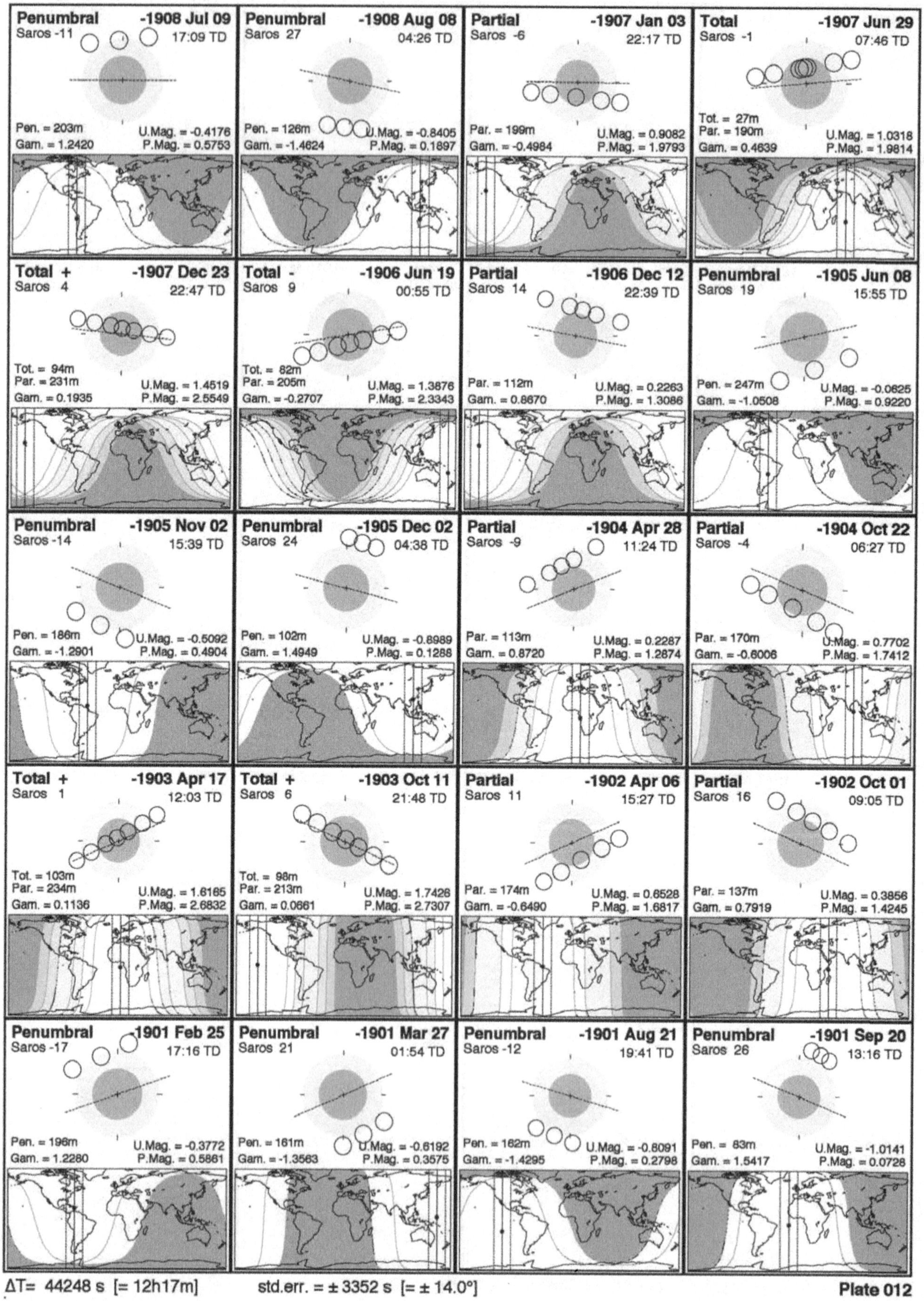

ΔT= 44248 s [= 12h17m] std.err. = ± 3352 s [= ± 14.0°] **Plate 012**

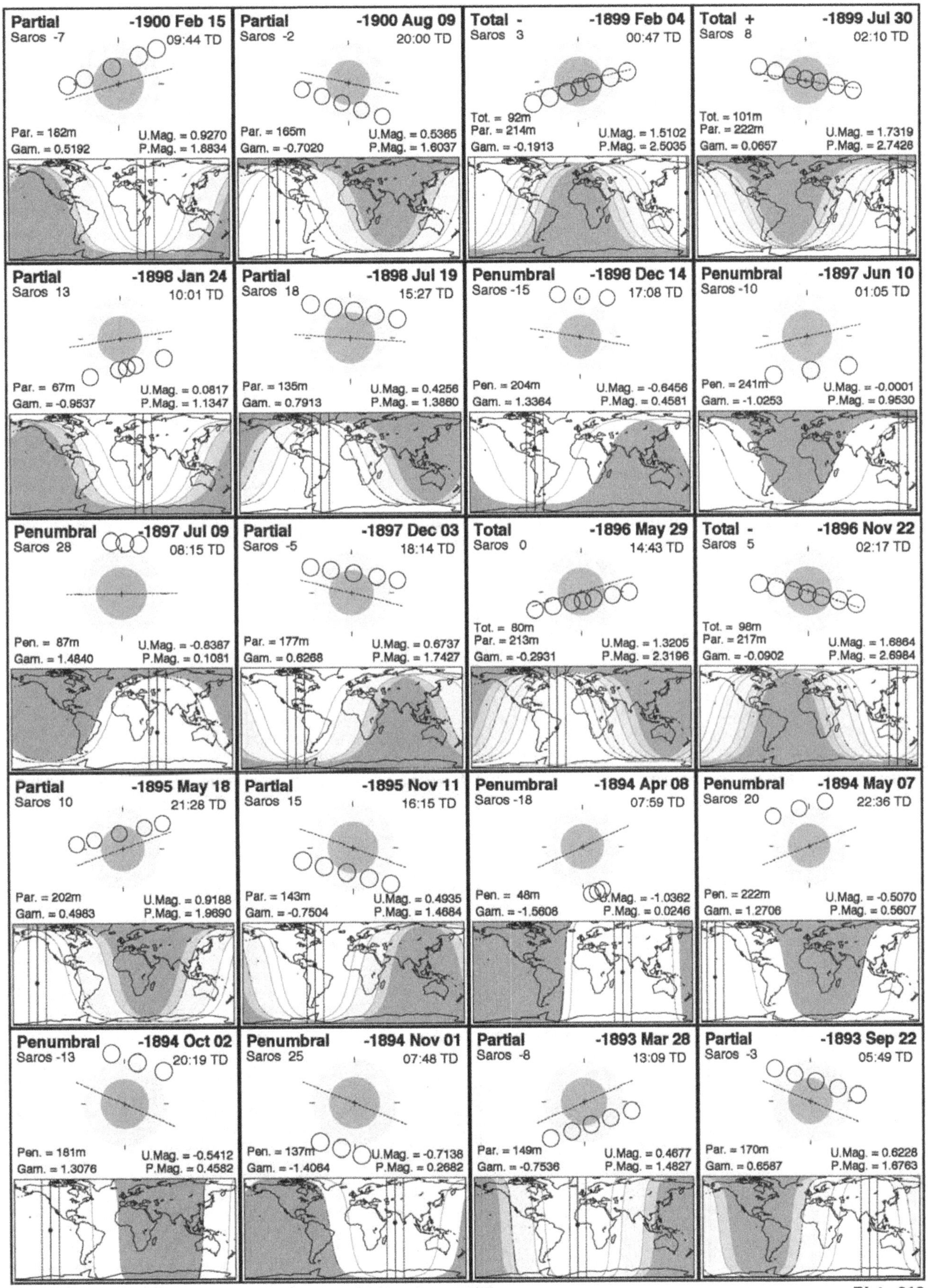

ΔT= 44068 s [= 12h14m] std.err. = ± 3322 s [= ± 13.8°] **Plate 013**

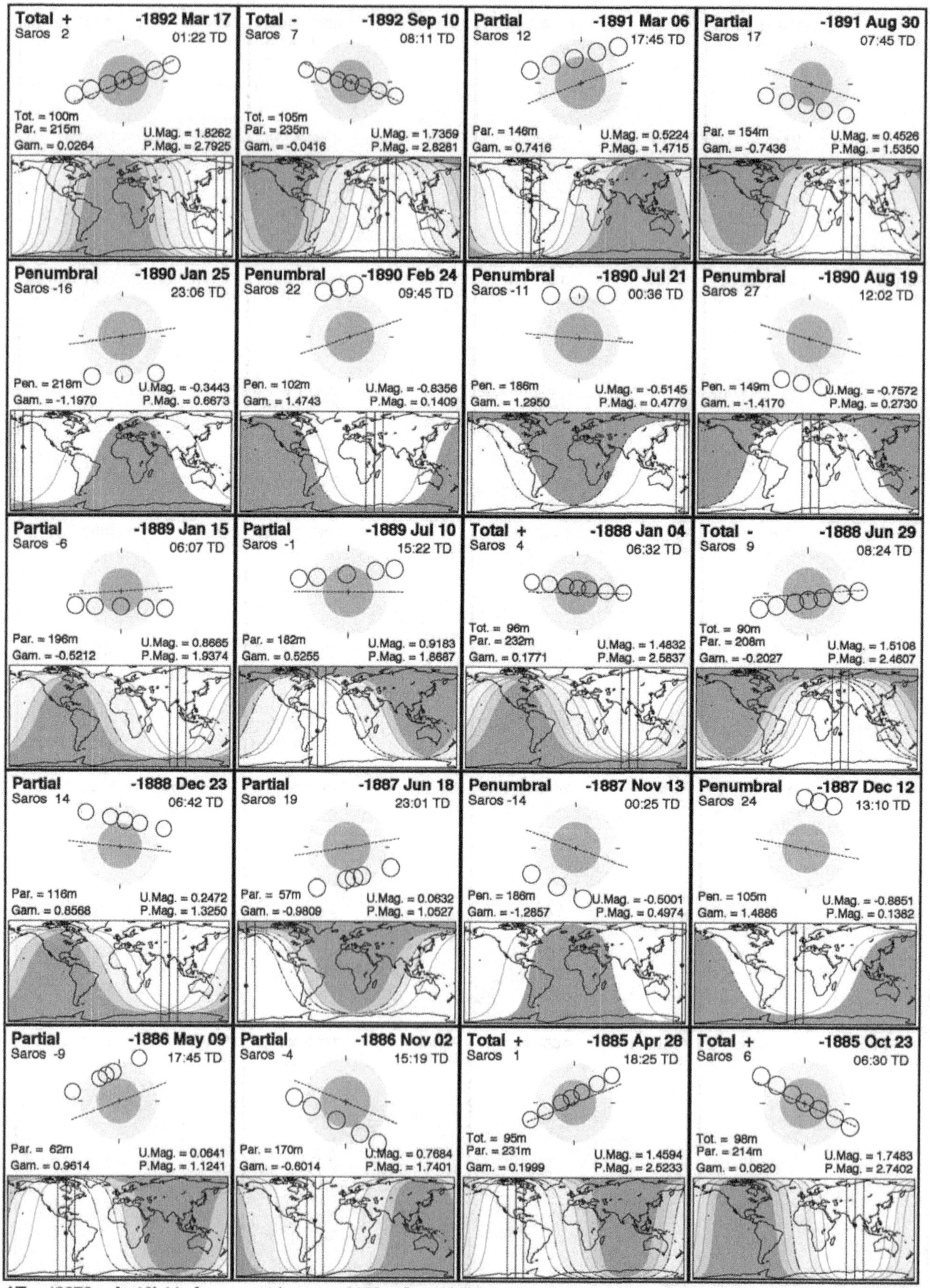

ΔT= 43876 s [= 12h11m] std.err. = ± 3290 s [= ± 13.7°] **Plate 014**

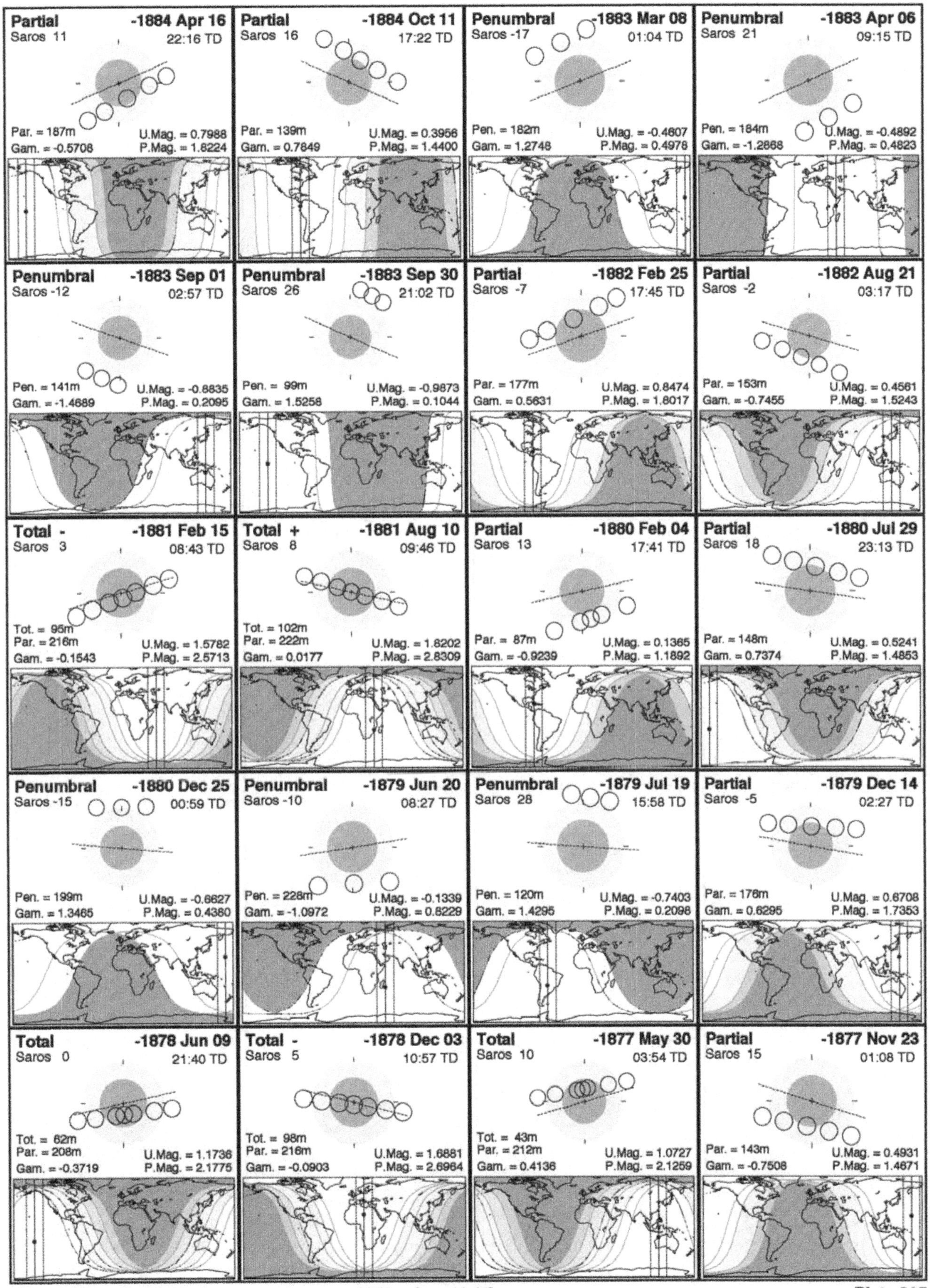

ΔT= 43685 s [= 12h08m] std.err. = ± 3258 s [= ± 13.6°] **Plate 015**

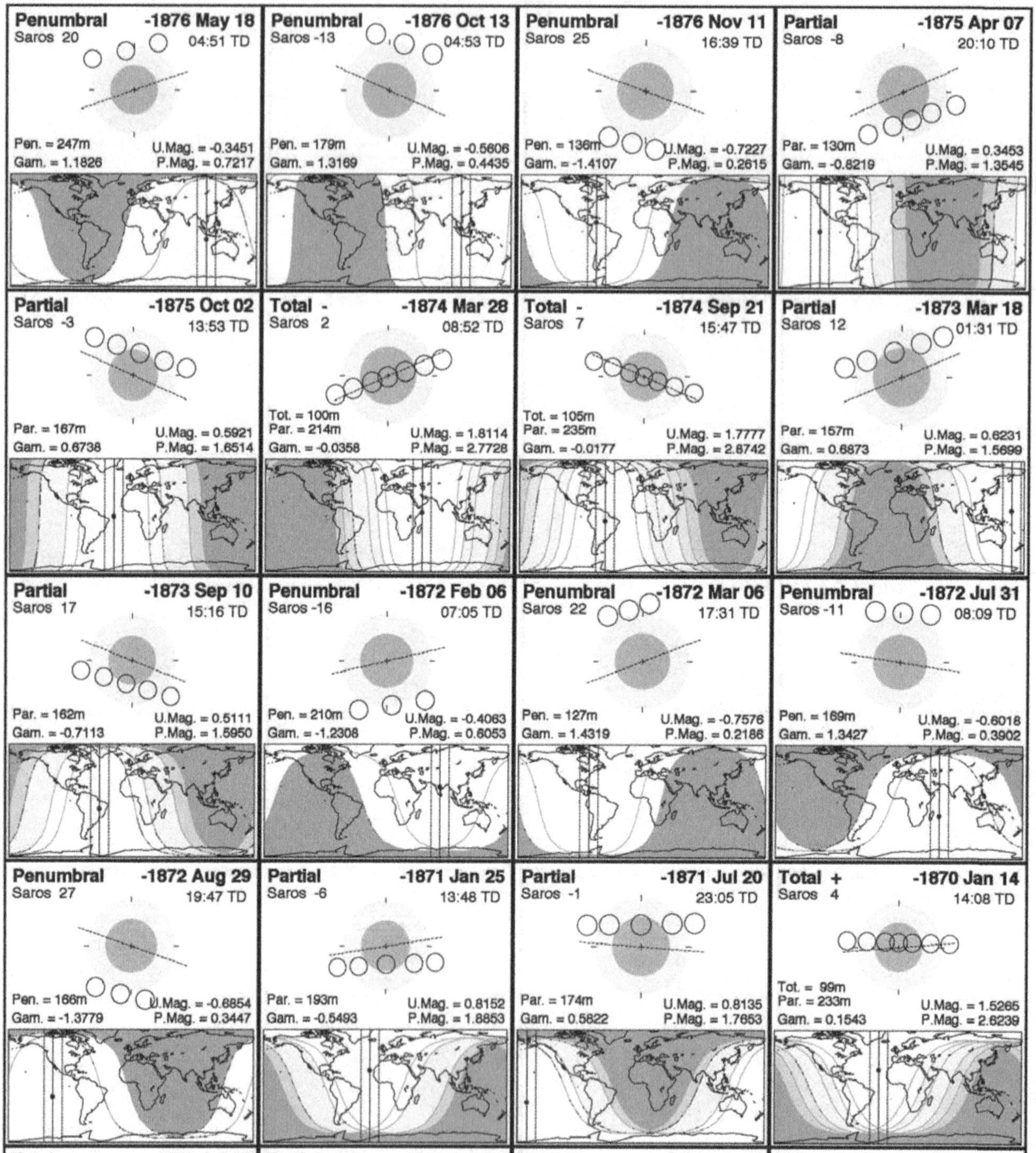

ΔT= 43495 s [= 12h05m] std.err. = ± 3226 s [= ± 13.4°] **Plate 016**

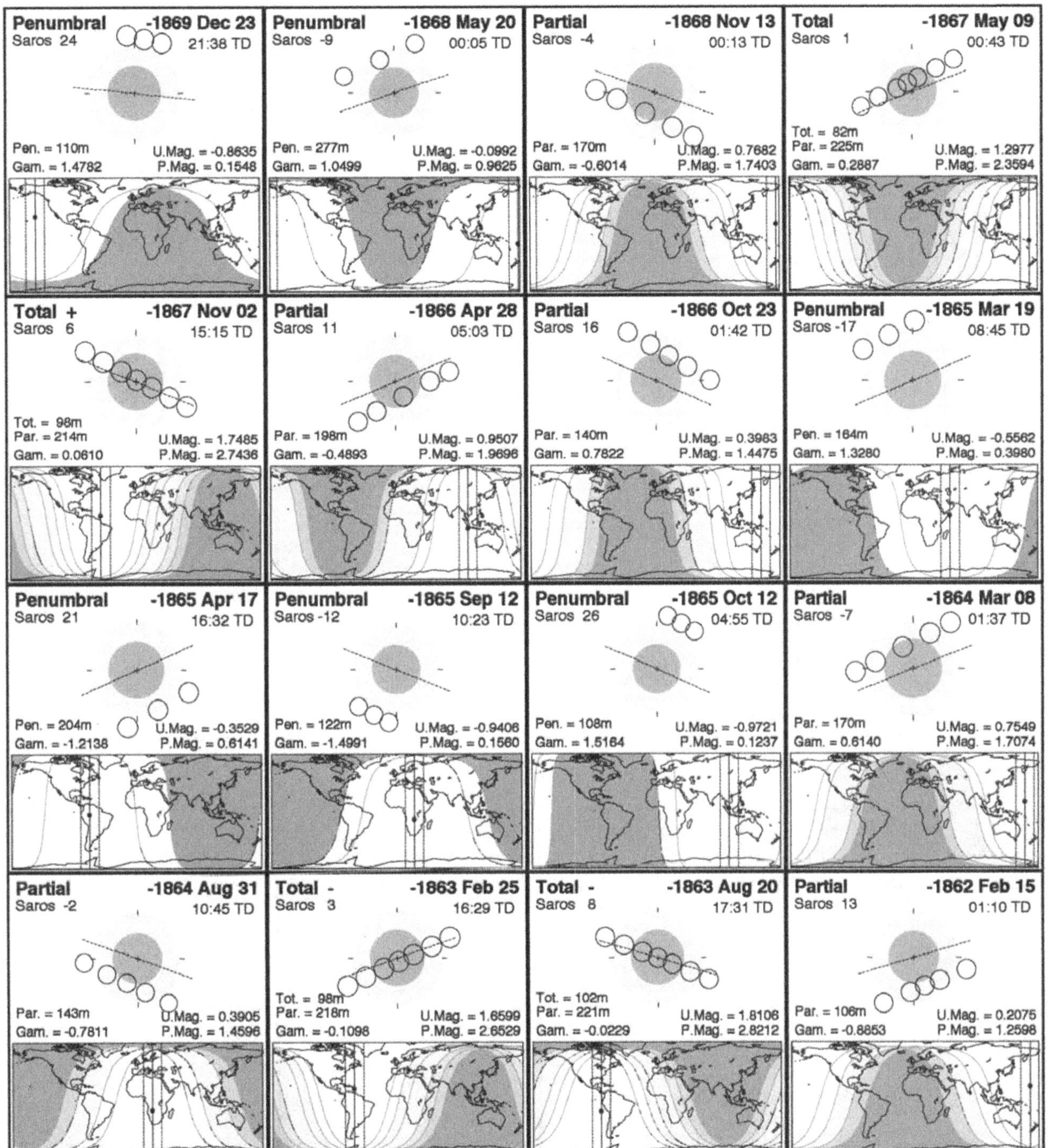

ΔT= 43316 s [= 12h02m]   std.err. = ± 3196 s [= ± 13.3°]   **Plate 017**

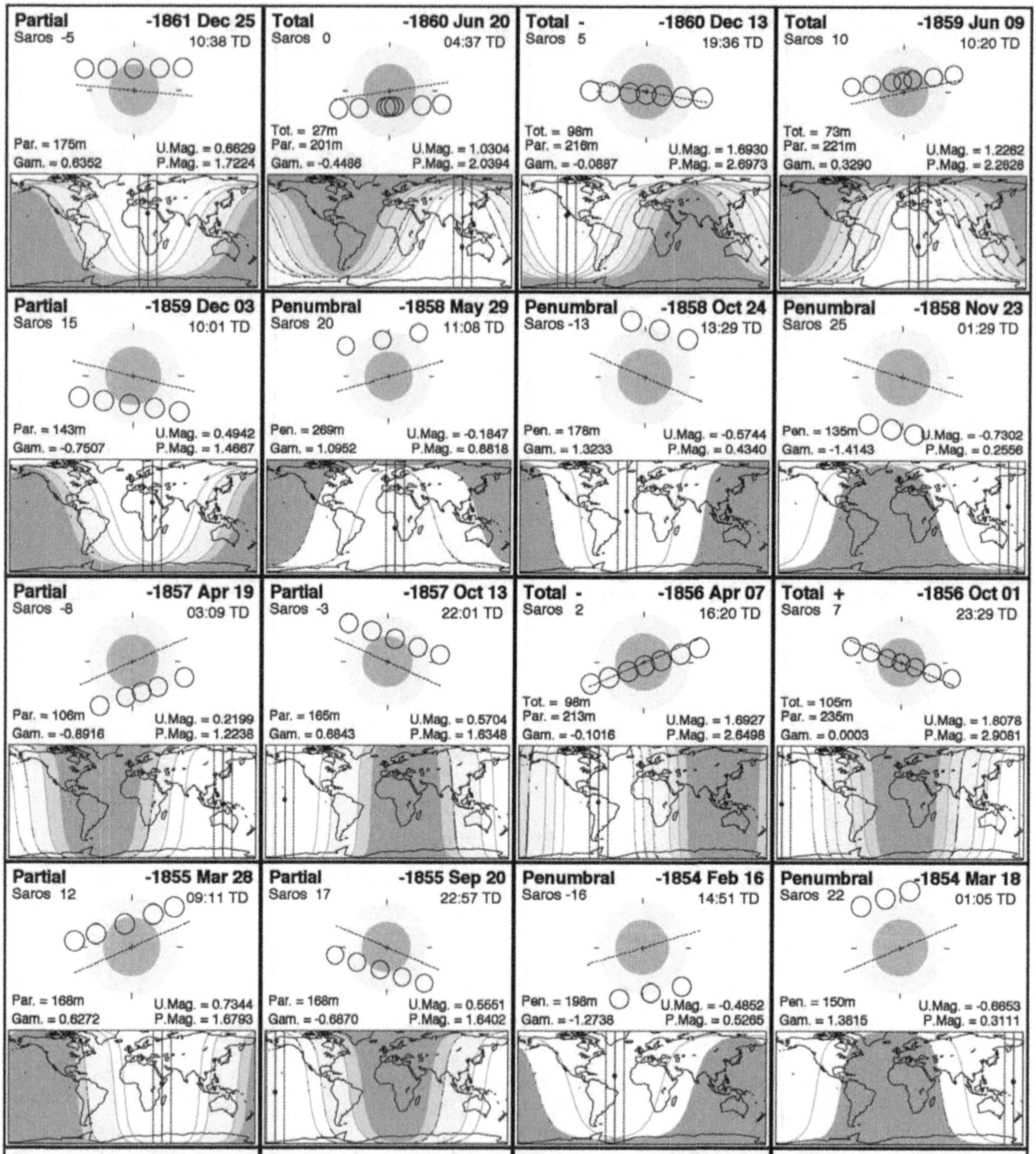

ΔT= 43128 s [= 11h59m] std.err. = ± 3165 s [= ± 13.2°] **Plate 018**

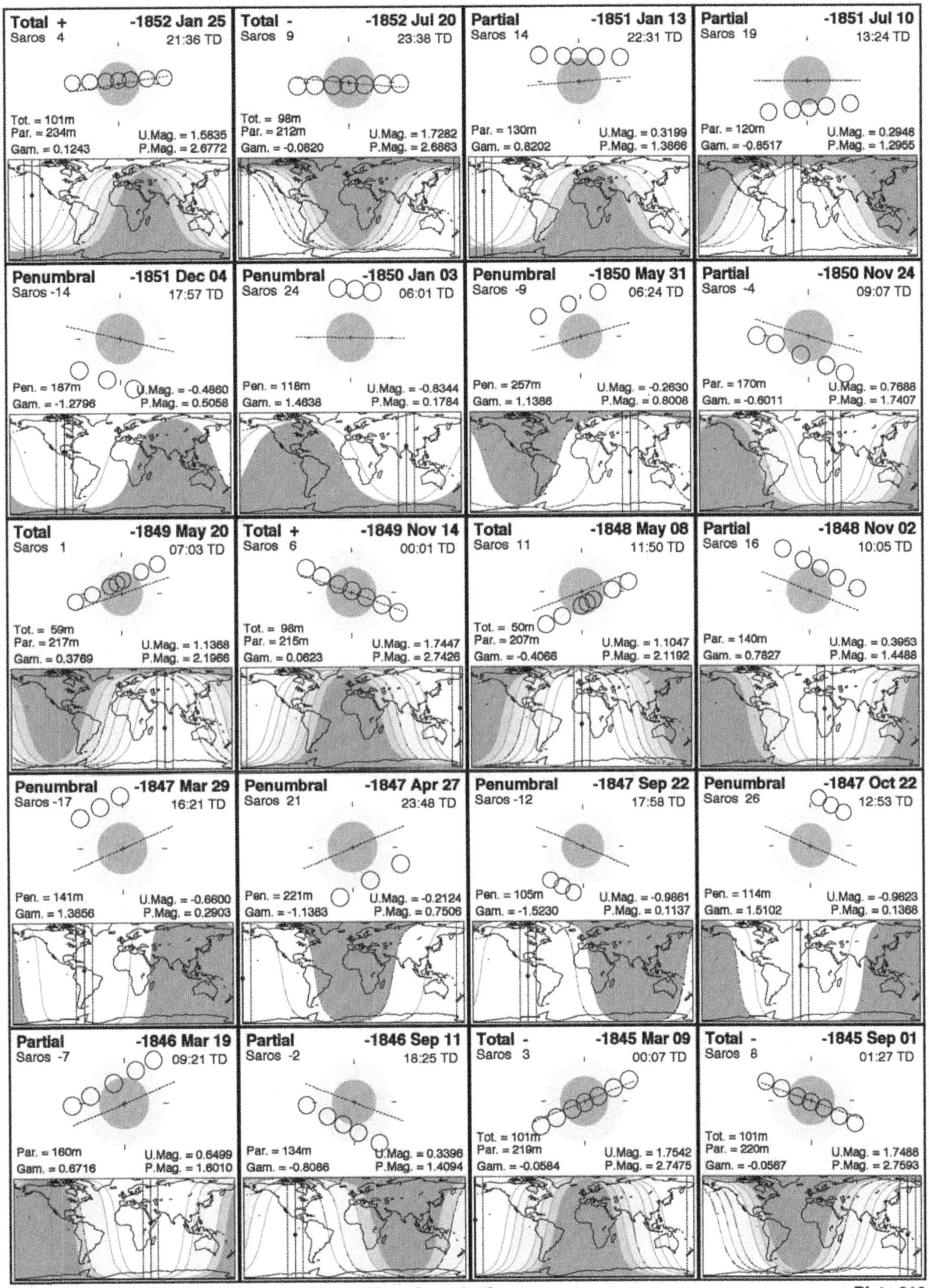

ΔT= 42938 s [= 11h56m] std.err. = ± 3133 s [= ± 13.1°] Plate 019

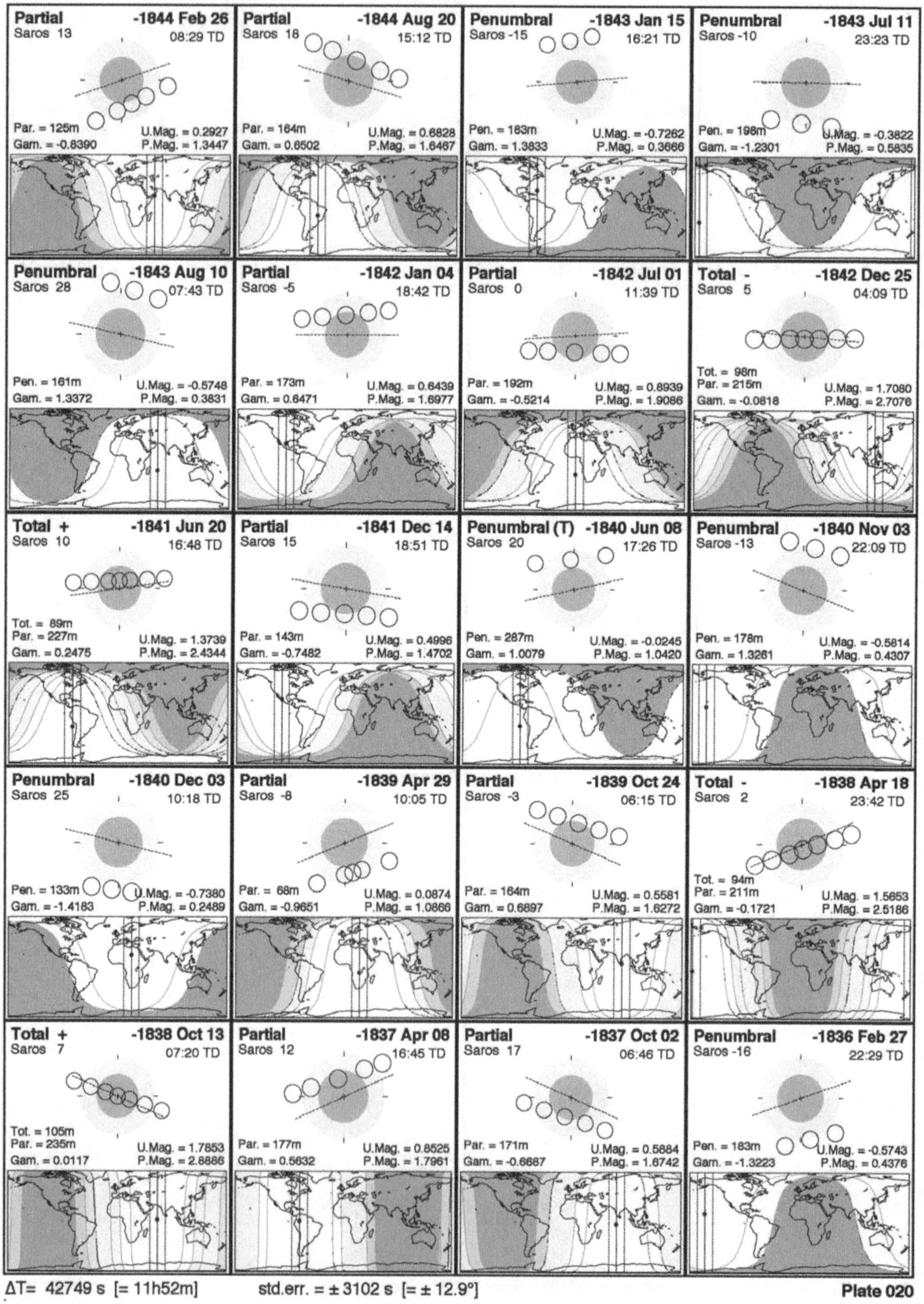

ΔT= 42749 s [= 11h52m] std.err. = ± 3102 s [= ± 12.9°] **Plate 020**

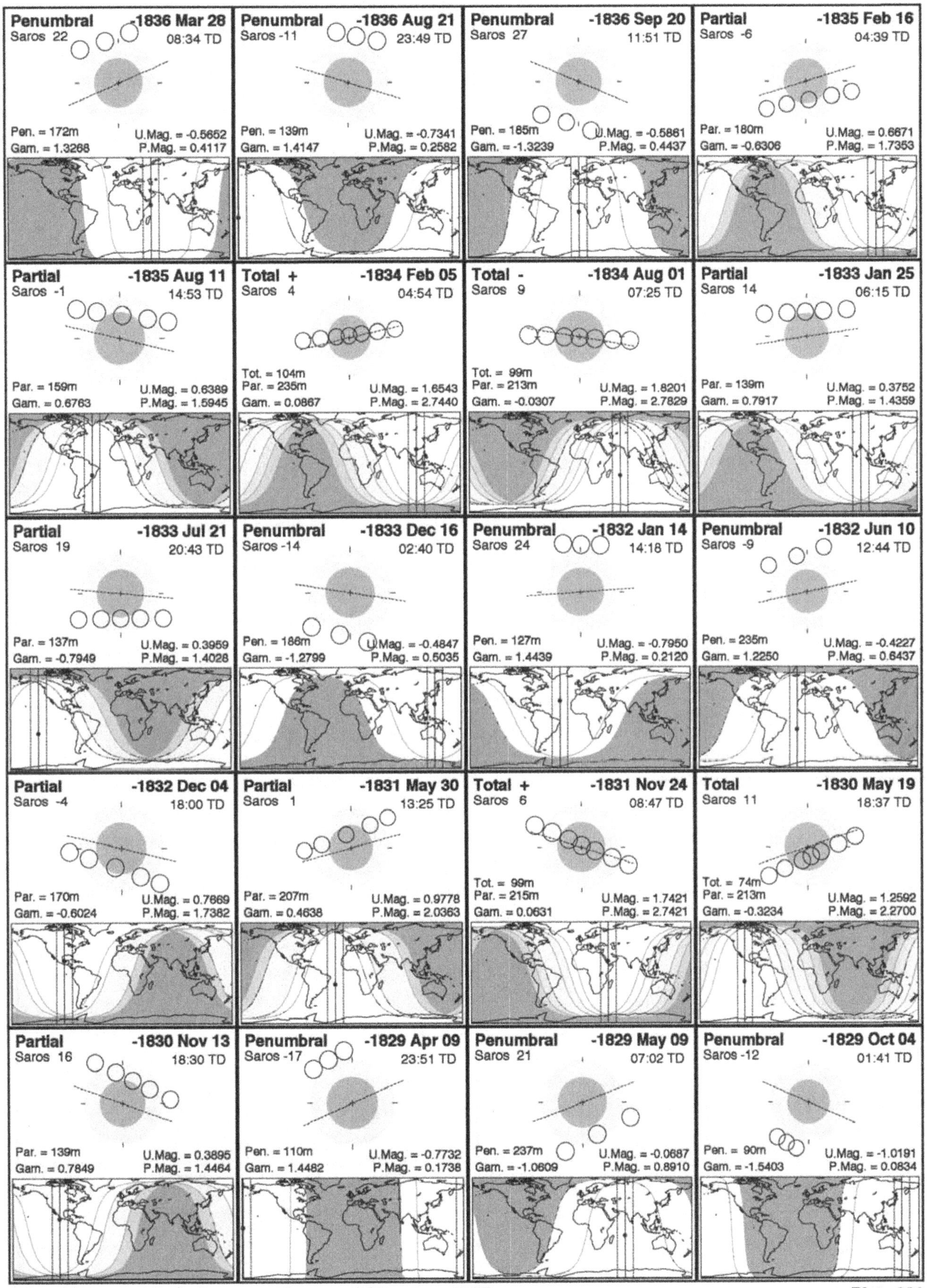

ΔT= 42561 s [= 11h49m] std.err. = ± 3071 s [= ± 12.8°] **Plate 021**

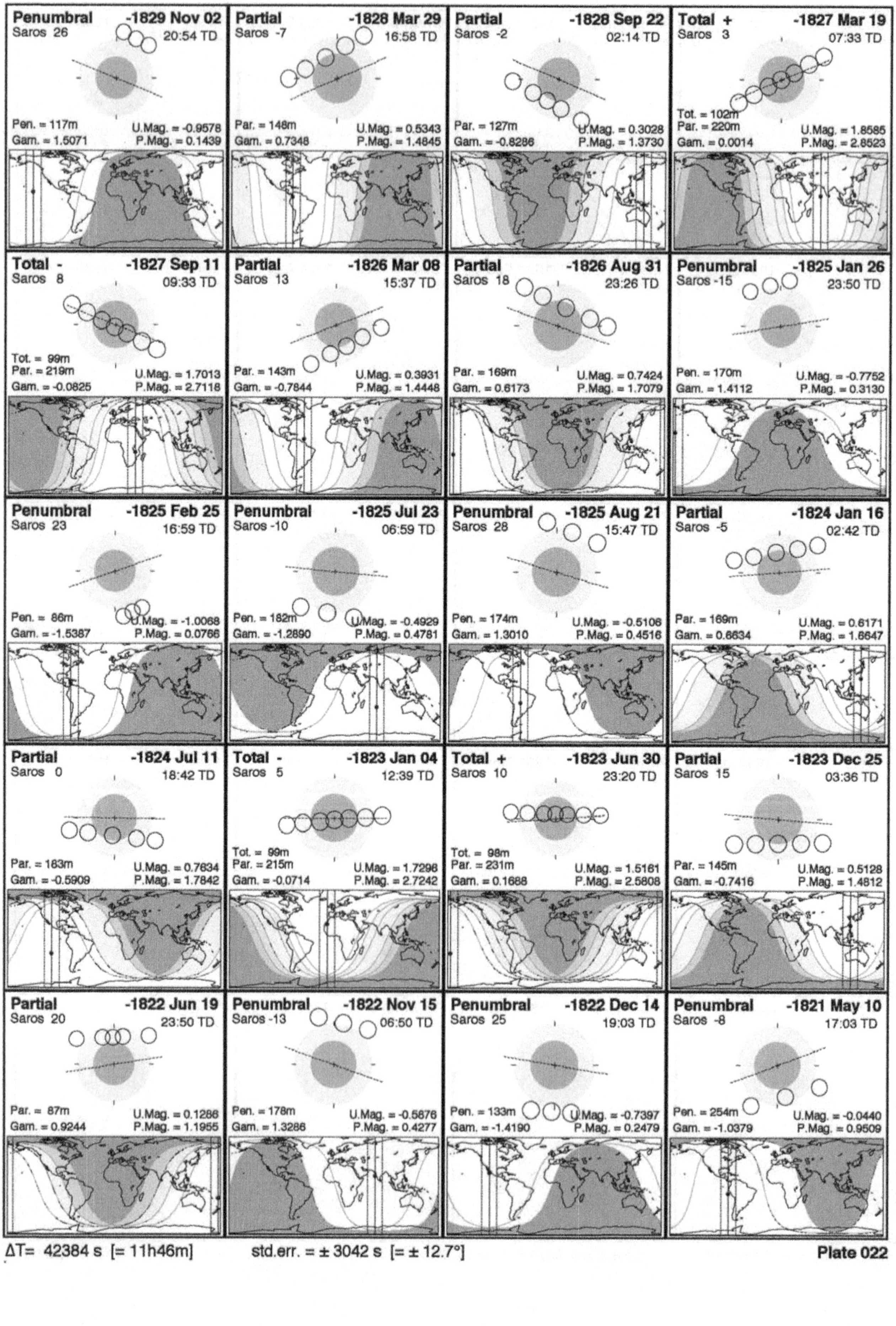

ΔT= 42384 s [= 11h46m] std.err. = ± 3042 s [= ± 12.7°] **Plate 022**

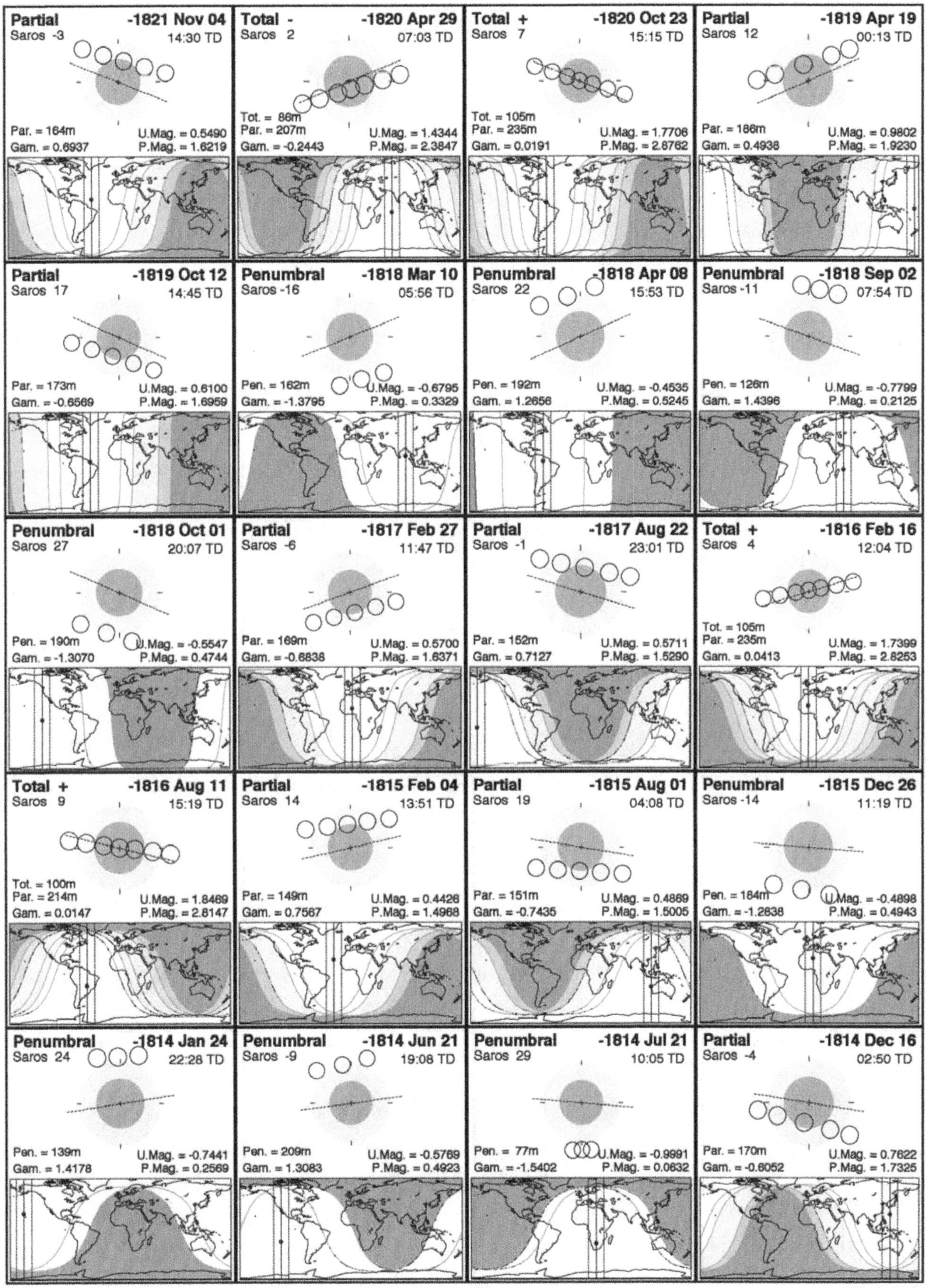

ΔT= 42198 s [= 11h43m] std.err. = ± 3011 s [= ± 12.5°] Plate 023

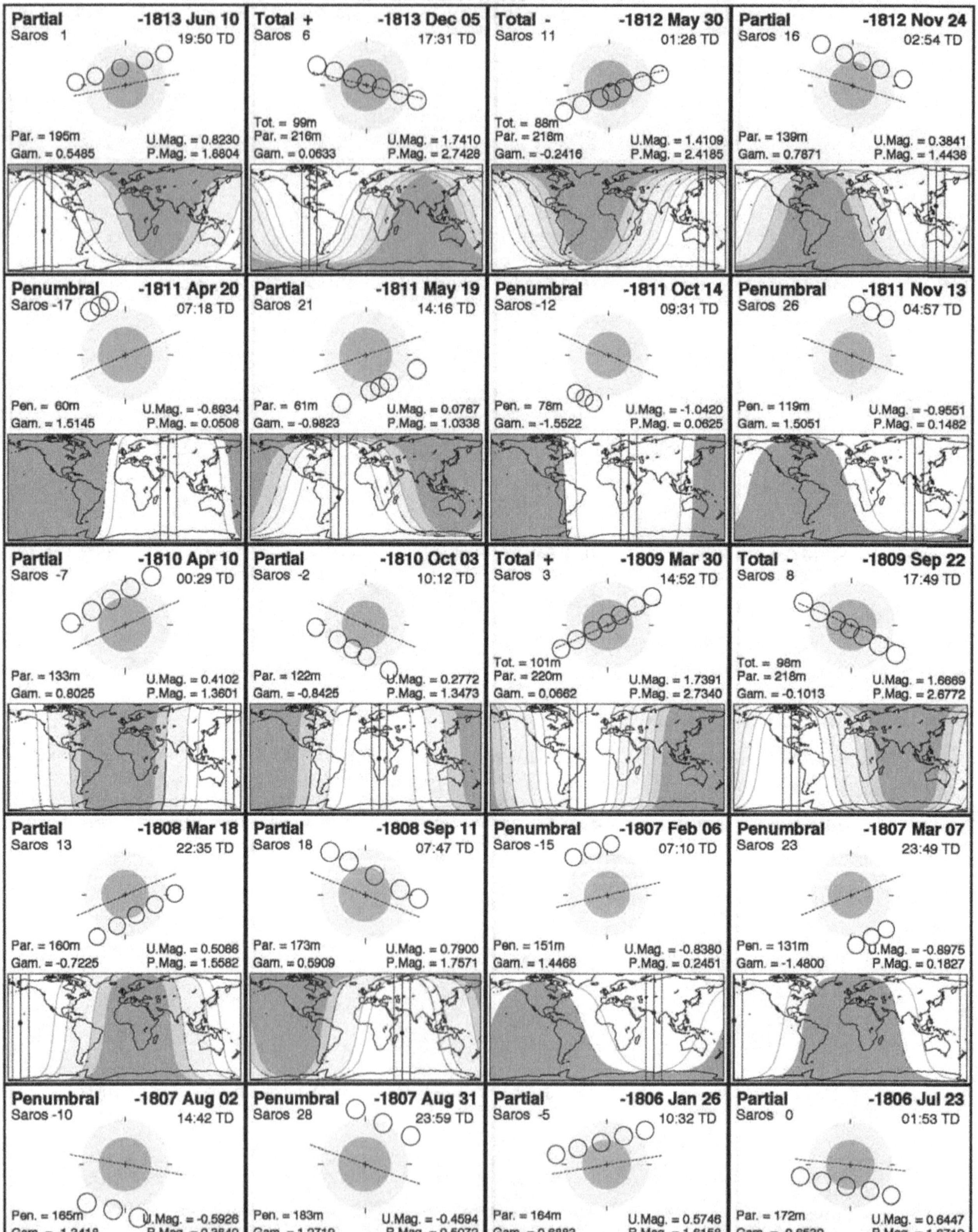

ΔT= 42022 s [= 11h40m] std.err. = ± 2982 s [= ± 12.4°] **Plate 024**

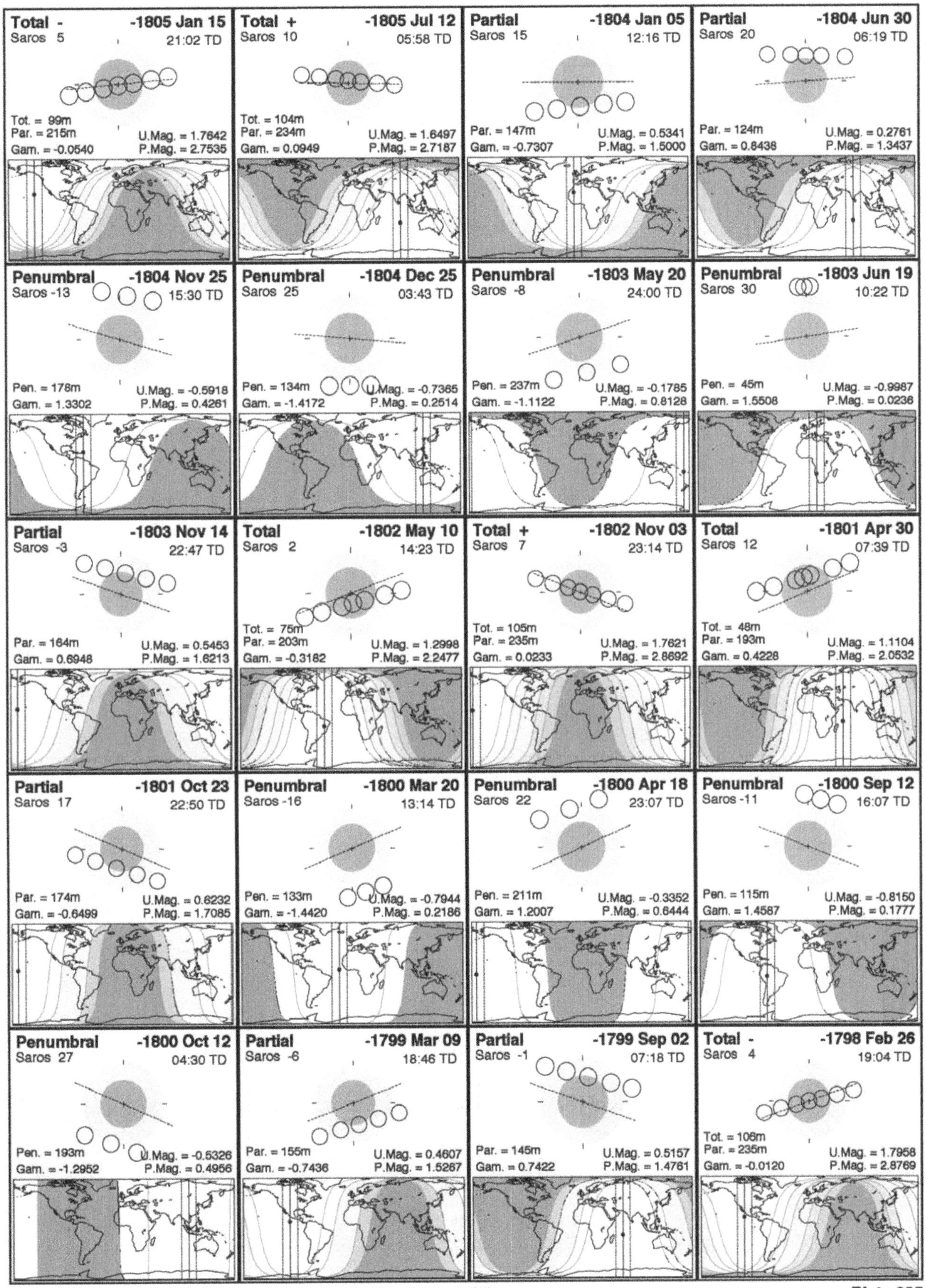

ΔT= 41846 s [= 11h37m] std.err. = ± 2953 s [= ± 12.3°] Plate 025

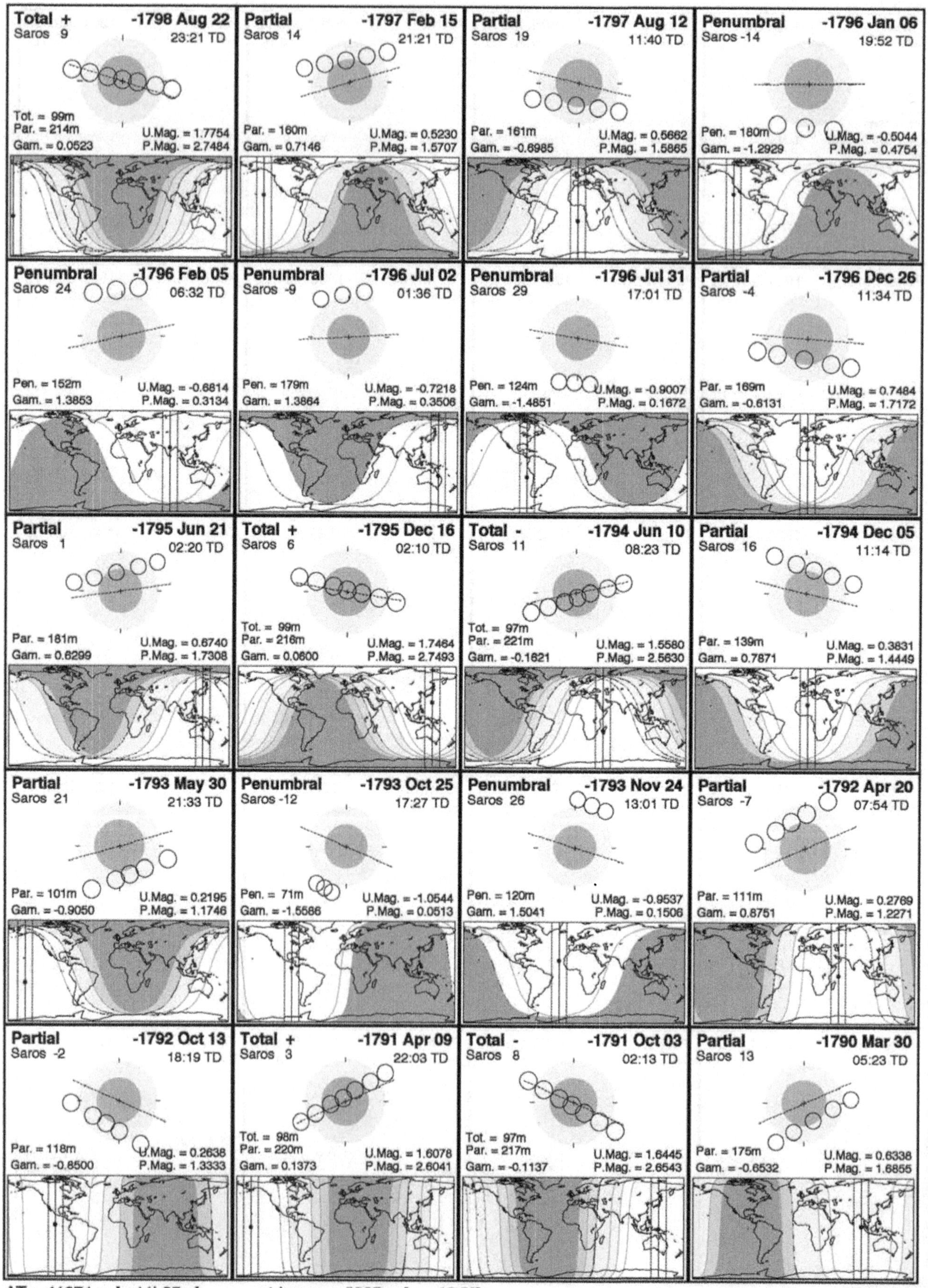

ΔT= 41671 s [= 11h35m] std.err. = ± 2925 s [= ± 12.2°] **Plate 026**

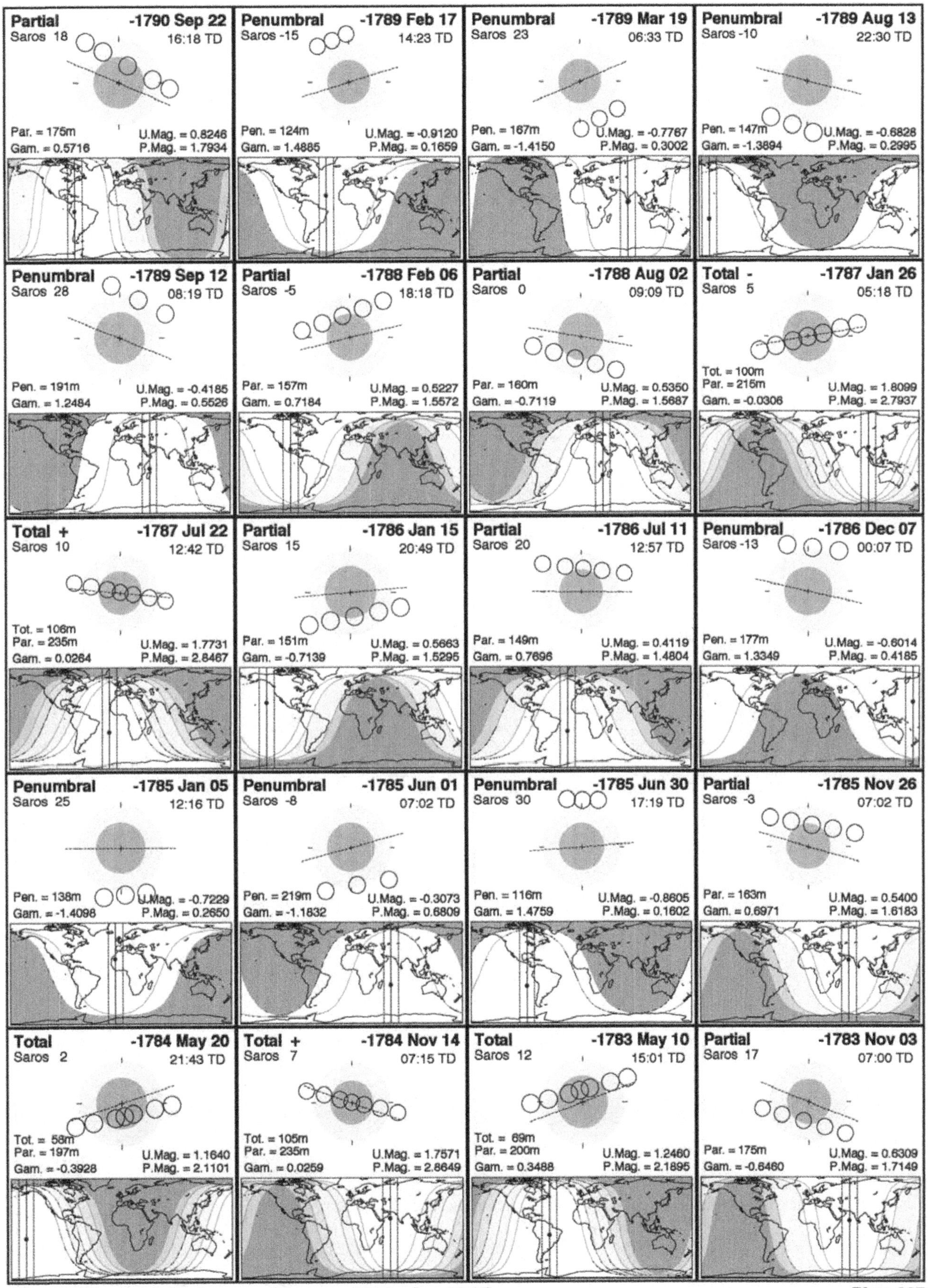

ΔT= 41485 s [= 11h31m] std.err. = ± 2894 s [= ± 12.1°] **Plate 027**

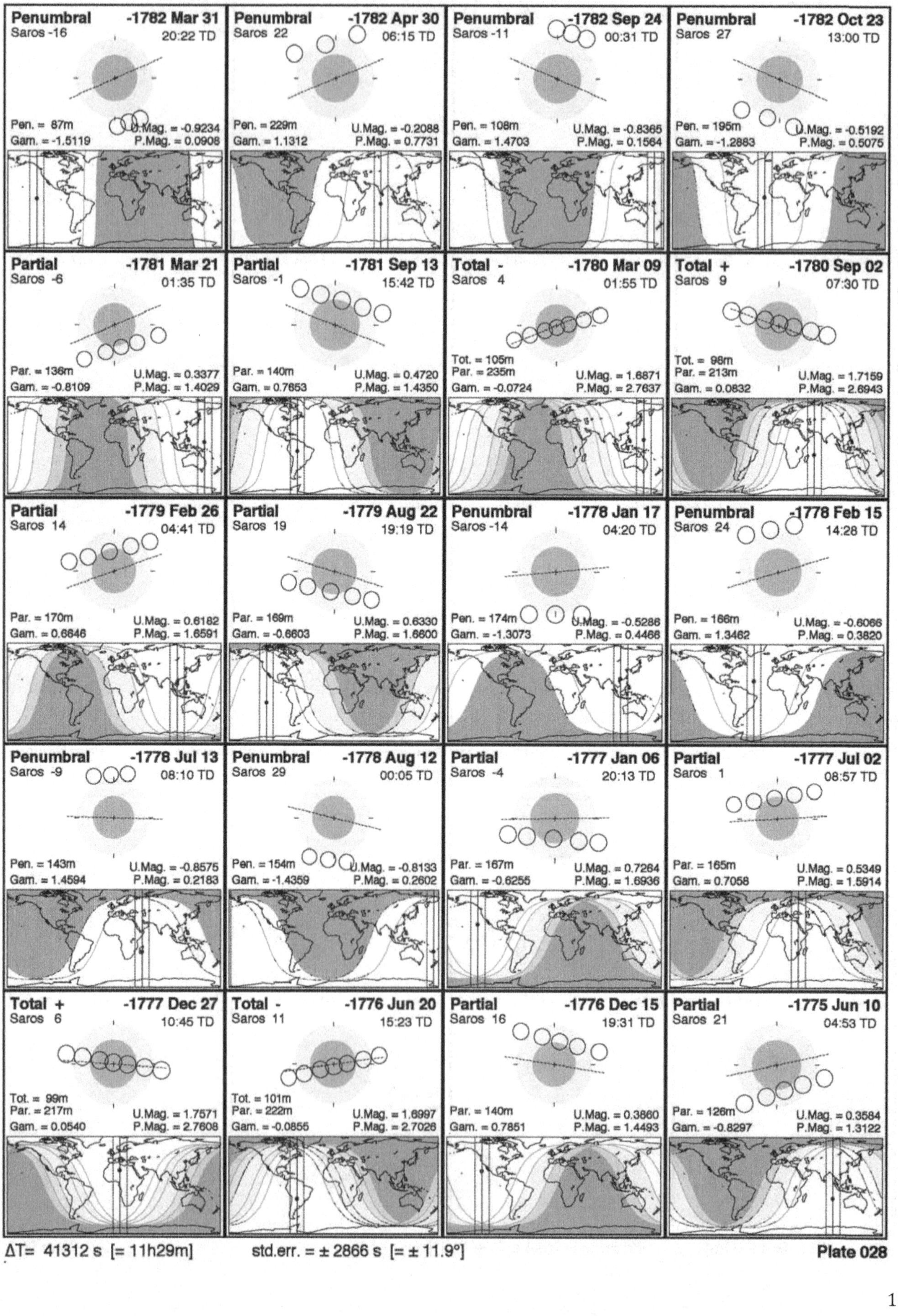

ΔT= 41312 s [= 11h29m] std.err. = ± 2866 s [= ± 11.9°] **Plate 028**

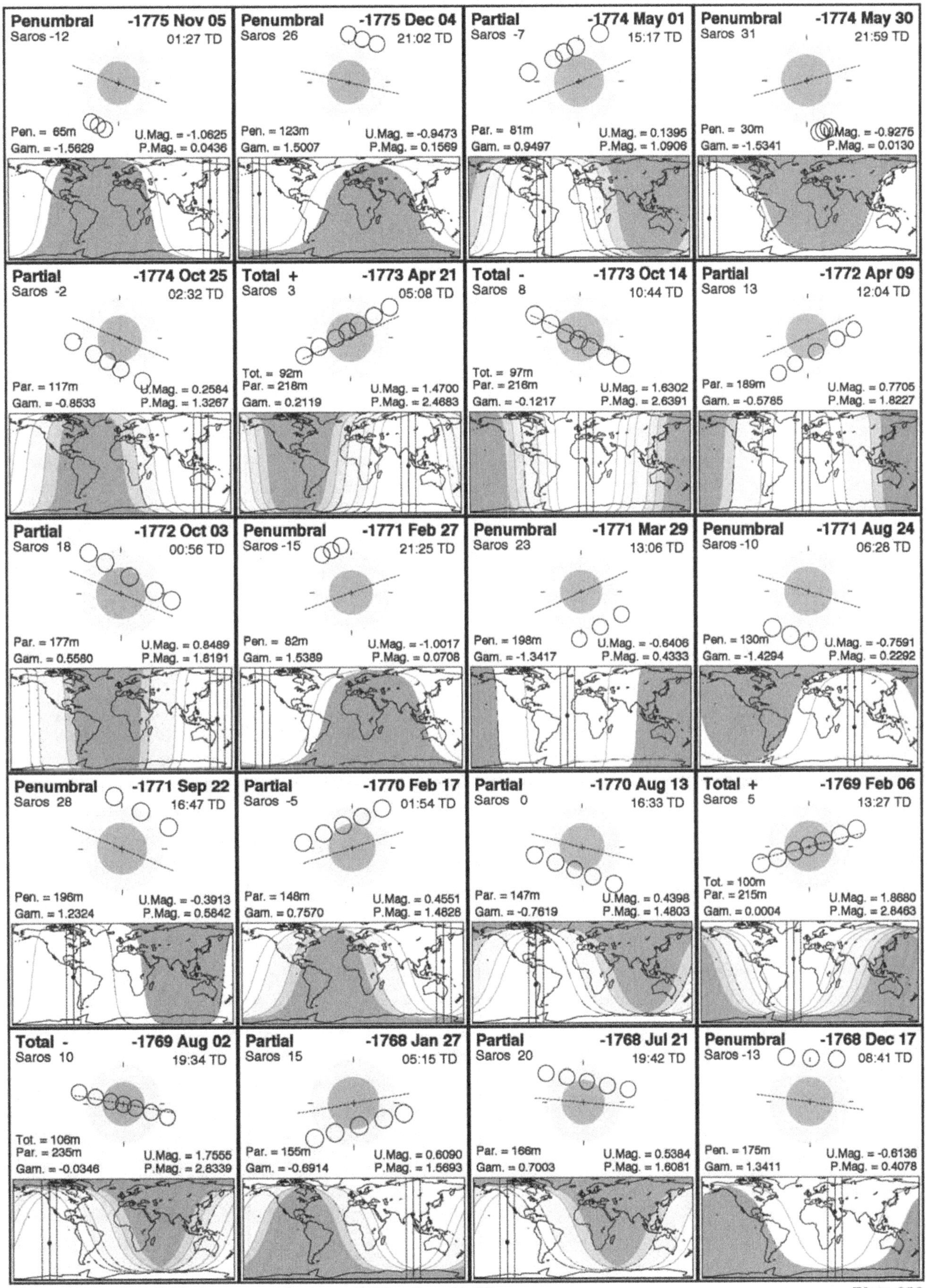

ΔT= 41138 s [= 11h26m] std.err. = ± 2838 s [= ± 11.8°] Plate 029

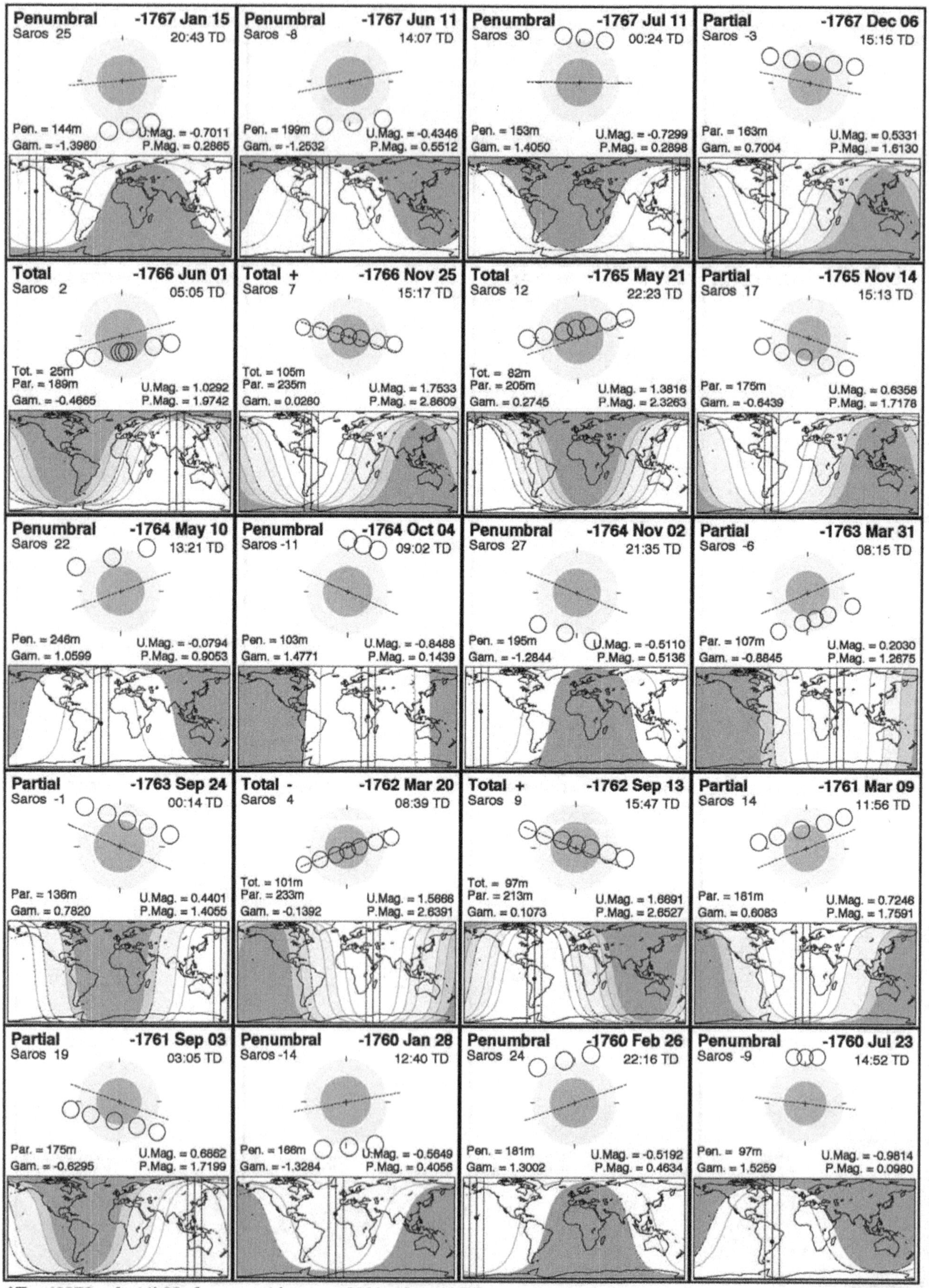

ΔT= 40973 s [= 11h23m] std.err. = ± 2811 s [= ± 11.7°] **Plate 030**

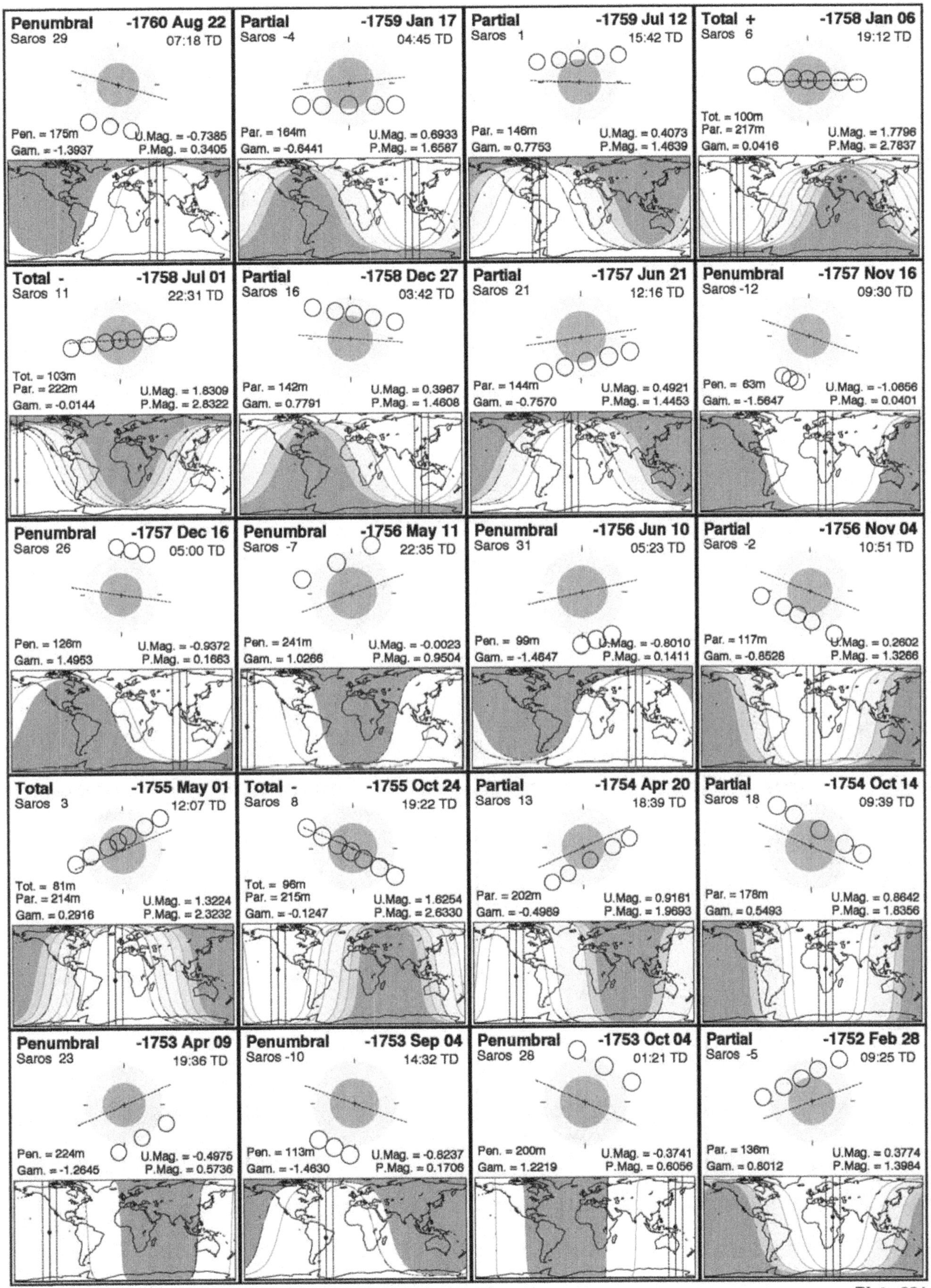

ΔT= 40799 s [= 11h20m] std.err. = ± 2783 s [= ± 11.6°] **Plate 031**

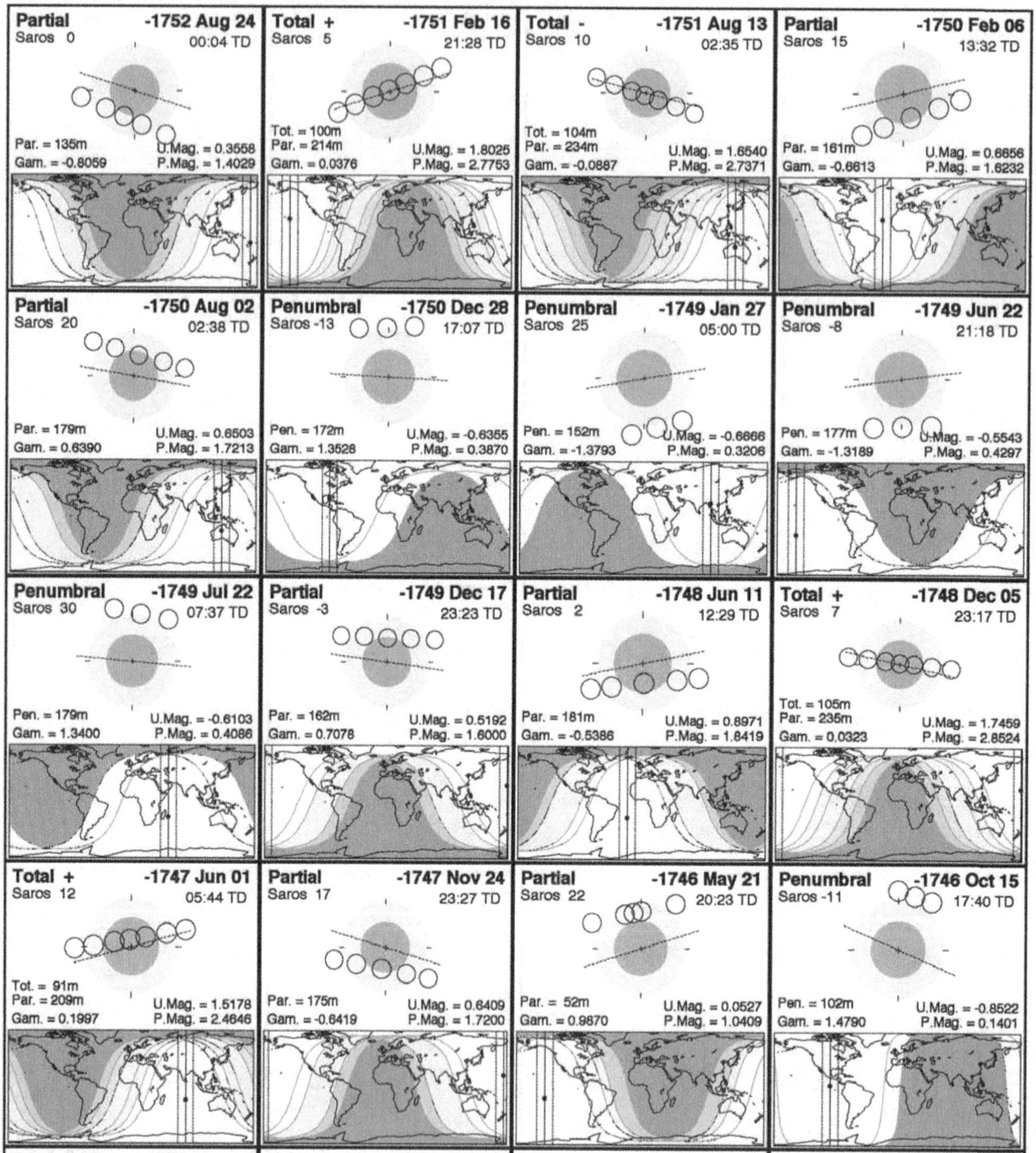

ΔT= 40617 s [= 11h17m] std.err. = ± 2754 s [= ± 11.5°] **Plate 032**

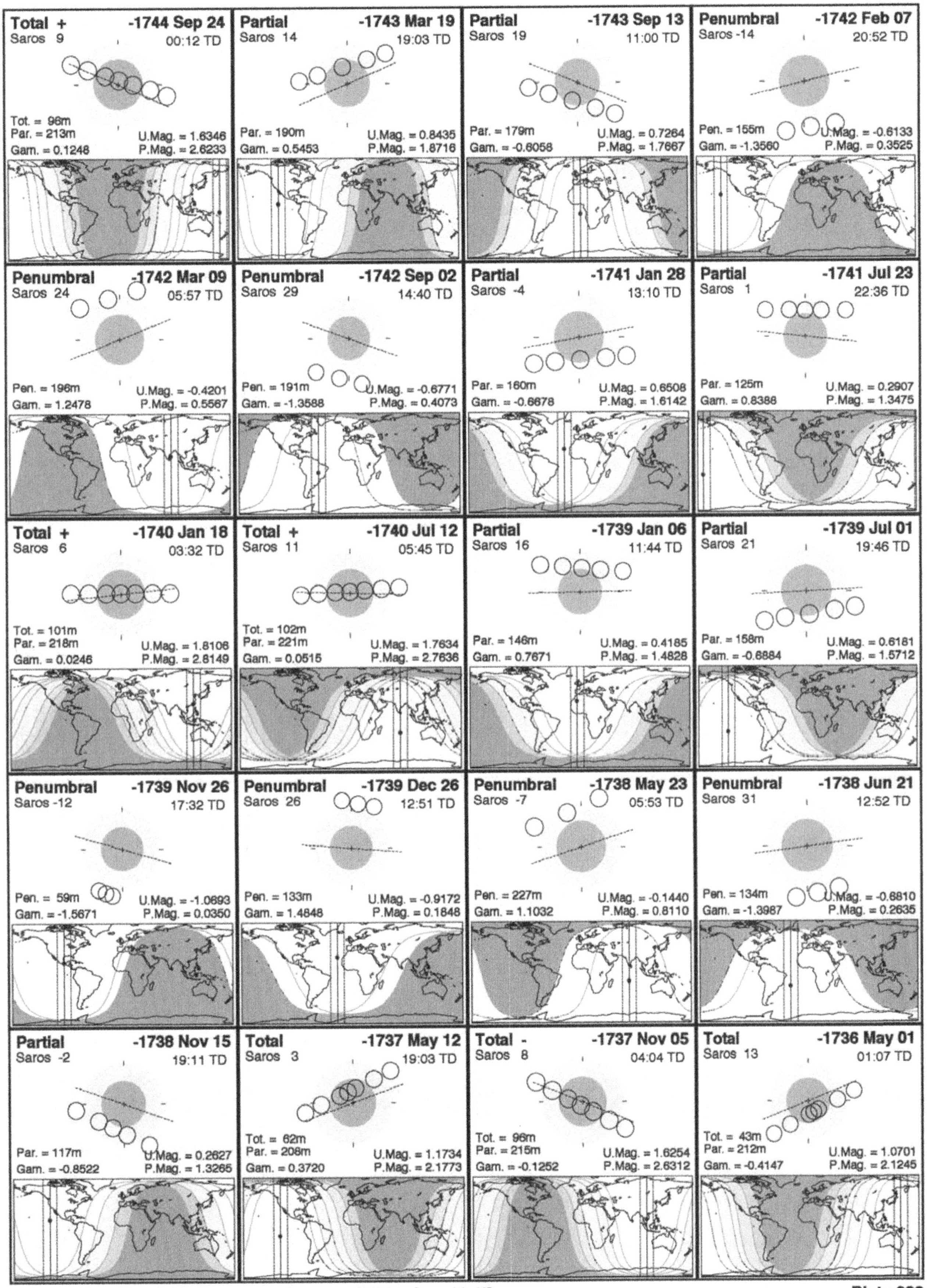

ΔT= 40433 s [= 11h14m] std.err. = ± 2724 s [= ± 11.4°] **Plate 033**

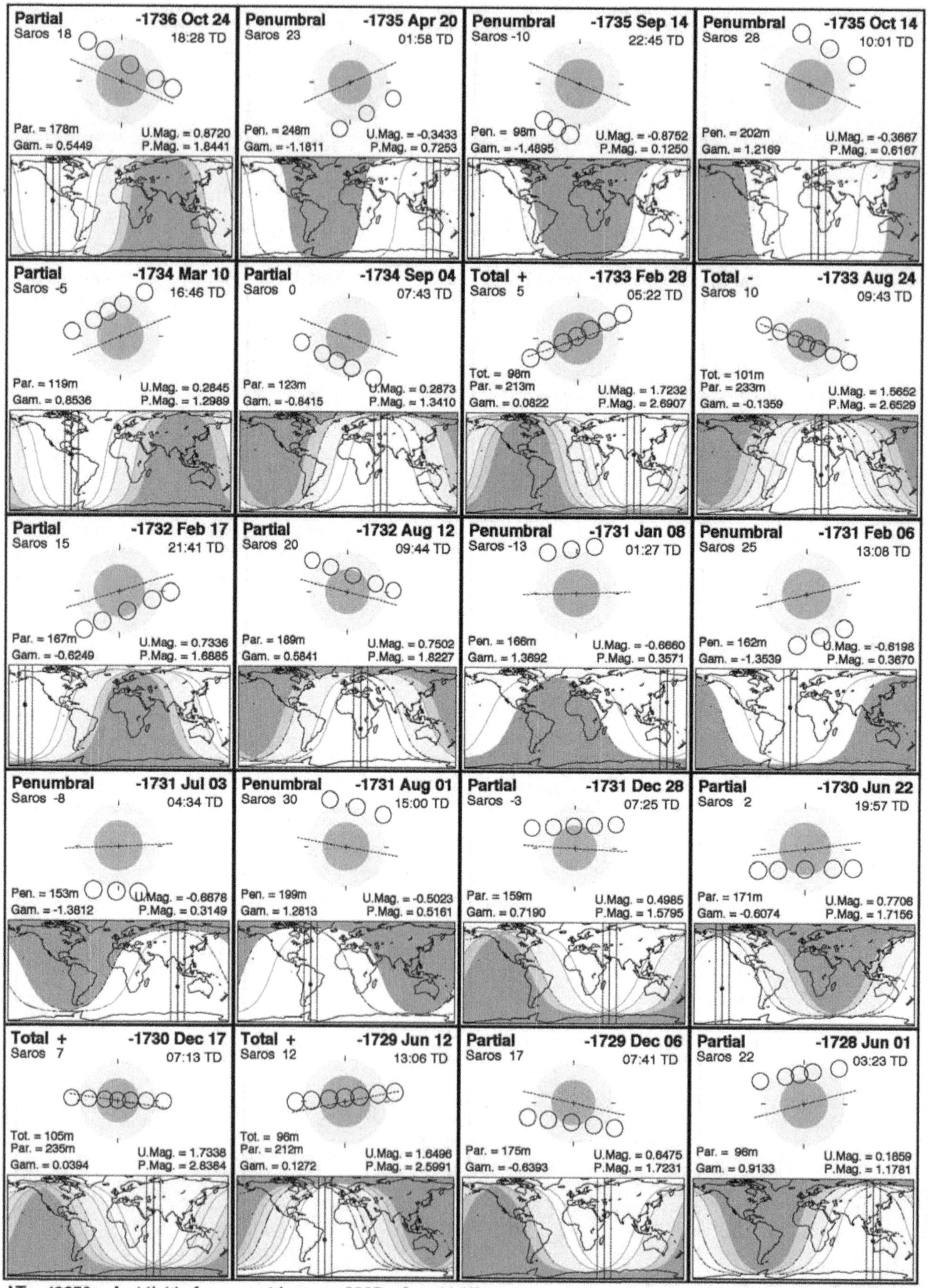

ΔT= 40250 s [= 11h11m] std.err. = ± 2695 s [= ± 11.2°] **Plate 034**

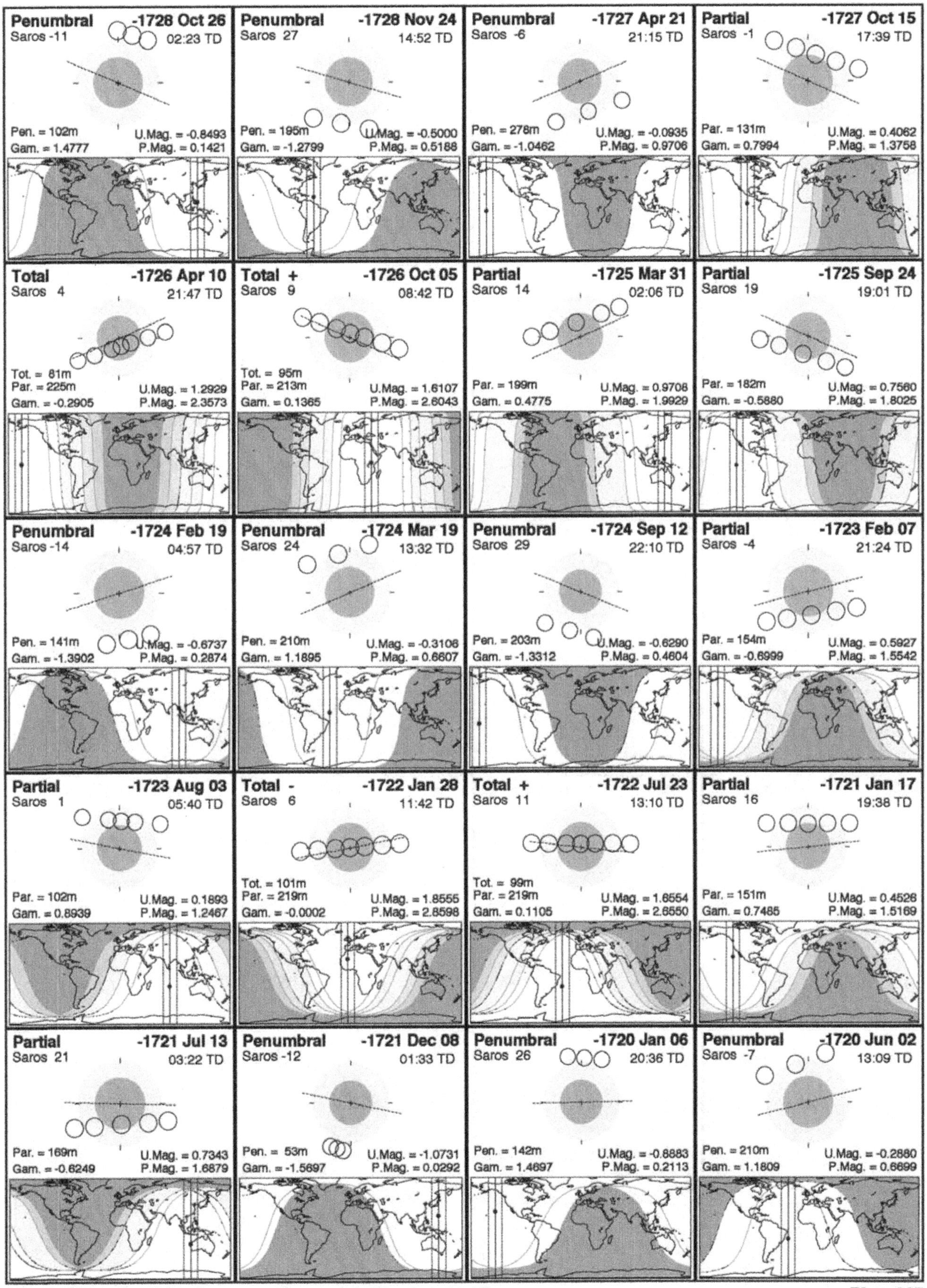

ΔT= 40069 s [= 11h08m]   std.err. = ± 2666 s [= ± 11.1°]   **Plate 035**

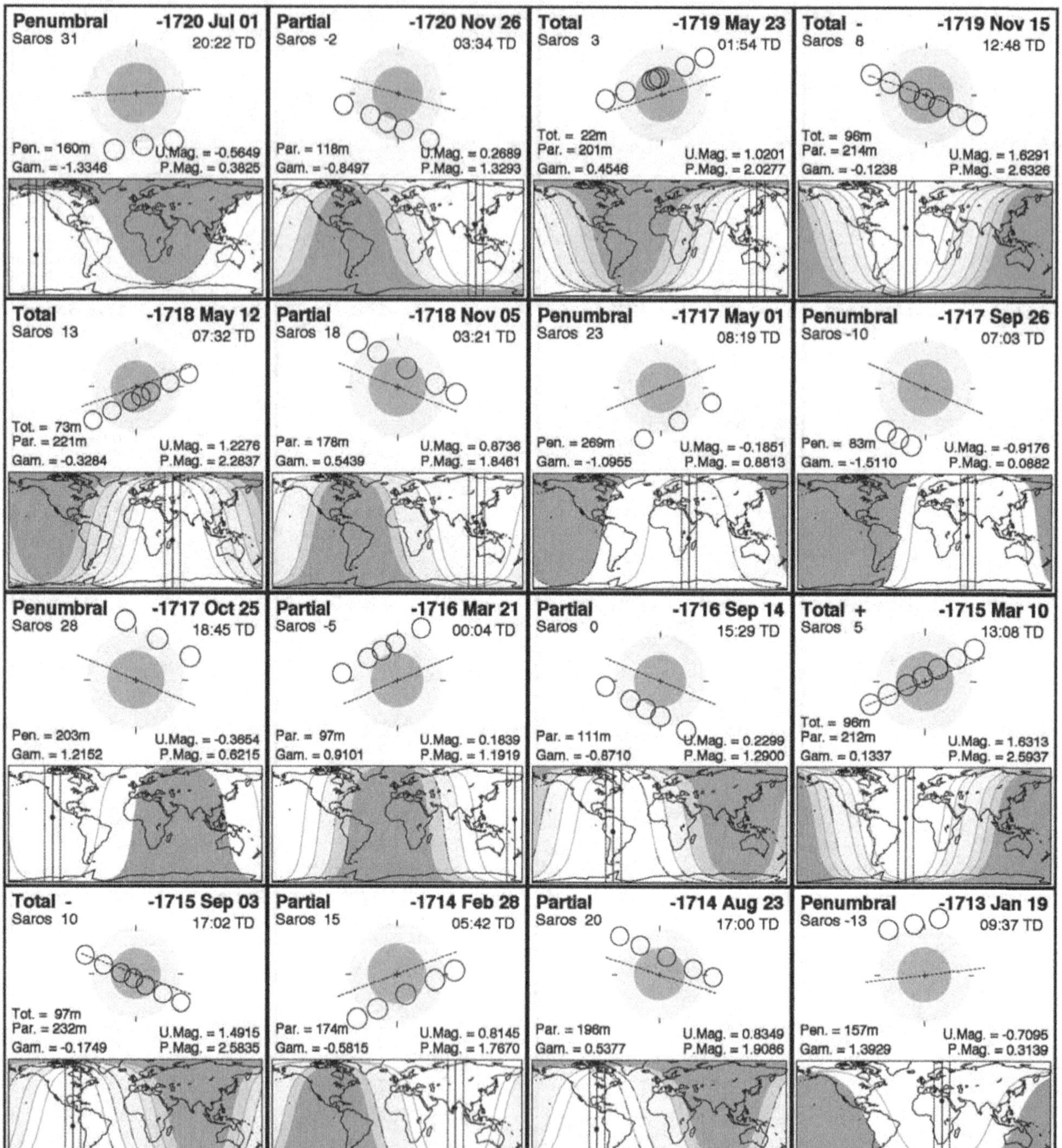

ΔT= 39895 s [= 11h05m] std.err. = ± 2638 s [= ± 11.0°] **Plate 036**

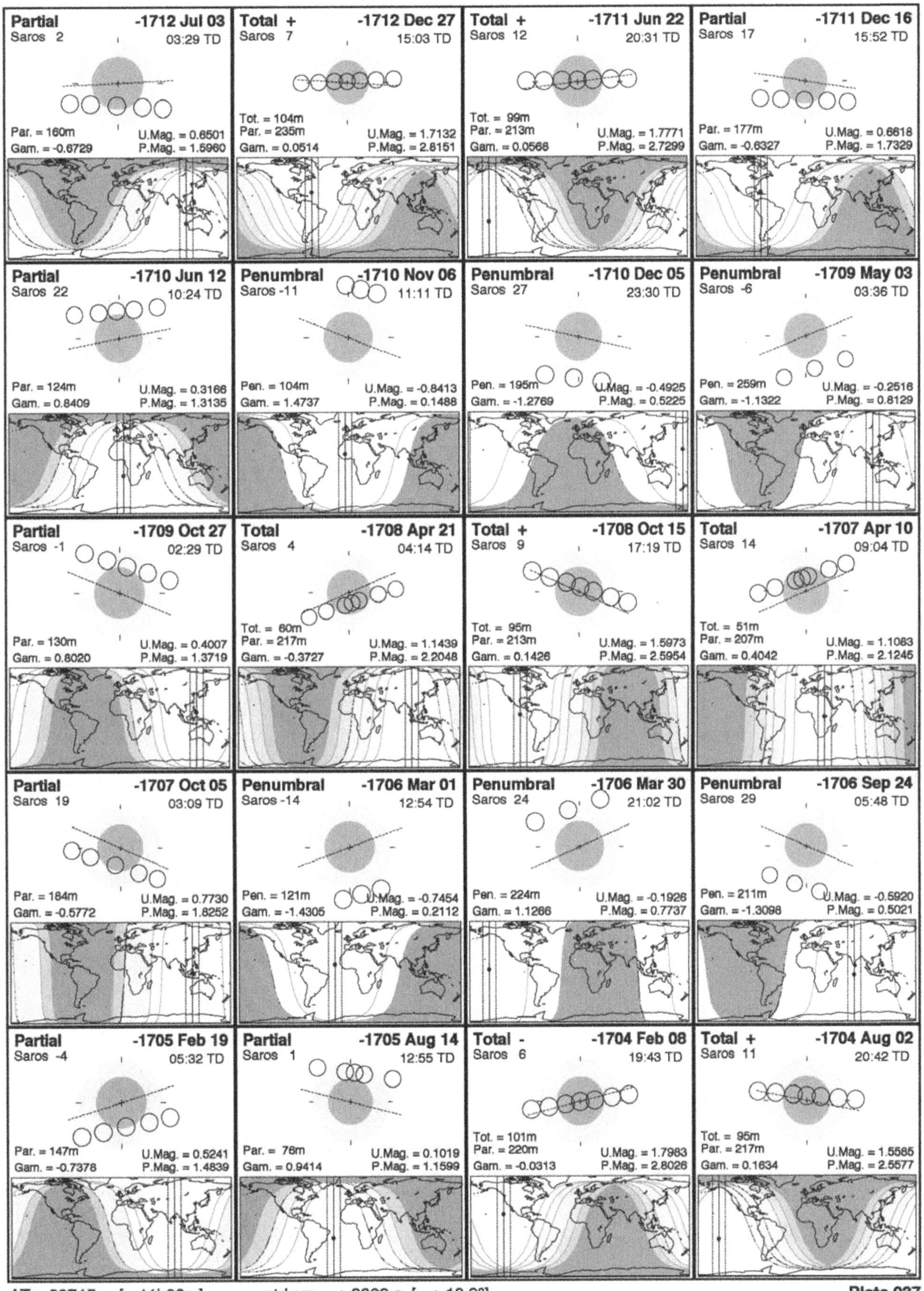

ΔT= 39715 s [= 11h02m] std.err. = ± 2609 s [= ± 10.9°] **Plate 037**

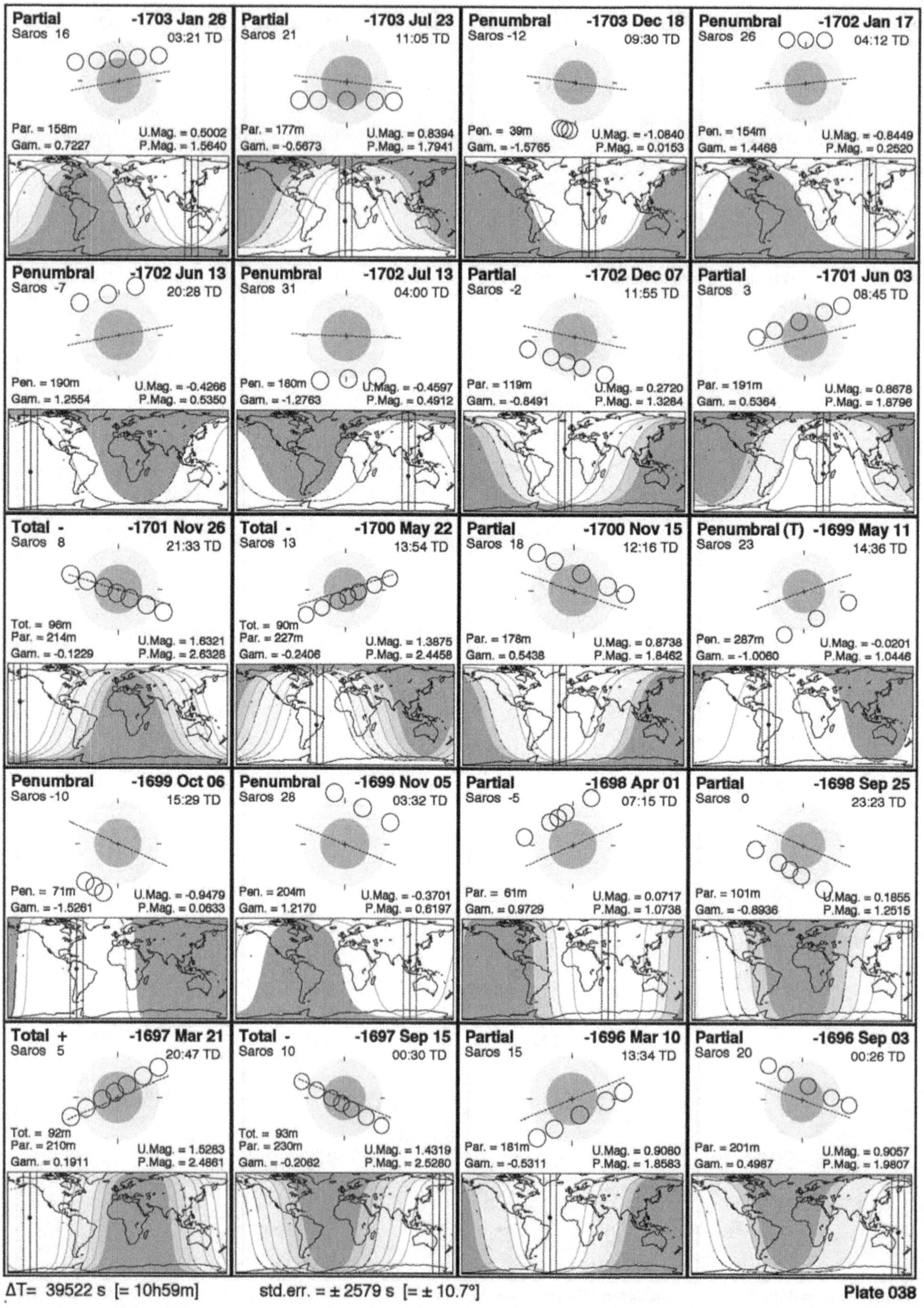

ΔT= 39522 s [= 10h59m] std.err. = ± 2579 s [= ± 10.7°] **Plate 038**

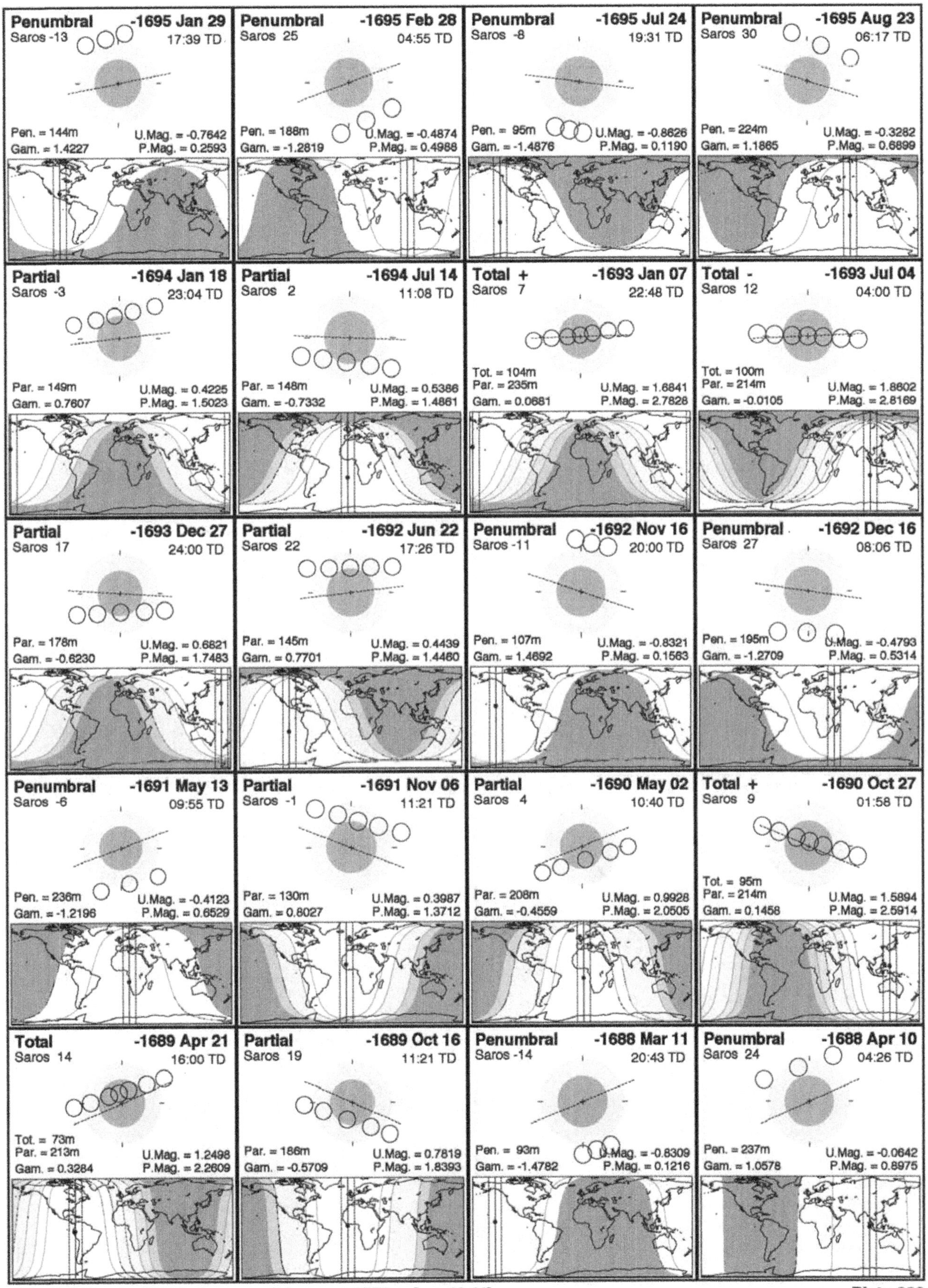

ΔT= 39343 s [= 10h56m] std.err. = ± 2550 s [= ± 10.6°] **Plate 039**

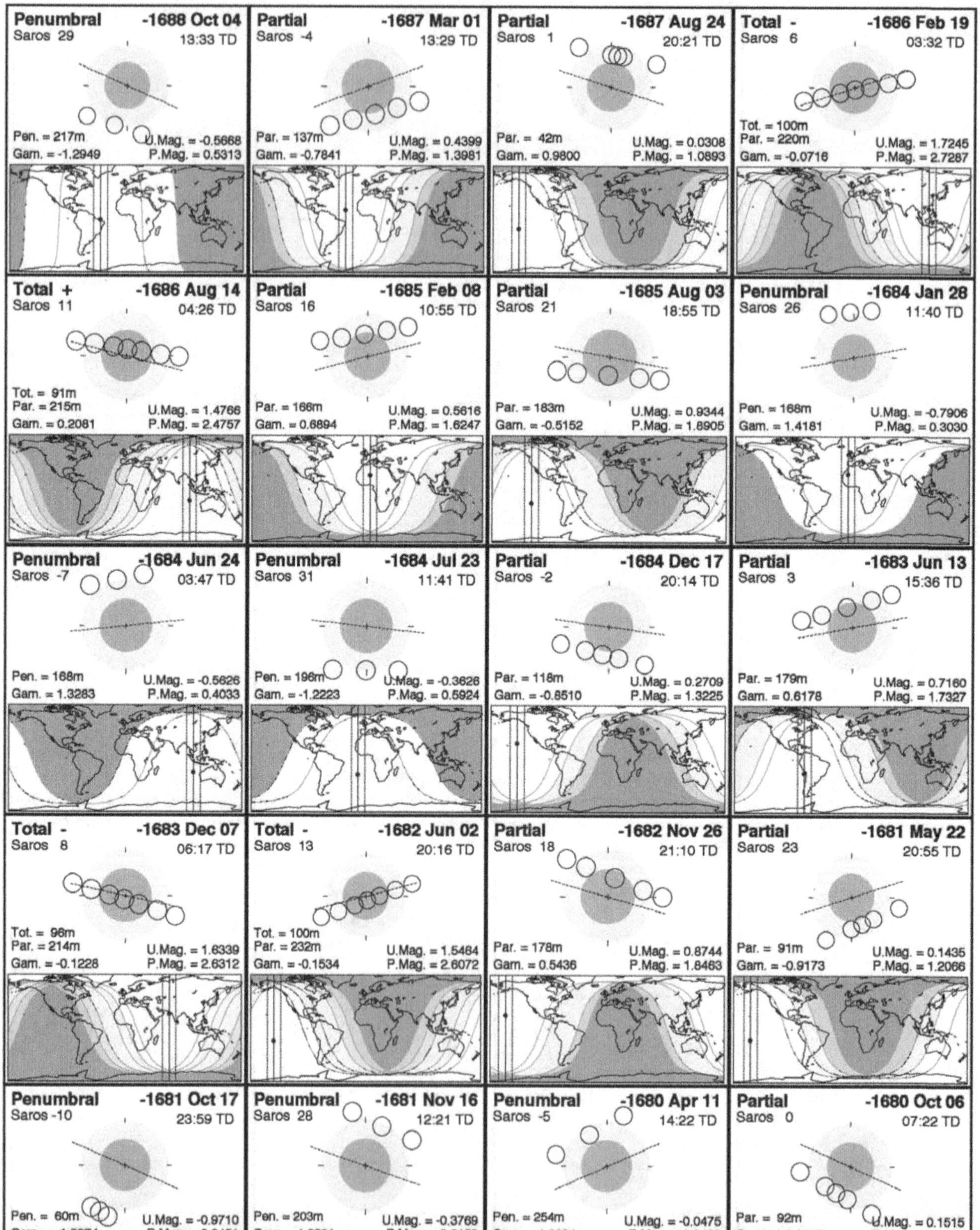

ΔT= 39171 s [= 10h53m] std.err. = ± 2523 s [= ± 10.5°] **Plate 040**

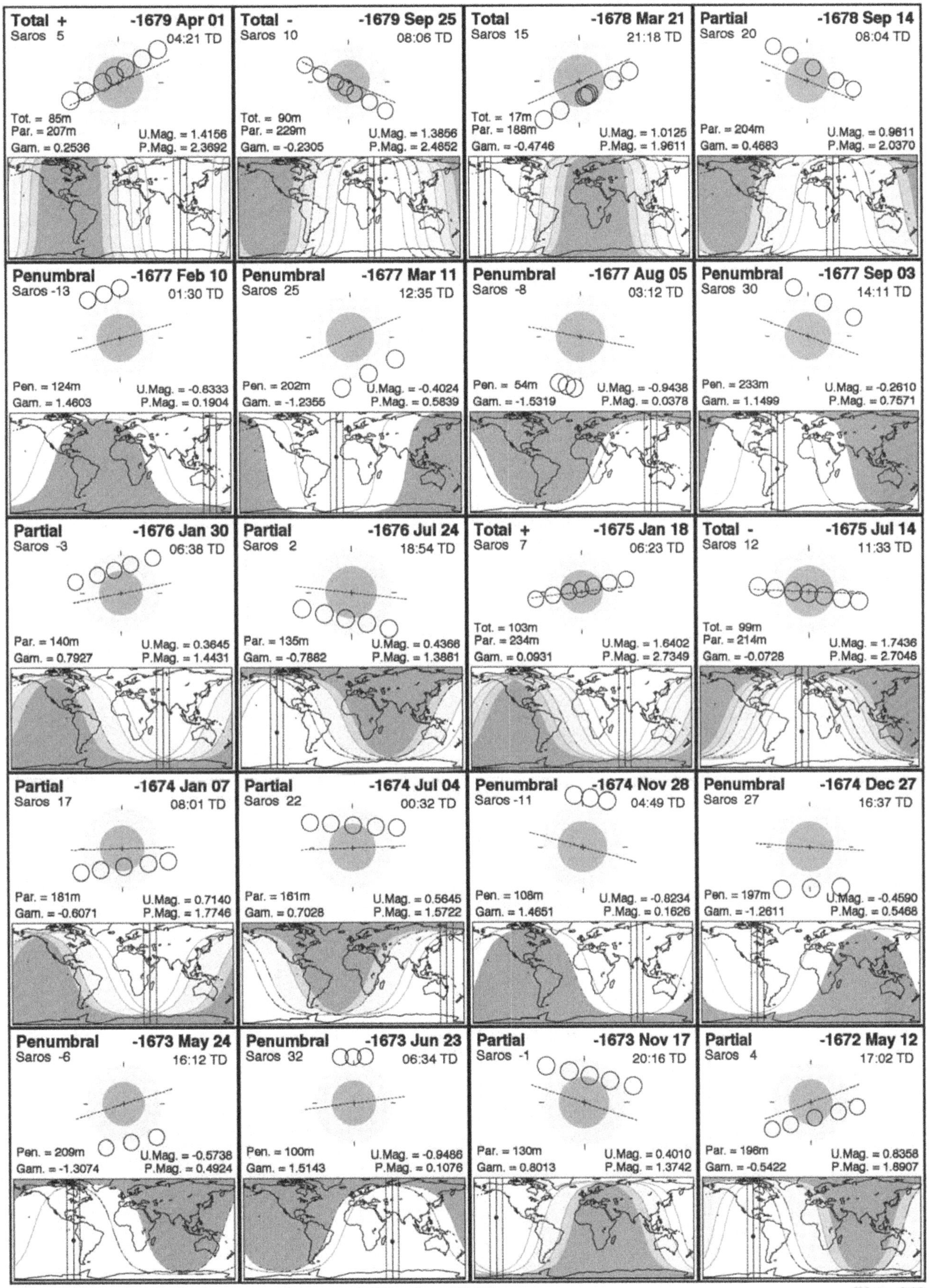

ΔT= 38981 s [= 10h50m] std.err. = ± 2493 s [= ± 10.4°] **Plate 041**

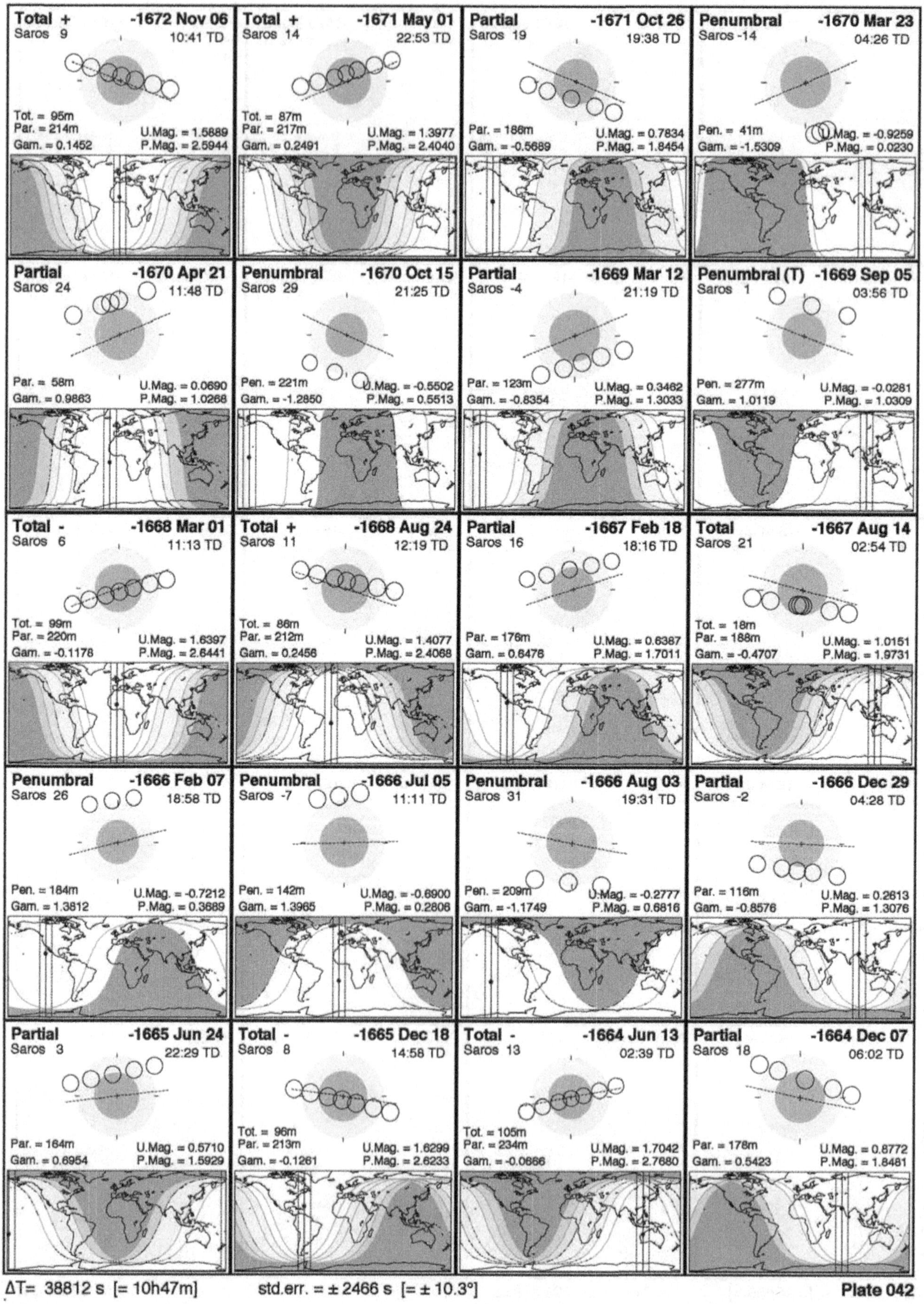

ΔT= 38812 s [= 10h47m] std.err. = ± 2466 s [= ± 10.3°] **Plate 042**

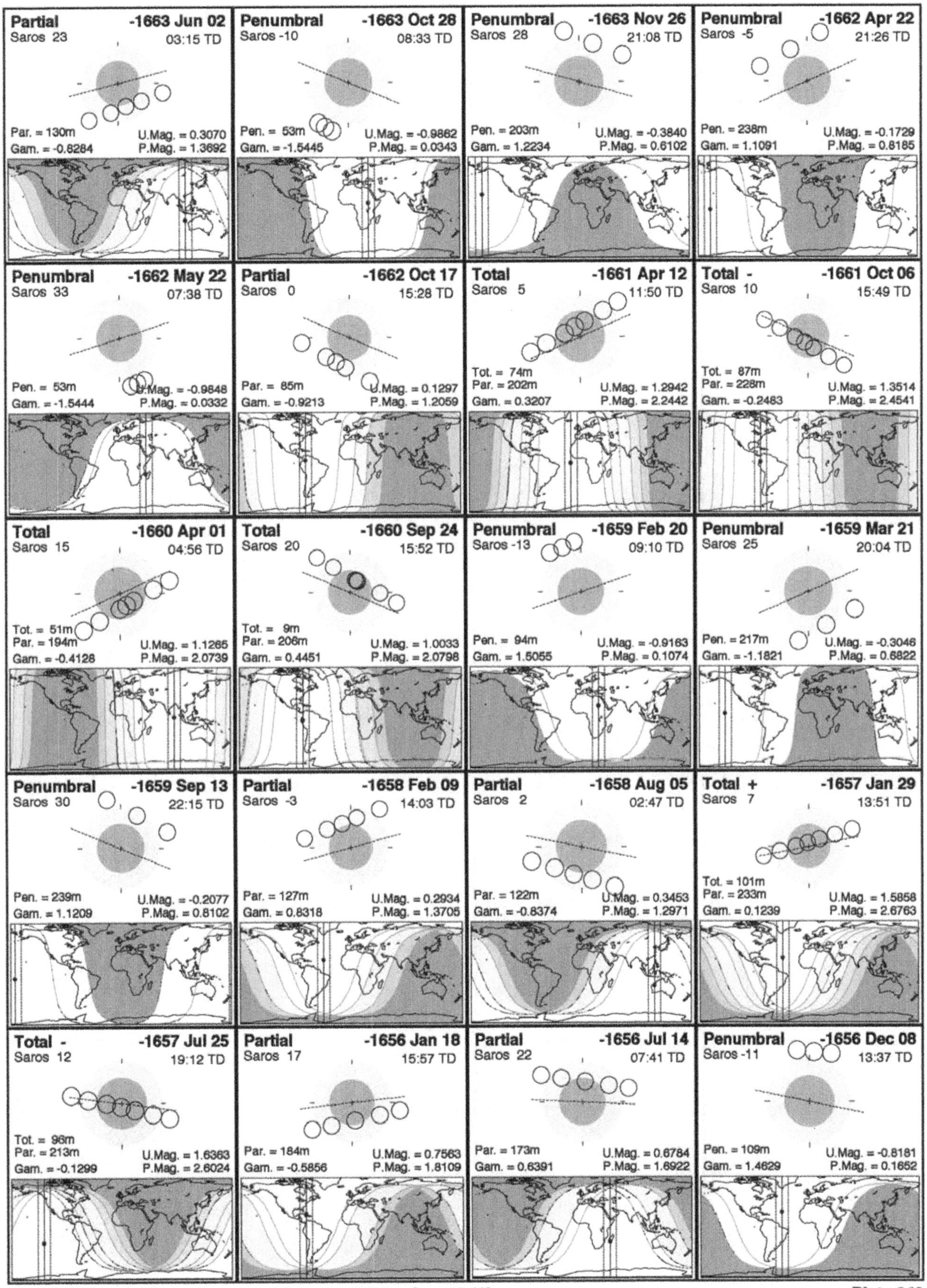

ΔT= 38622 s [= 10h44m] std.err. = ± 2437 s [= ± 10.2°] Plate 043

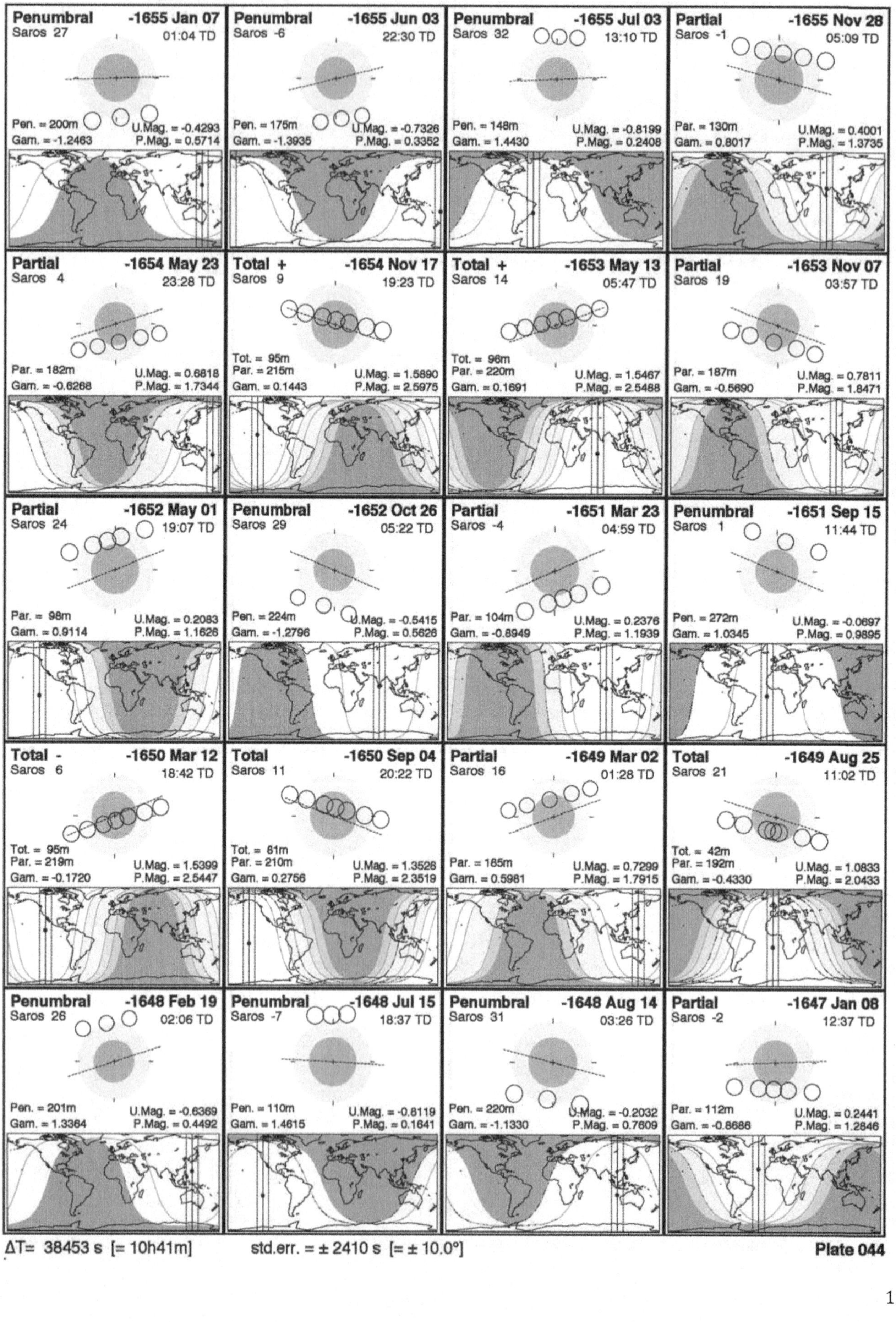

ΔT= 38453 s [= 10h41m] std.err. = ± 2410 s [= ± 10.0°] **Plate 044**

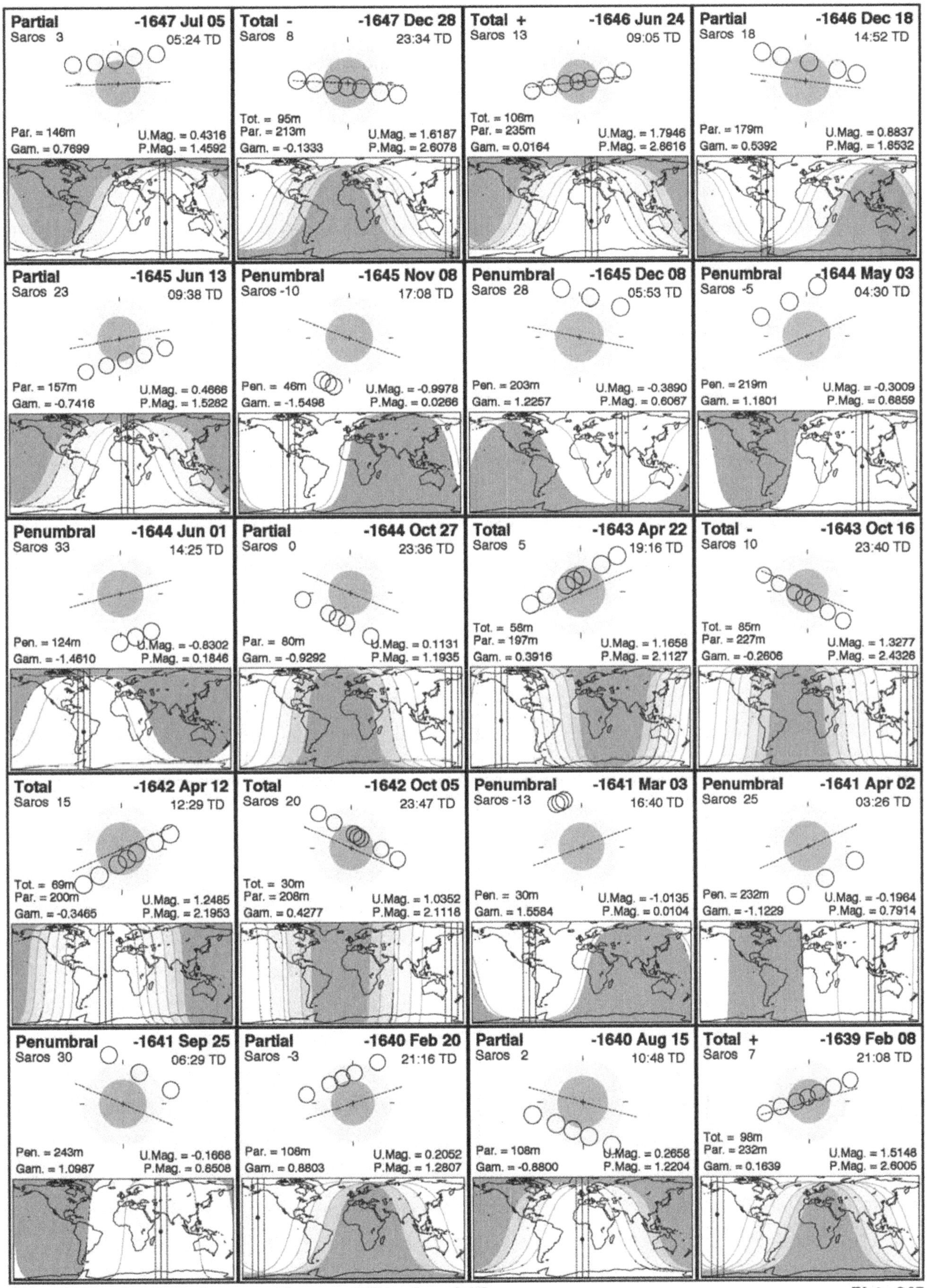

ΔT= 38265 s [= 10h38m] std.err. = ± 2381 s [= ± 9.9°] Plate 045

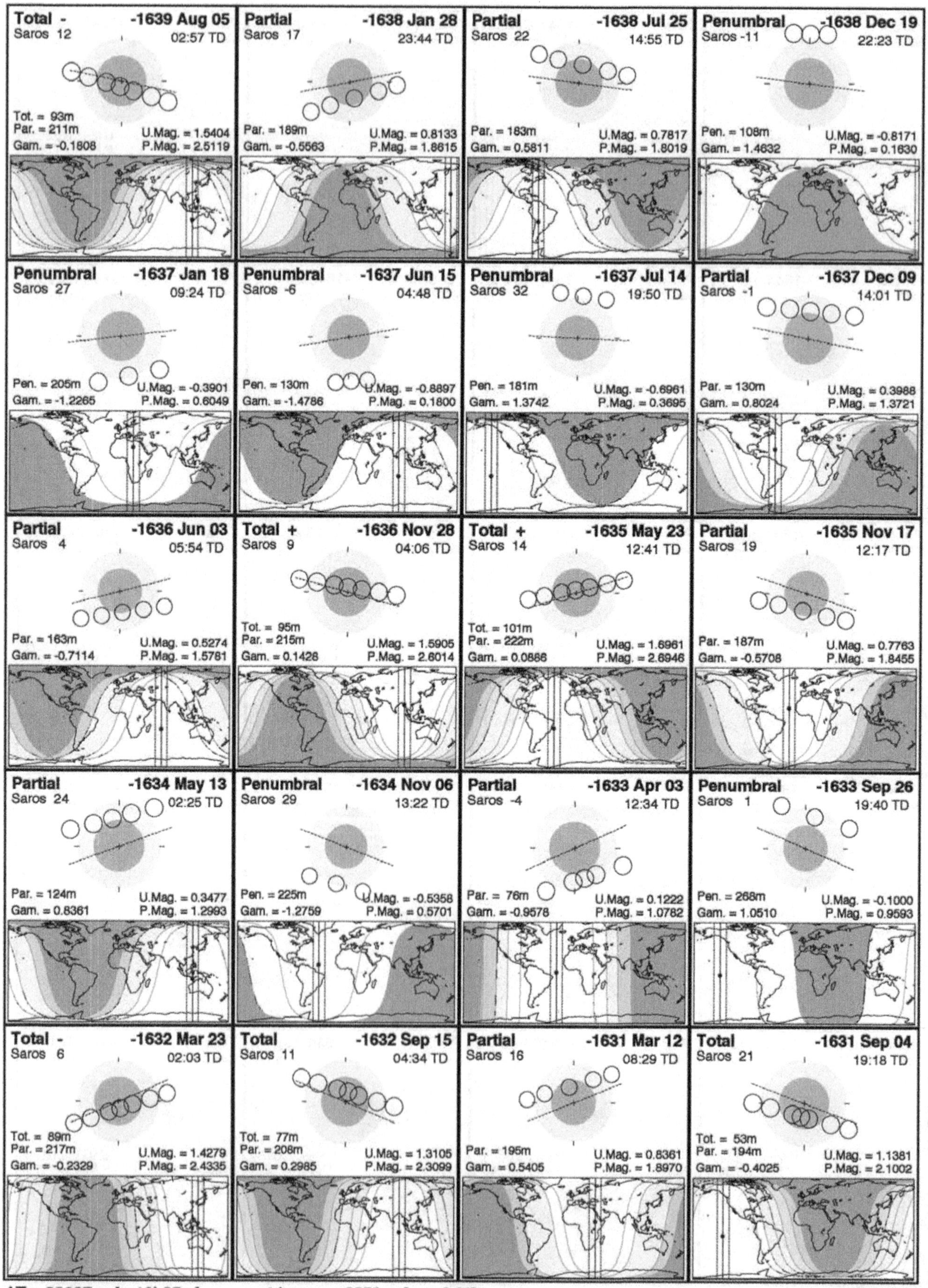

ΔT= 38087 s [= 10h35m] std.err. = ± 2353 s [= ± 9.8°] **Plate 046**

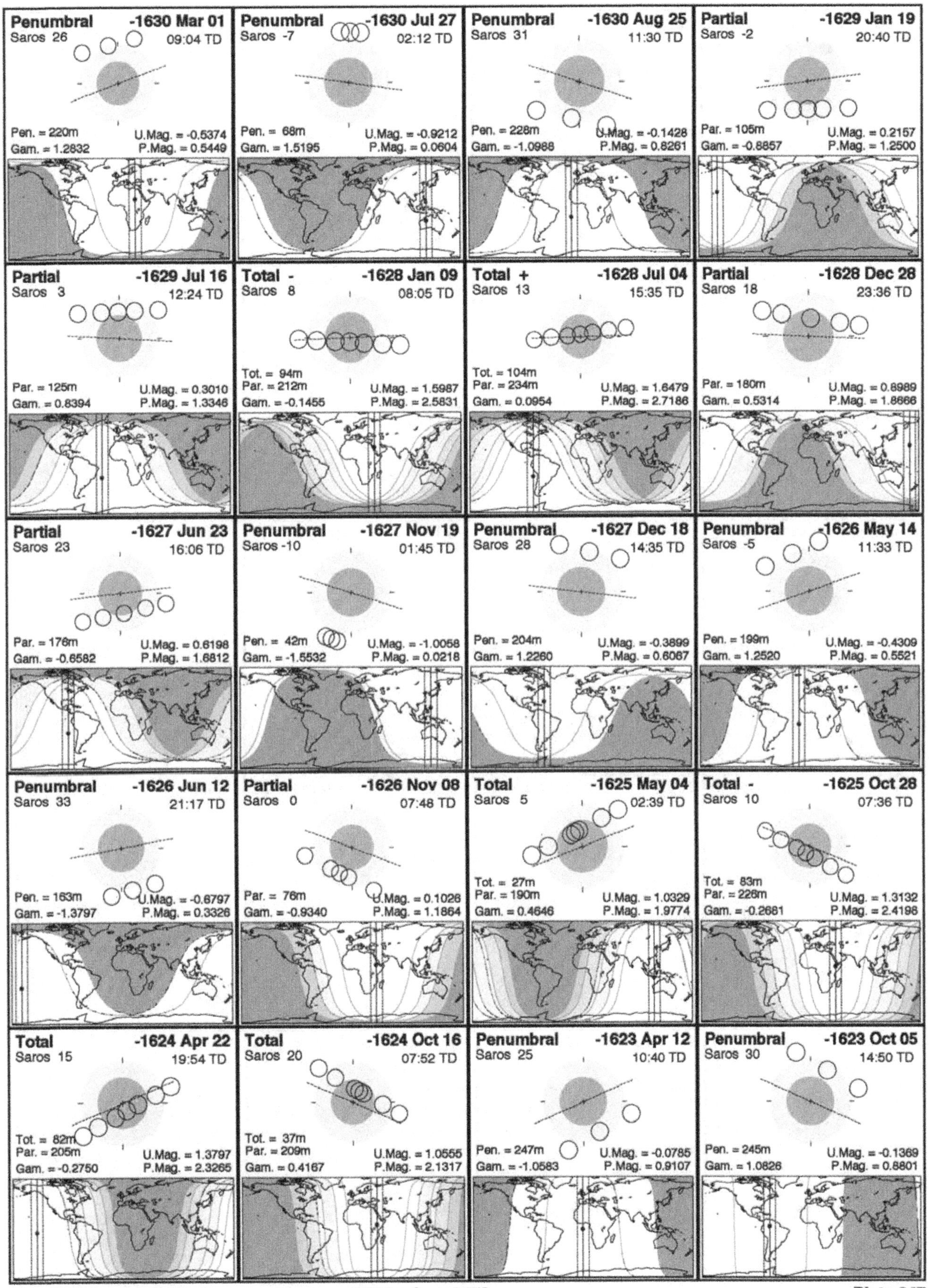

ΔT= 37898 s [= 10h32m] std.err. = ± 2324 s [= ± 9.7°] **Plate 047**

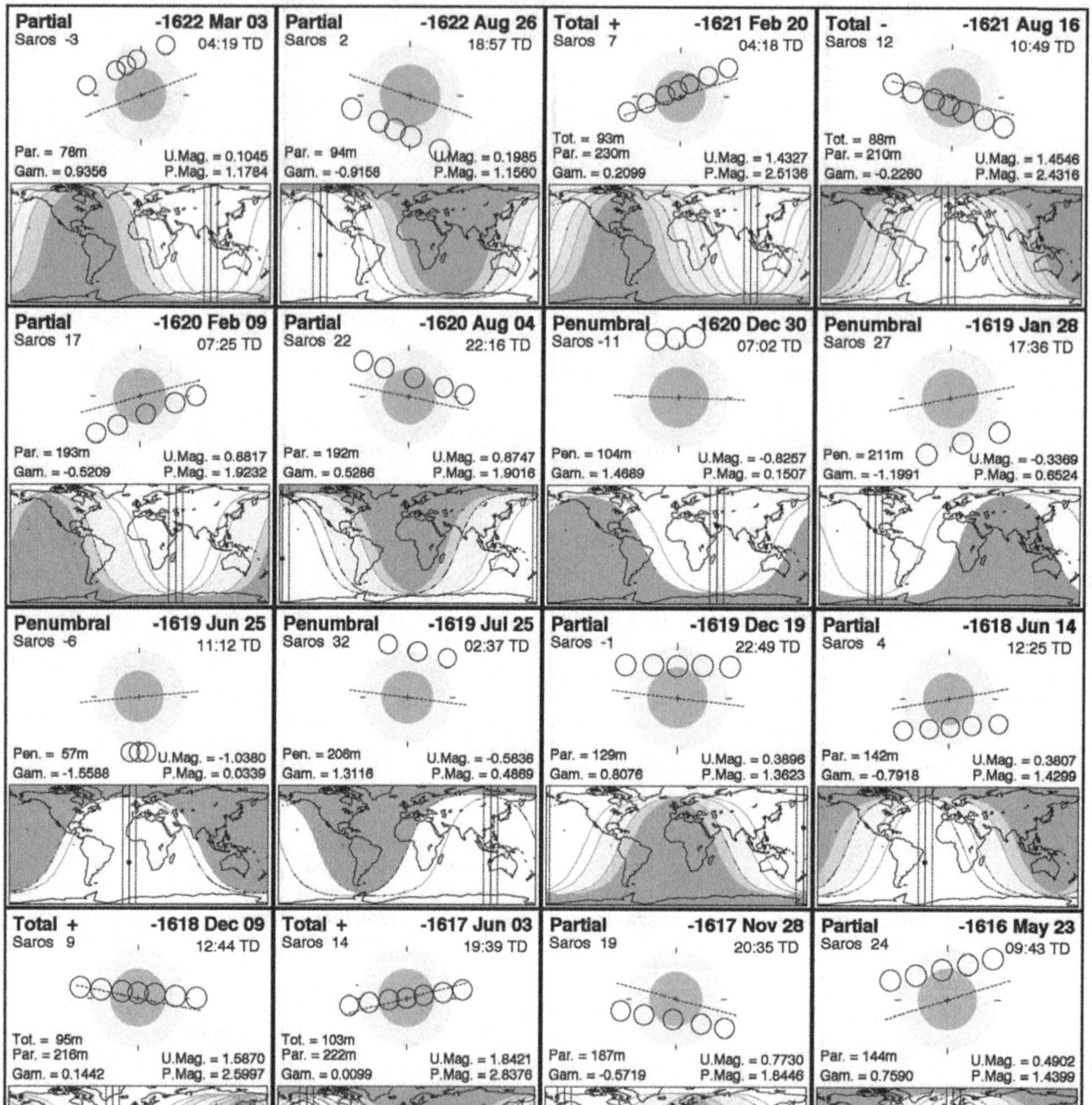

ΔT= 37722 s [= 10h29m] std.err. = ± 2297 s [= ± 9.6°] **Plate 048**

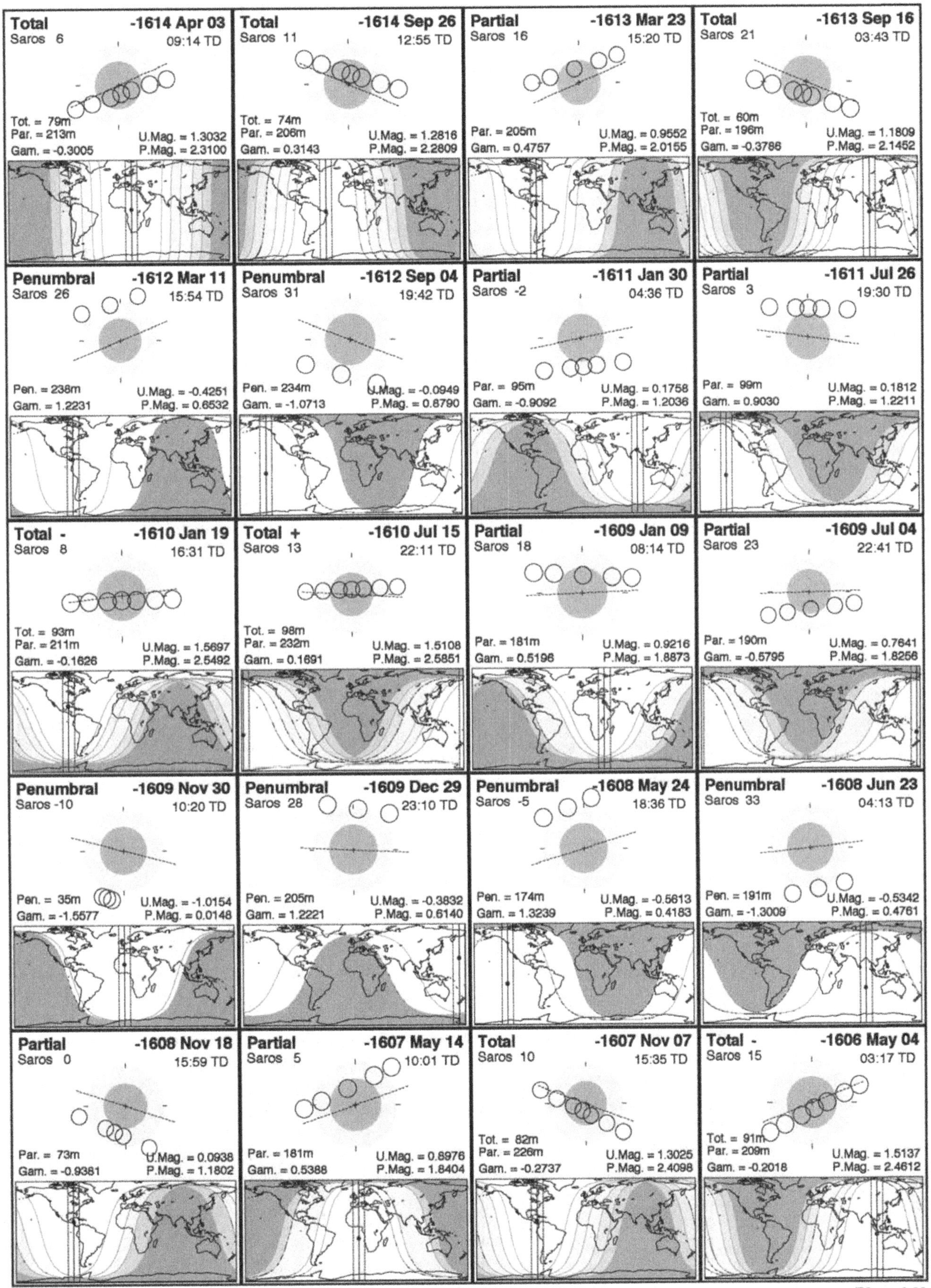

ΔT= 37545 s [= 10h26m] std.err. = ± 2269 s [= ± 9.5°] Plate 049

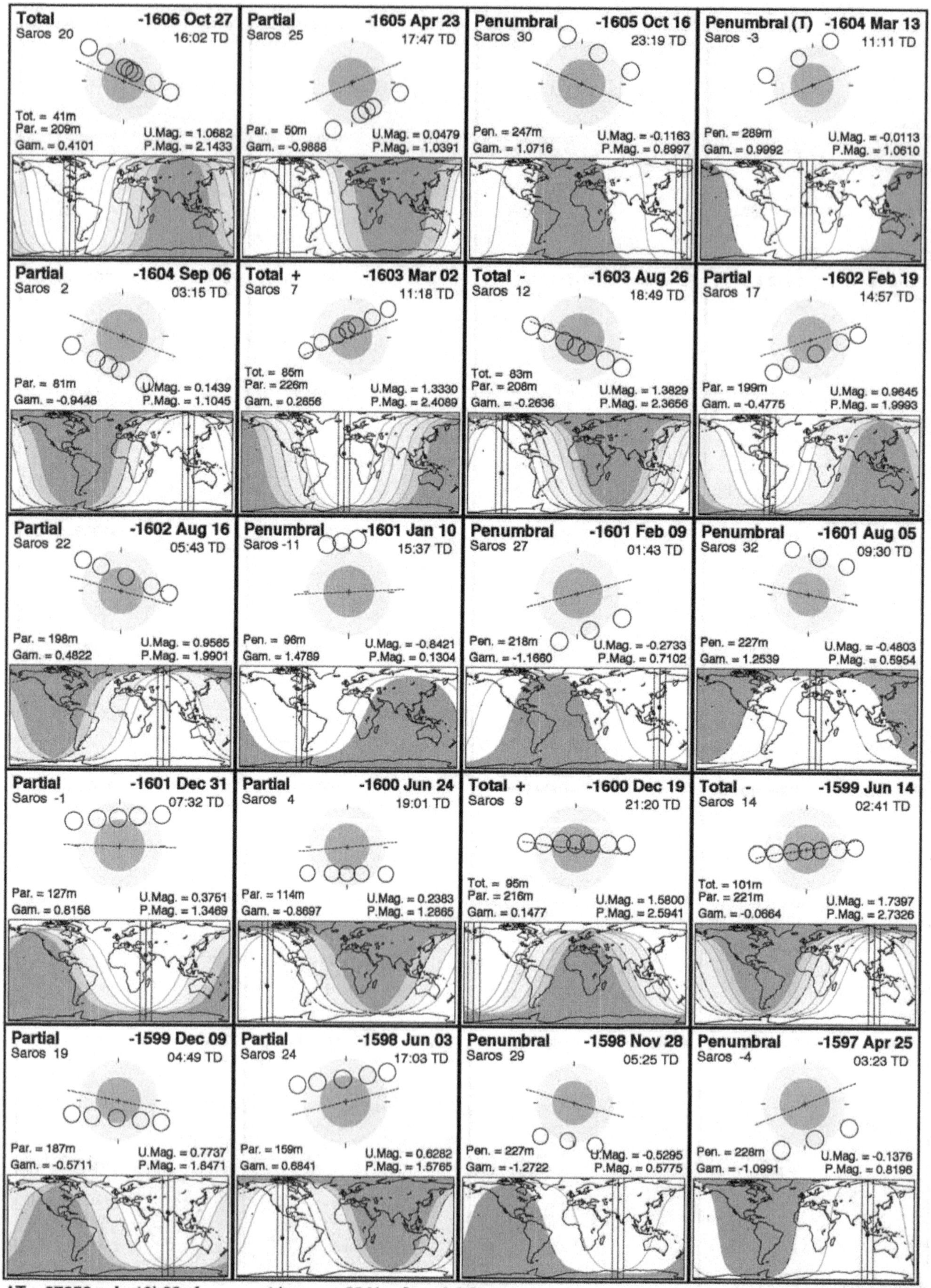

ΔT= 37358 s [= 10h23m] std.err. = ± 2241 s [= ± 9.3°] **Plate 050**

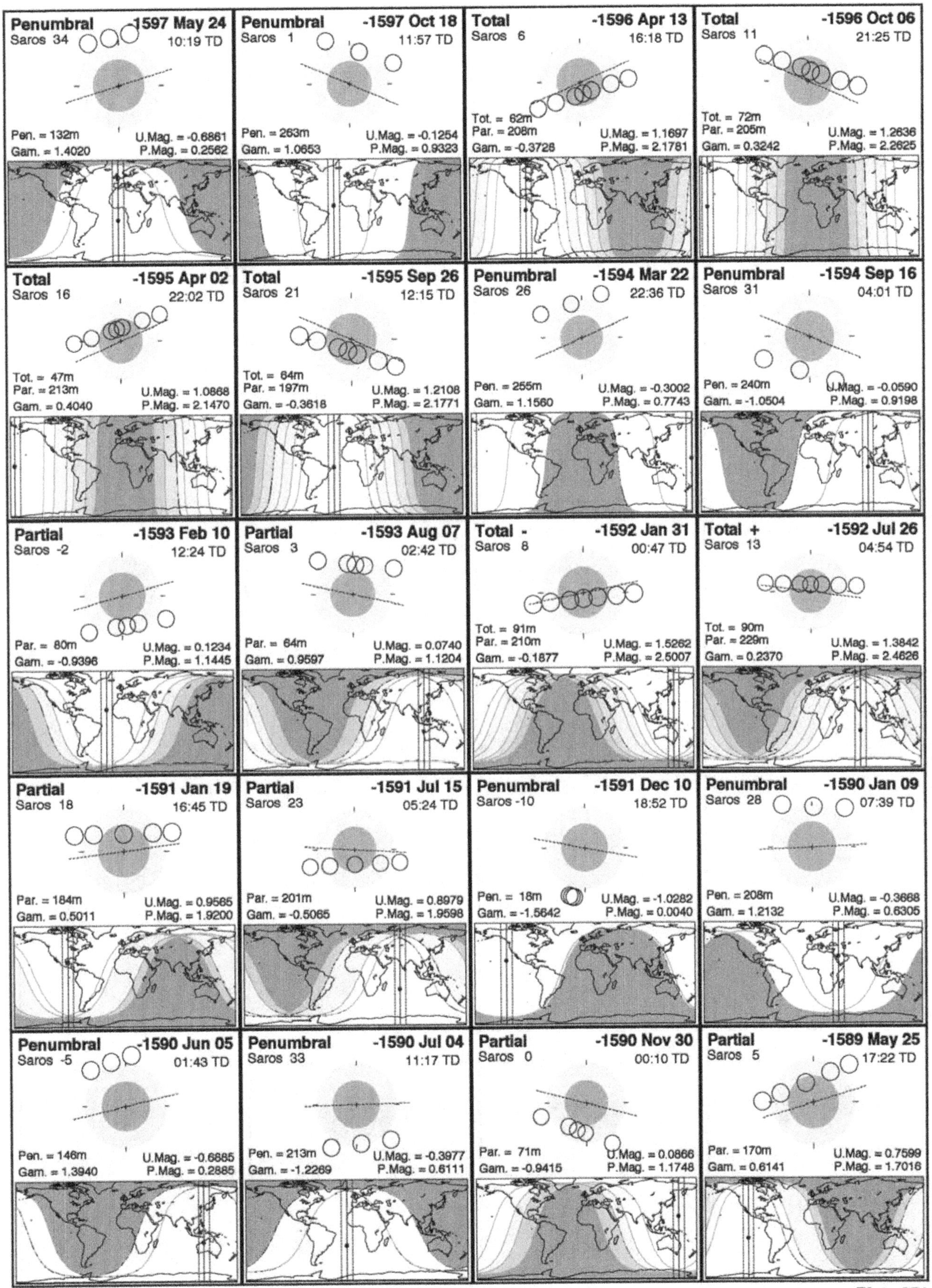

ΔT= 37171 s [= 10h20m] std.err. = ± 2212 s [= ± 9.2°] Plate 051

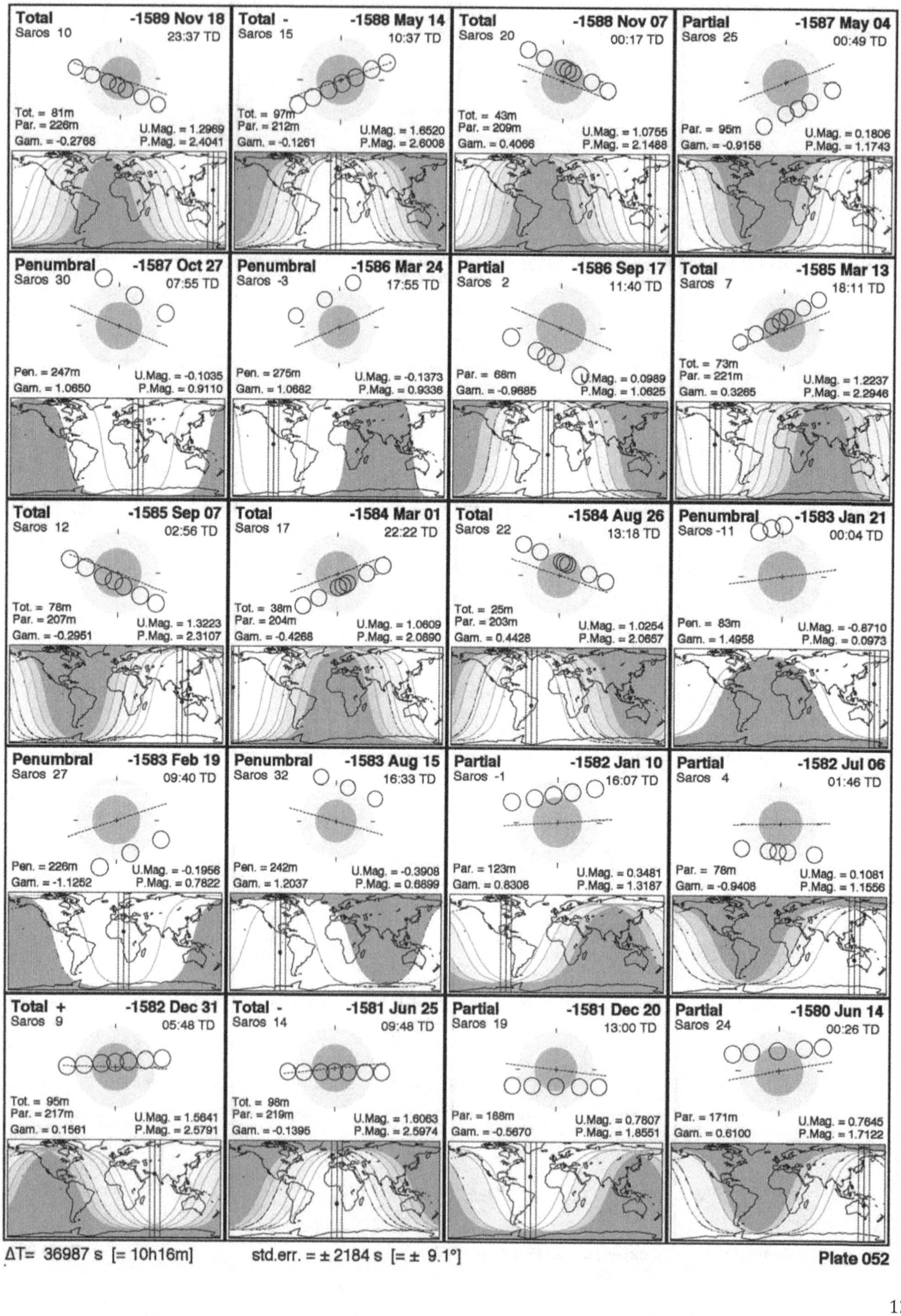

ΔT= 36987 s [= 10h16m] std.err. = ± 2184 s [= ± 9.1°] **Plate 052**

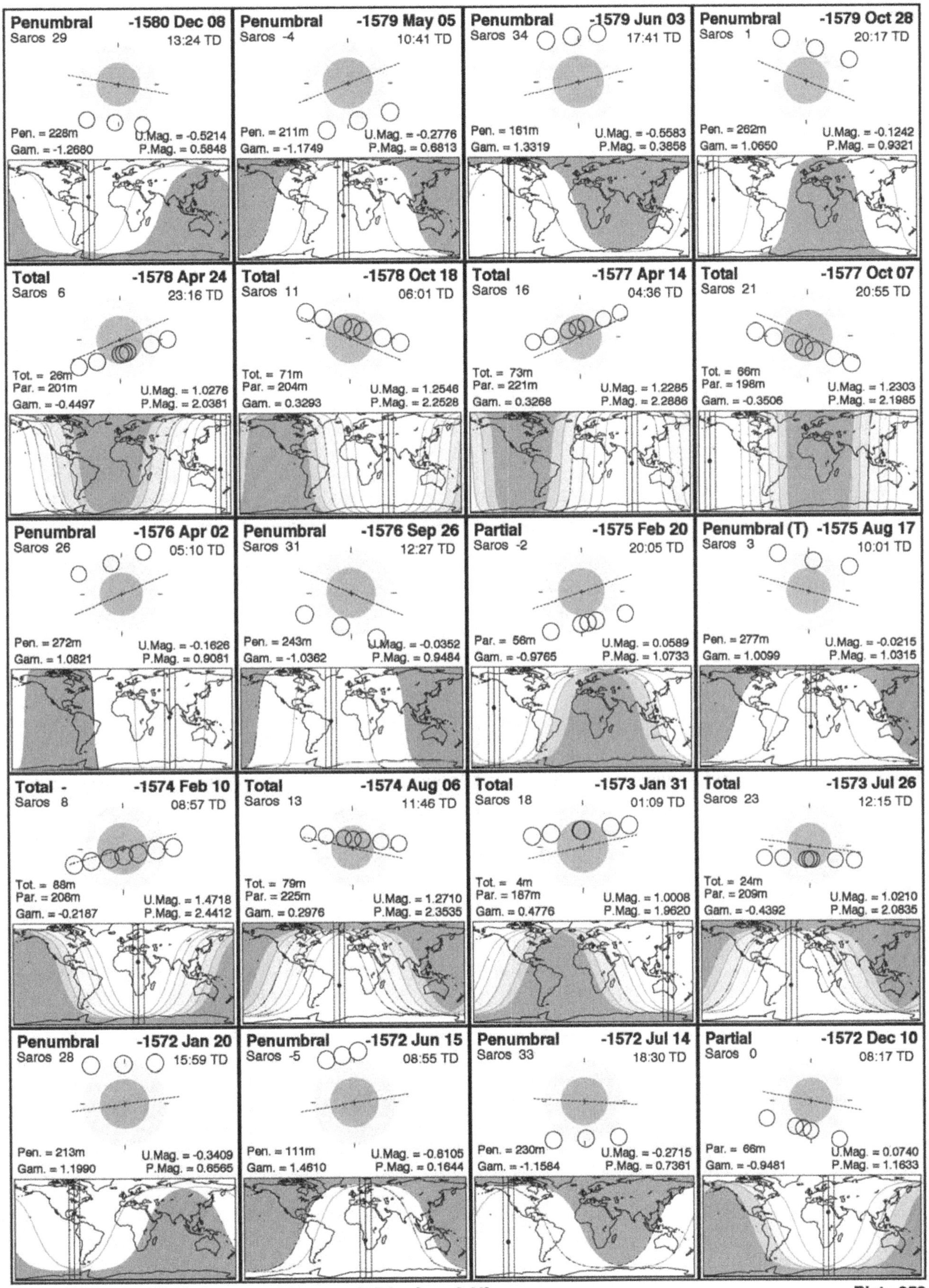

ΔT= 36790 s [= 10h13m] std.err. = ± 2154 s [= ± 9.0°] Plate 053

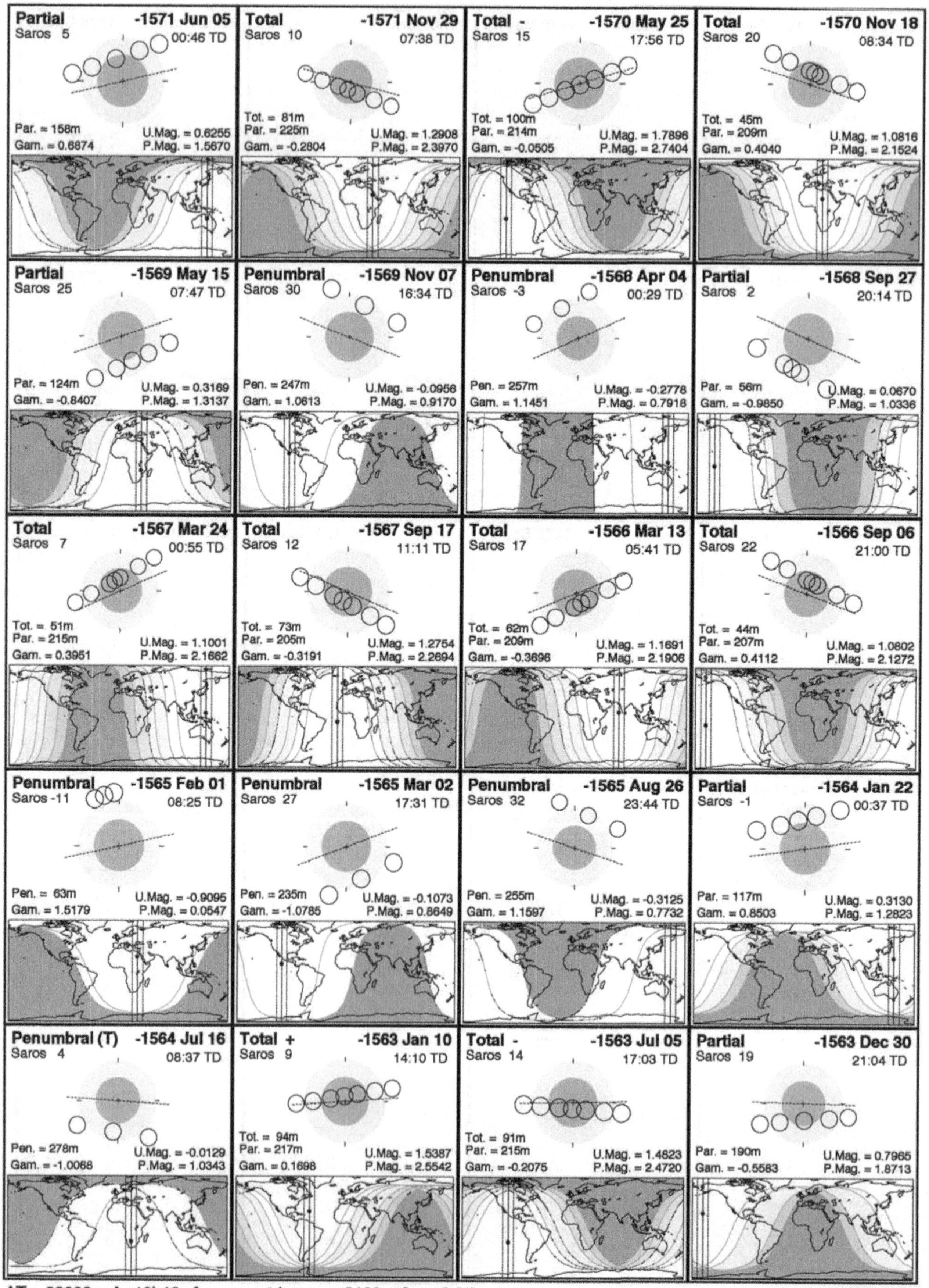

ΔT= 36606 s [= 10h10m] std.err. = ± 2126 s [= ± 8.9°] **Plate 054**

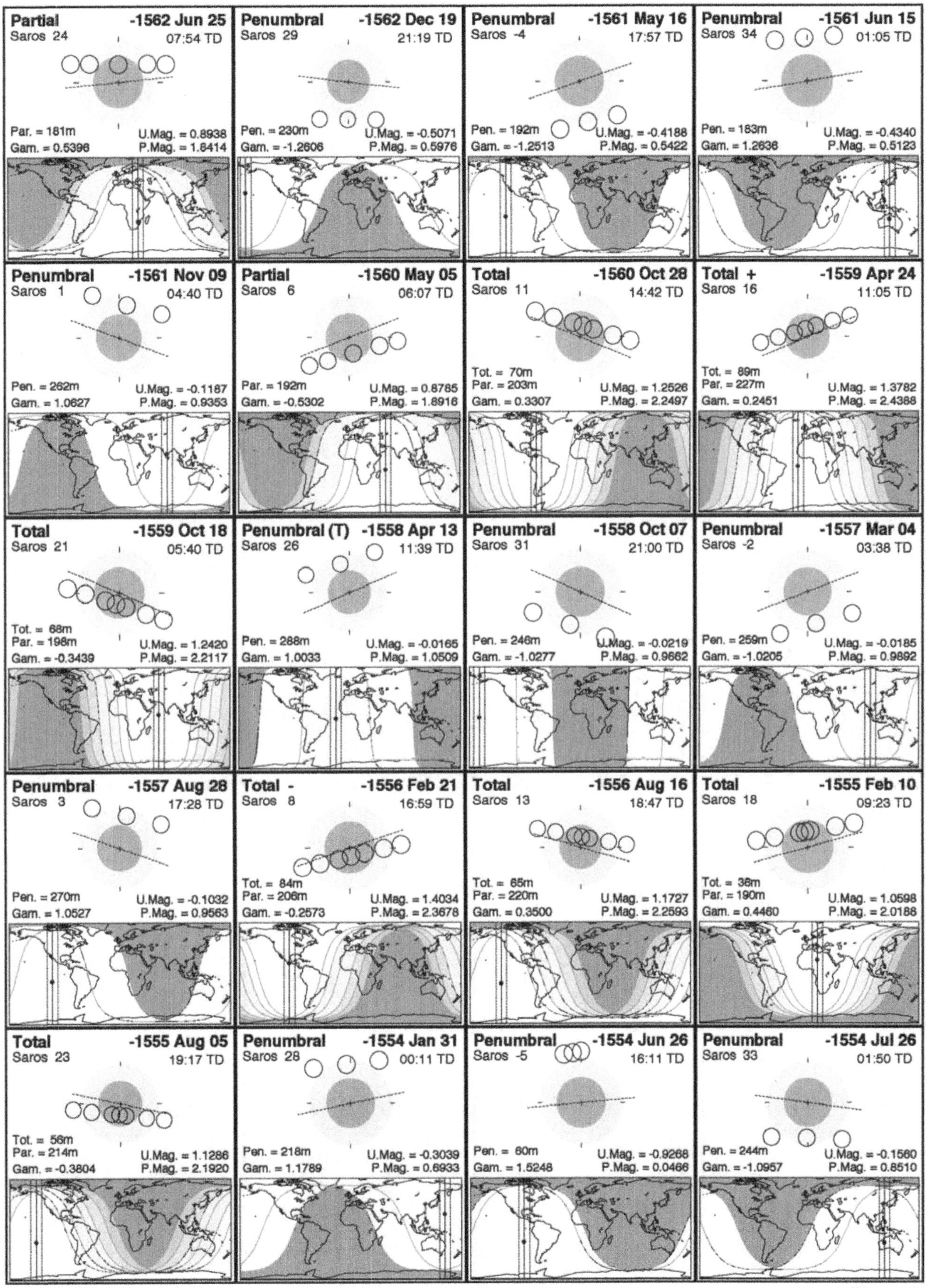

ΔT= 36411 s [= 10h07m] std.err. = ± 2096 s [= ± 8.7°] **Plate 055**

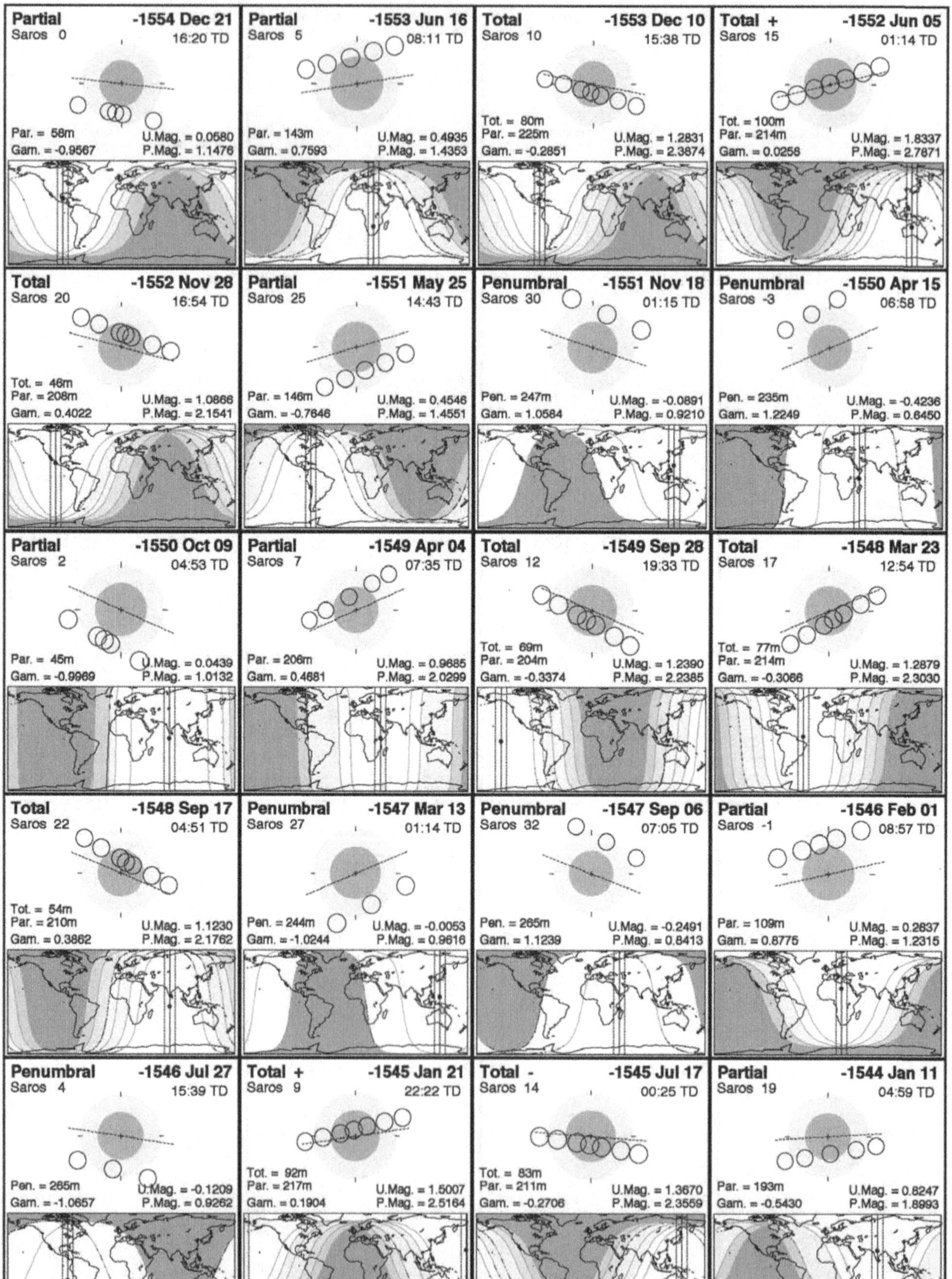

ΔT= 36228 s [= 10h04m] std.err. = ± 2069 s [= ± 8.6°] Plate 056

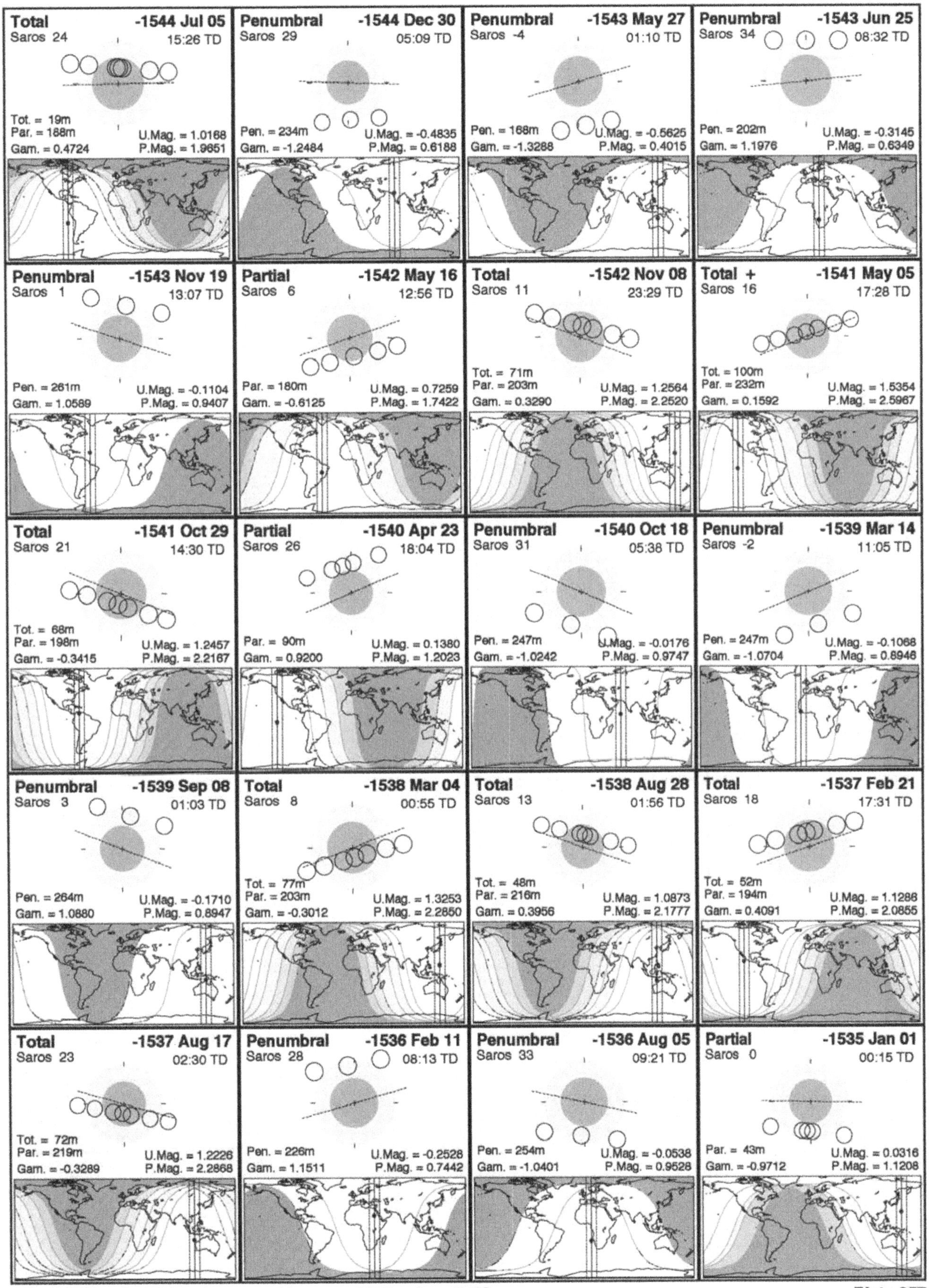

ΔT= 36023 s [= 10h00m] std.err. = ± 2038 s [= ± 8.5°] Plate 057

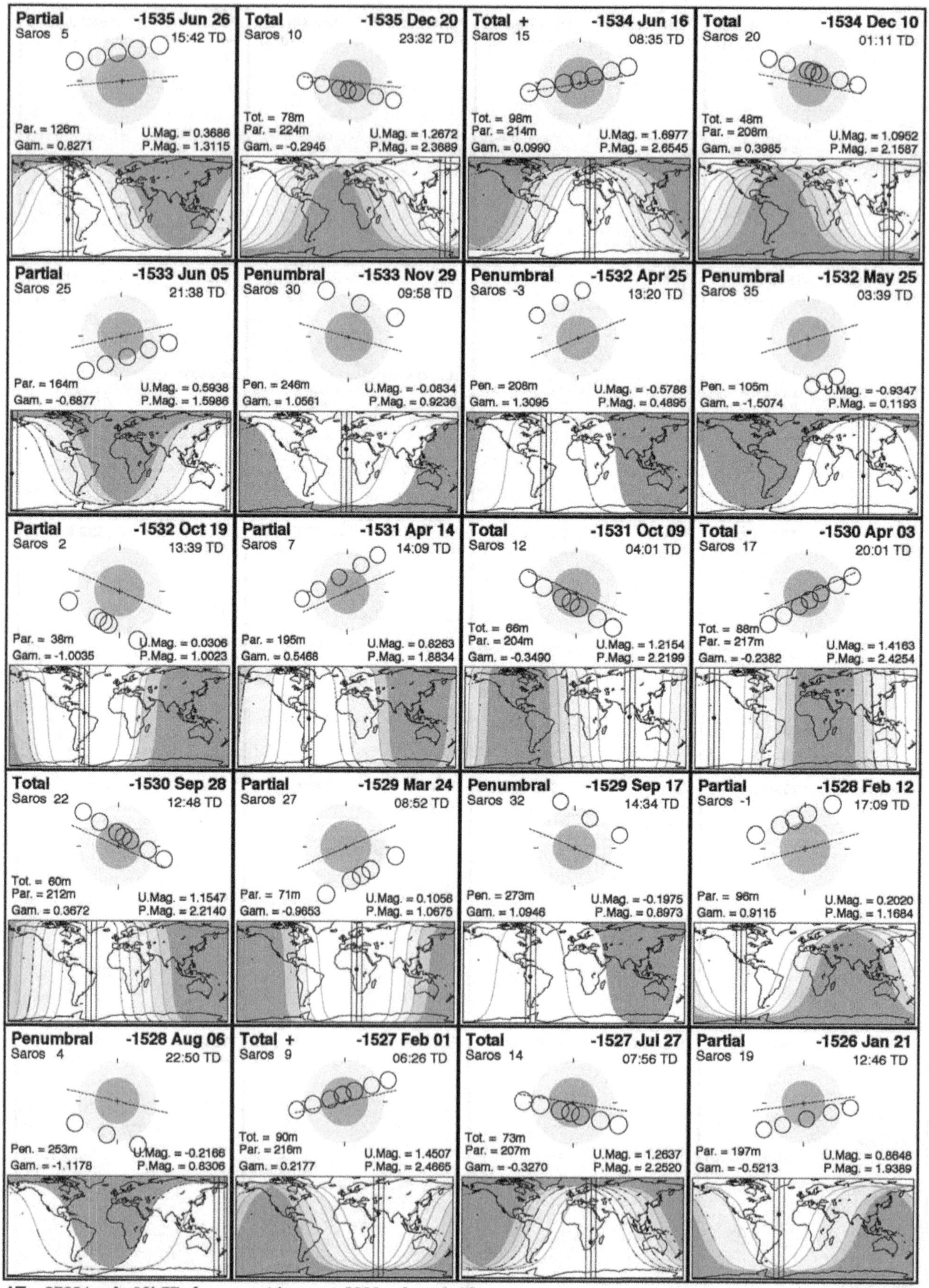

ΔT= 35831 s [= 09h57m] std.err. = ± 2009 s [= ± 8.4°] **Plate 058**

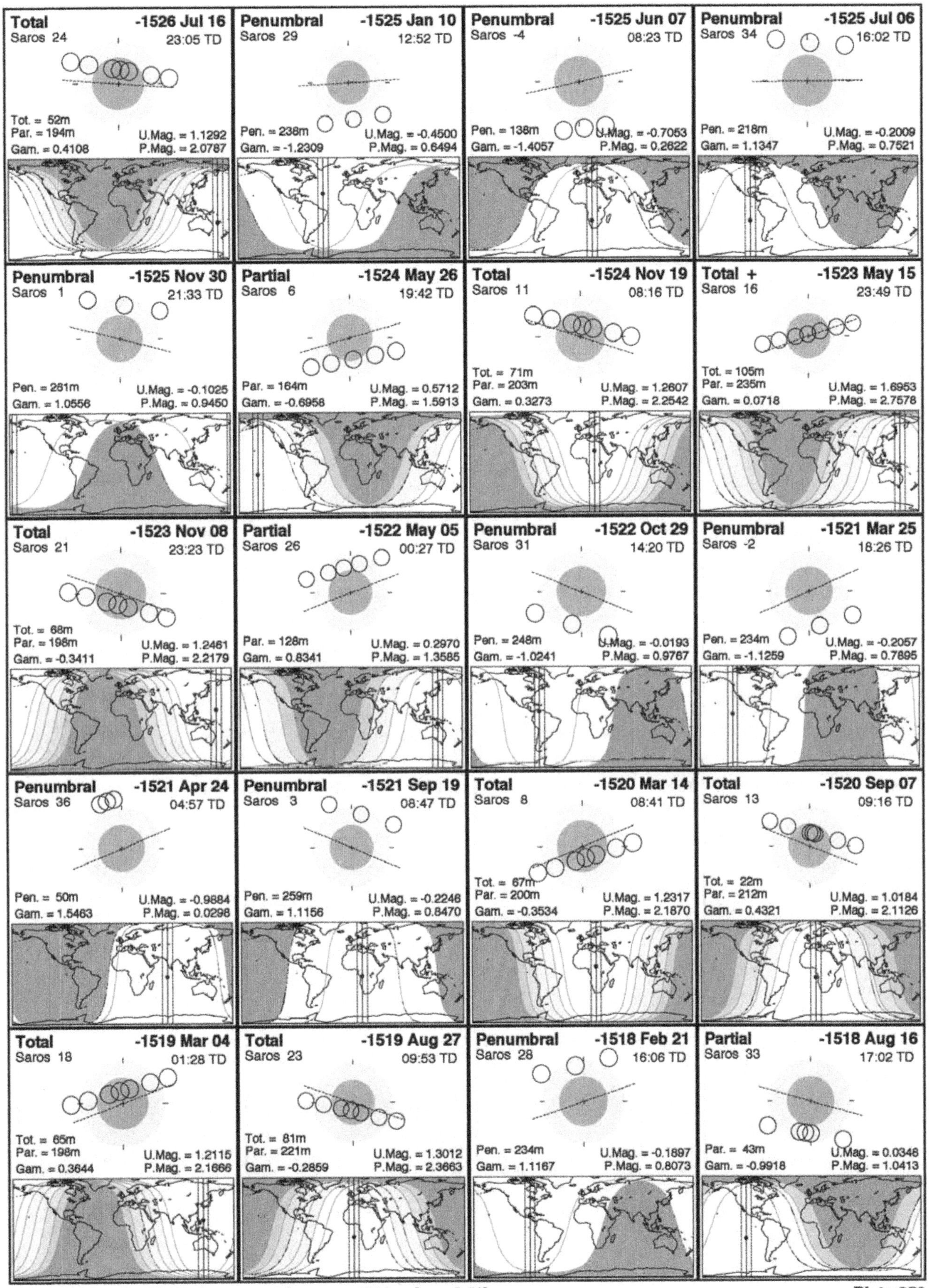

ΔT= 35638 s [= 09h54m] std.err. = ± 1980 s [= ± 8.3°] **Plate 059**

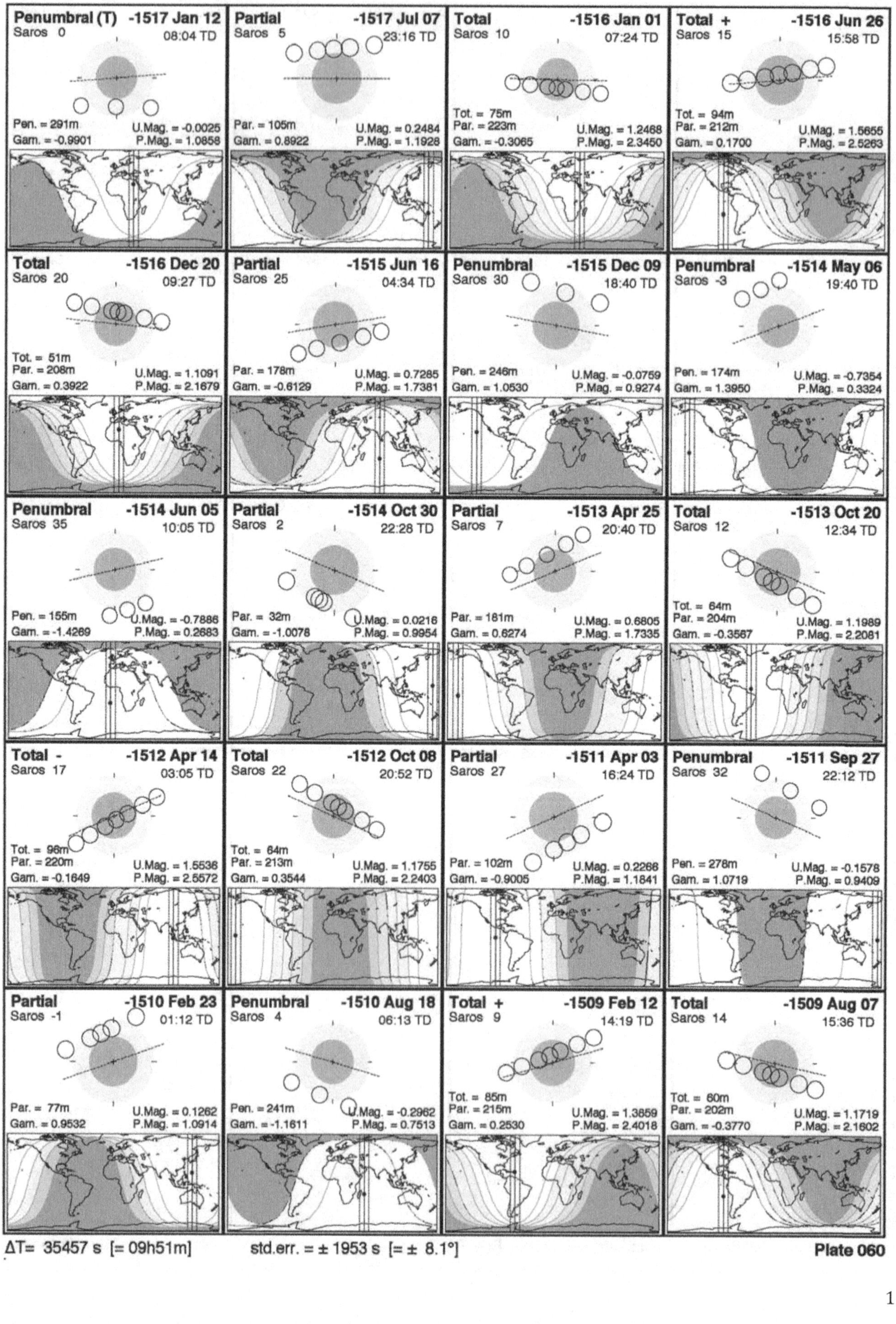

ΔT= 35457 s [= 09h51m] std.err. = ± 1953 s [= ± 8.1°] **Plate 060**

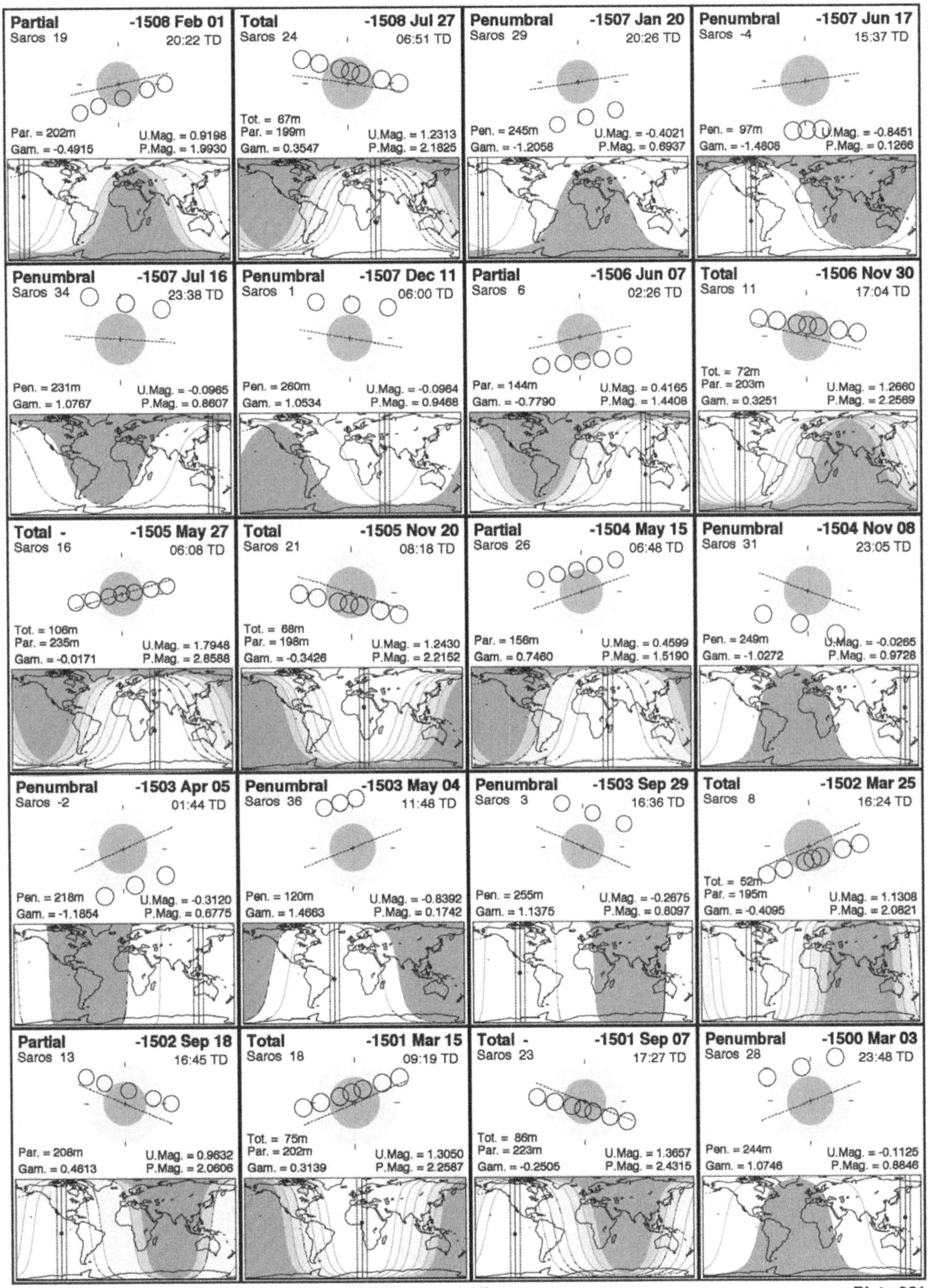

ΔT= 35265 s [= 09h48m] std.err. = ± 1925 s [= ± 8.0°] **Plate 061**

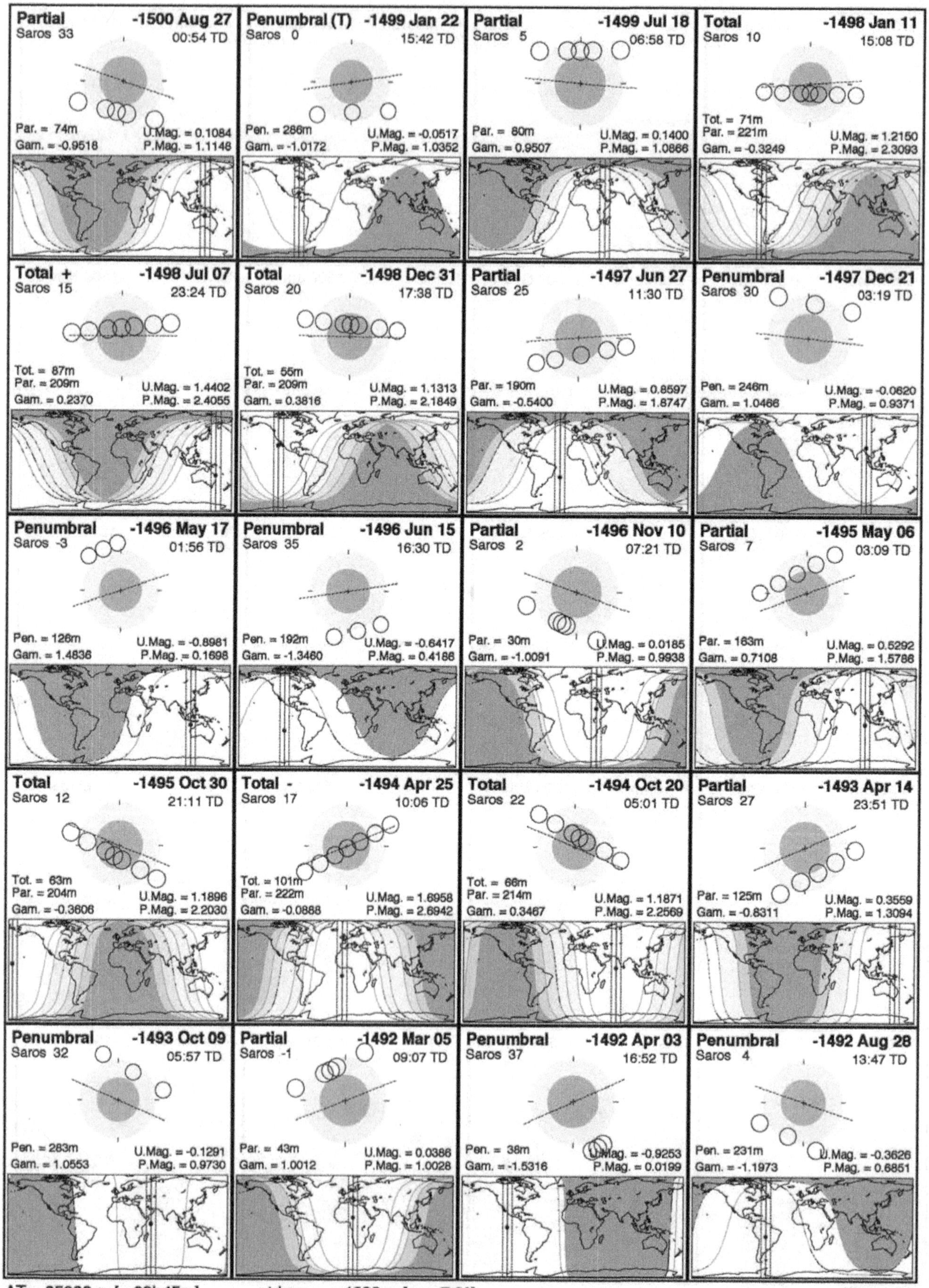

ΔT= 35083 s [= 09h45m] std.err. = ± 1898 s [= ± 7.9°] **Plate 062**

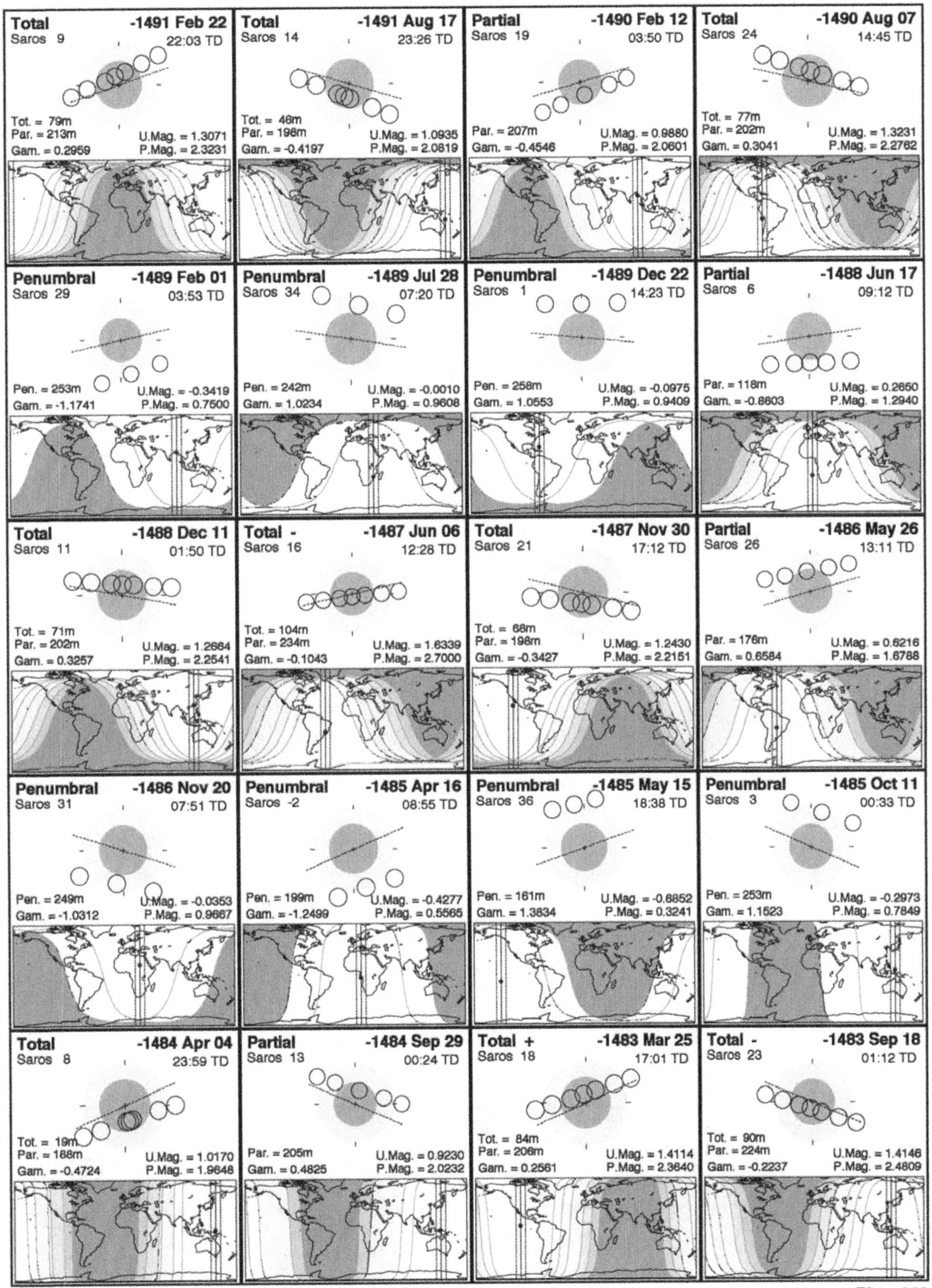

ΔT= 34904 s [= 09h42m] std.err. = ± 1872 s [= ± 7.8°] **Plate 063**

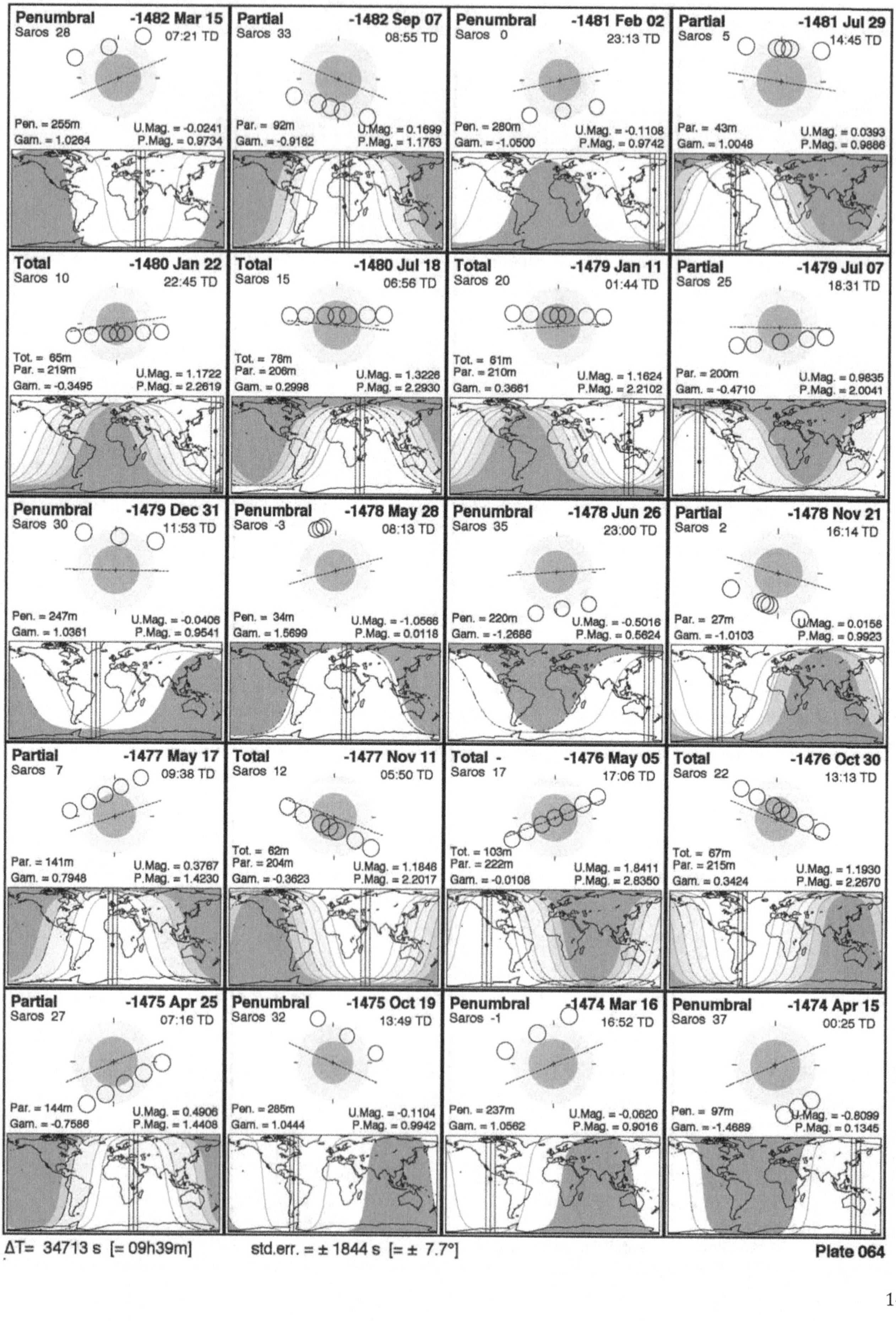

ΔT= 34713 s [= 09h39m] std.err. = ± 1844 s [= ± 7.7°] **Plate 064**

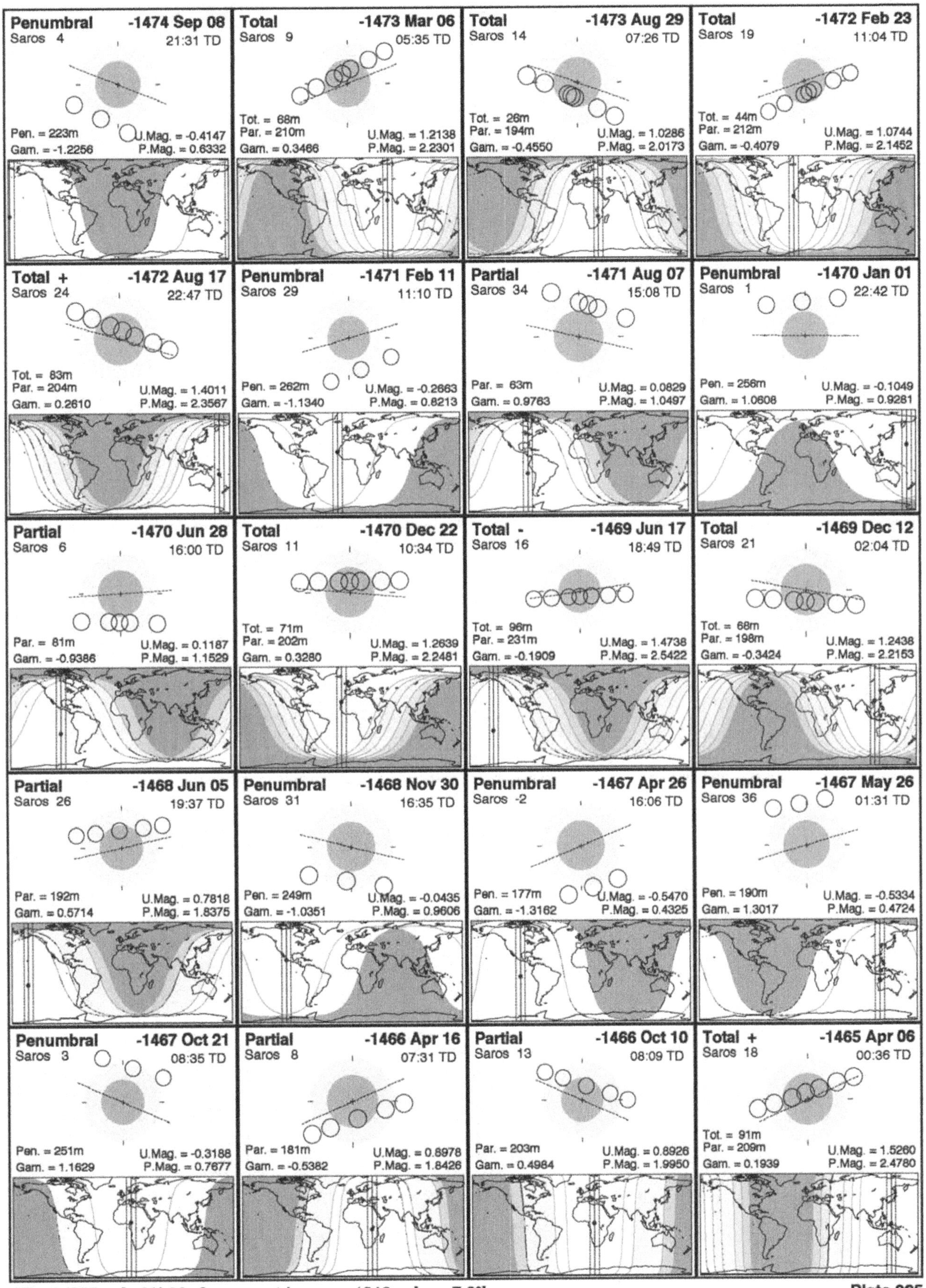

ΔT= 34535 s [= 09h36m] std.err. = ± 1818 s [= ± 7.6°] Plate 065

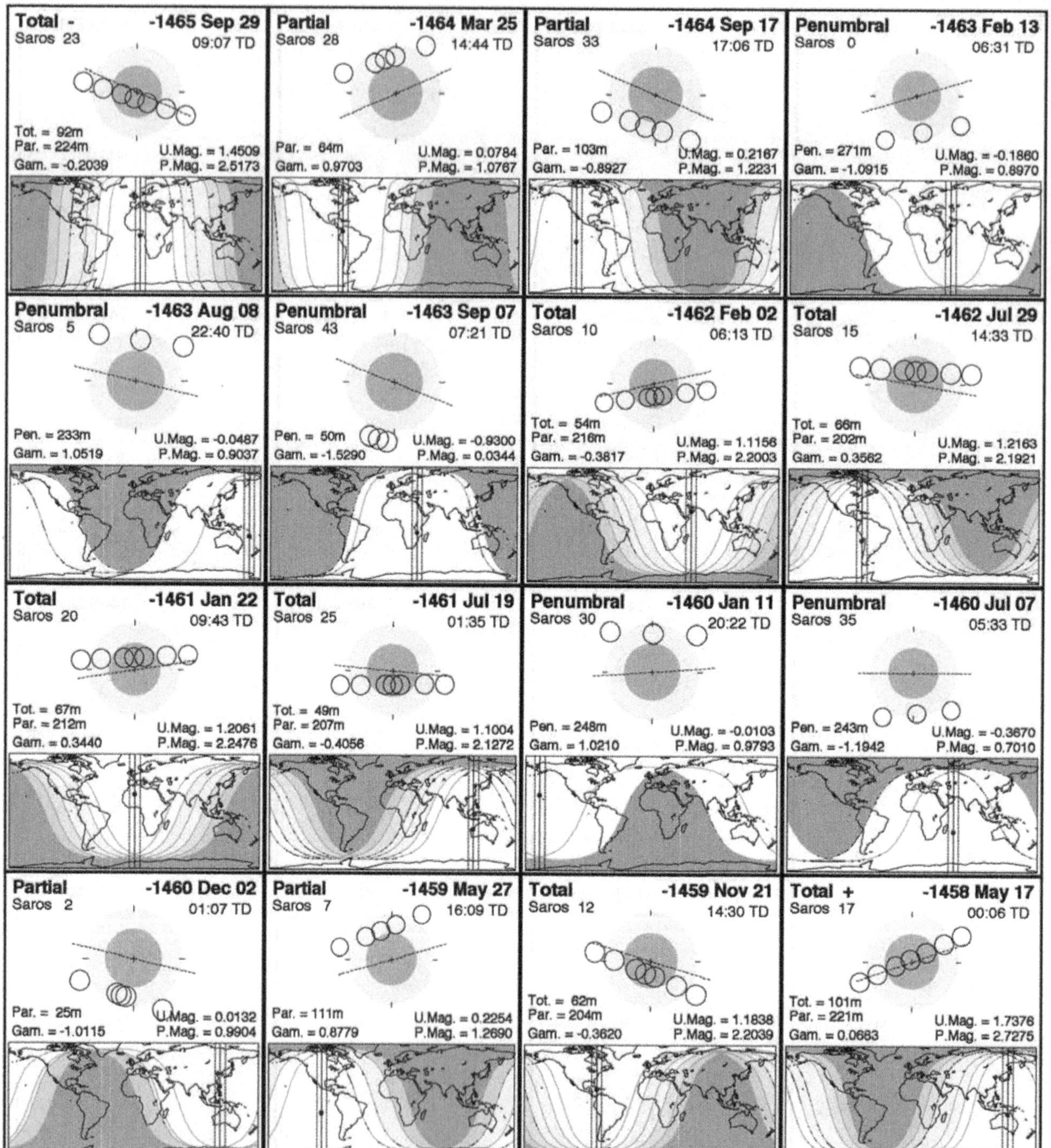

ΔT= 34345 s [= 09h32m] std.err. = ± 1790 s [= ± 7.5°] **Plate 066**

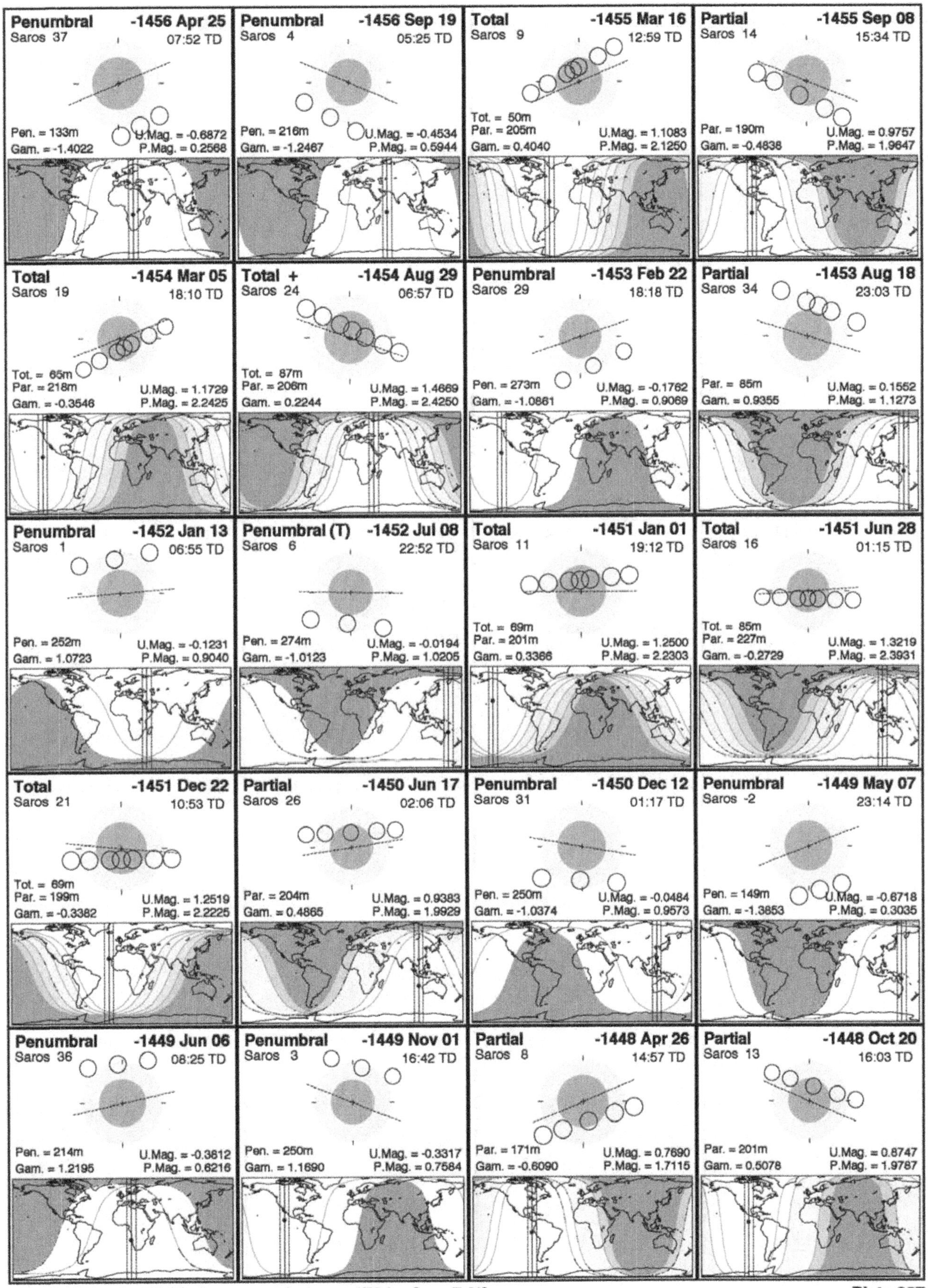

ΔT= 34166 s [= 09h29m] std.err. = ± 1764 s [= ± 7.3°] **Plate 067**

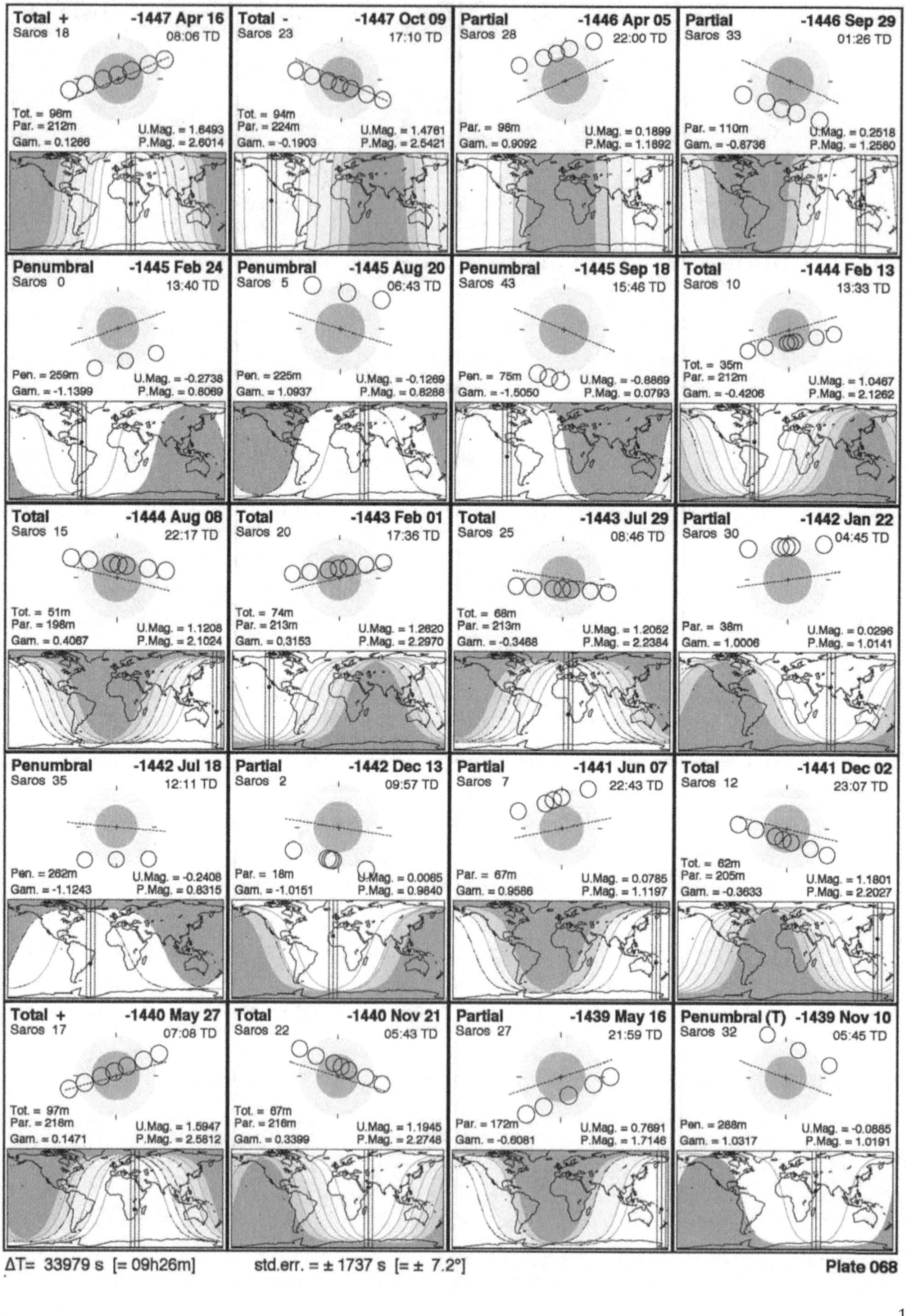

ΔT= 33979 s [= 09h26m] std.err. = ± 1737 s [= ± 7.2°] **Plate 068**

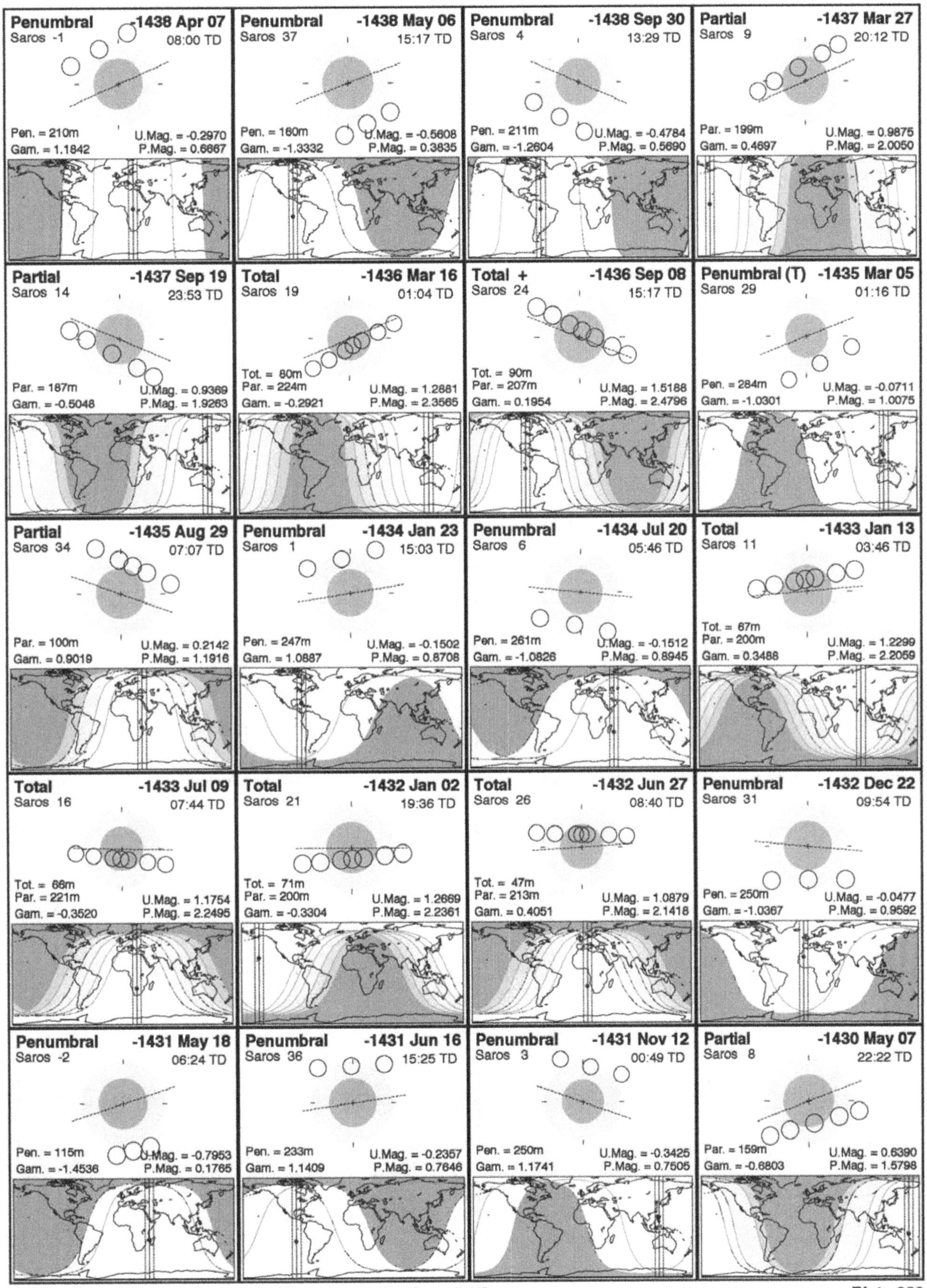

ΔT= 33792 s [= 09h23m] std.err. = ± 1710 s [= ± 7.1°] **Plate 069**

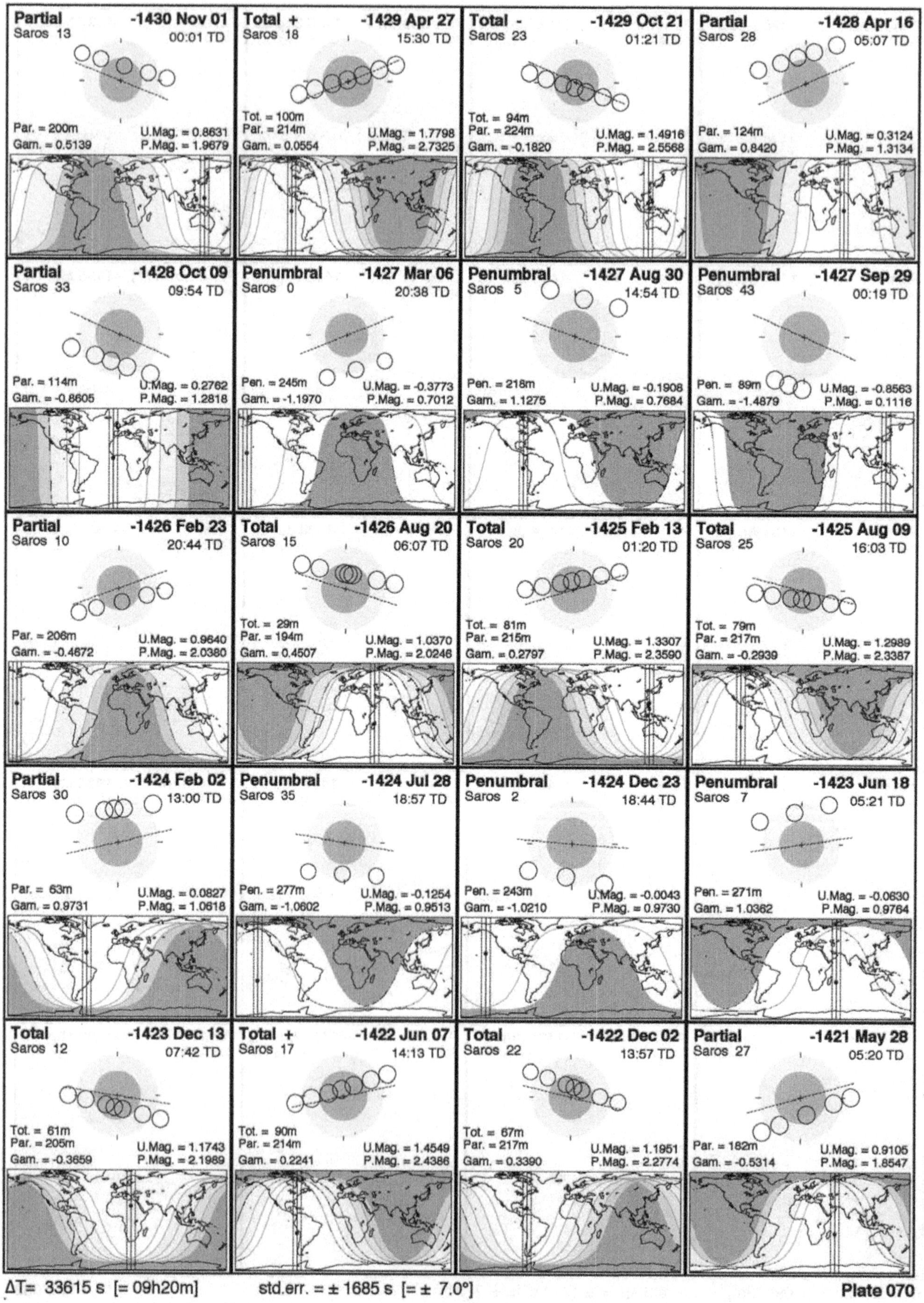

ΔT= 33615 s [= 09h20m] std.err. = ± 1685 s [= ± 7.0°] **Plate 070**

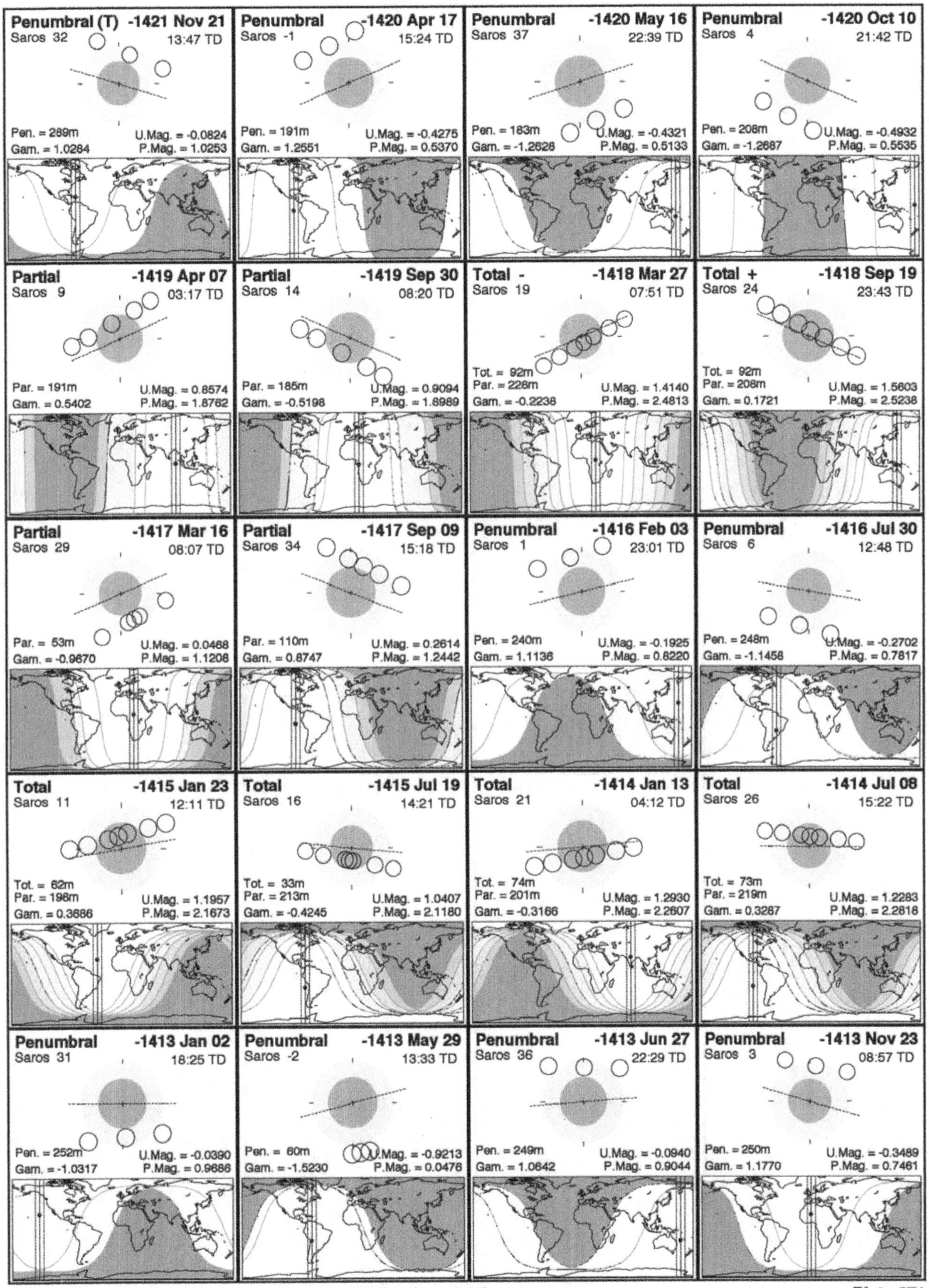

ΔT= 33427 s [= 09h17m] std.err. = ± 1658 s [= ± 6.9°] **Plate 071**

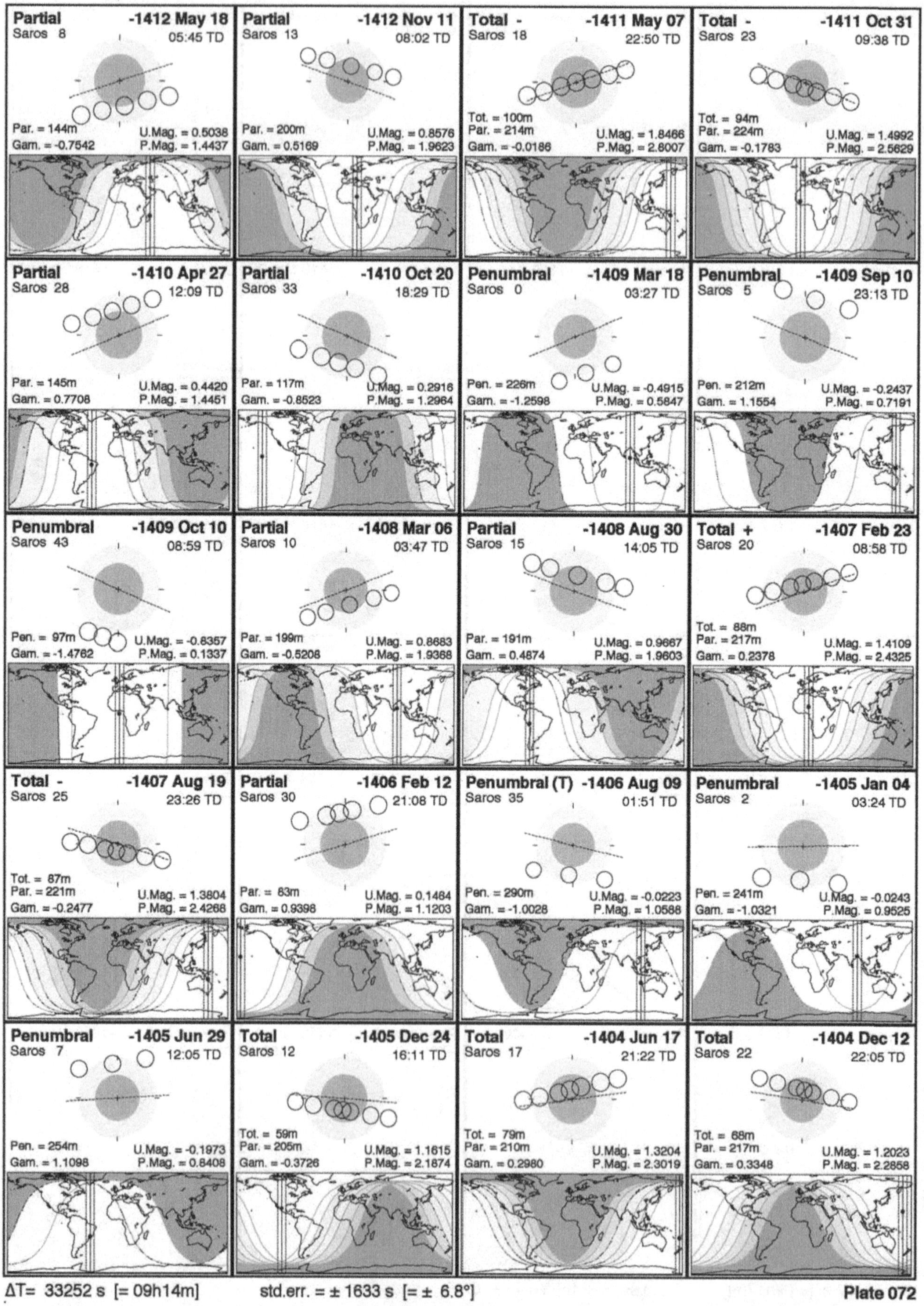

ΔT= 33252 s [= 09h14m] std.err. = ± 1633 s [= ± 6.8°] Plate 072

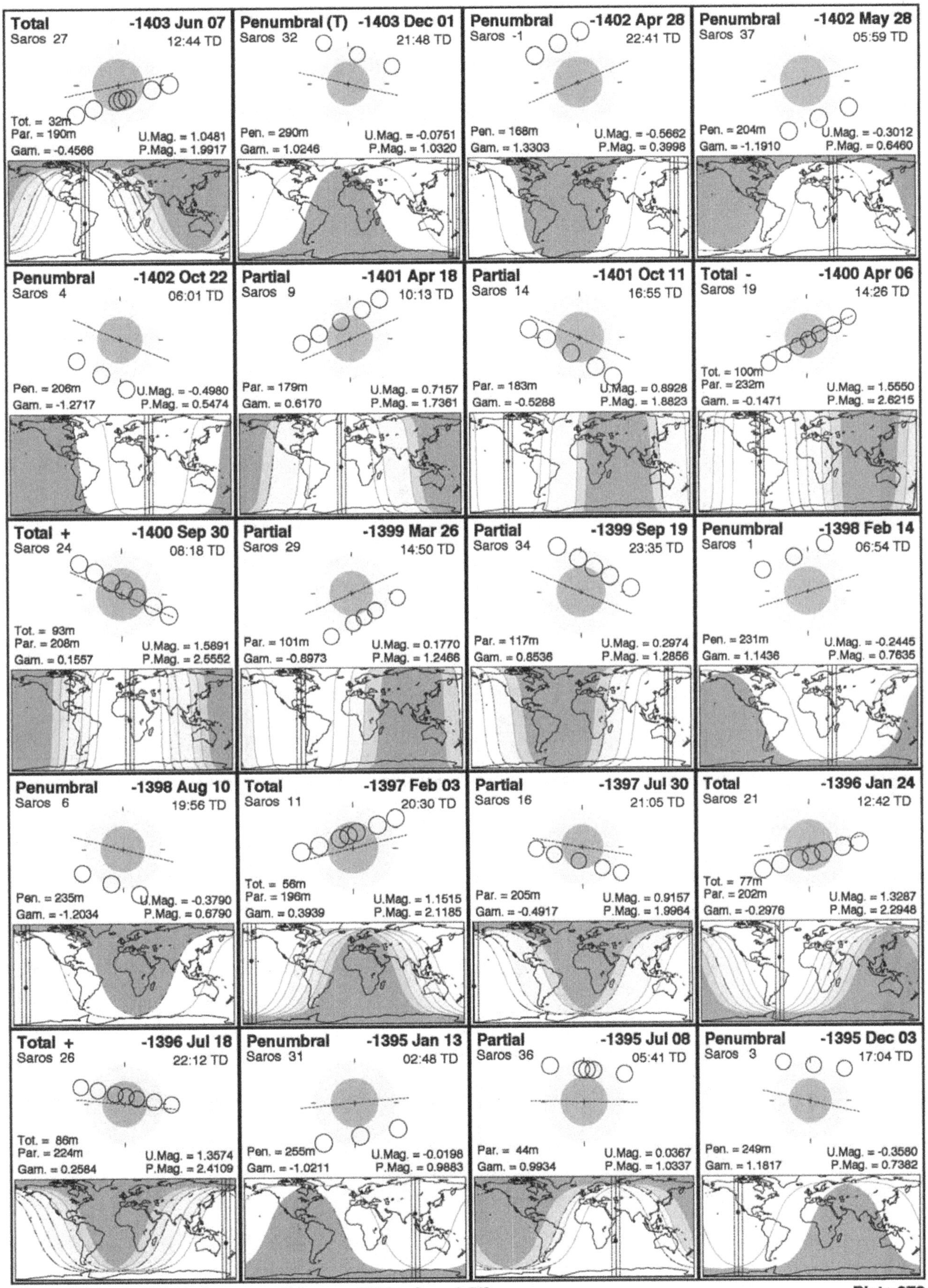

ΔT= 33066 s [= 09h11m] std.err. = ± 1606 s [= ± 6.7°] **Plate 073**

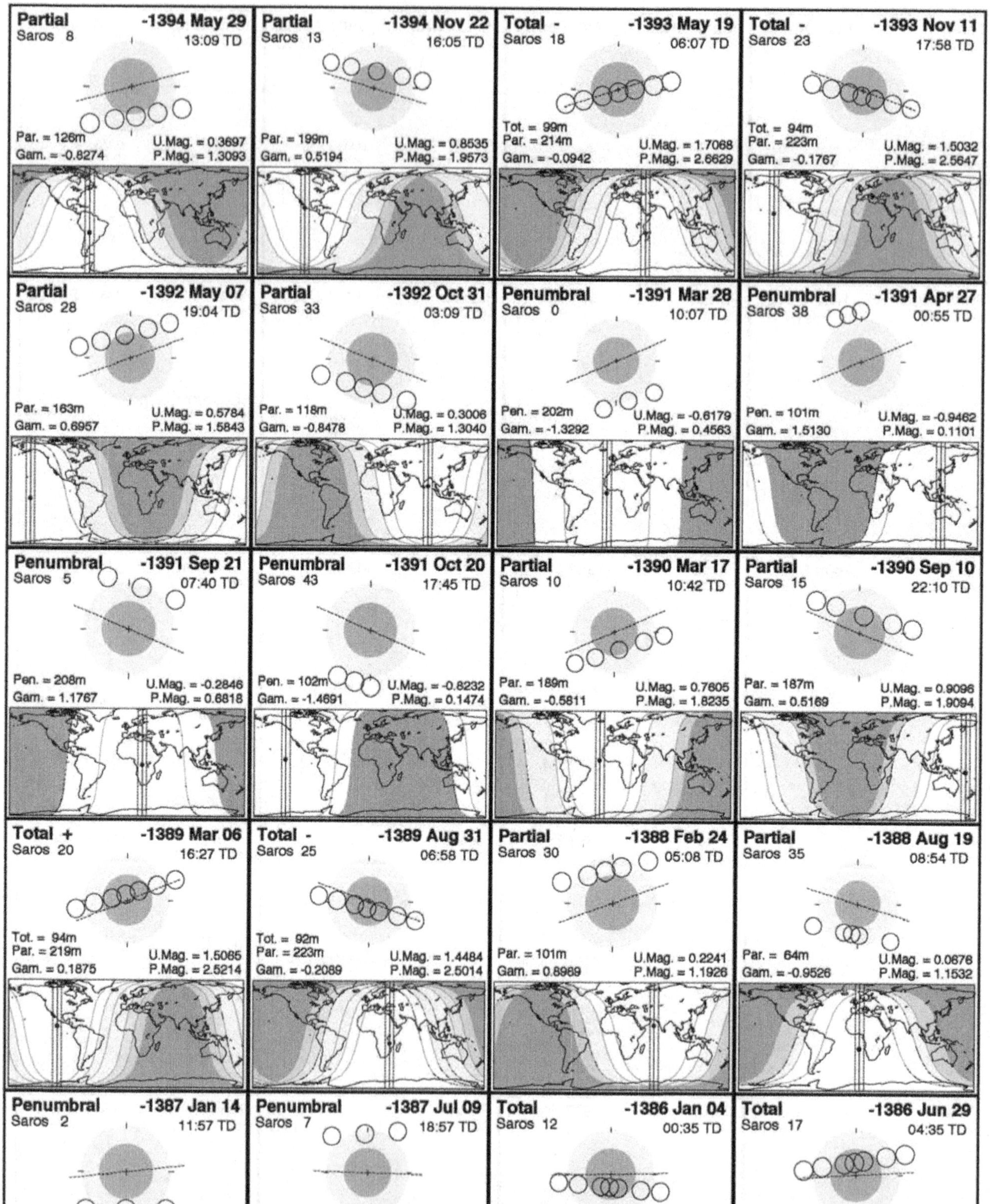

ΔT= 32882 s [= 09h08m] std.err. = ± 1580 s [= ± 6.6°] **Plate 074**

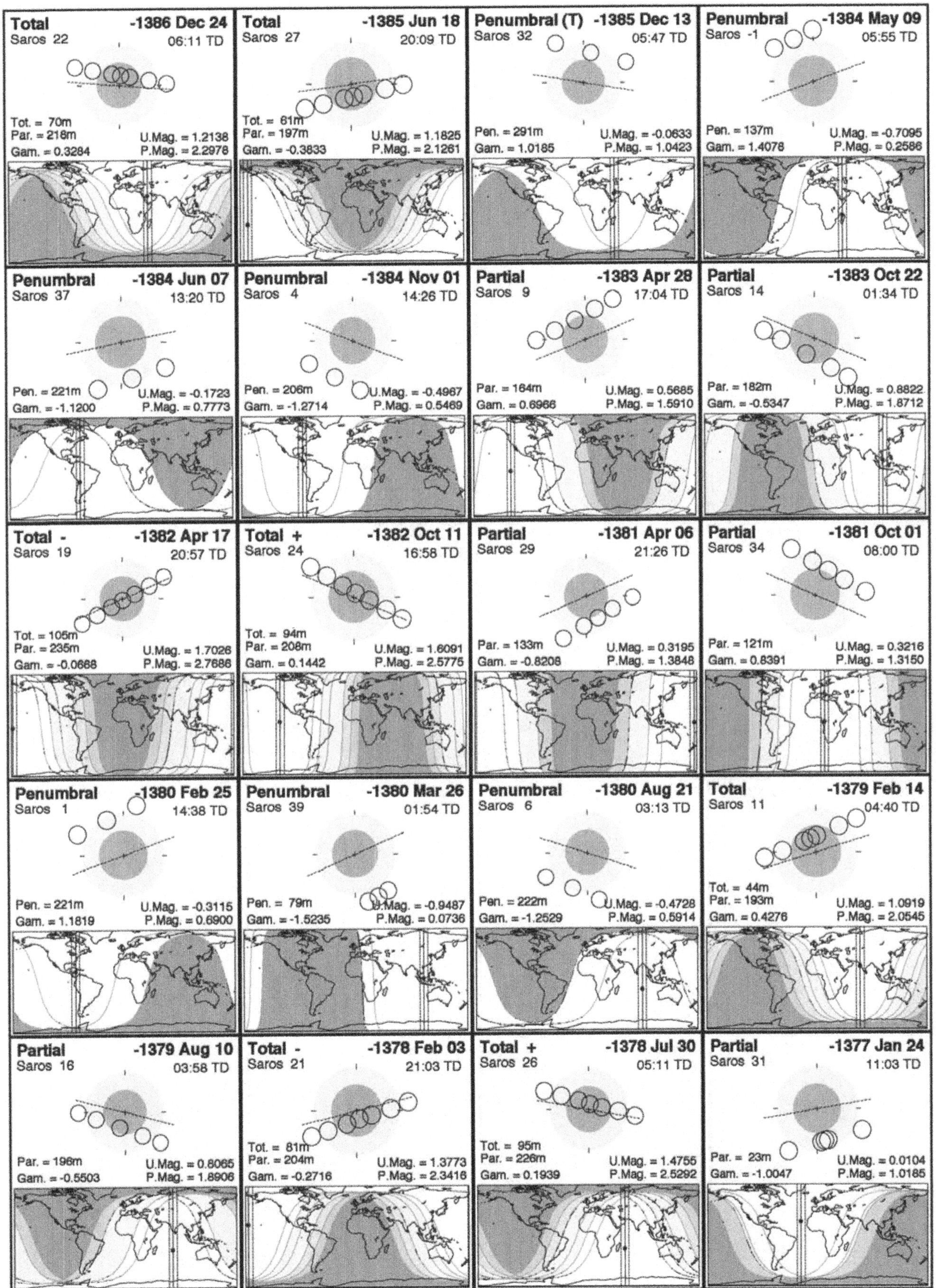

ΔT= 32707 s [= 09h05m] std.err. = ± 1556 s [= ± 6.5°] **Plate 075**

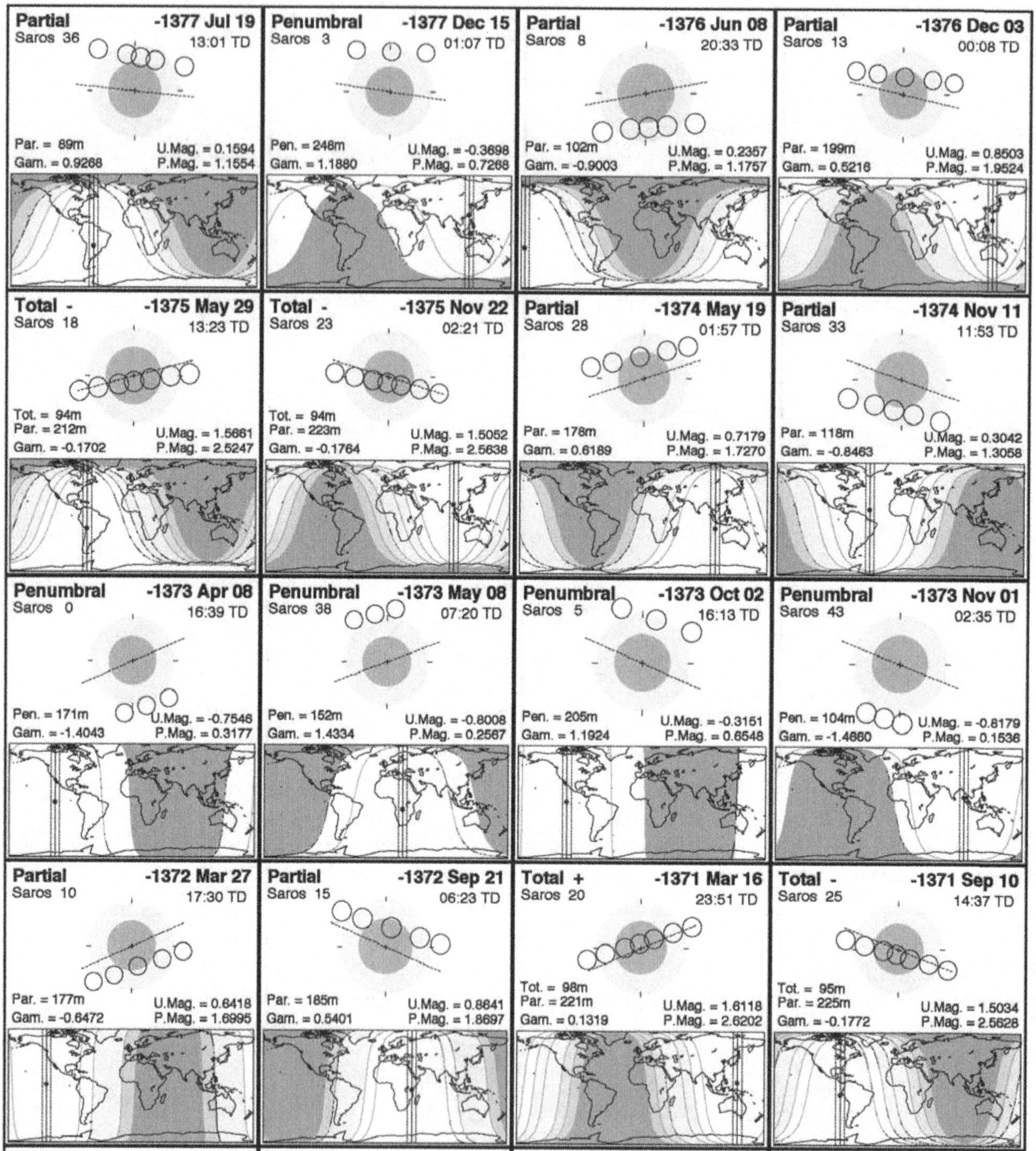

ΔT= 32532 s [= 09h02m] std.err. = ± 1531 s [= ± 6.4°] **Plate 076**

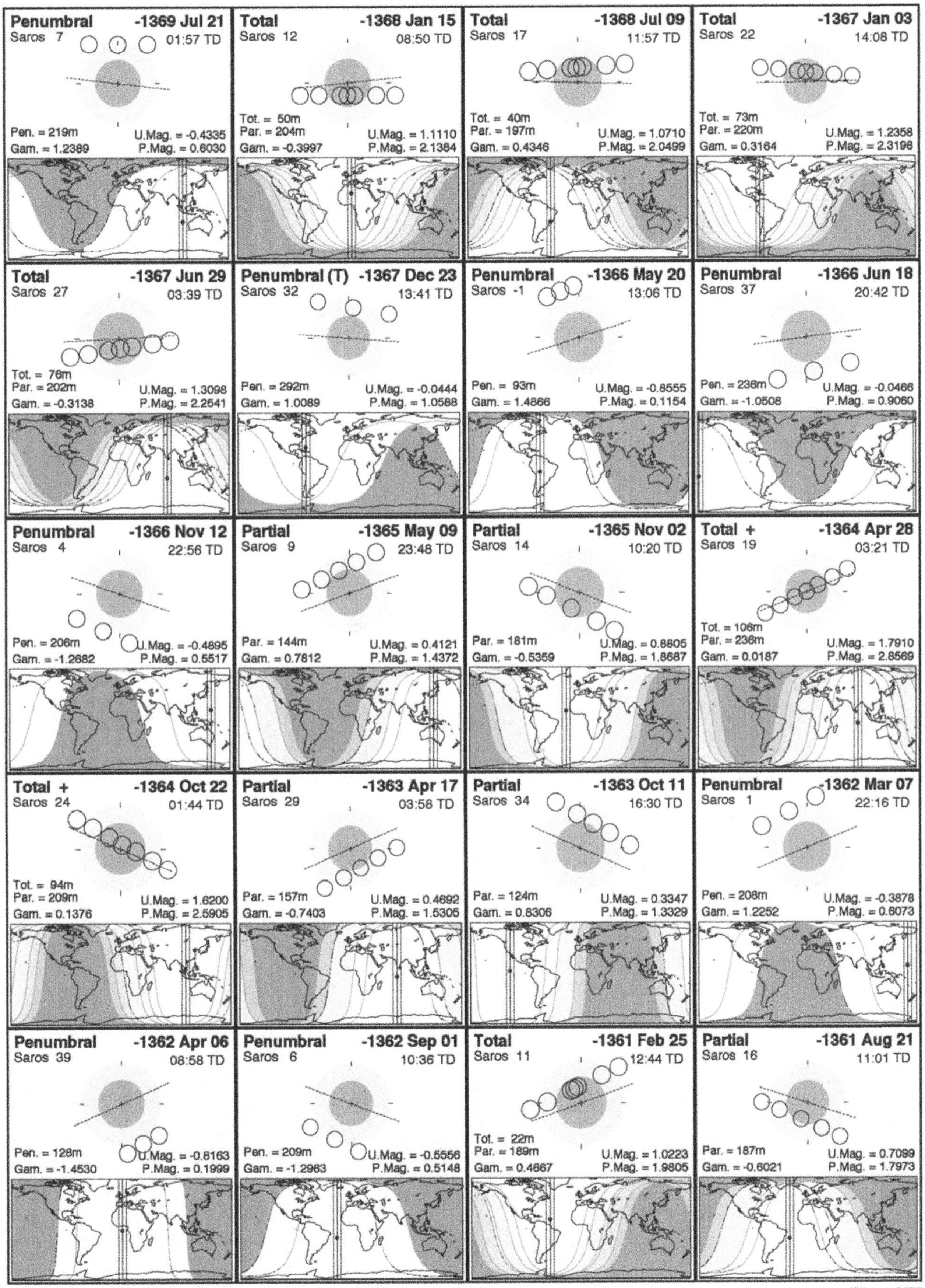

ΔT= 32369 s [= 08h59m] std.err. = ± 1509 s [= ± 6.3°] Plate 077

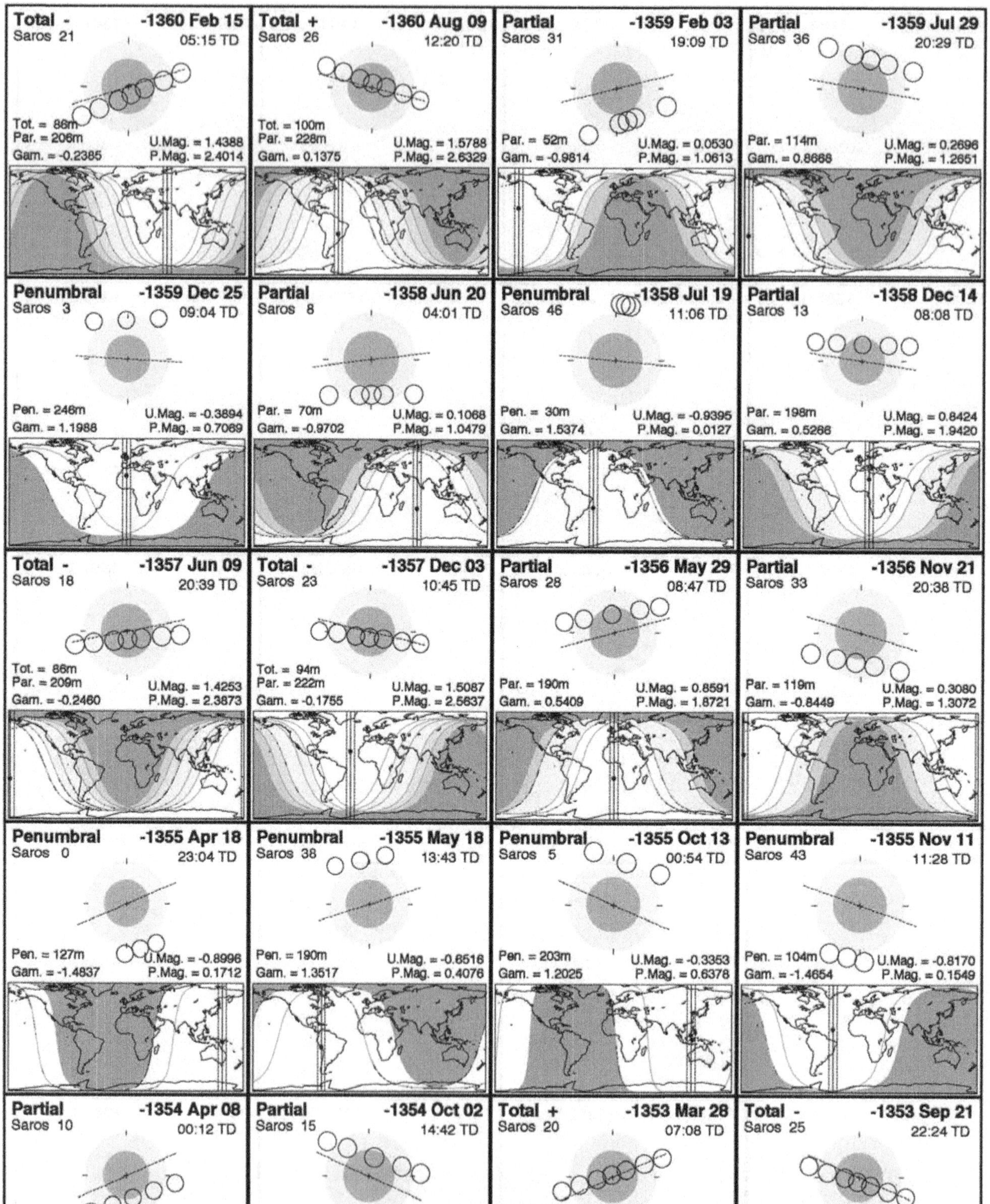

ΔT= 32195 s [= 08h57m] std.err. = ± 1484 s [= ± 6.2°] **Plate 078**

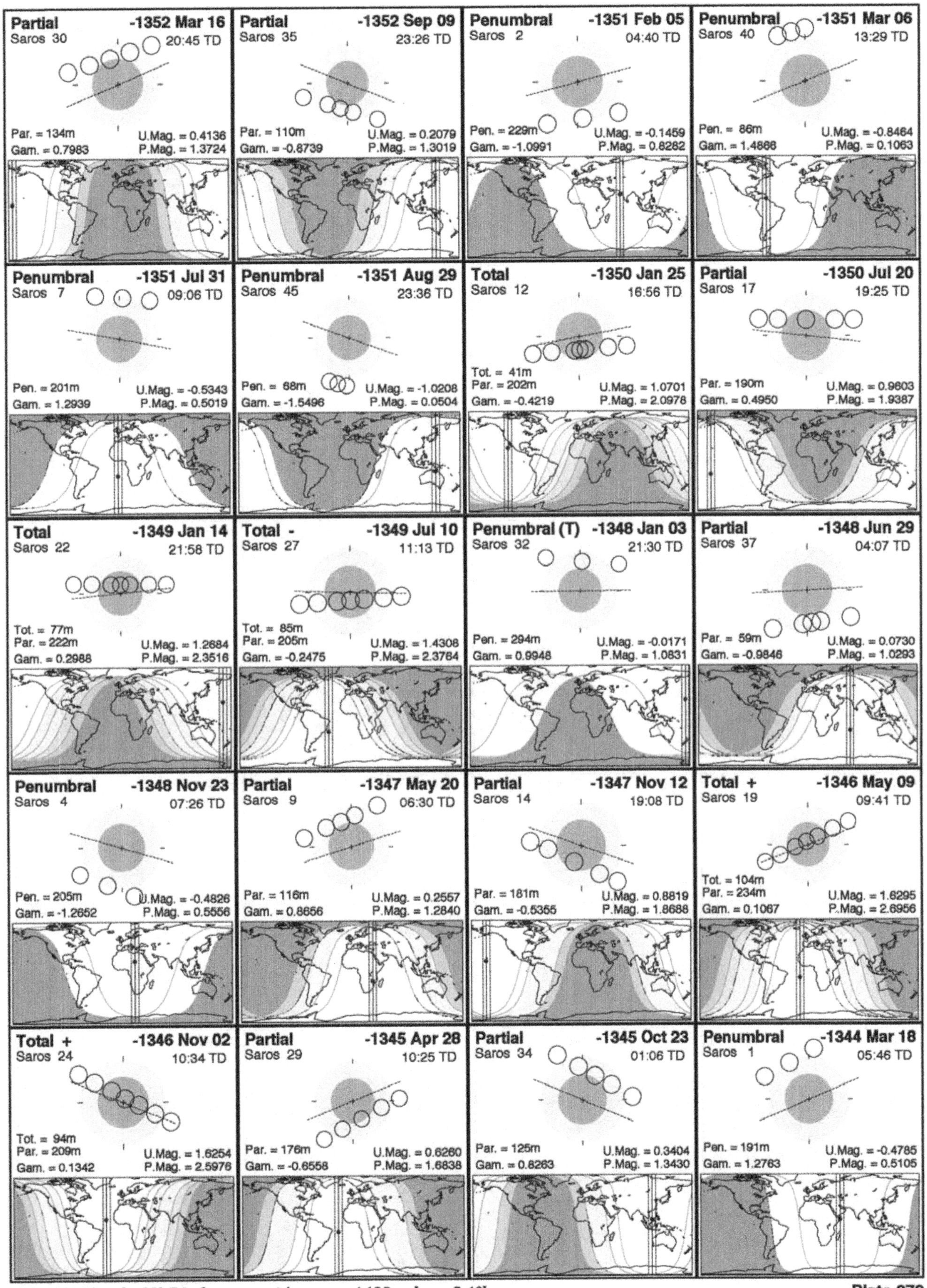

ΔT= 32031 s [= 08h54m] std.err. = ± 1462 s [= ± 6.1°] **Plate 079**

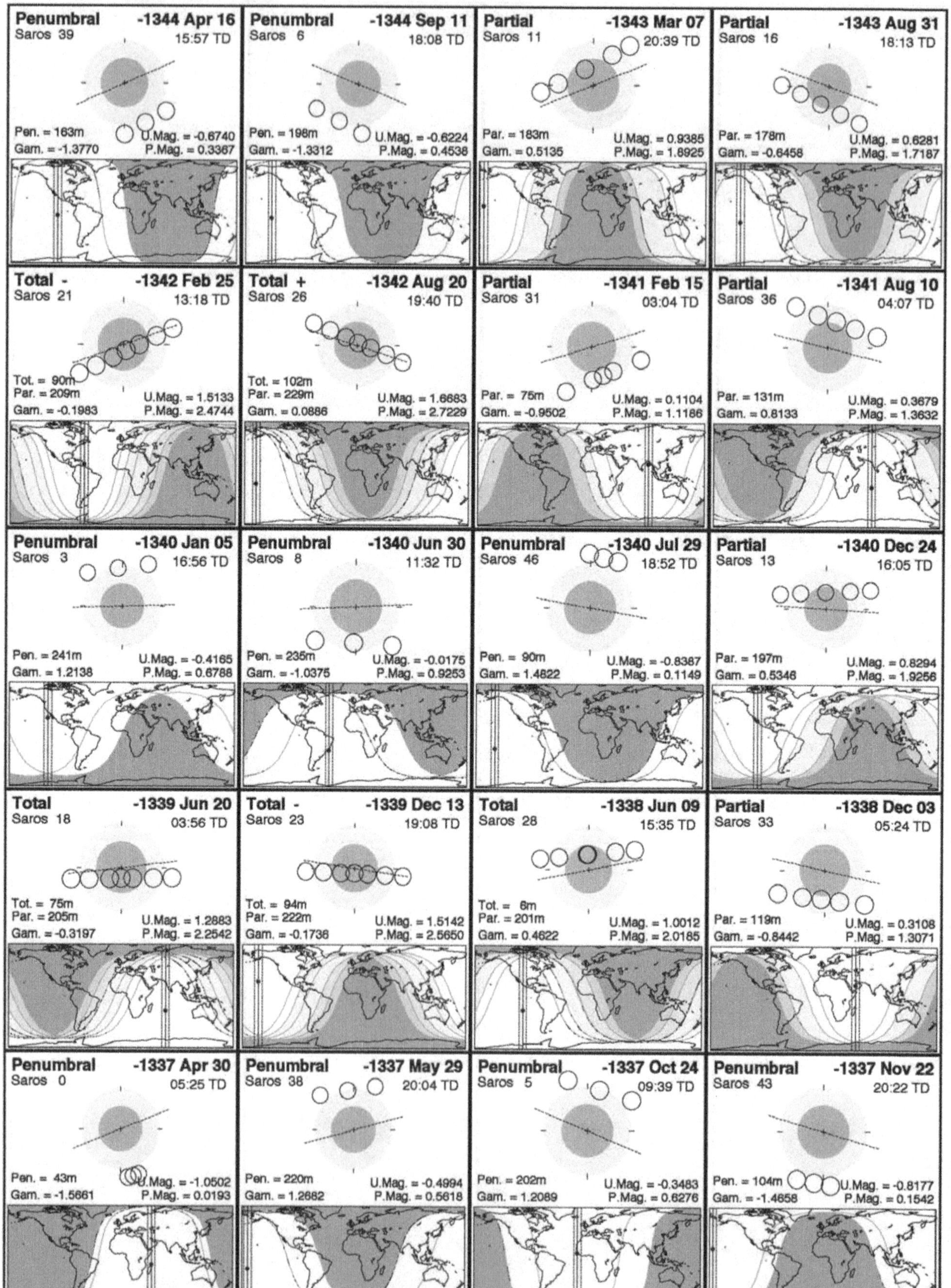

ΔT= 31868 s [= 08h51m] std.err. = ± 1439 s [= ± 6.0°] **Plate 080**

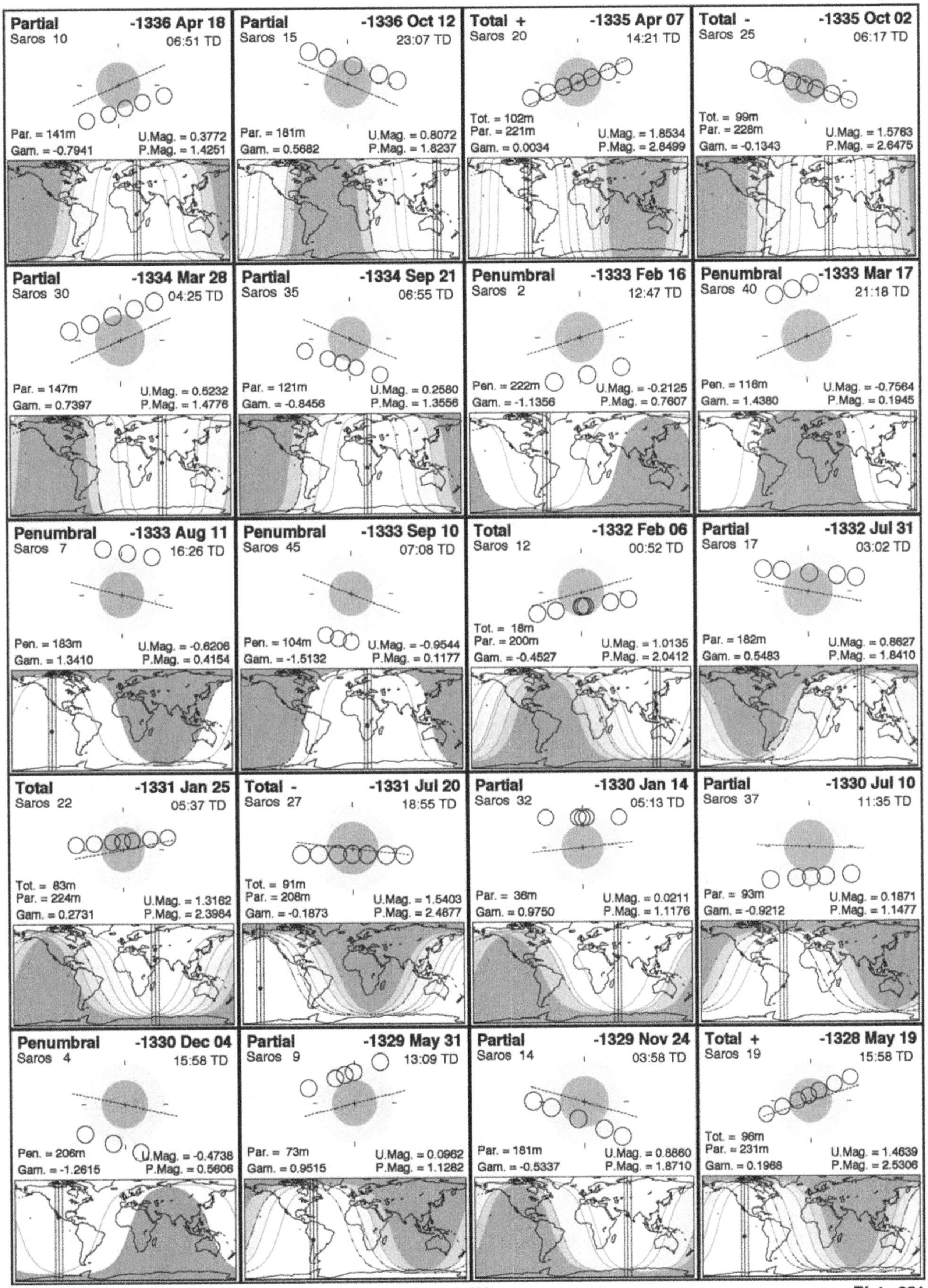

ΔT= 31707 s [= 08h48m] std.err. = ± 1417 s [= ± 5.9°] **Plate 081**

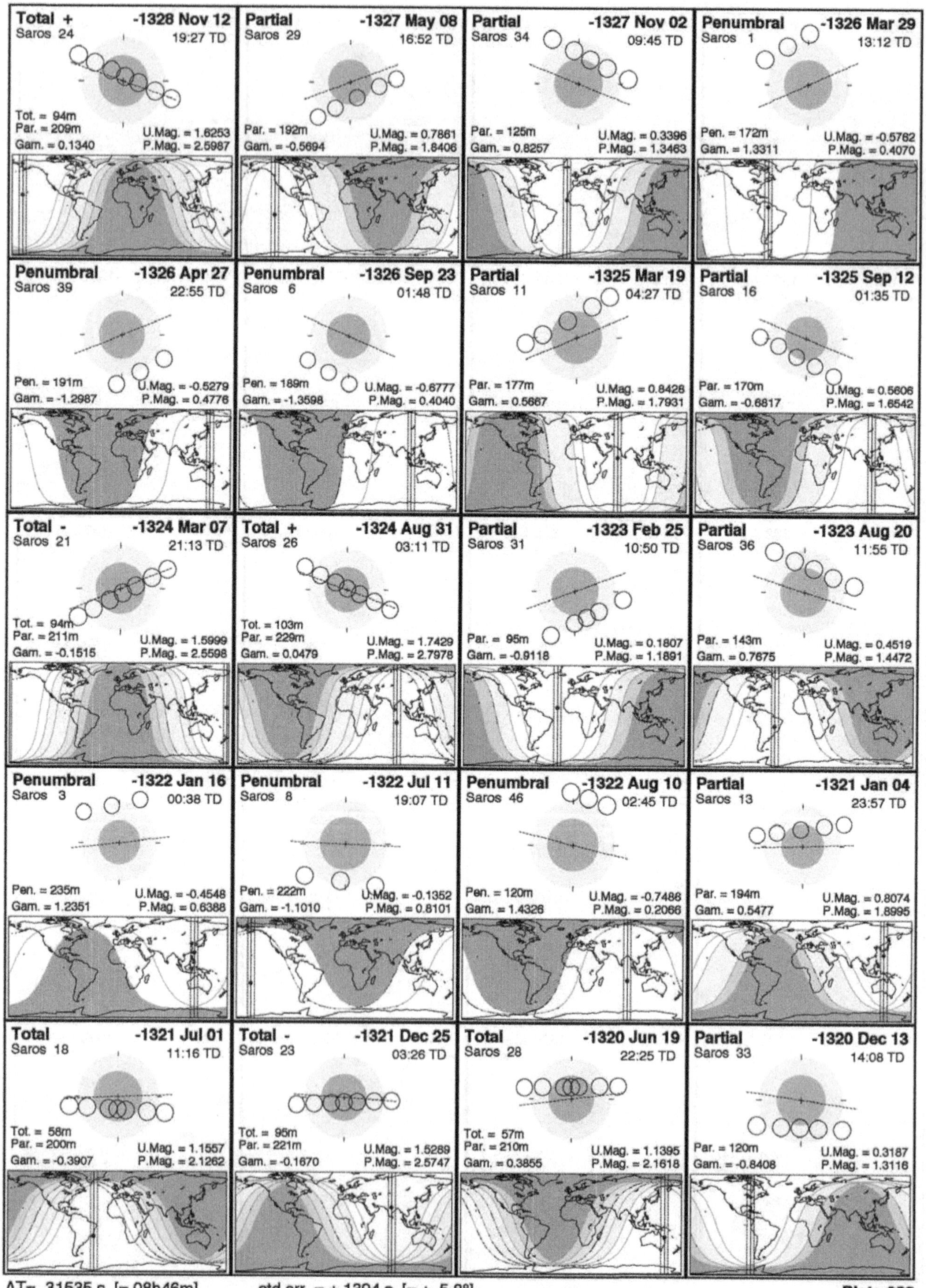

ΔT= 31535 s [= 08h46m] std.err. = ± 1394 s [= ± 5.8°] **Plate 082**

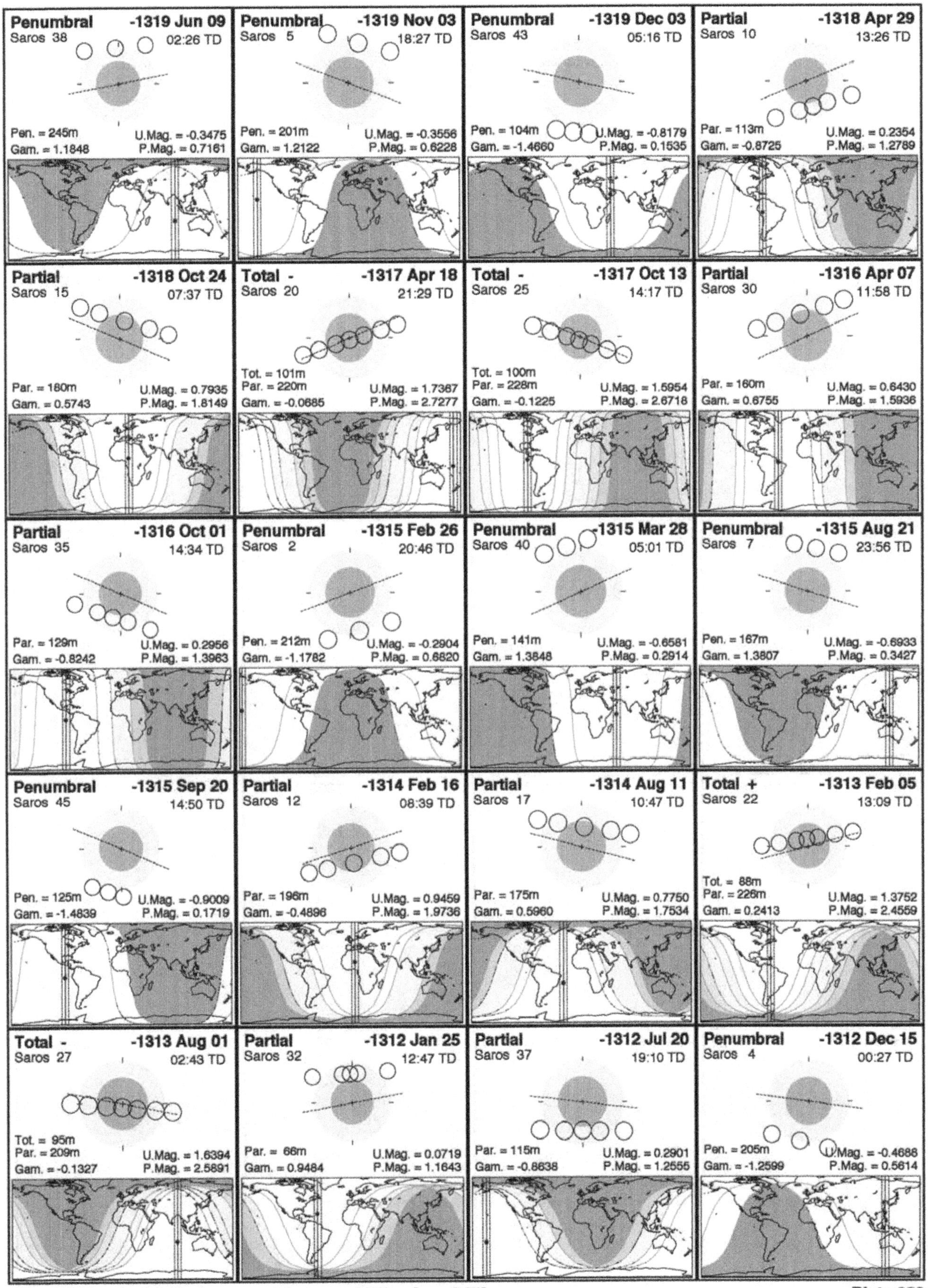

ΔT= 31363 s [= 08h43m] std.err. = ± 1370 s [= ± 5.7°] **Plate 083**

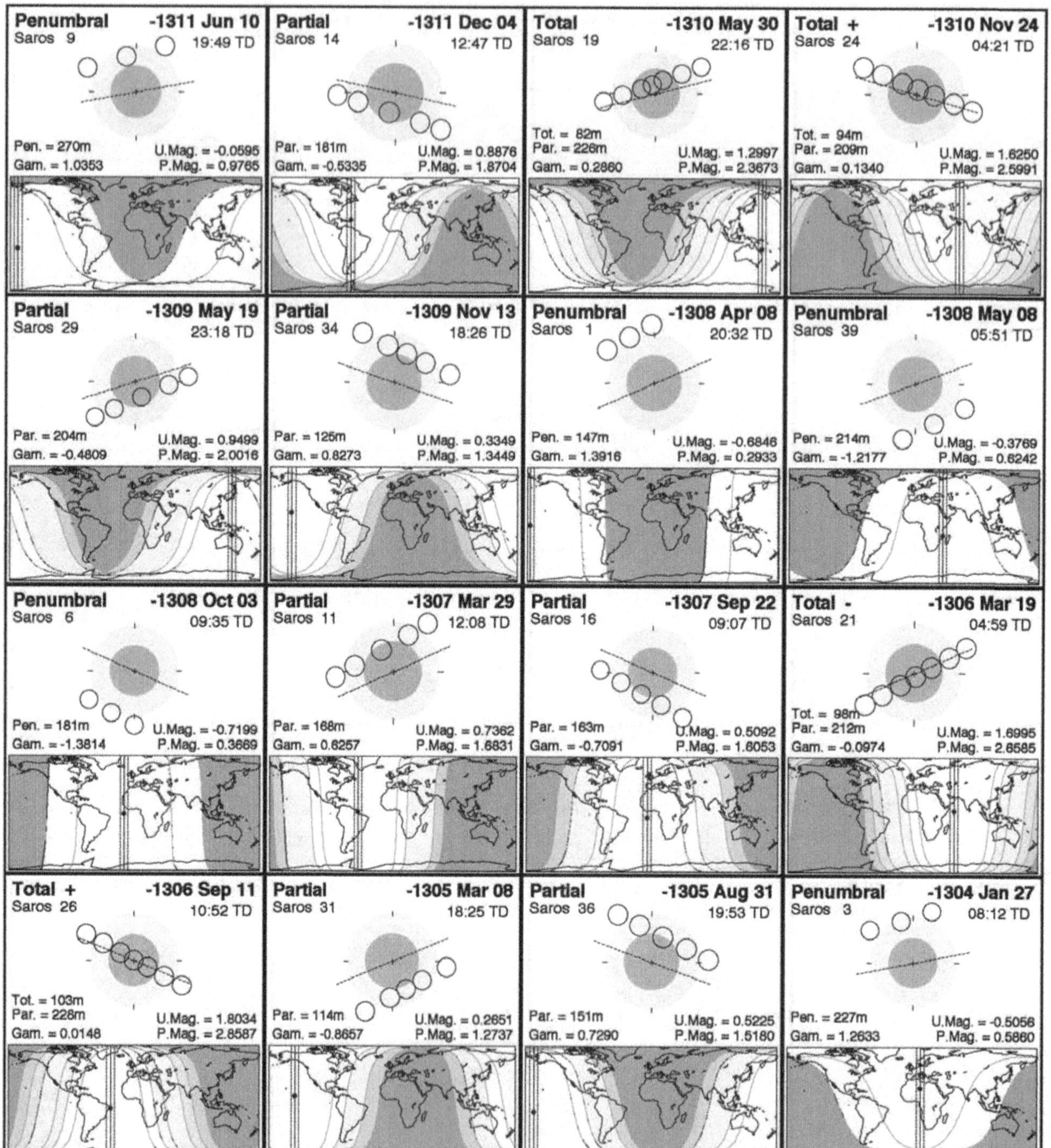

ΔT= 31203 s [= 08h40m] std.err. = ± 1349 s [= ± 5.6°] **Plate 084**

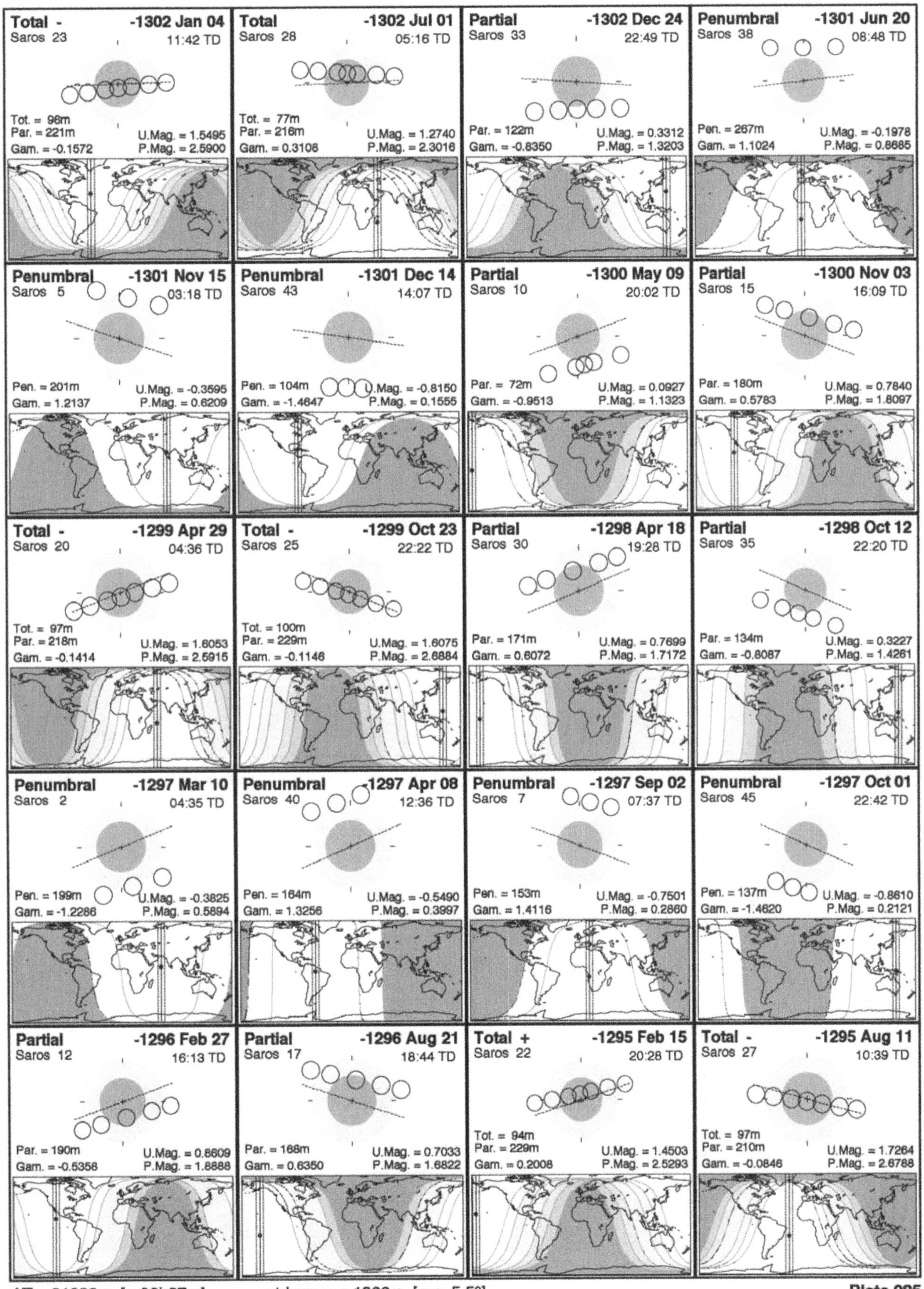

ΔT= 31033 s [= 08h37m] std.err. = ± 1326 s [= ± 5.5°] **Plate 085**

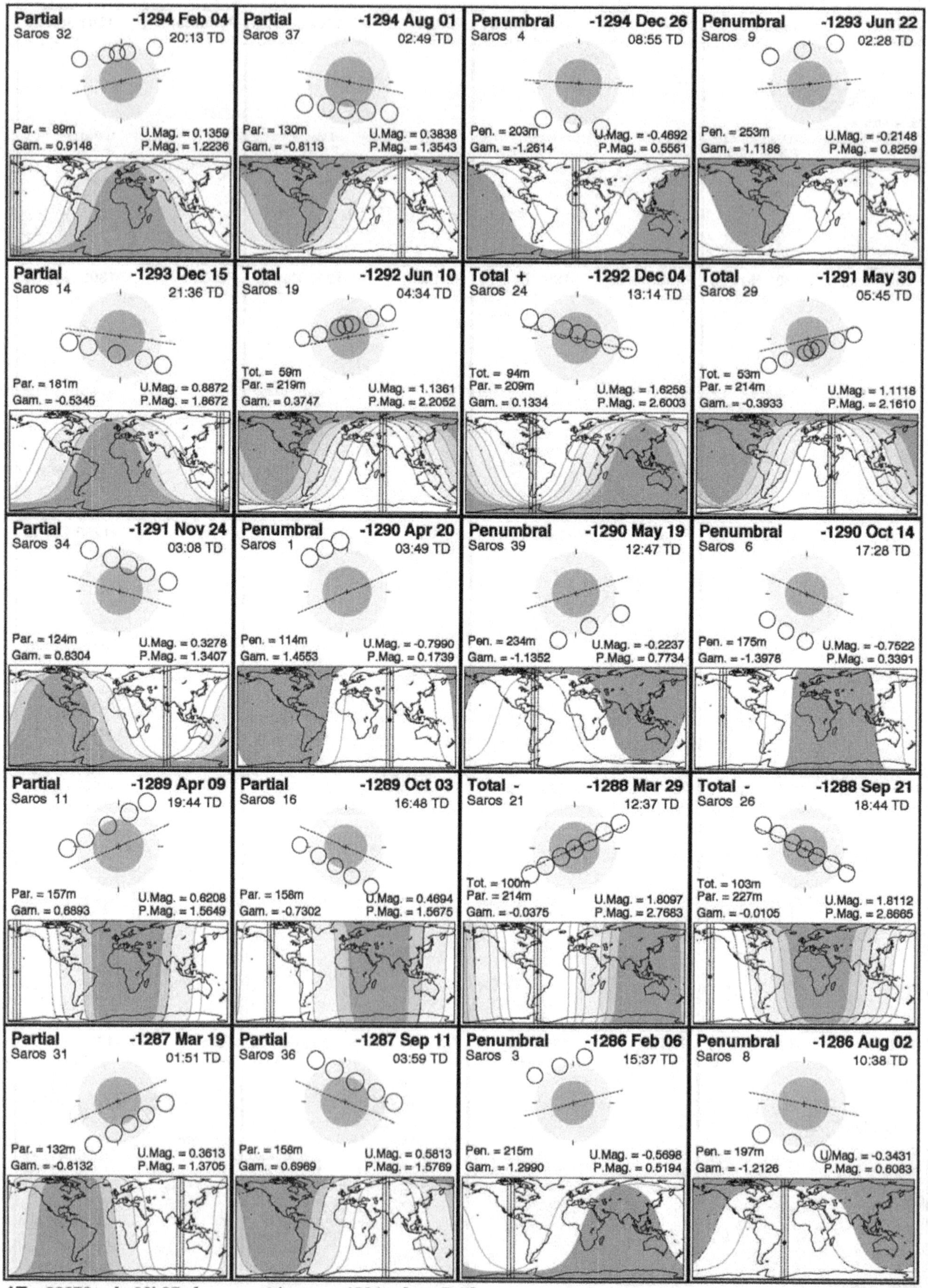

ΔT= 30872 s [= 08h35m] std.err. = ± 1304 s [= ± 5.4°] **Plate 086**

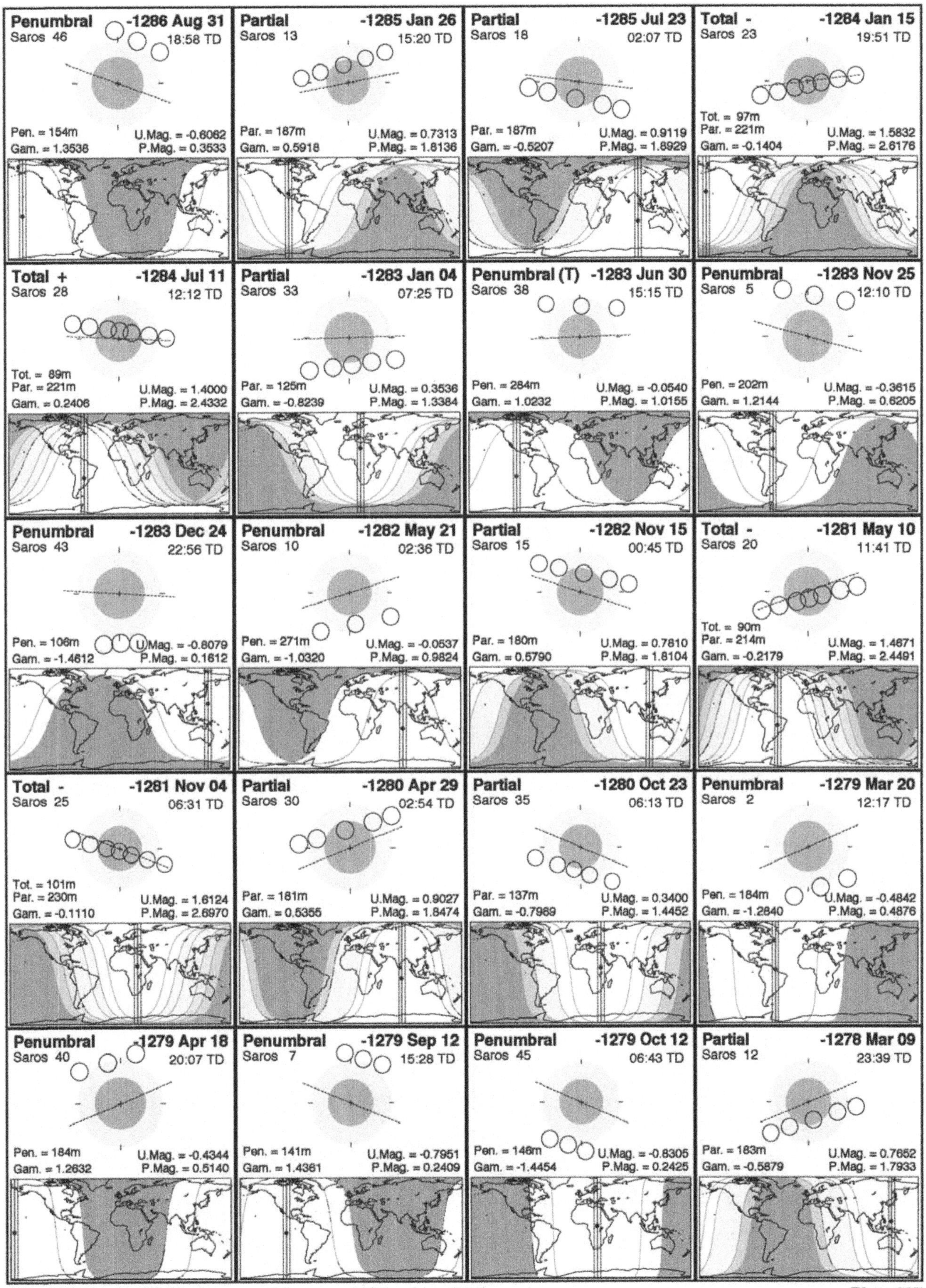

ΔT= 30702 s [= 08h32m] std.err. = ± 1281 s [= ± 5.3°] Plate 087

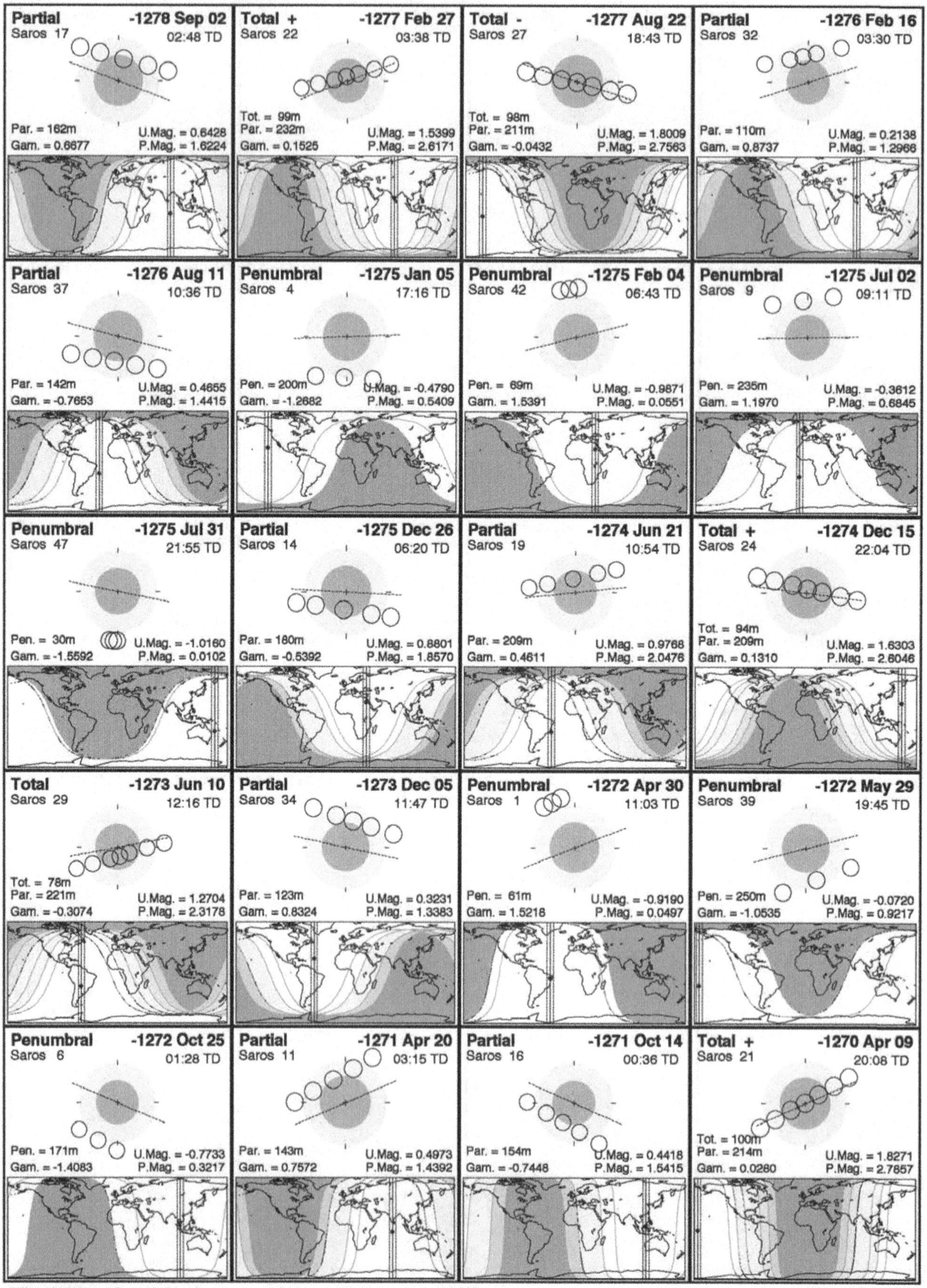

ΔT= 30544 s [= 08h29m] std.err. = ± 1260 s [= ± 5.3°] **Plate 088**

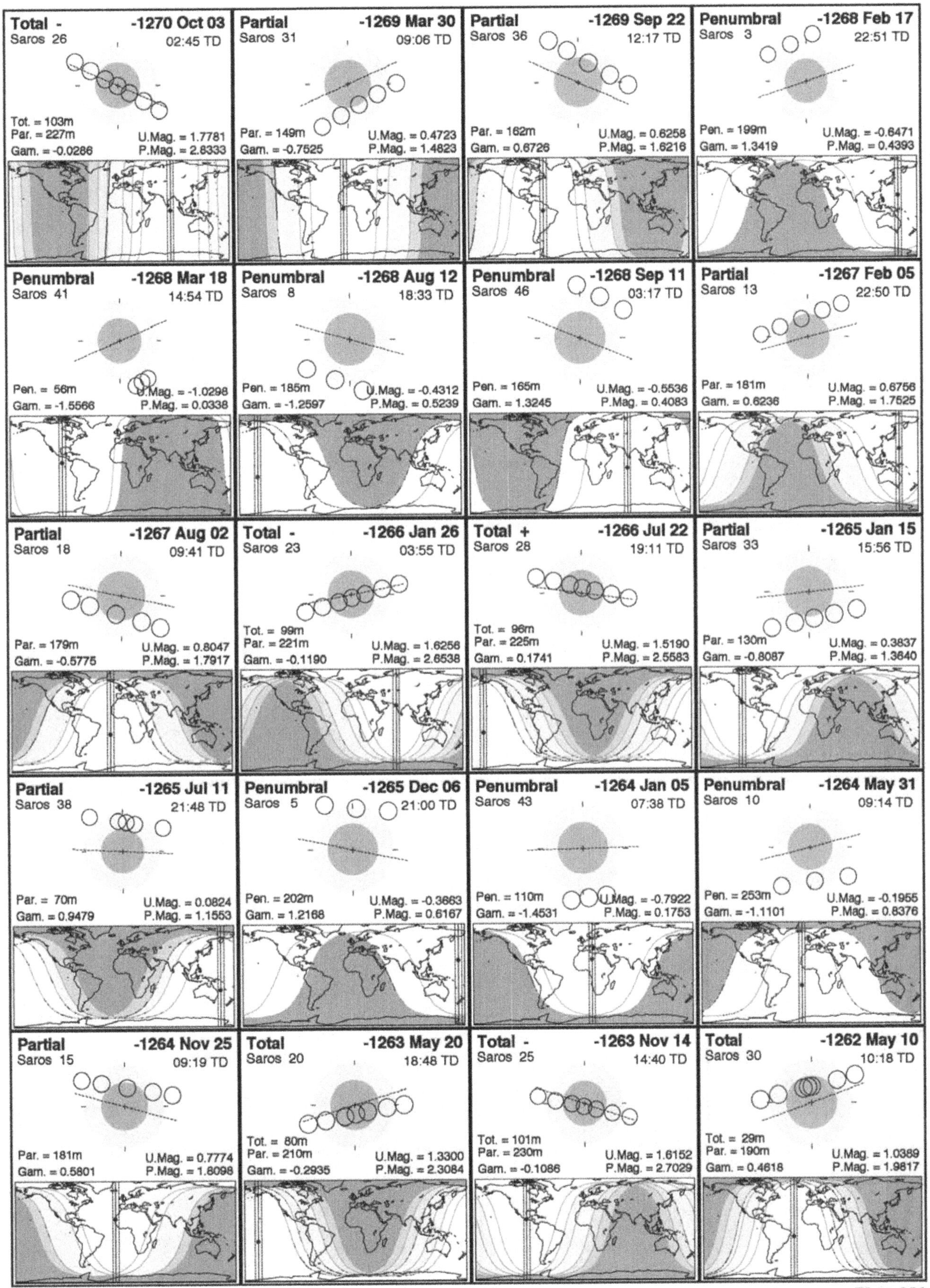

ΔT= 30385 s [= 08h26m] std.err. = ± 1239 s [= ± 5.2°] Plate 089

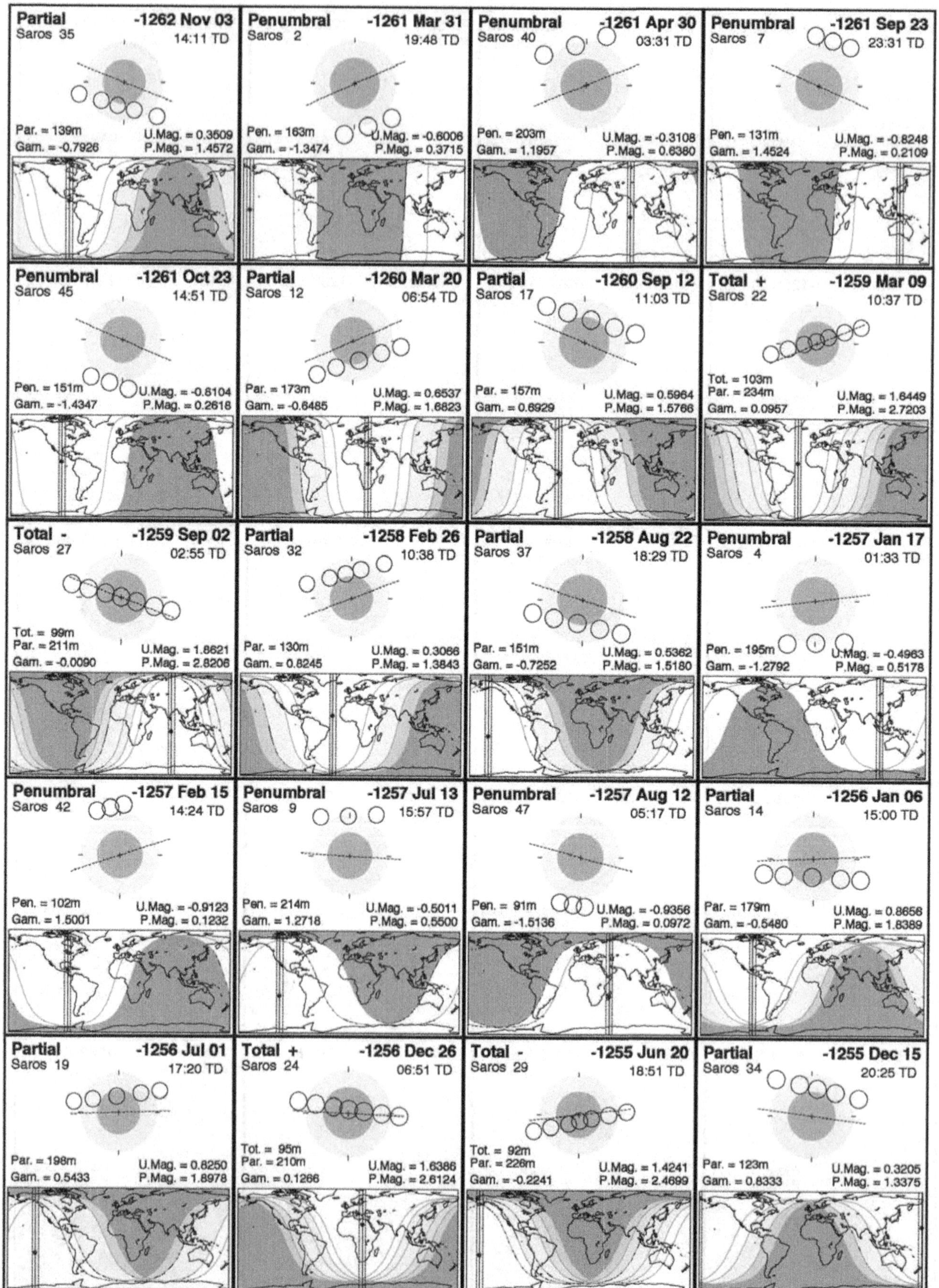

ΔT= 30226 s [= 08h24m] std.err. = ± 1218 s [= ± 5.1°] **Plate 090**

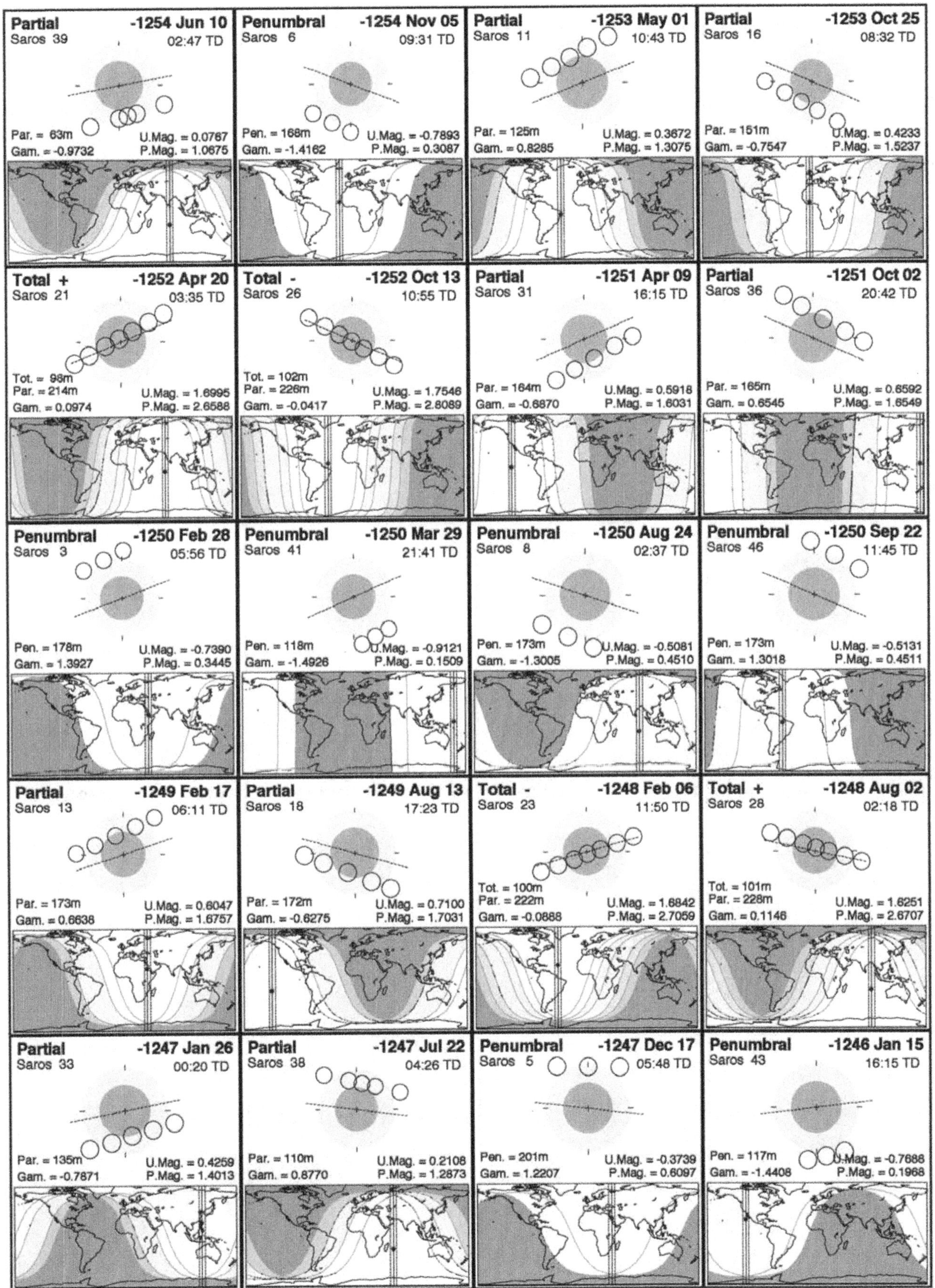

ΔT= 30077 s [= 08h21m] std.err. = ± 1198 s [= ± 5.0°] **Plate 091**

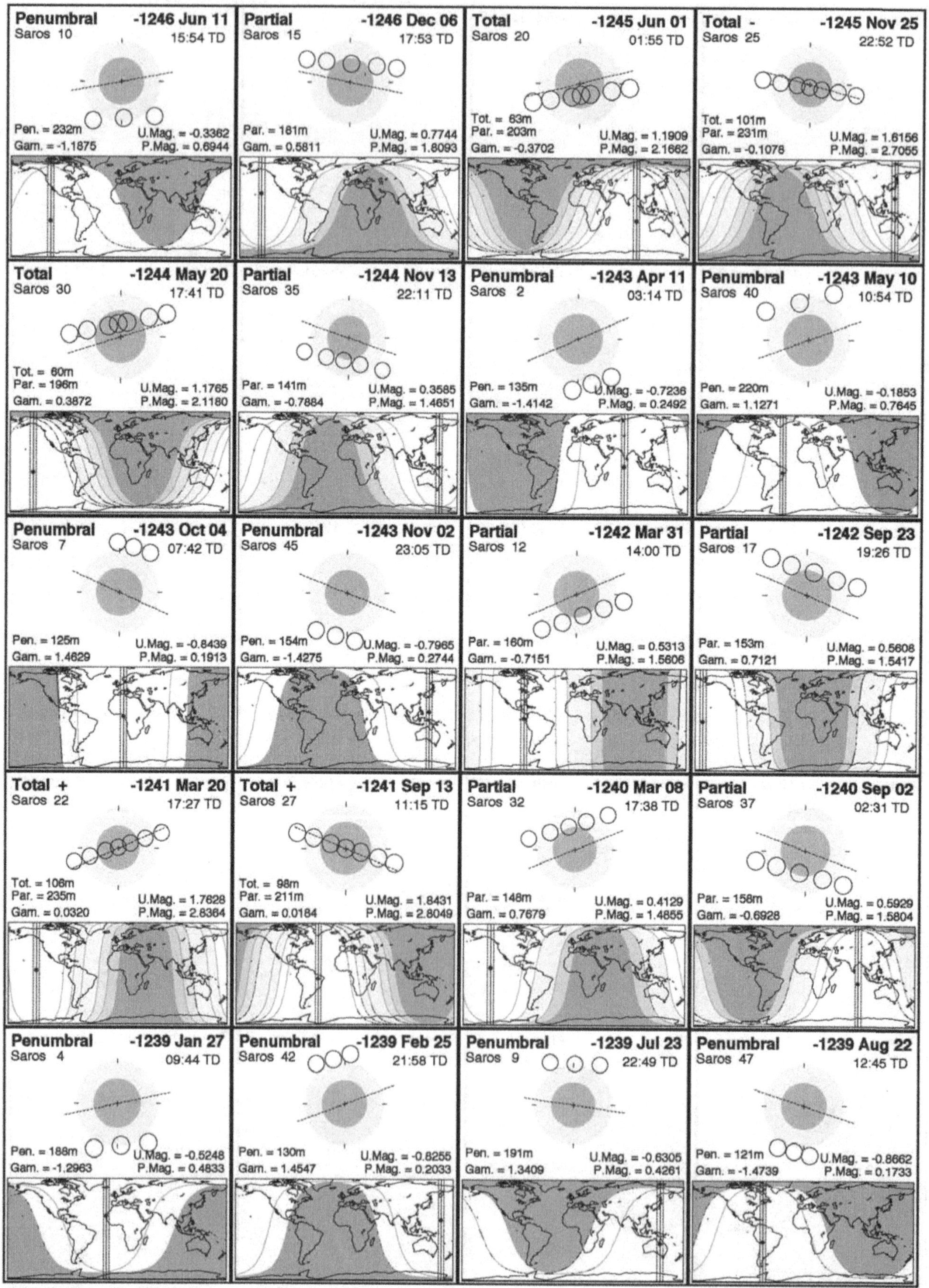

ΔT= 29920 s [= 08h19m] std.err. = ± 1178 s [= ± 4.9°] **Plate 092**

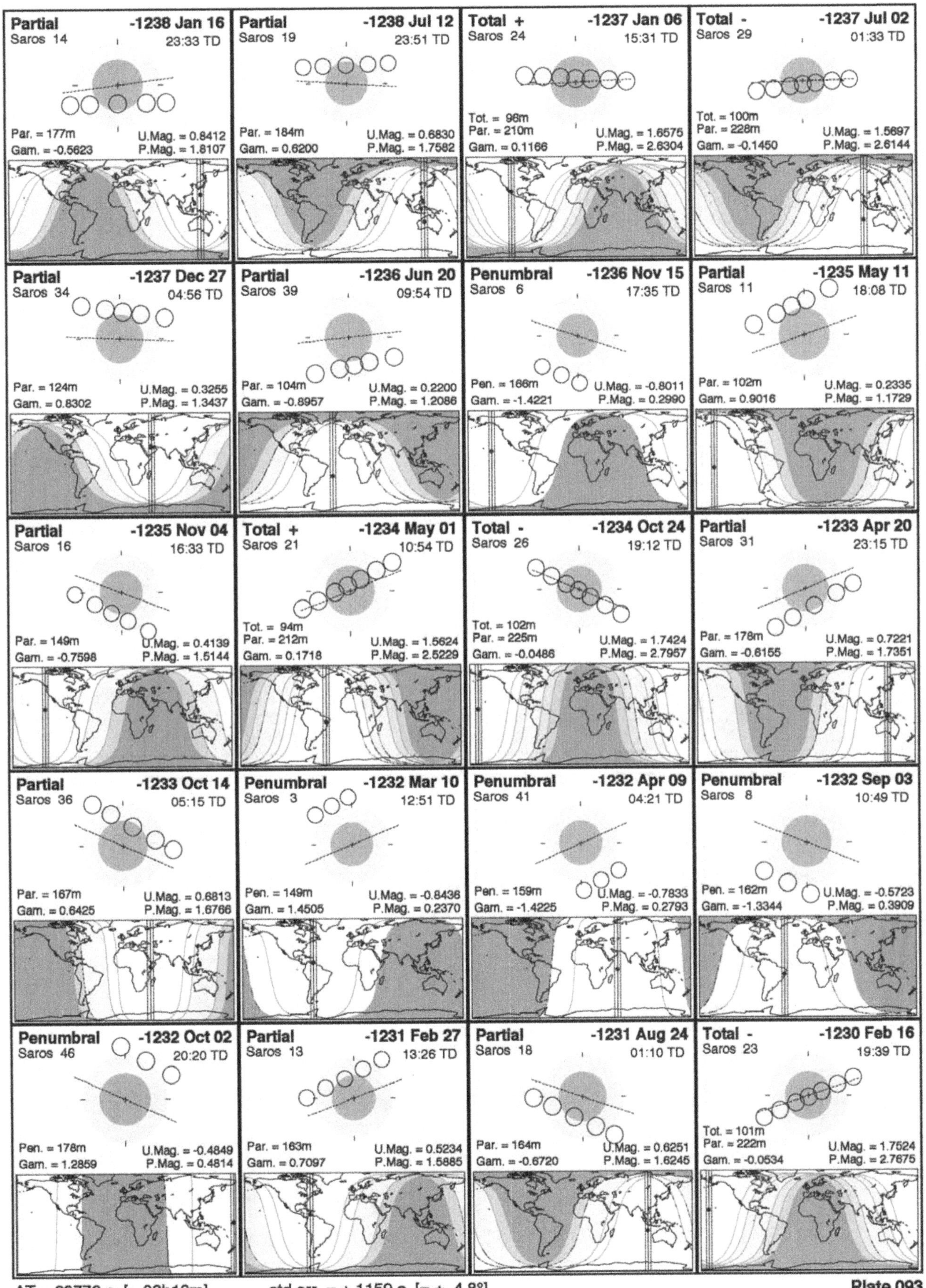

ΔT= 29772 s [= 08h16m] std.err. = ± 1159 s [= ± 4.8°] **Plate 093**

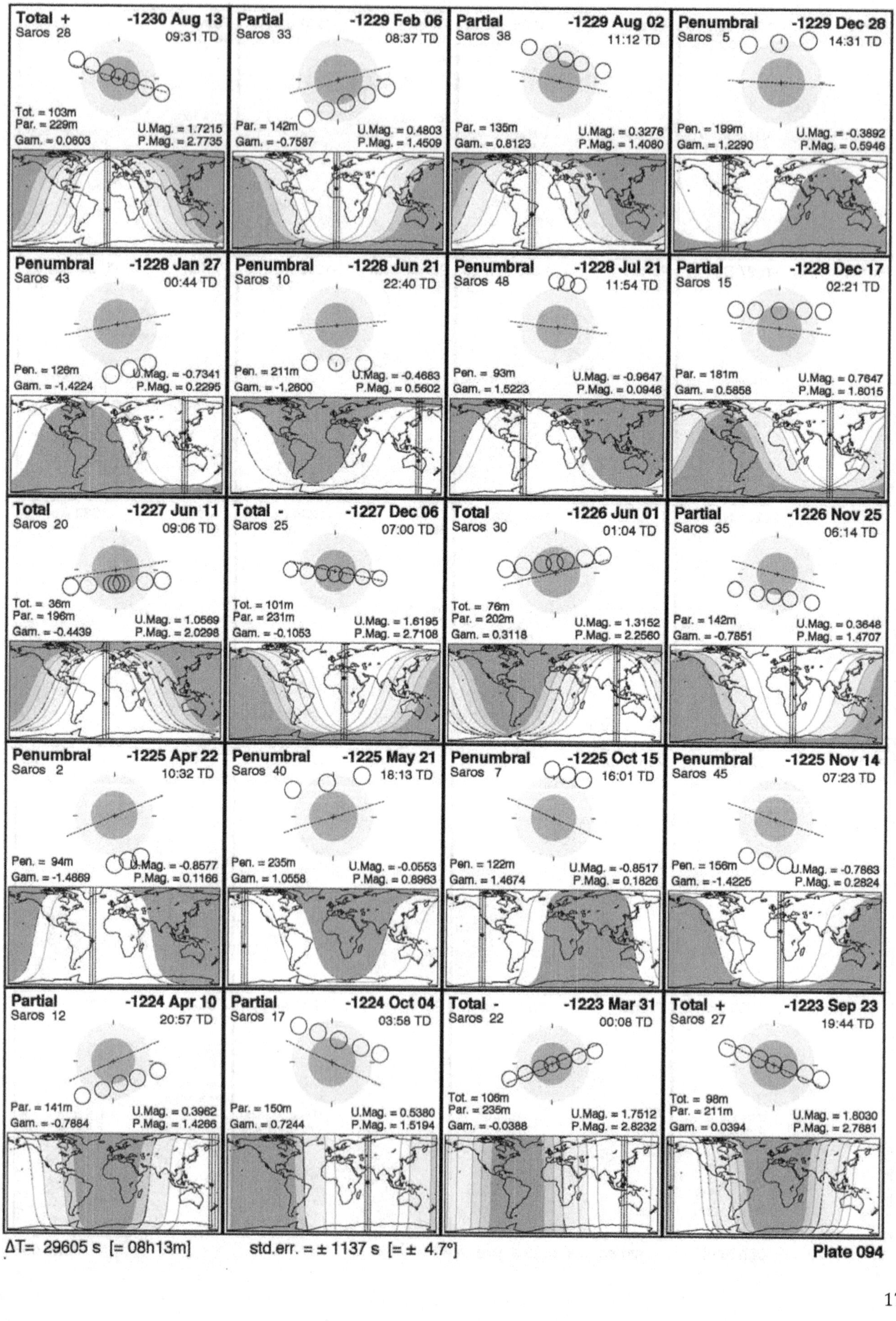

ΔT= 29605 s [= 08h13m] std.err. = ± 1137 s [= ± 4.7°] **Plate 094**

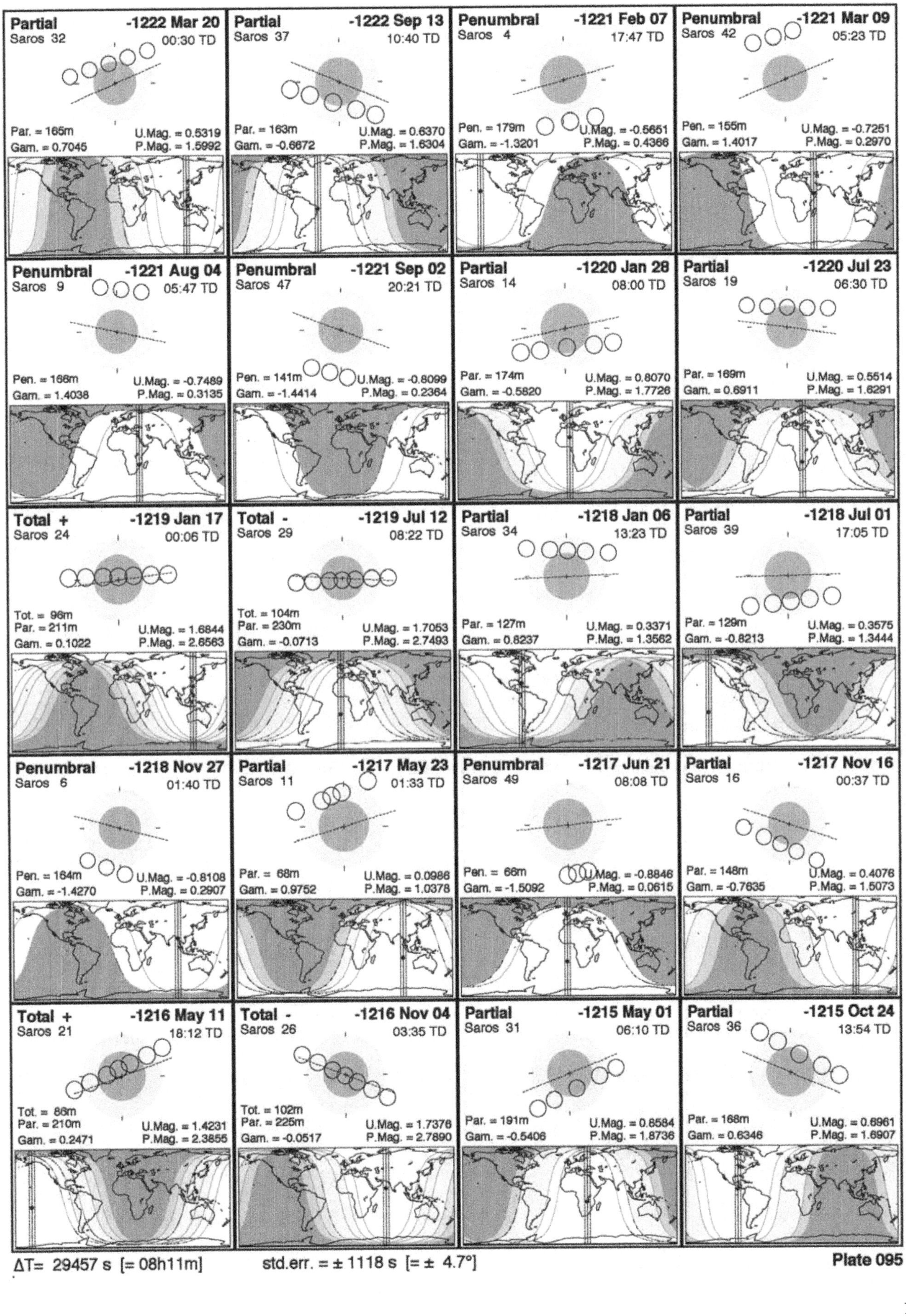

ΔT= 29457 s [= 08h11m] std.err. = ± 1118 s [= ± 4.7°] **Plate 095**

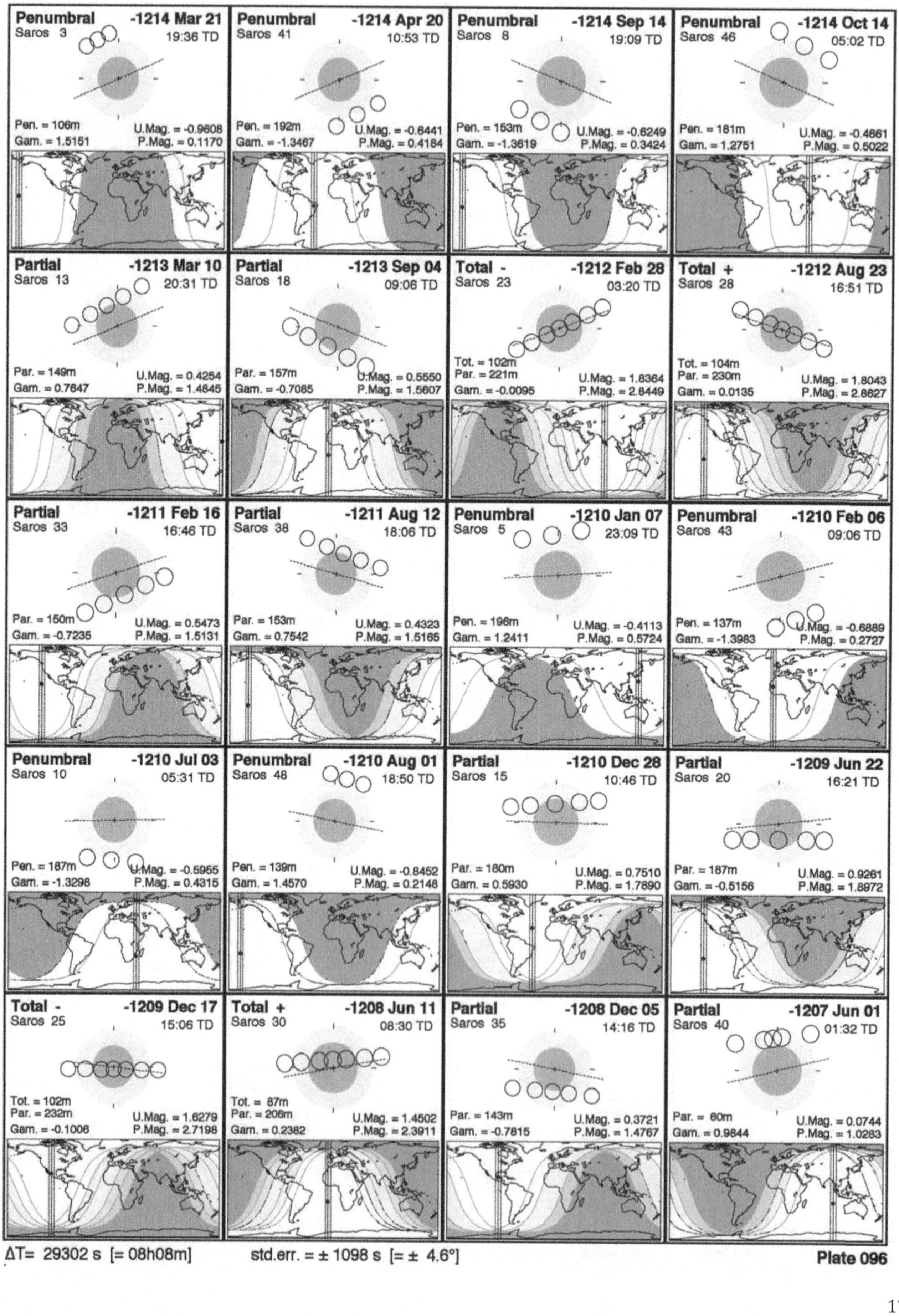

ΔT= 29302 s [= 08h08m] std.err. = ± 1098 s [= ± 4.6°] **Plate 096**

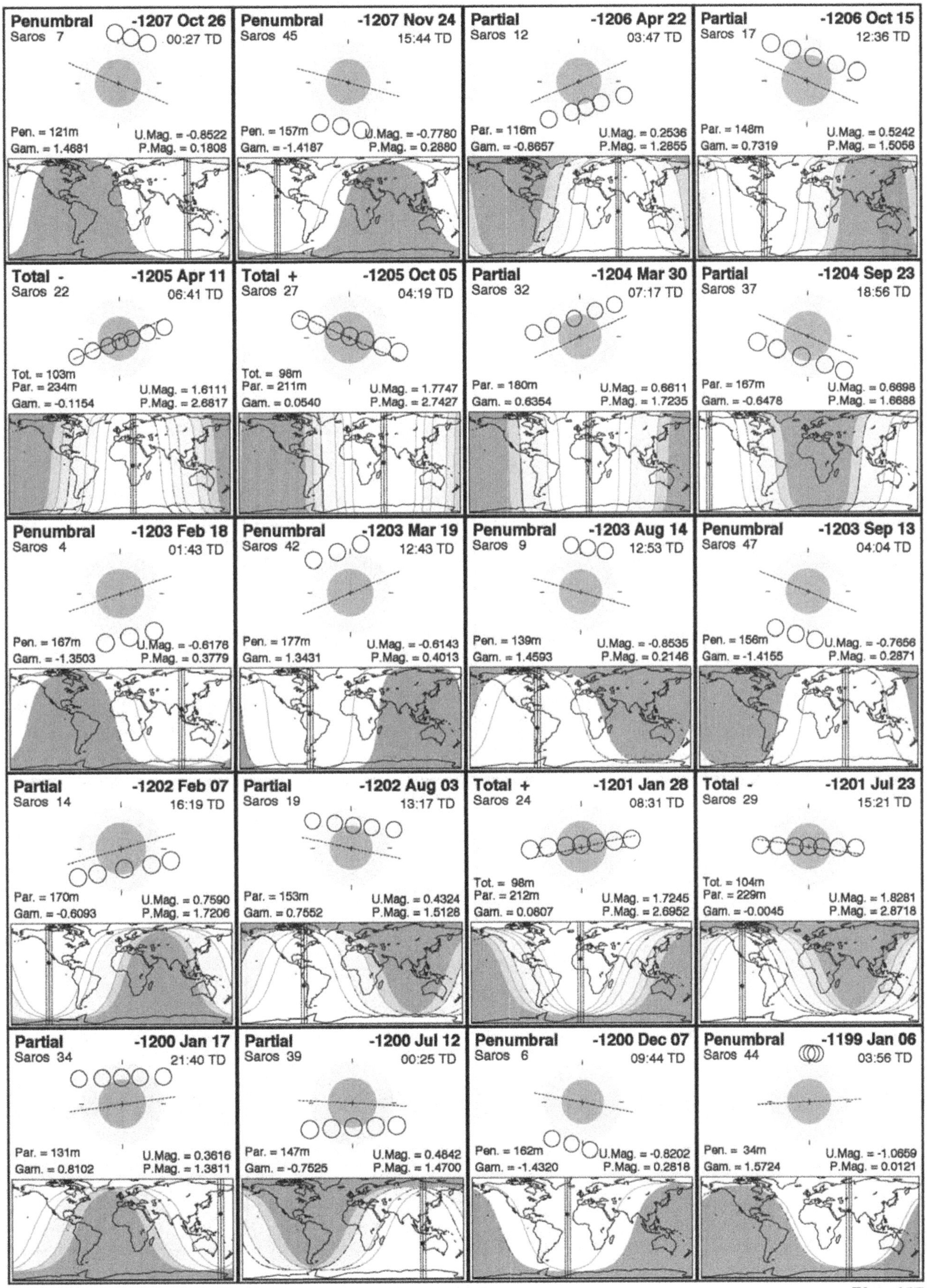

ΔT= 29156 s [= 08h06m] std.err. = ± 1079 s [= ± 4.5°] Plate 097

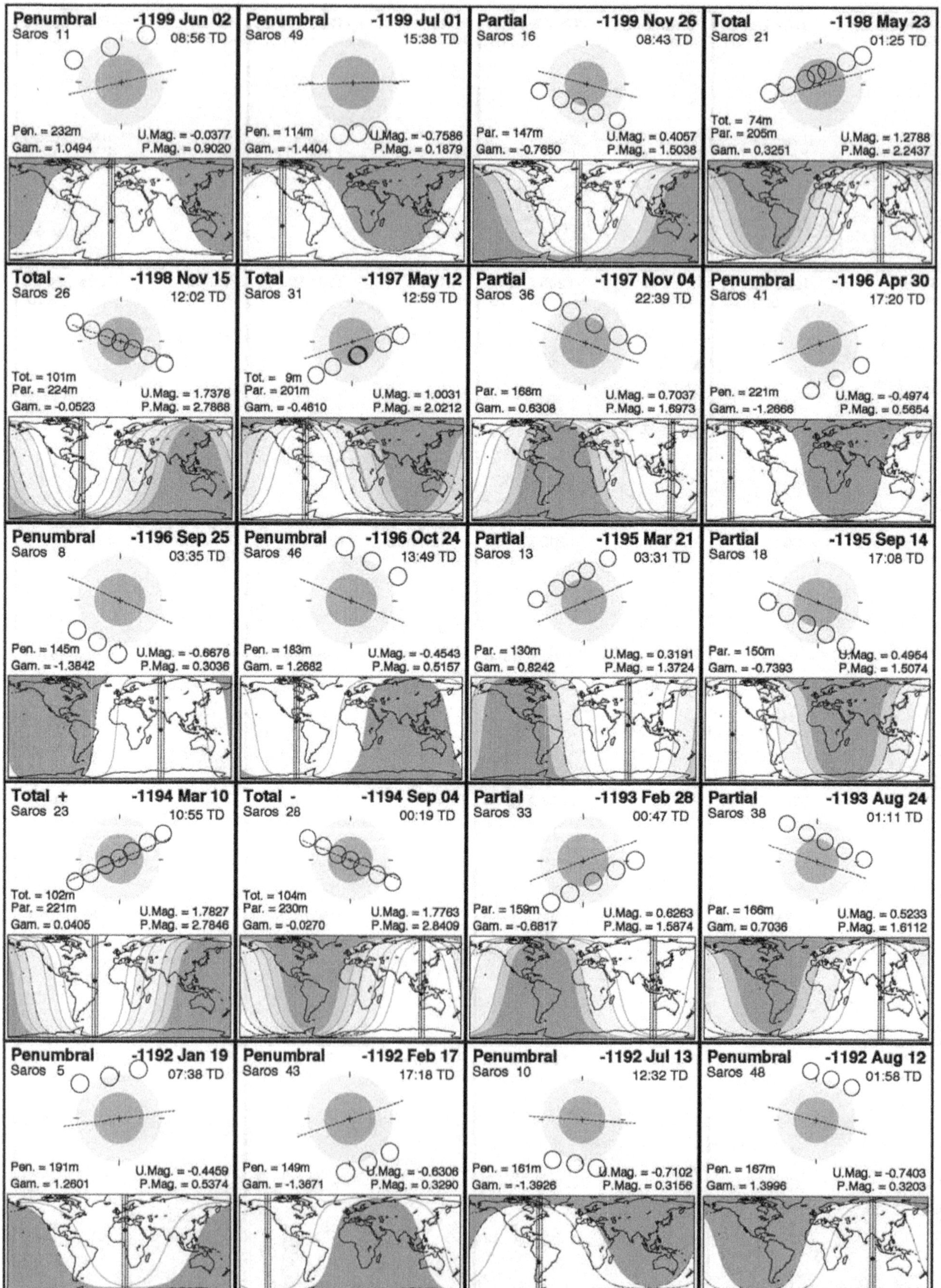

ΔT= 29009 s [= 08h03m] std.err. = ± 1061 s [= ± 4.4°] **Plate 098**

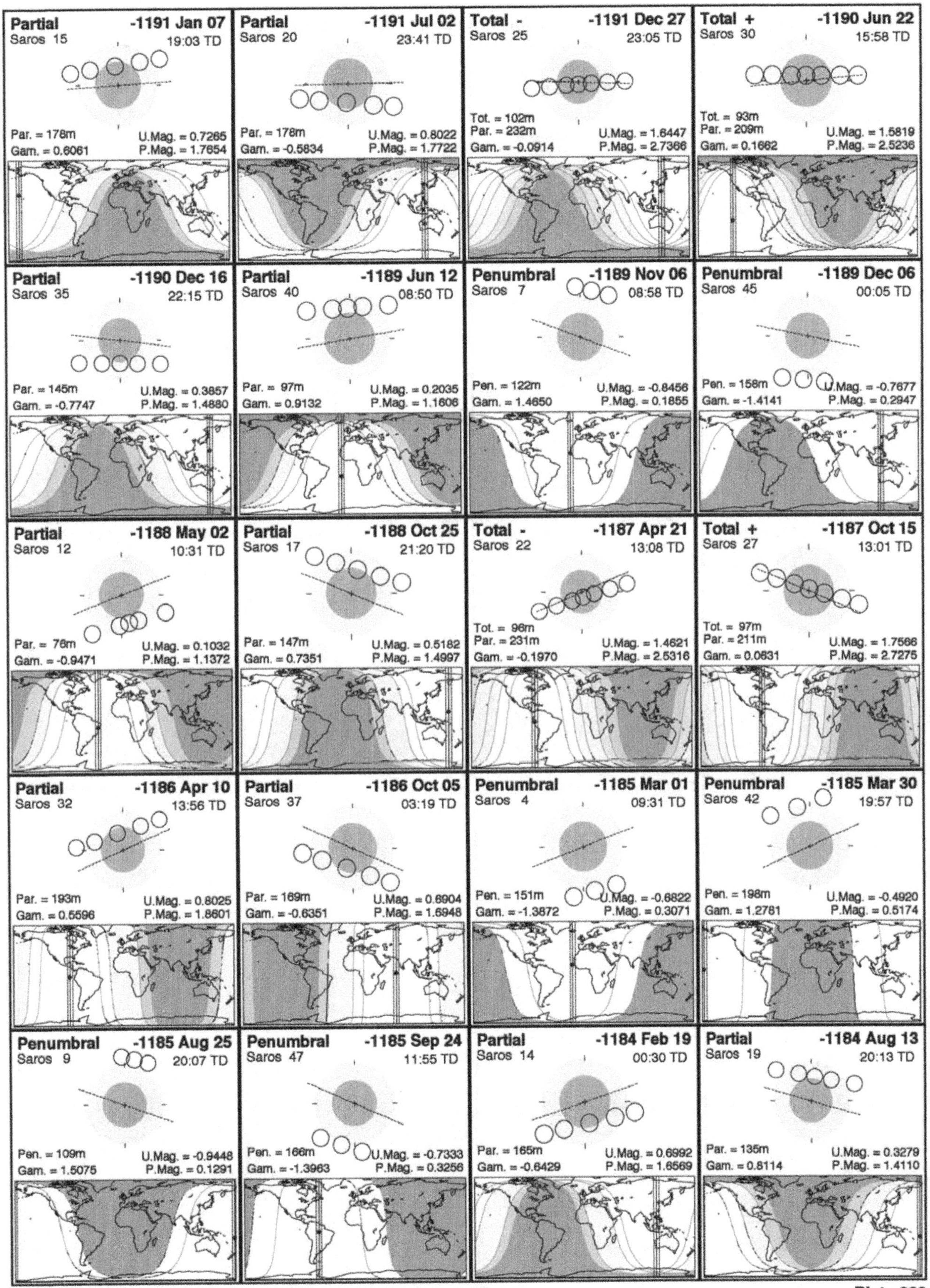

ΔT= 28863 s [= 08h01m] std.err. = ± 1042 s [= ± 4.3°] **Plate 099**

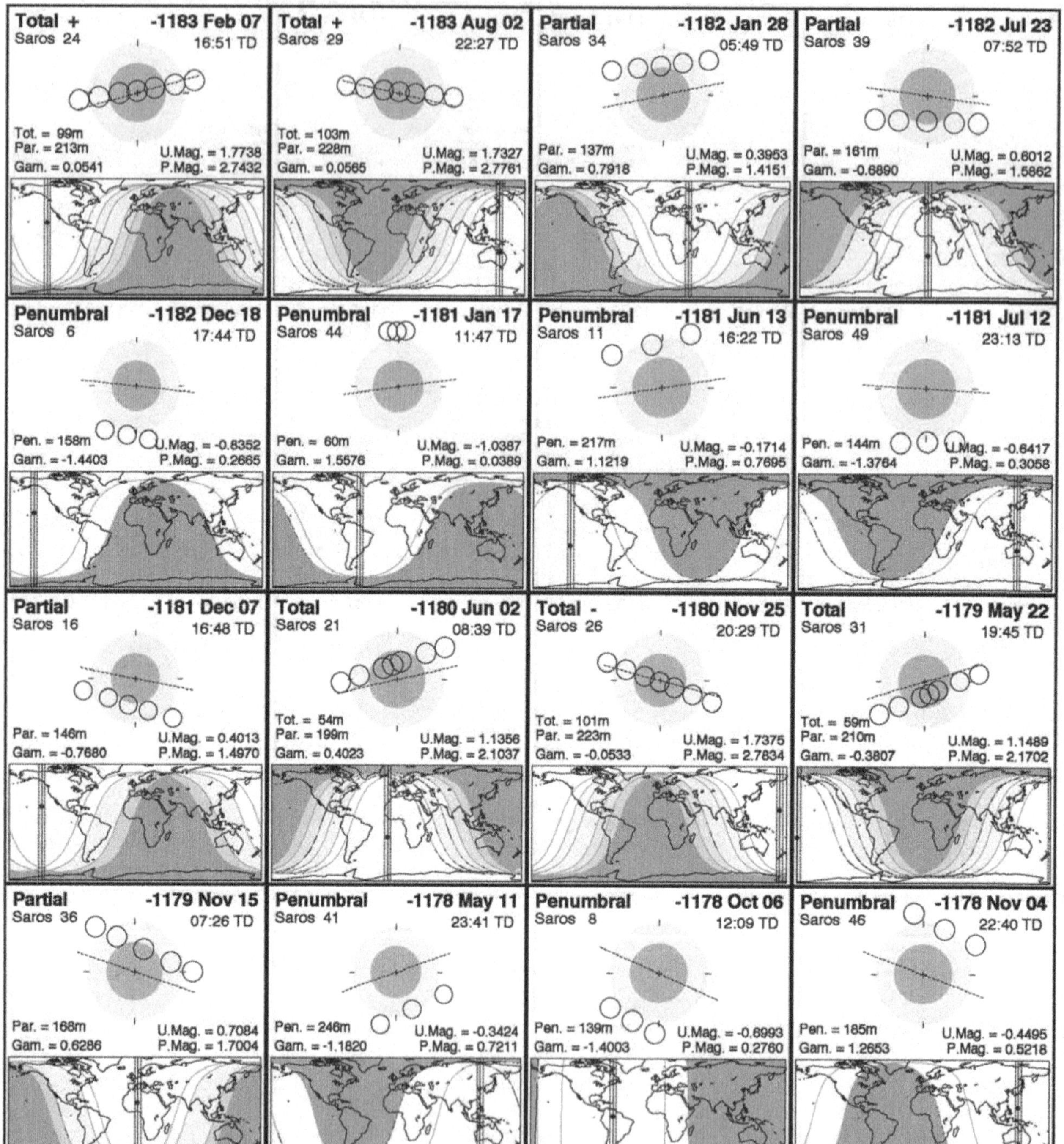

ΔT= 28708 s [= 07h58m] std.err. = ± 1023 s [= ± 4.3°] **Plate 100**

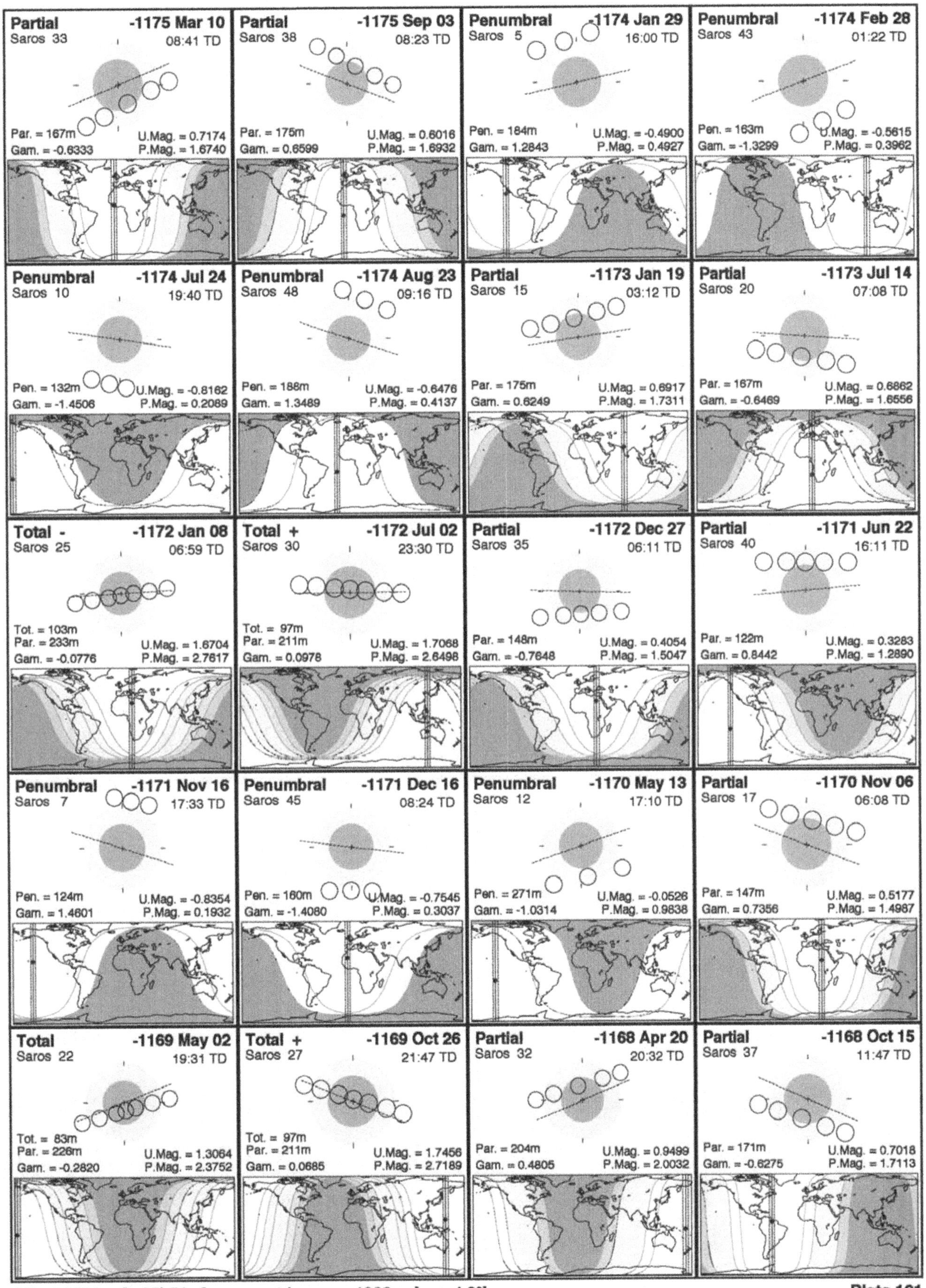

ΔT= 28554 s [= 07h56m] std.err. = ± 1003 s [= ± 4.2°] **Plate 101**

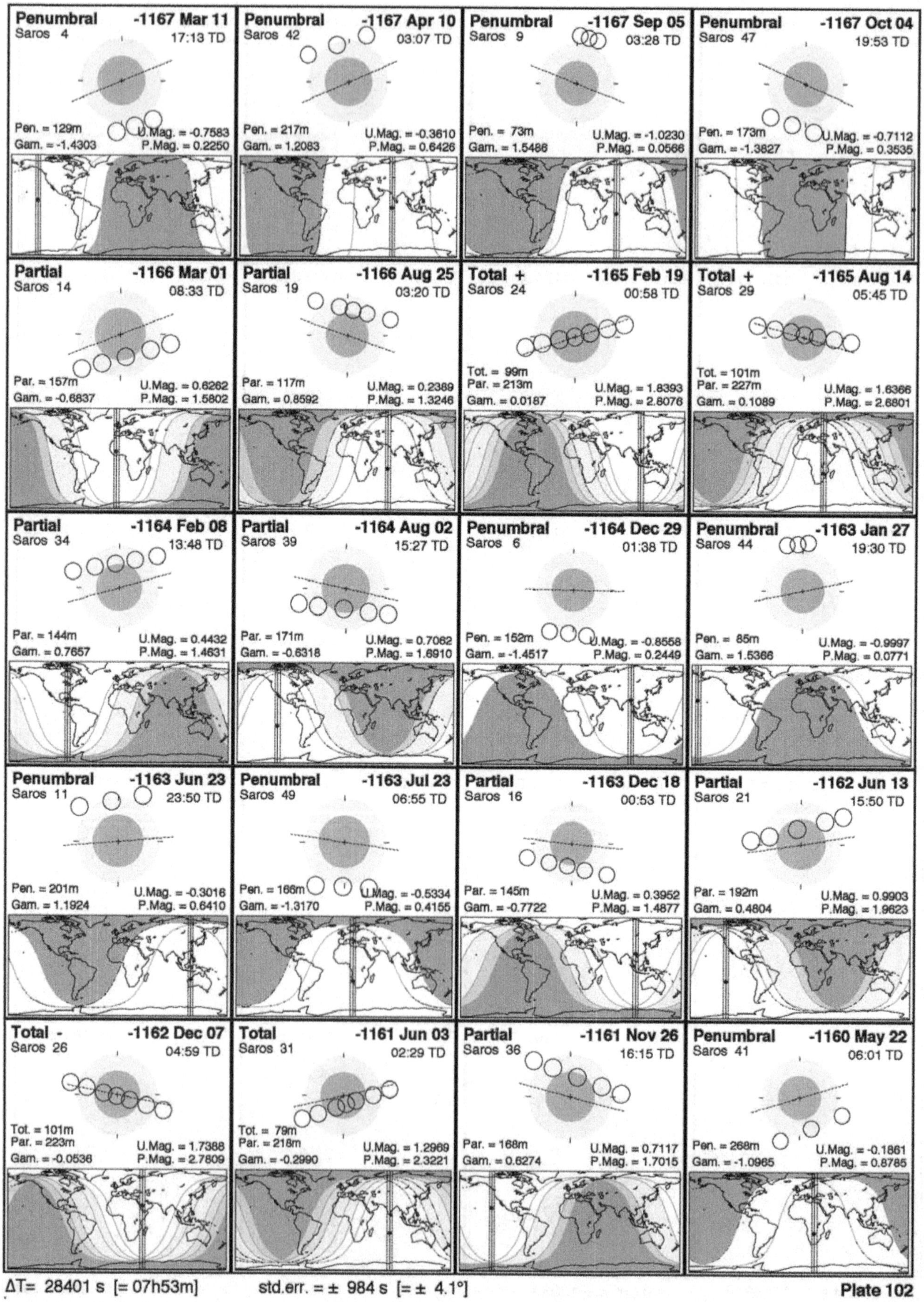

ΔT= 28401 s [= 07h53m] std.err. = ± 984 s [= ± 4.1°] **Plate 102**

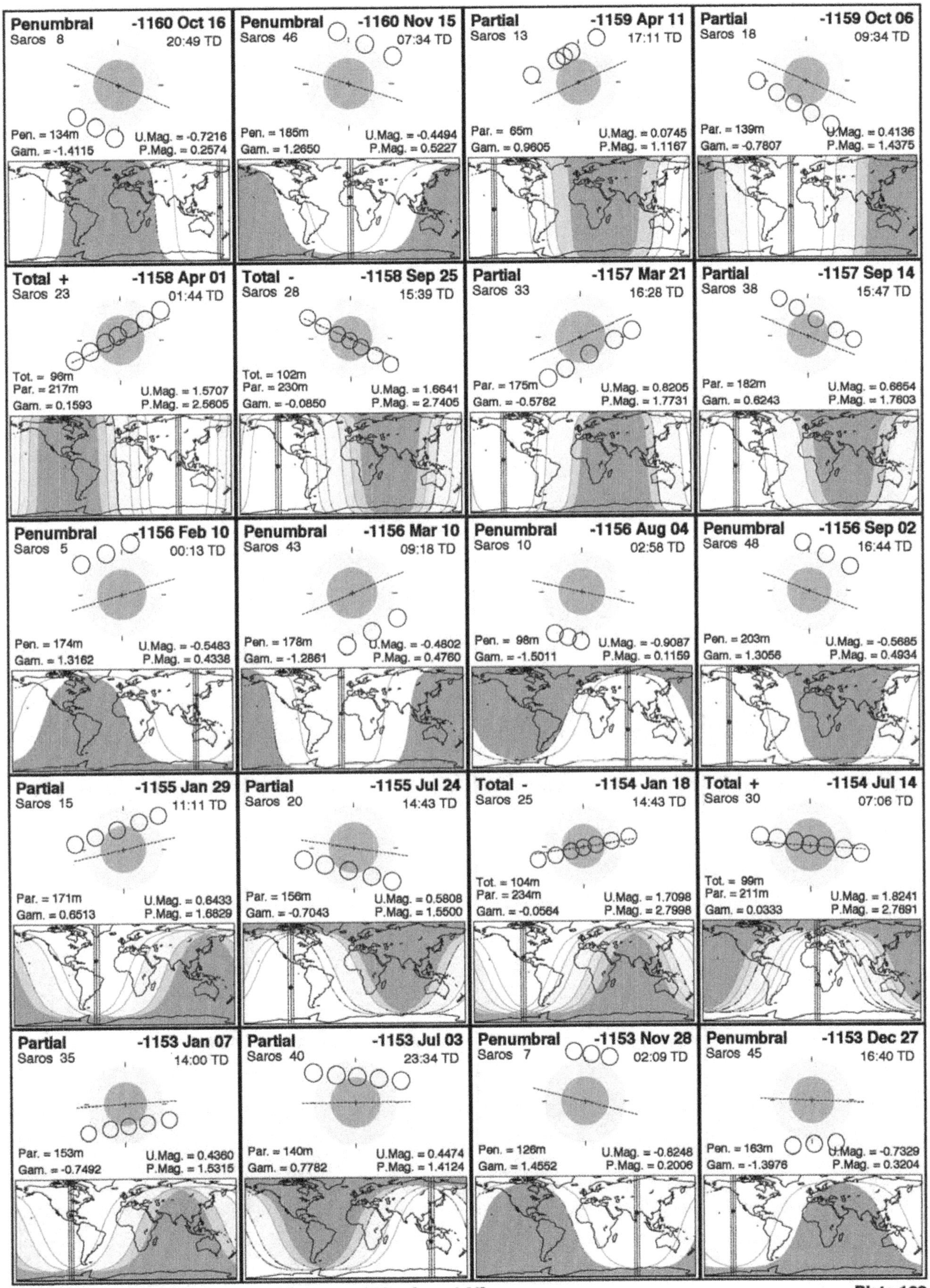

ΔT= 28257 s [= 07h51m] std.err. = ± 966 s [= ± 4.0°] **Plate 103**

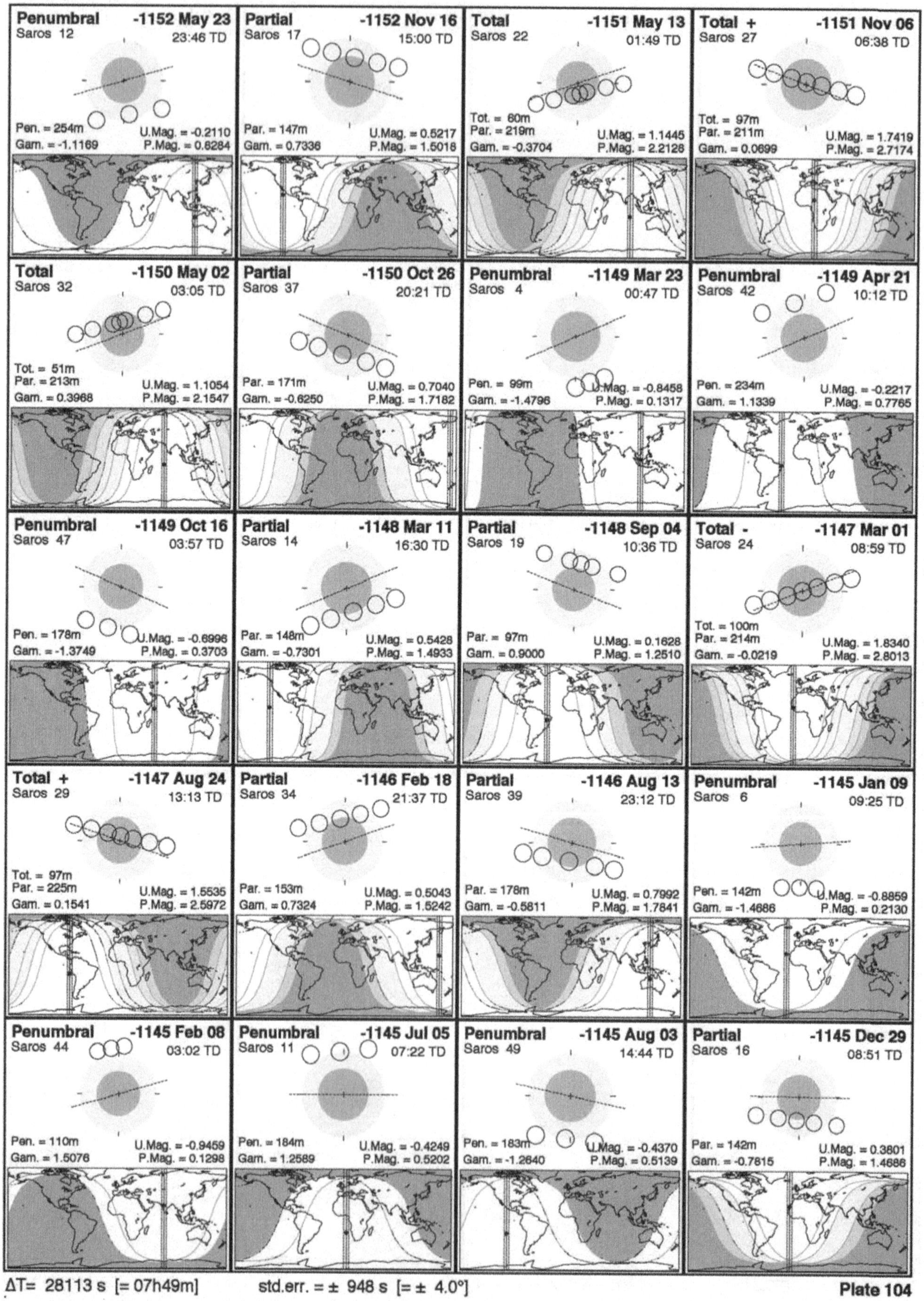

ΔT= 28113 s [= 07h49m] std.err. = ± 948 s [= ± 4.0°] **Plate 104**

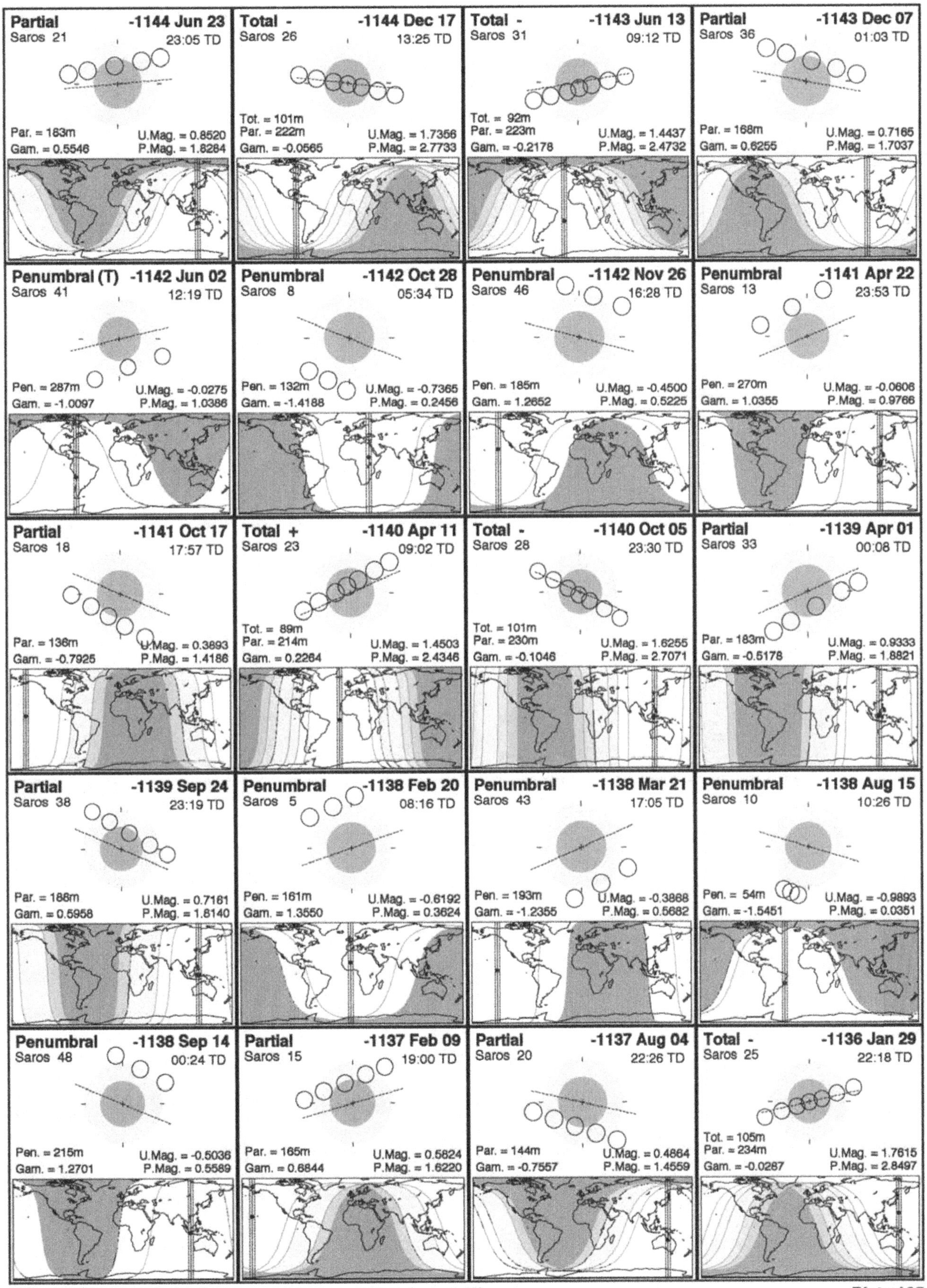

ΔT= 27960 s [= 07h46m] std.err. = ± 930 s [= ± 3.9°] **Plate 105**

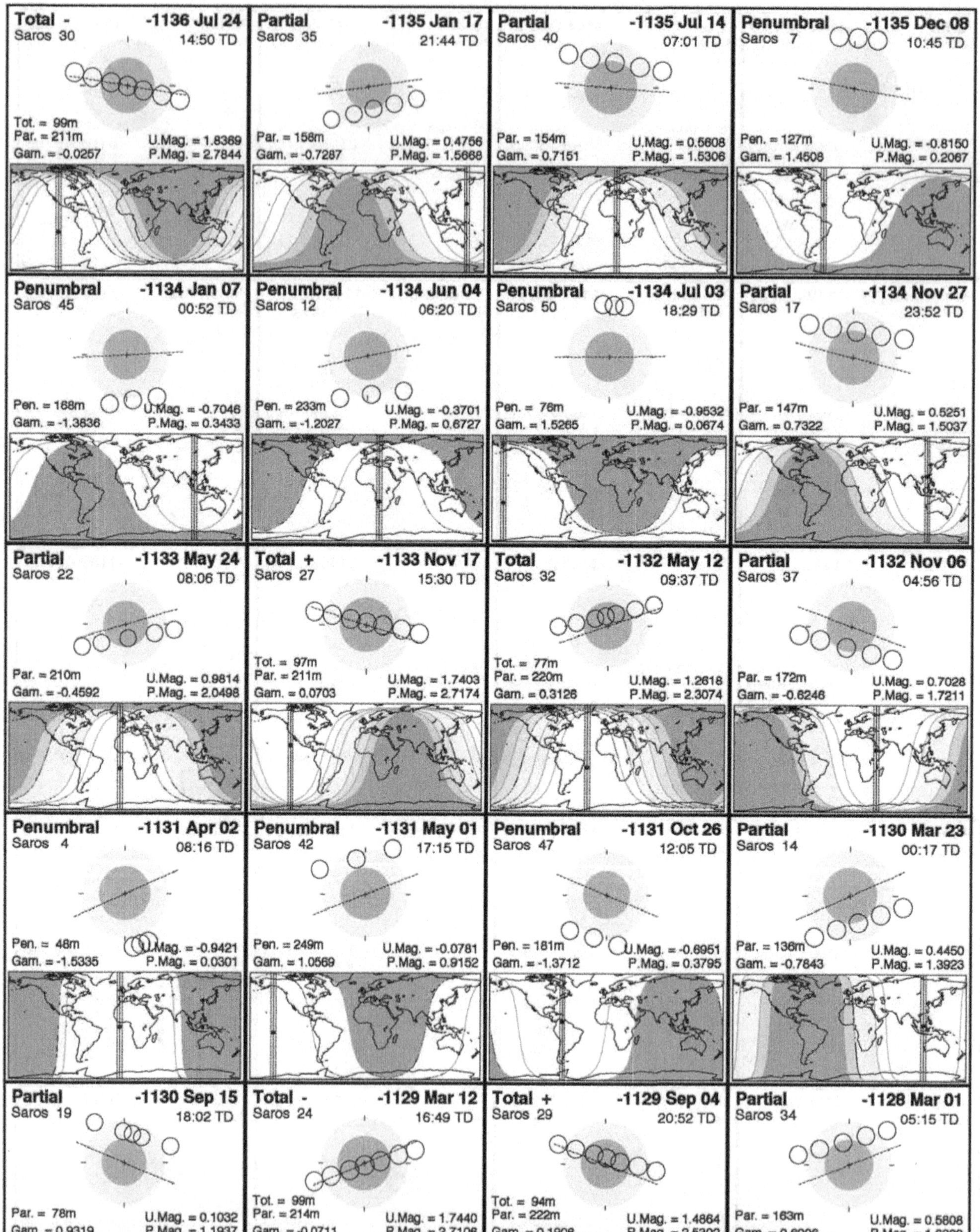

ΔT= 27807 s [= 07h43m] std.err. = ± 911 s [= ± 3.8°] **Plate 106**

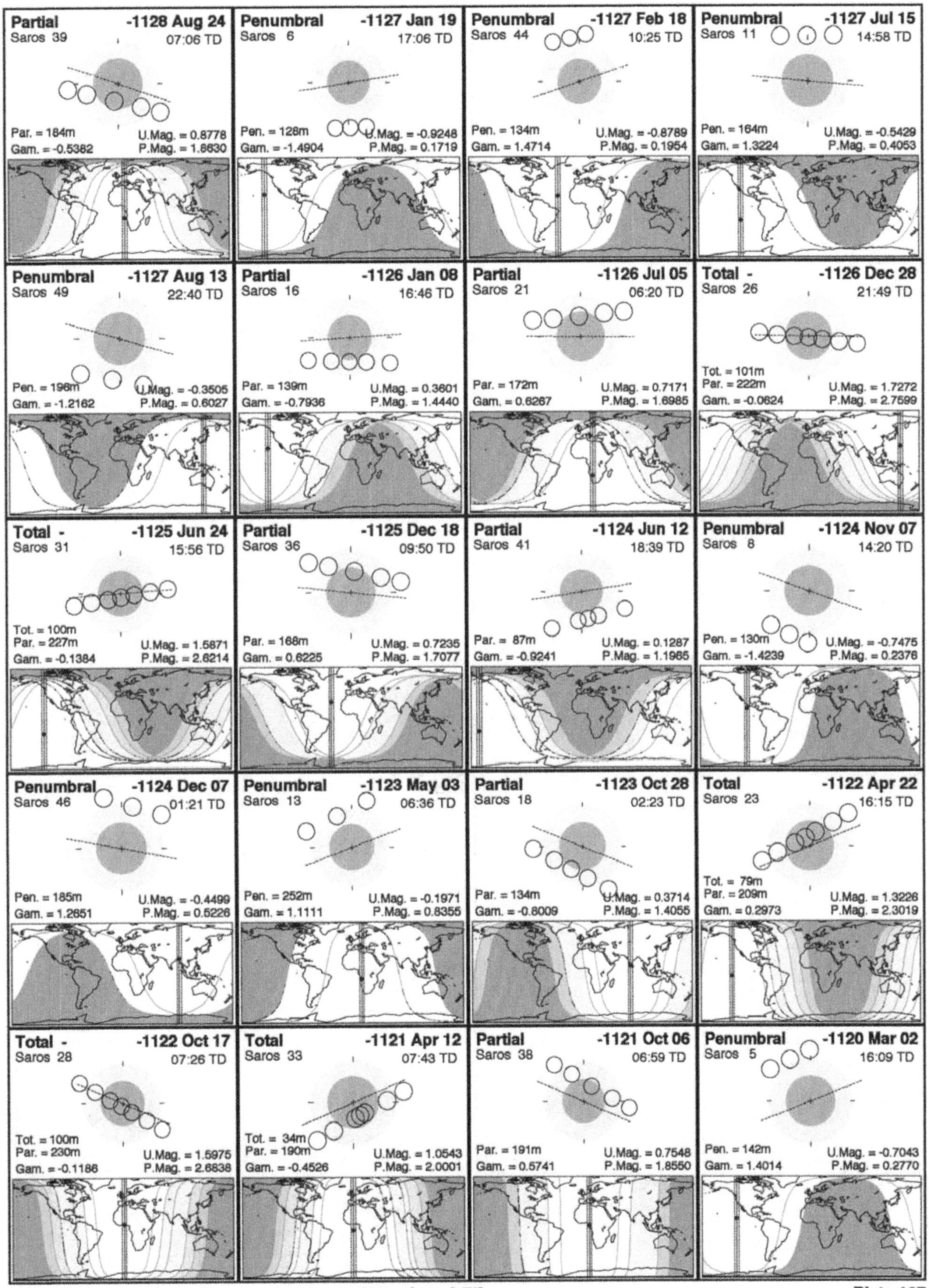

ΔT= 27655 s [= 07h41m] std.err. = ± 892 s [= ± 3.7°] **Plate 107**

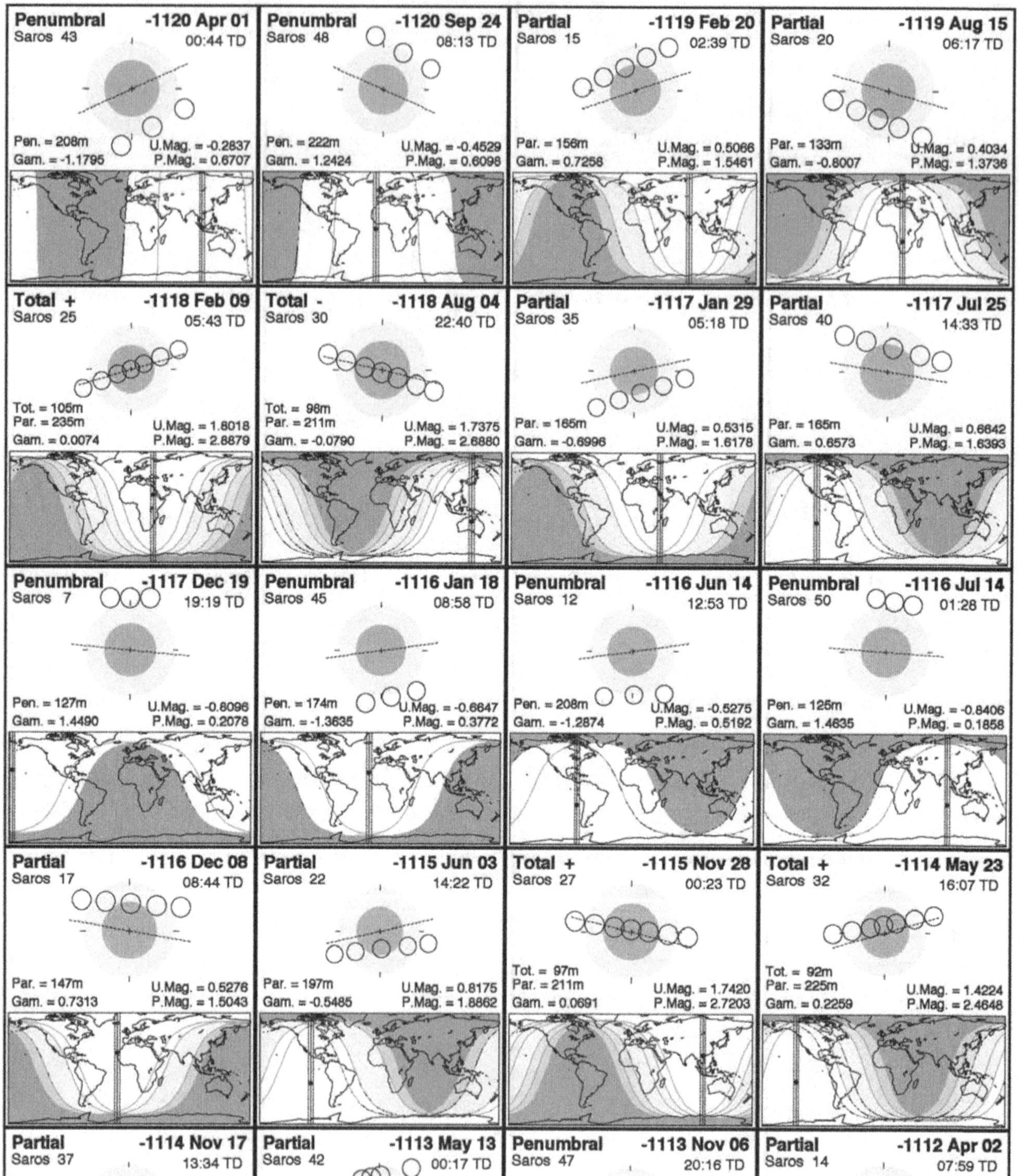

ΔT= 27513 s [= 07h39m] std.err. = ± 875 s [= ± 3.6°] **Plate 108**

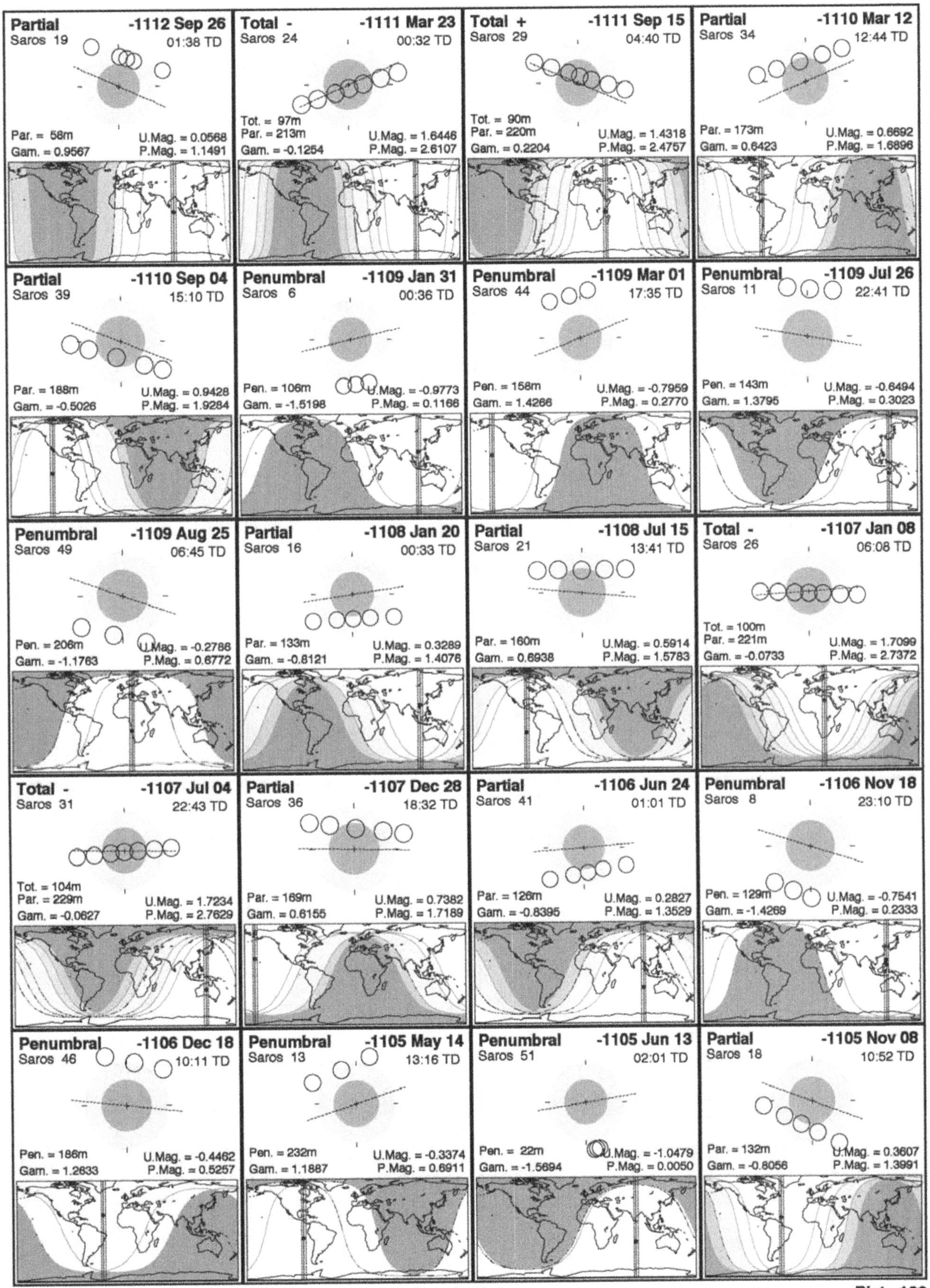

ΔT= 27354 s [= 07h36m] std.err. = ± 856 s [= ± 3.6°] Plate 109

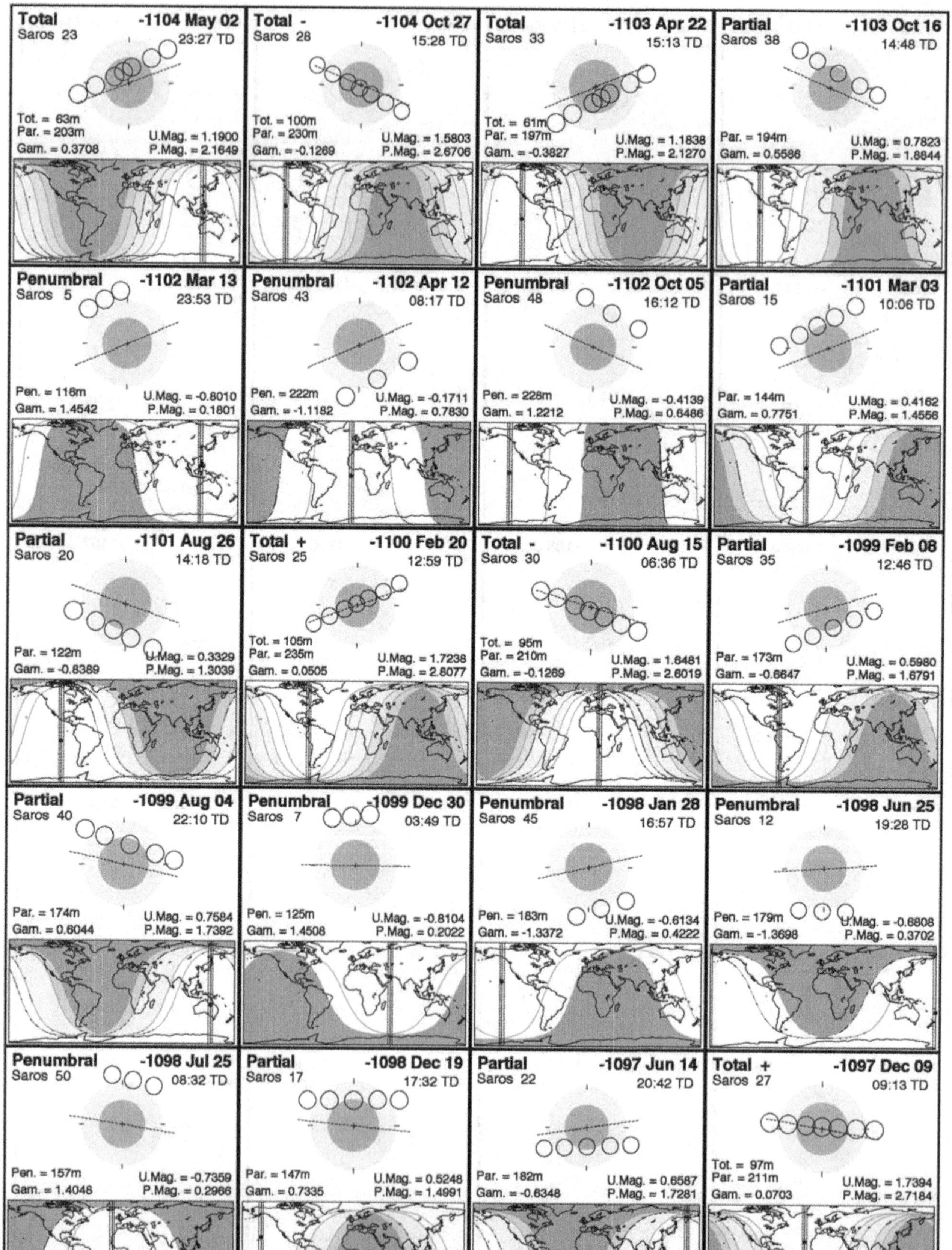

ΔT= 27212 s [= 07h34m] std.err. = ± 839 s [= ± 3.5°] **Plate 110**

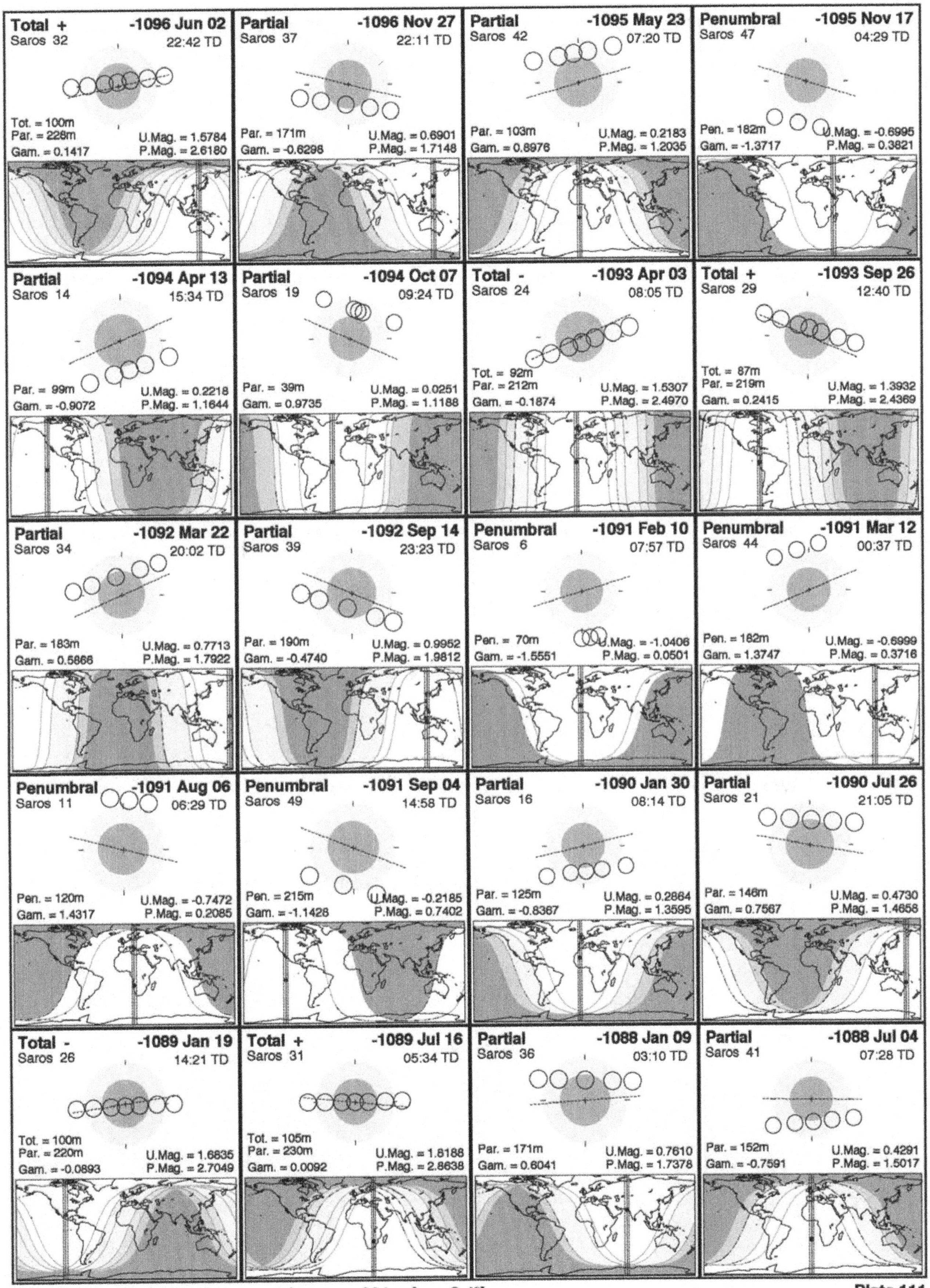

ΔT= 27062 s [= 07h31m] std.err. = ± 821 s [= ± 3.4°] **Plate 111**

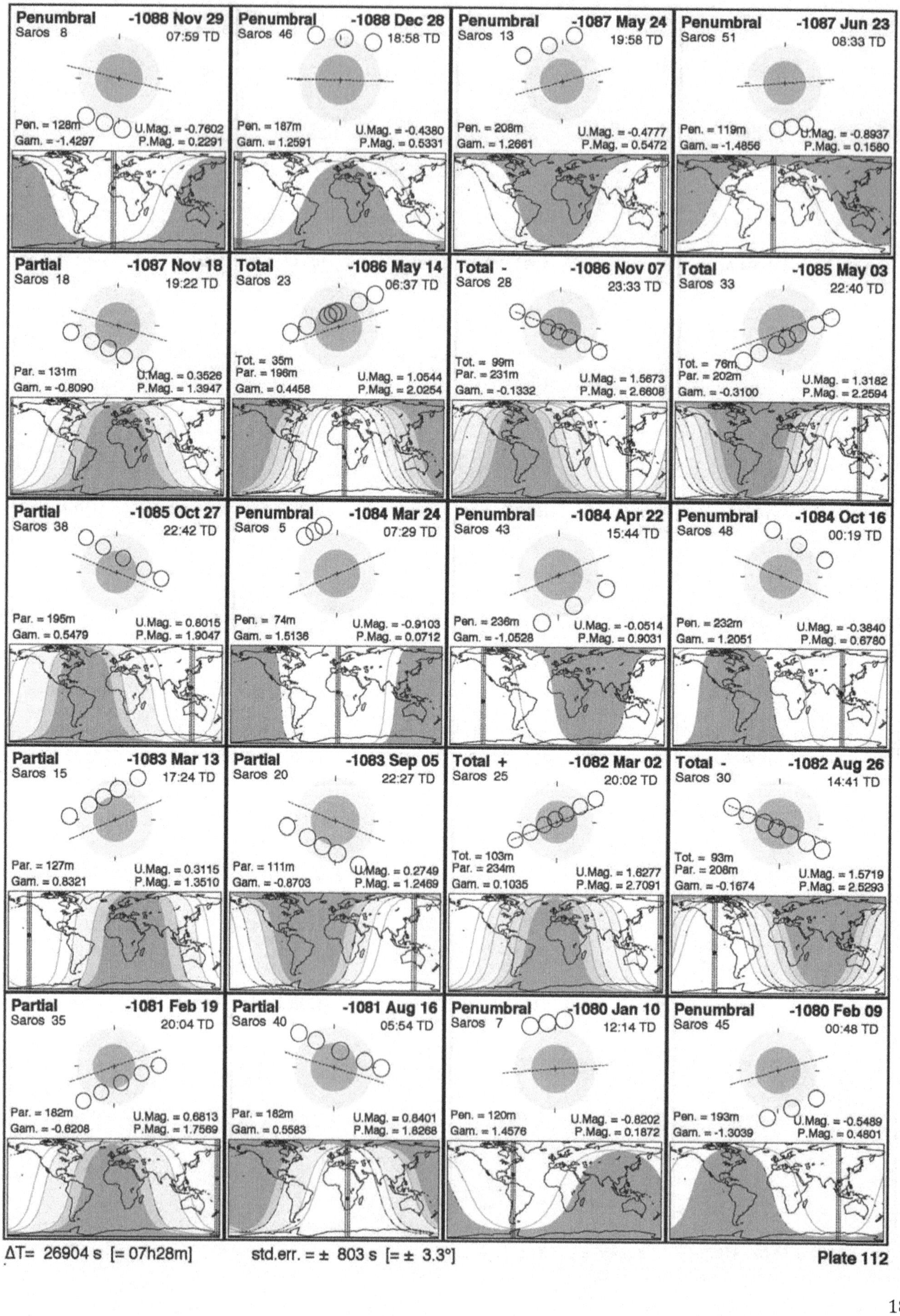

ΔT= 26904 s [= 07h28m] std.err. = ± 803 s [= ± 3.3°] **Plate 112**

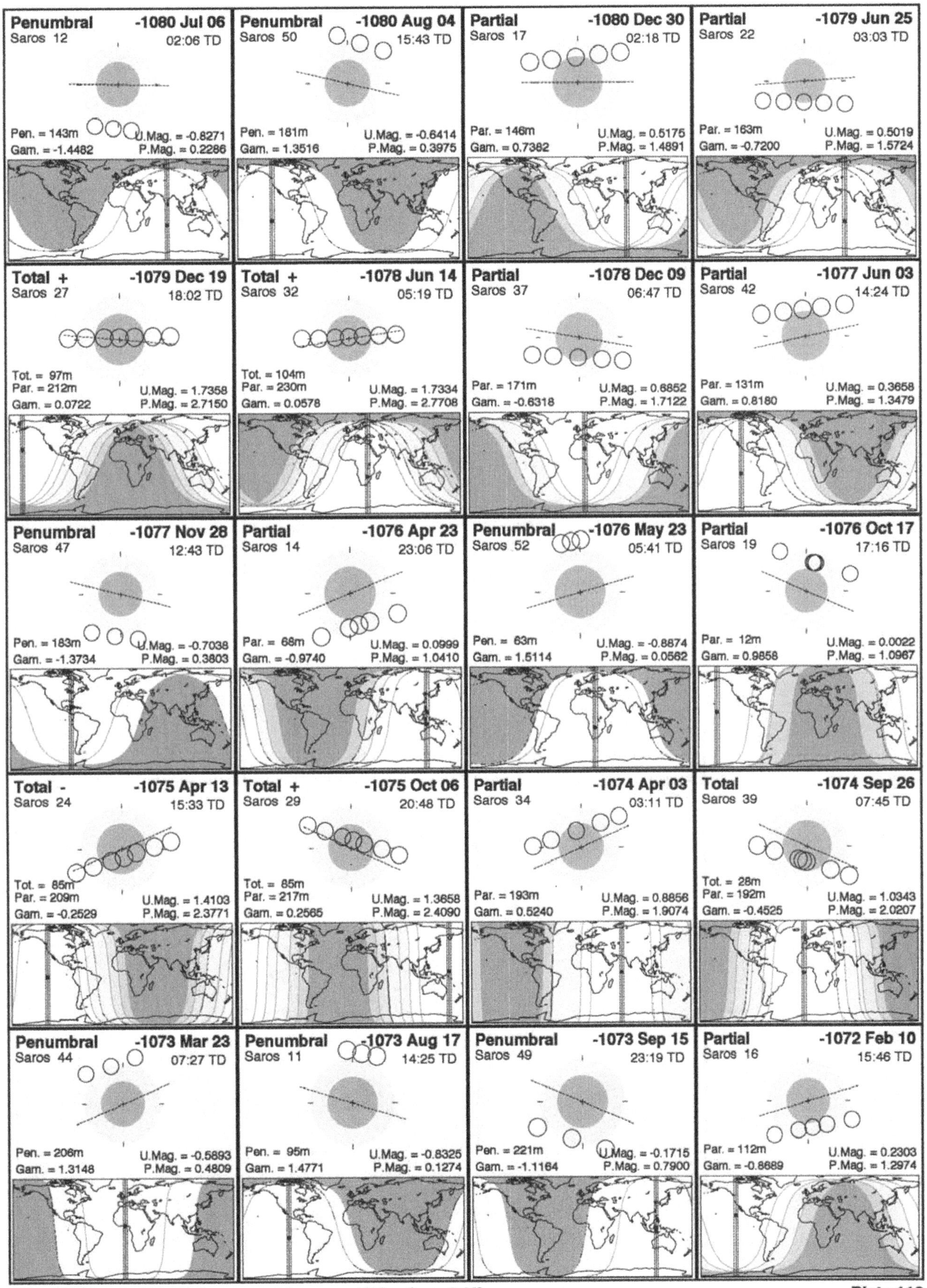

ΔT= 26763 s [= 07h26m] std.err. = ± 786 s [= ± 3.3°] **Plate 113**

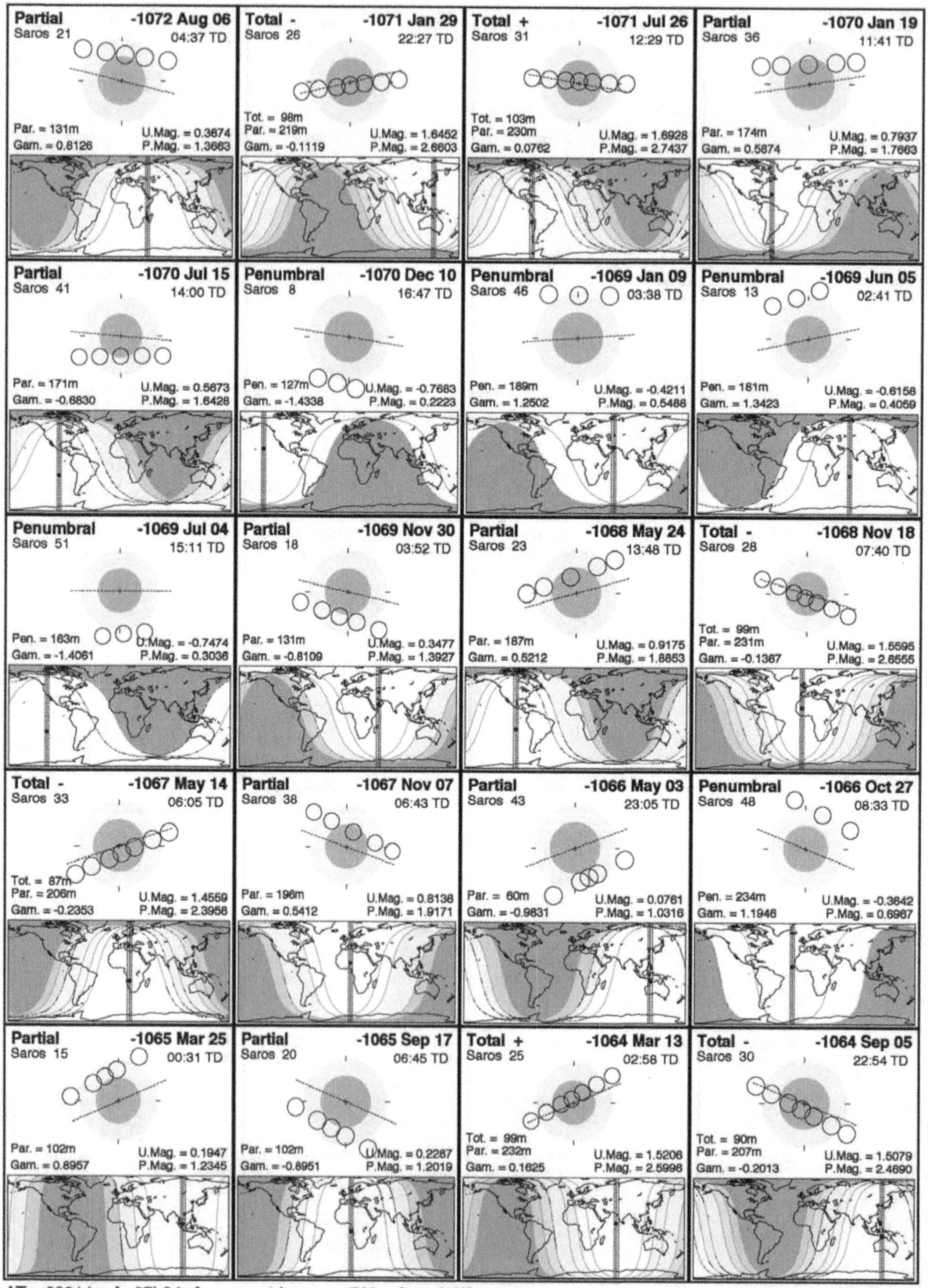

ΔT= 26614 s [= 07h24m] std.err. = ± 769 s [= ± 3.2°] **Plate 114**

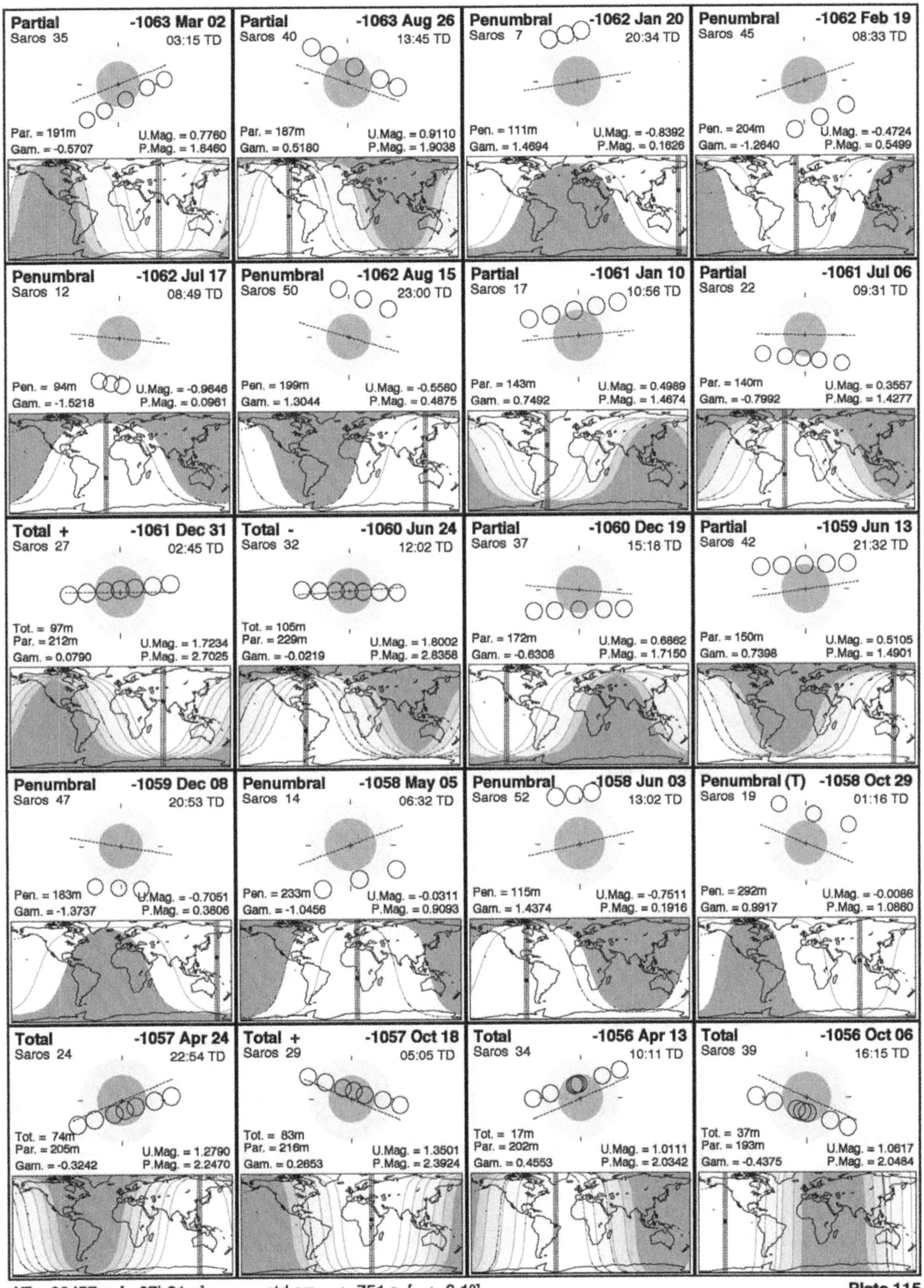

ΔT= 26457 s [= 07h21m] std.err. = ± 751 s [= ± 3.1°] **Plate 115**

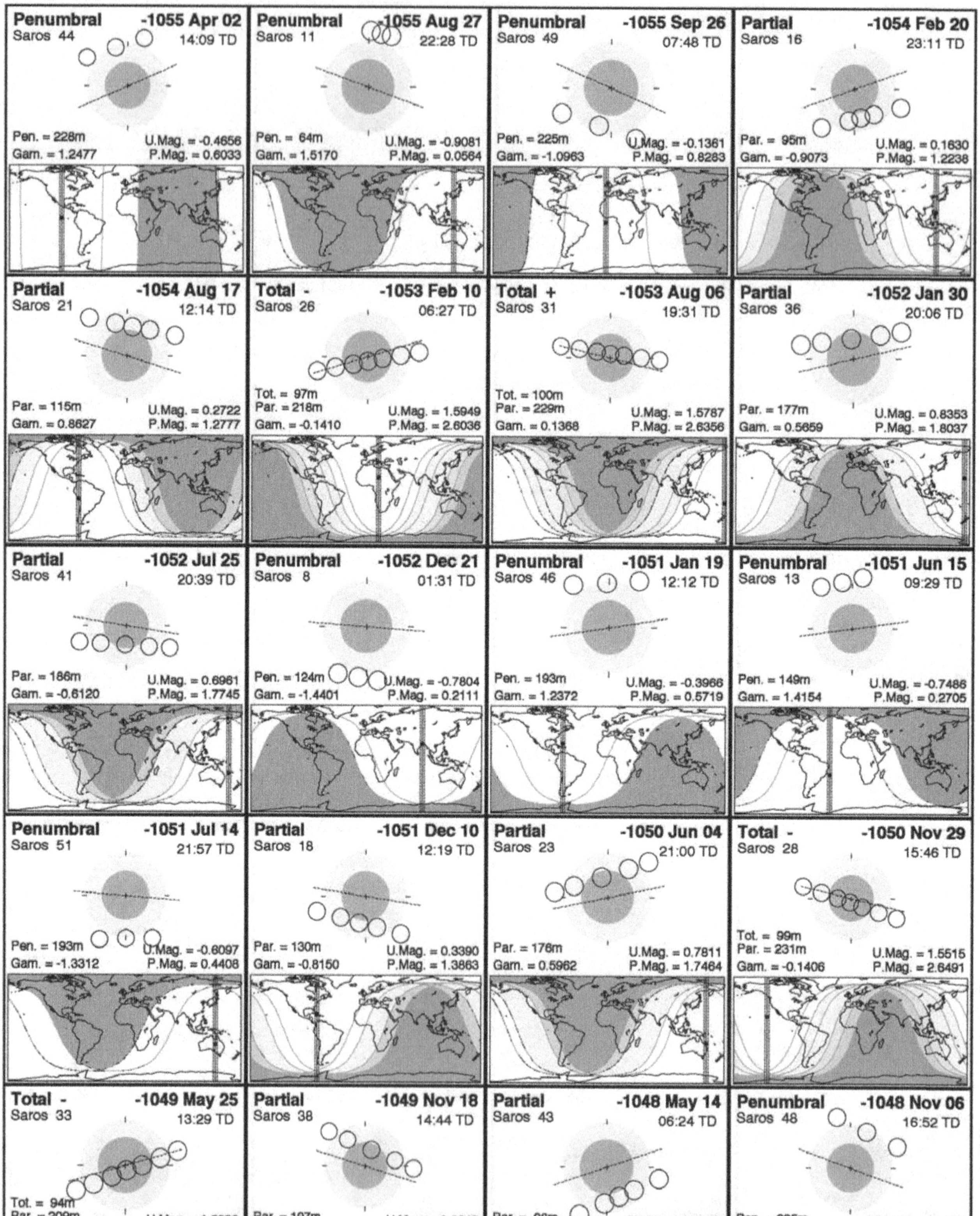

ΔT= 26308 s [= 07h18m] std.err. = ± 733 s [= ± 3.1°] **Plate 116**

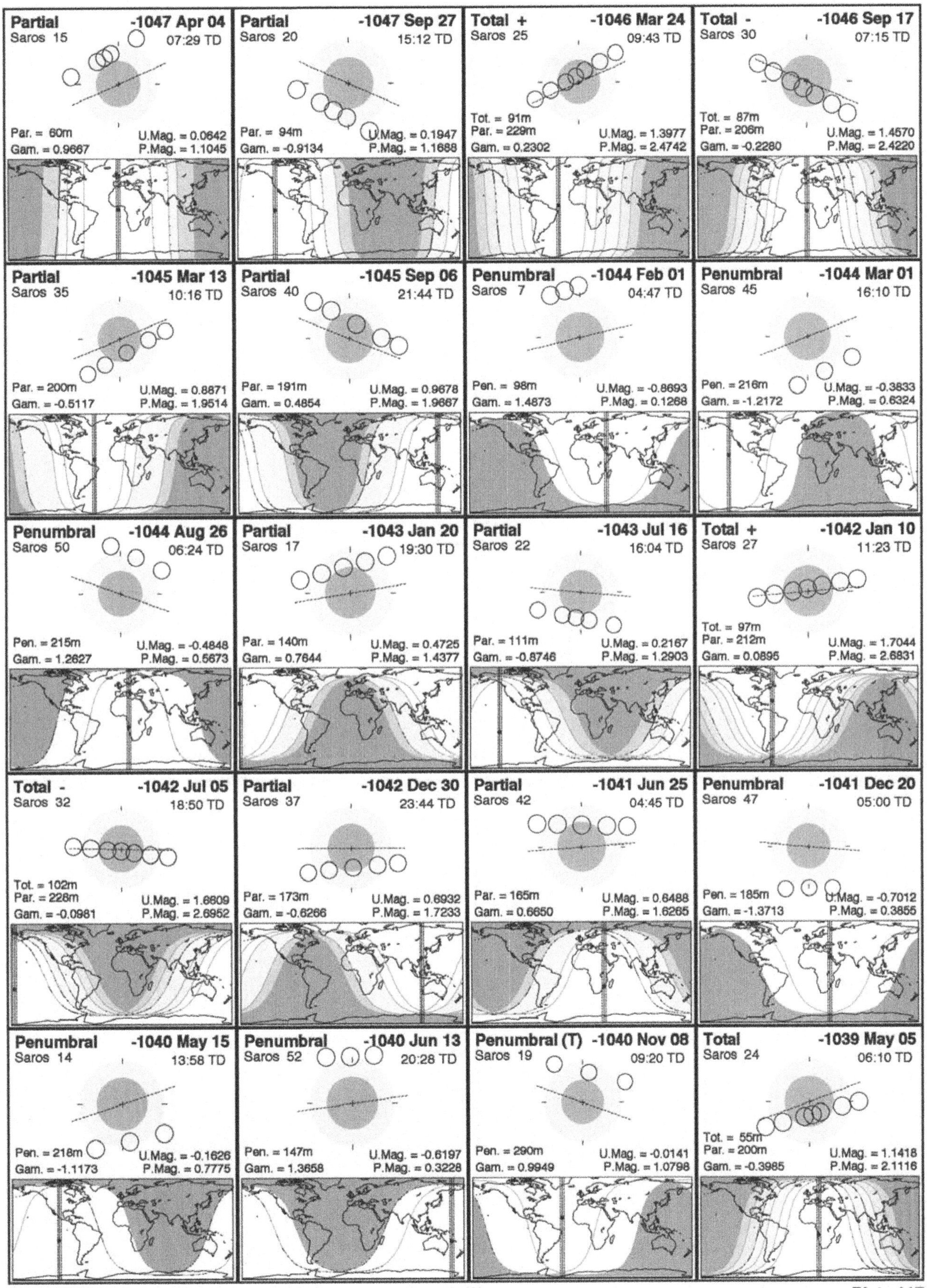

ΔT= 26162 s [= 07h16m] std.err. = ± 717 s [= ± 3.0°] Plate 117

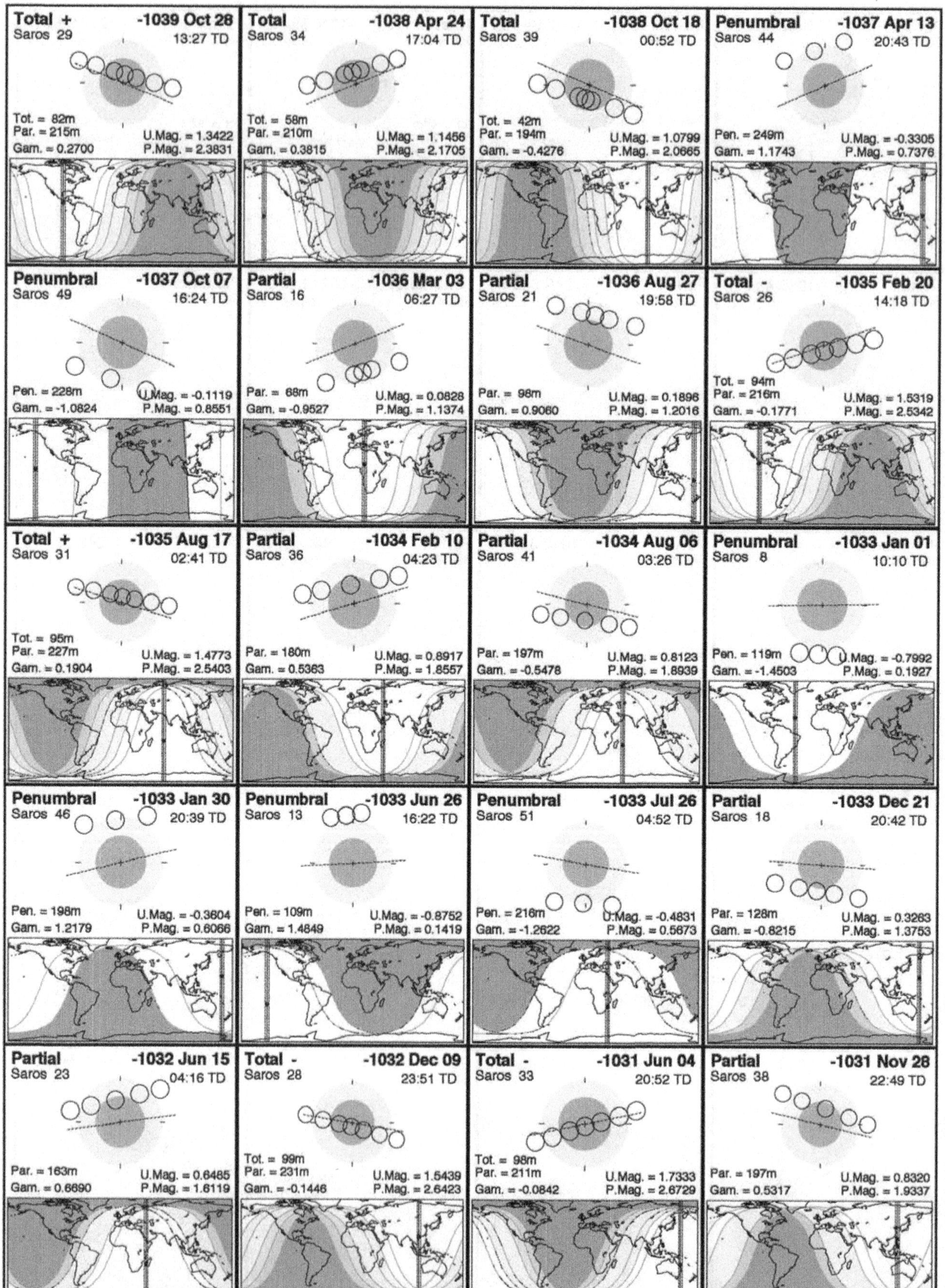

ΔT= 26005 s [= 07h13m] std.err. = ± 699 s [= ± 2.9°] **Plate 118**

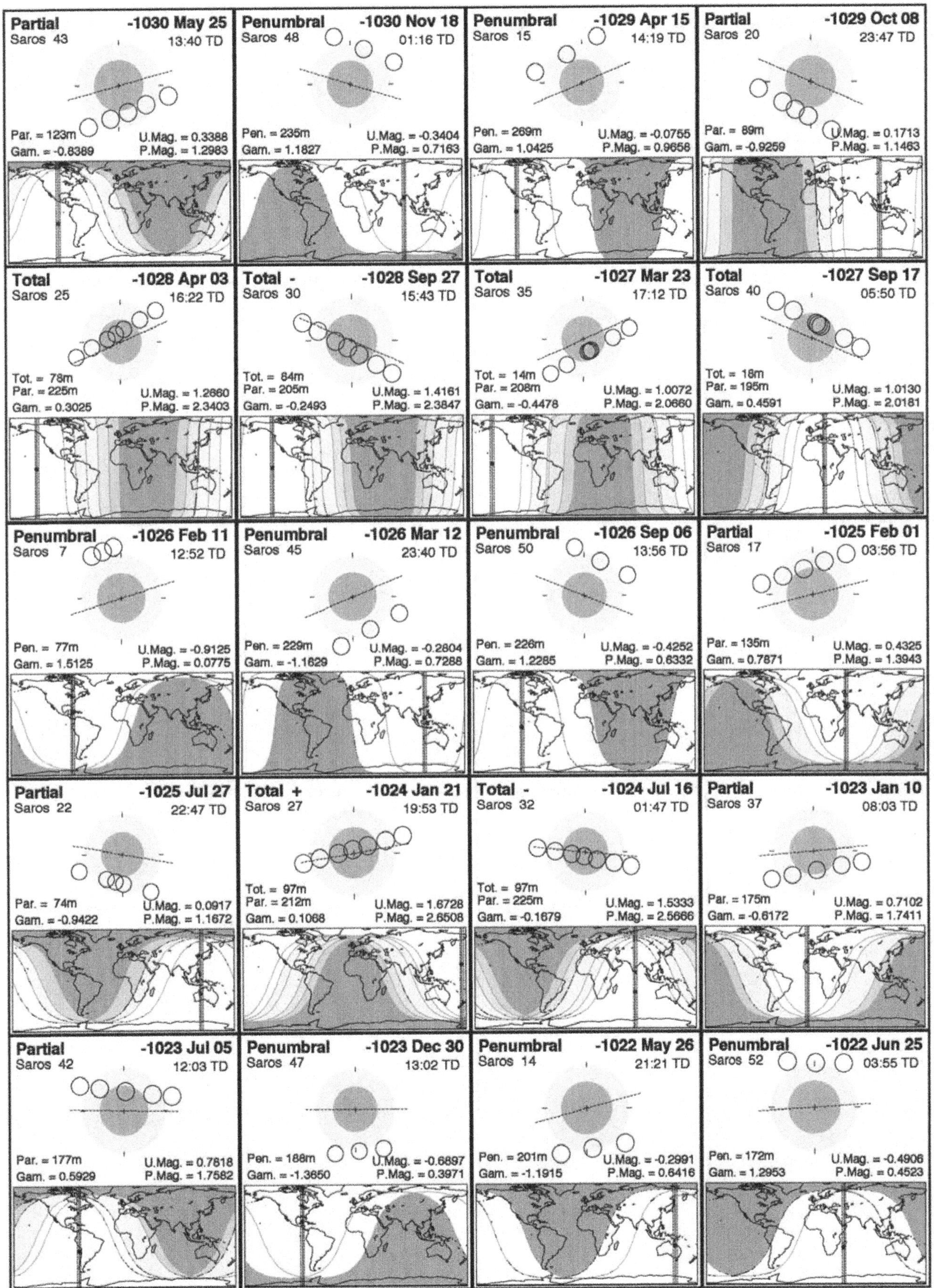

ΔT= 25850 s [= 07h11m] std.err. = ± 681 s [= ± 2.8°] **Plate 119**

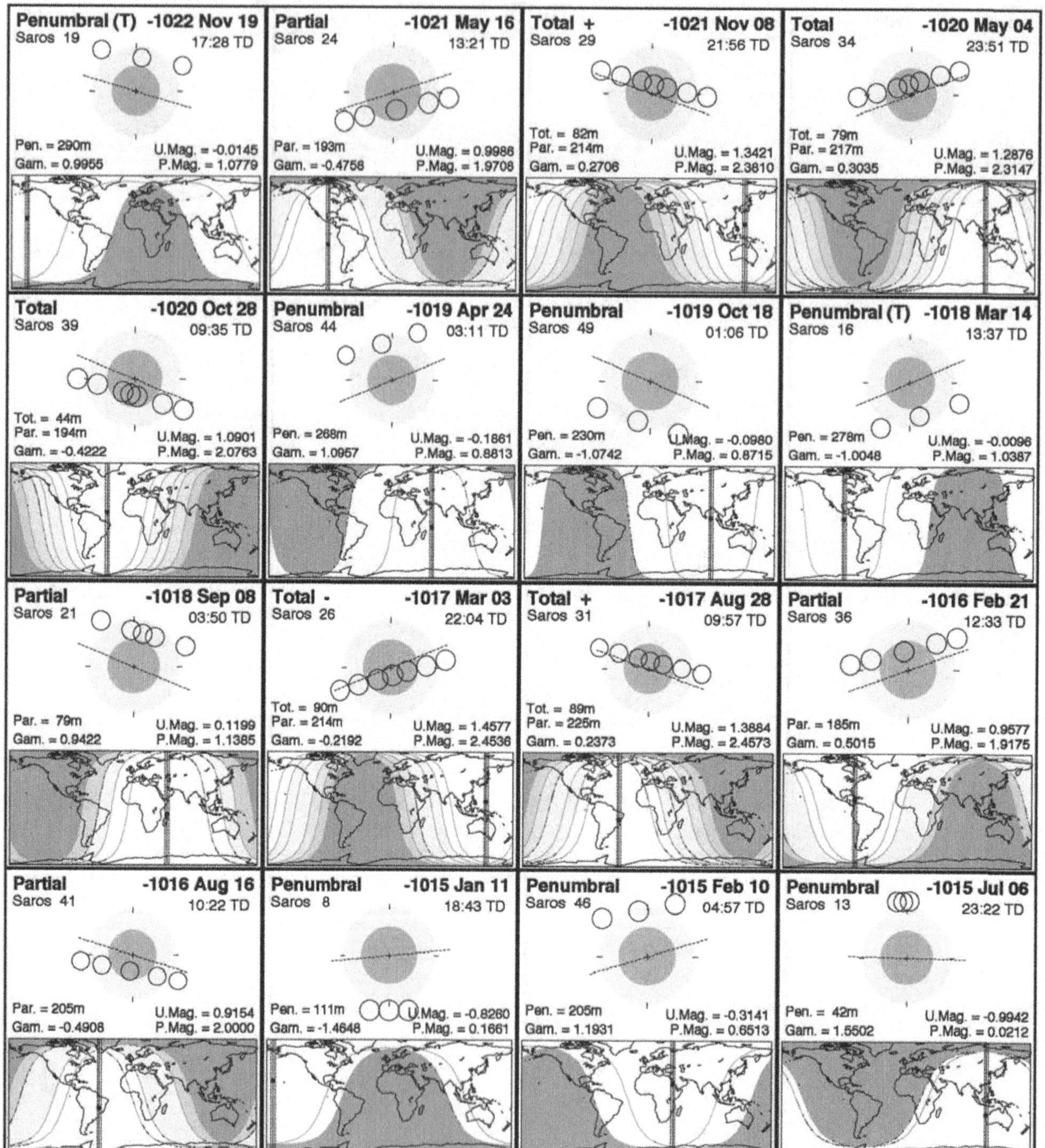

ΔT= 25696 s [= 07h08m] std.err. = ± 664 s [= ± 2.8°] **Plate 120**

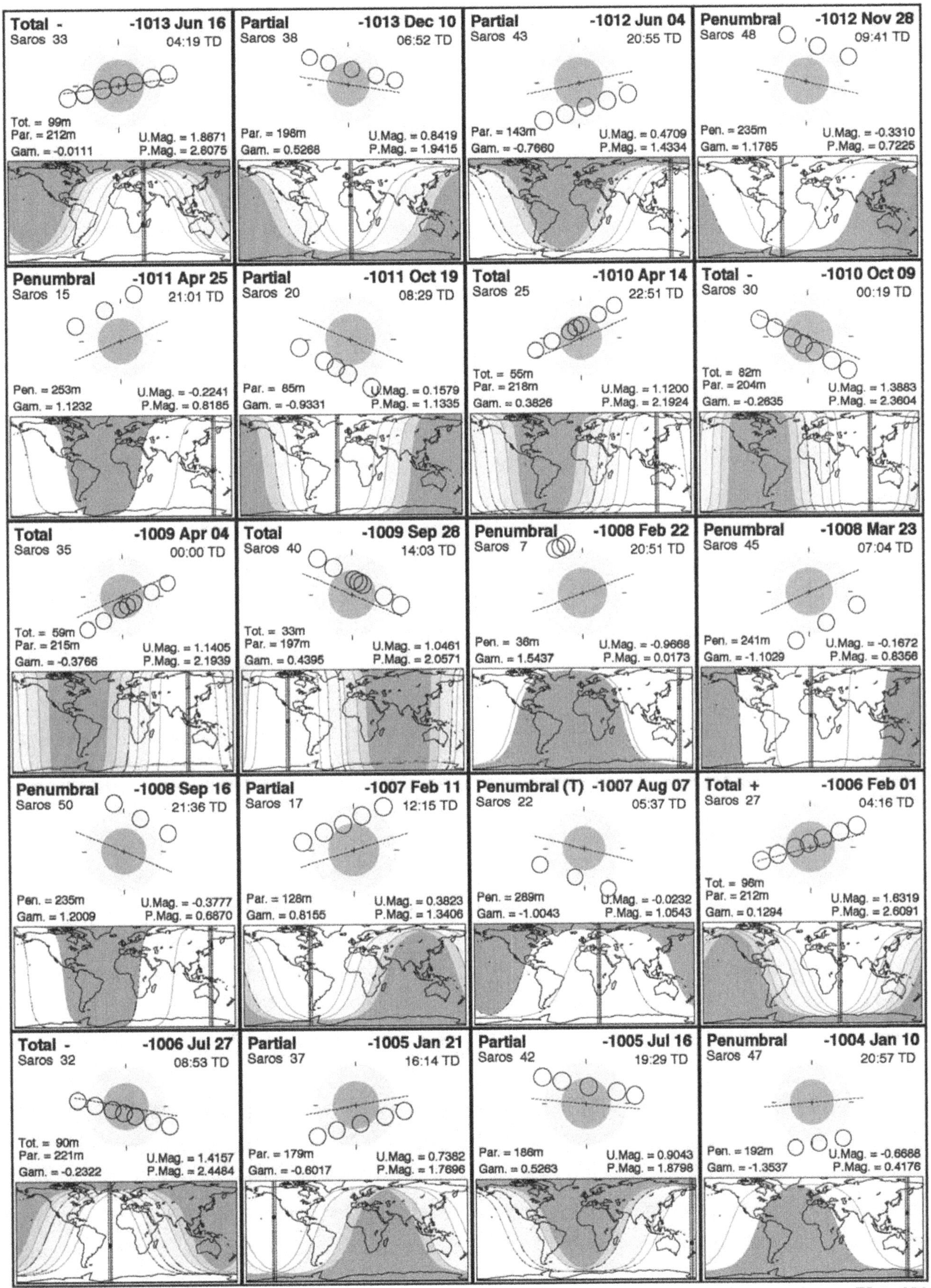

ΔT= 25541 s [= 07h06m] std.err. = ± 647 s [= ± 2.7°] **Plate 121**

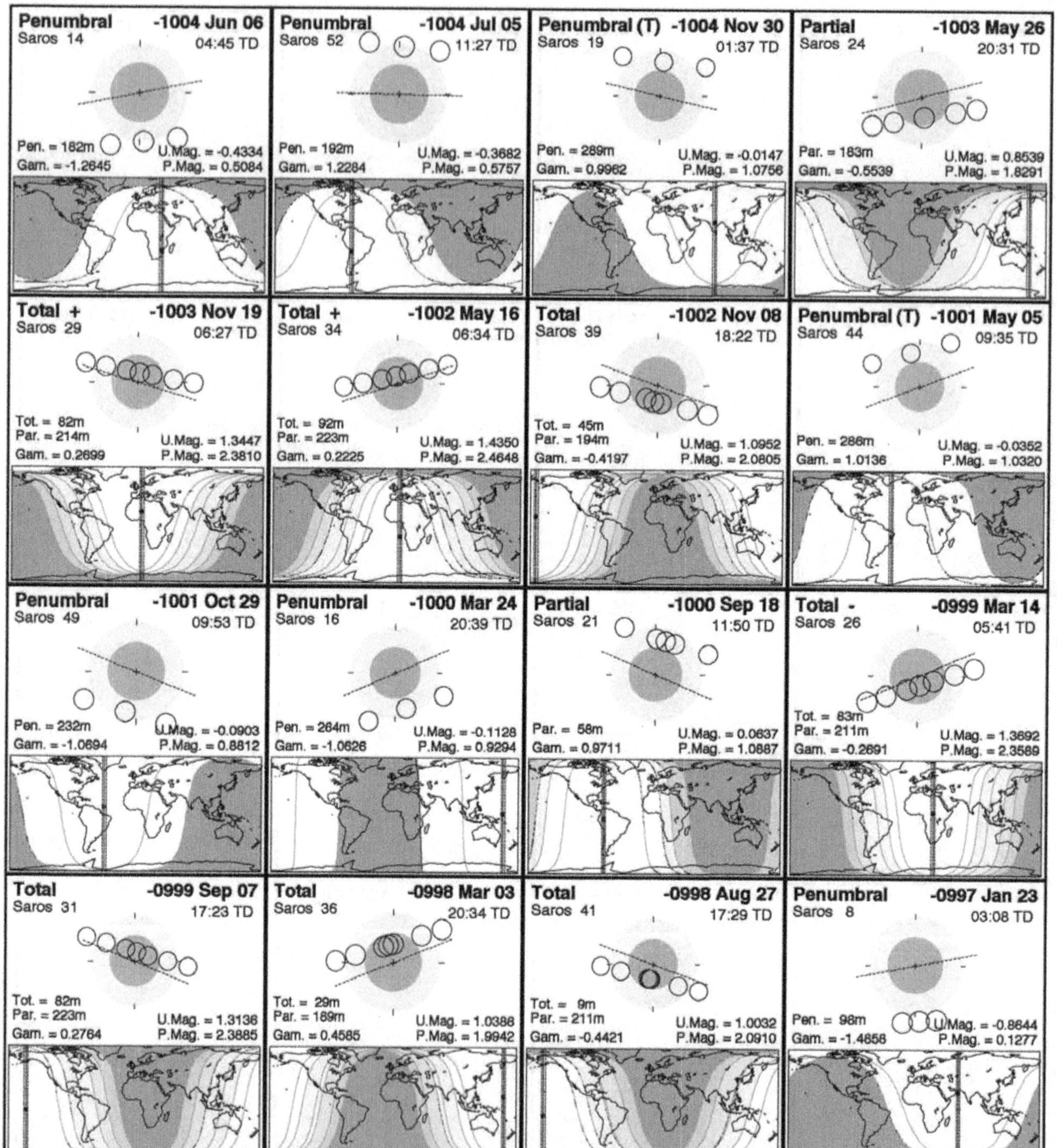

ΔT= 25379 s [= 07h03m] std.err. = ± 629 s [= ± 2.6°] **Plate 122**

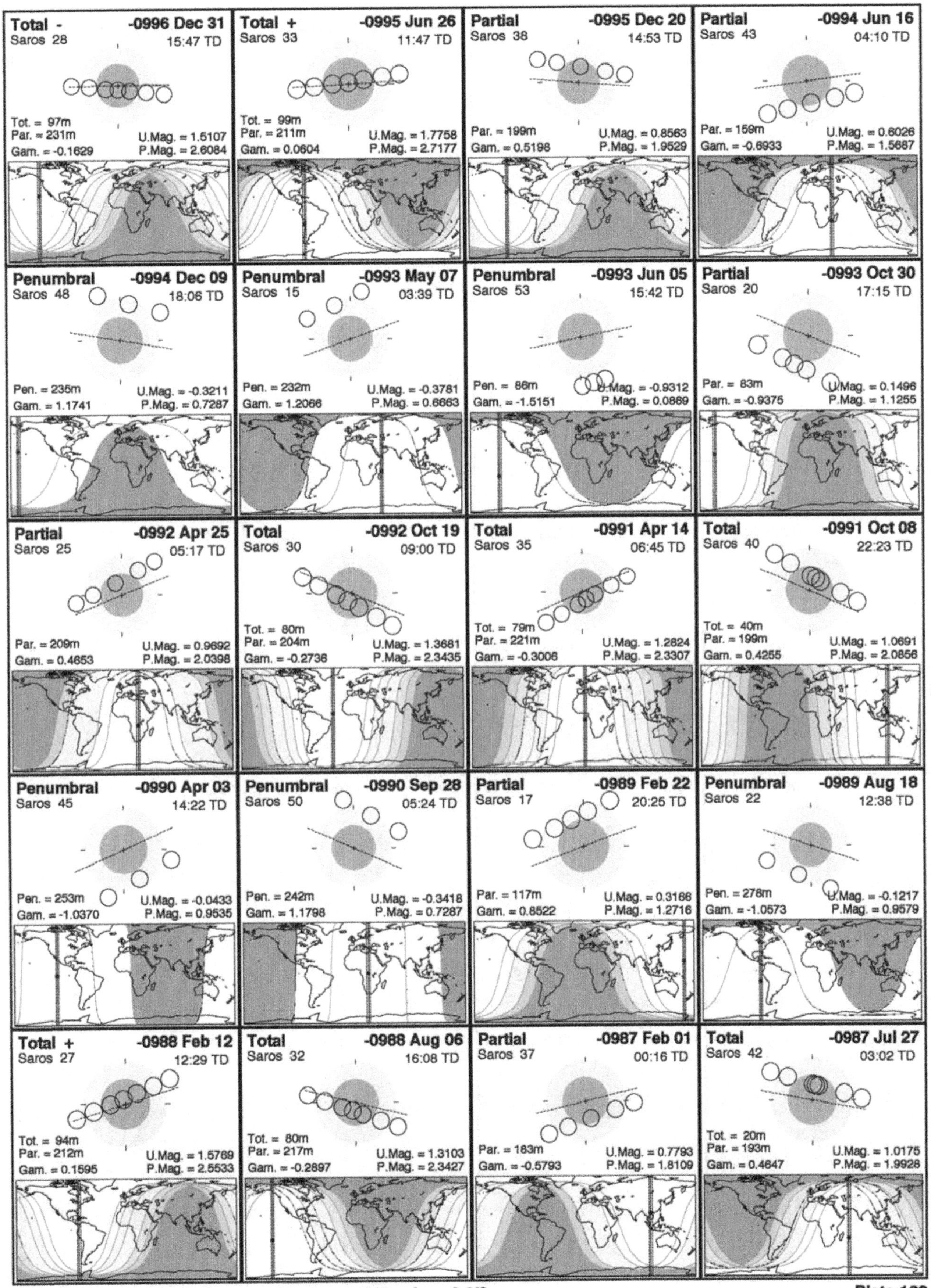

ΔT= 25225 s [= 07h00m] std.err. = ± 620 s [= ± 2.6°] **Plate 123**

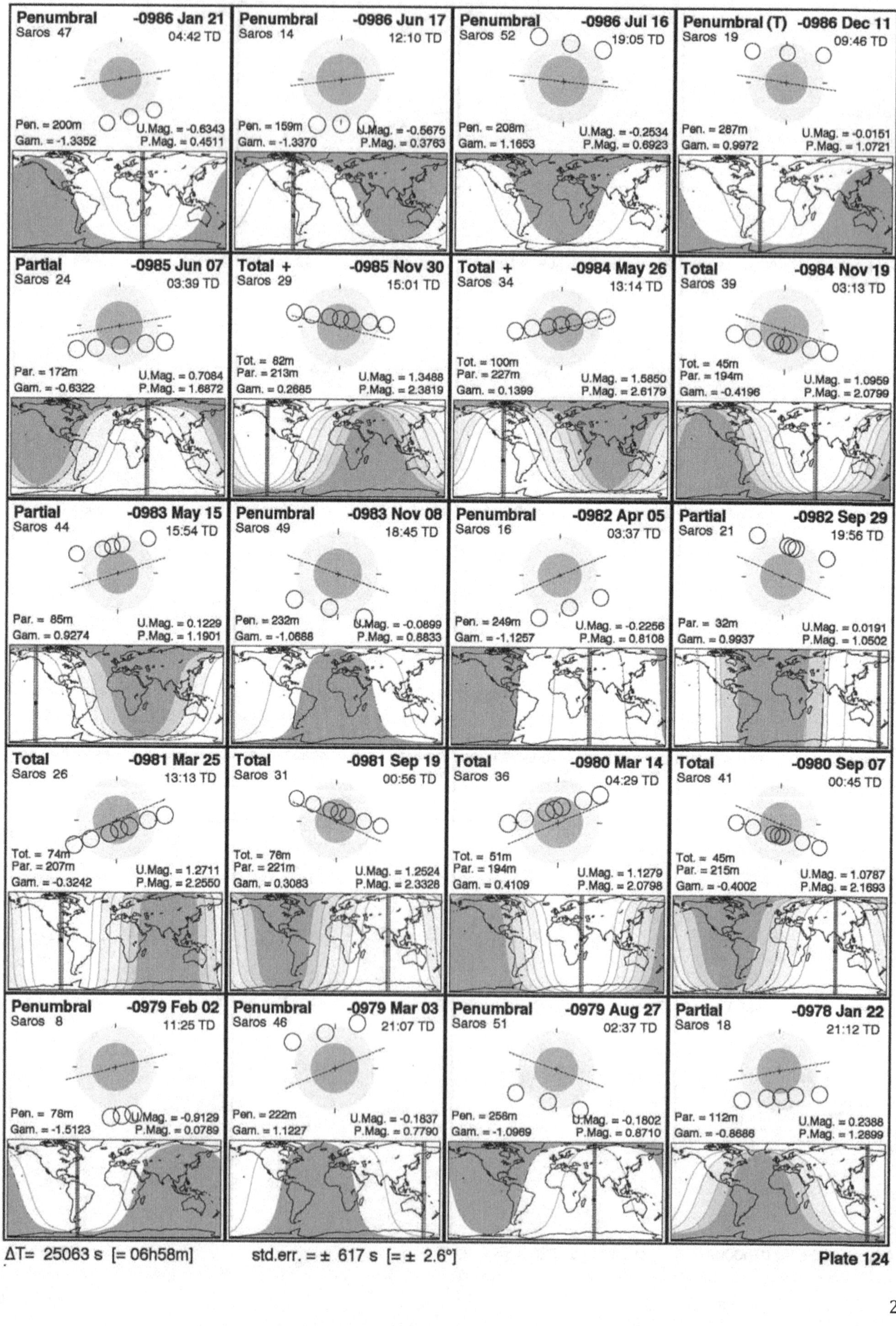

ΔT= 25063 s [= 06h58m] std.err. = ± 617 s [= ± 2.6°] **Plate 124**

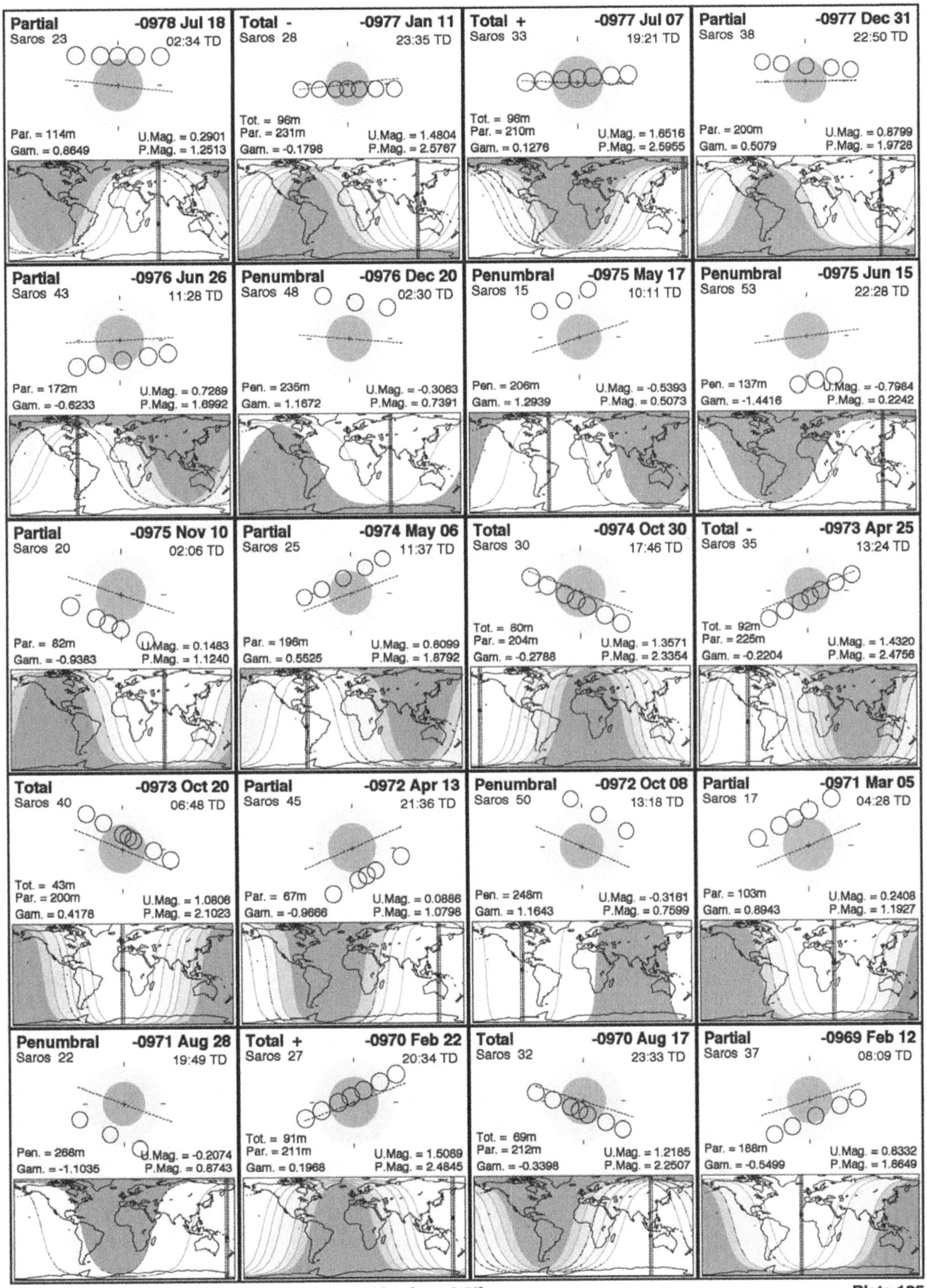

ΔT= 24911 s [= 06h55m] std.err. = ± 613 s [= ± 2.6°] **Plate 125**

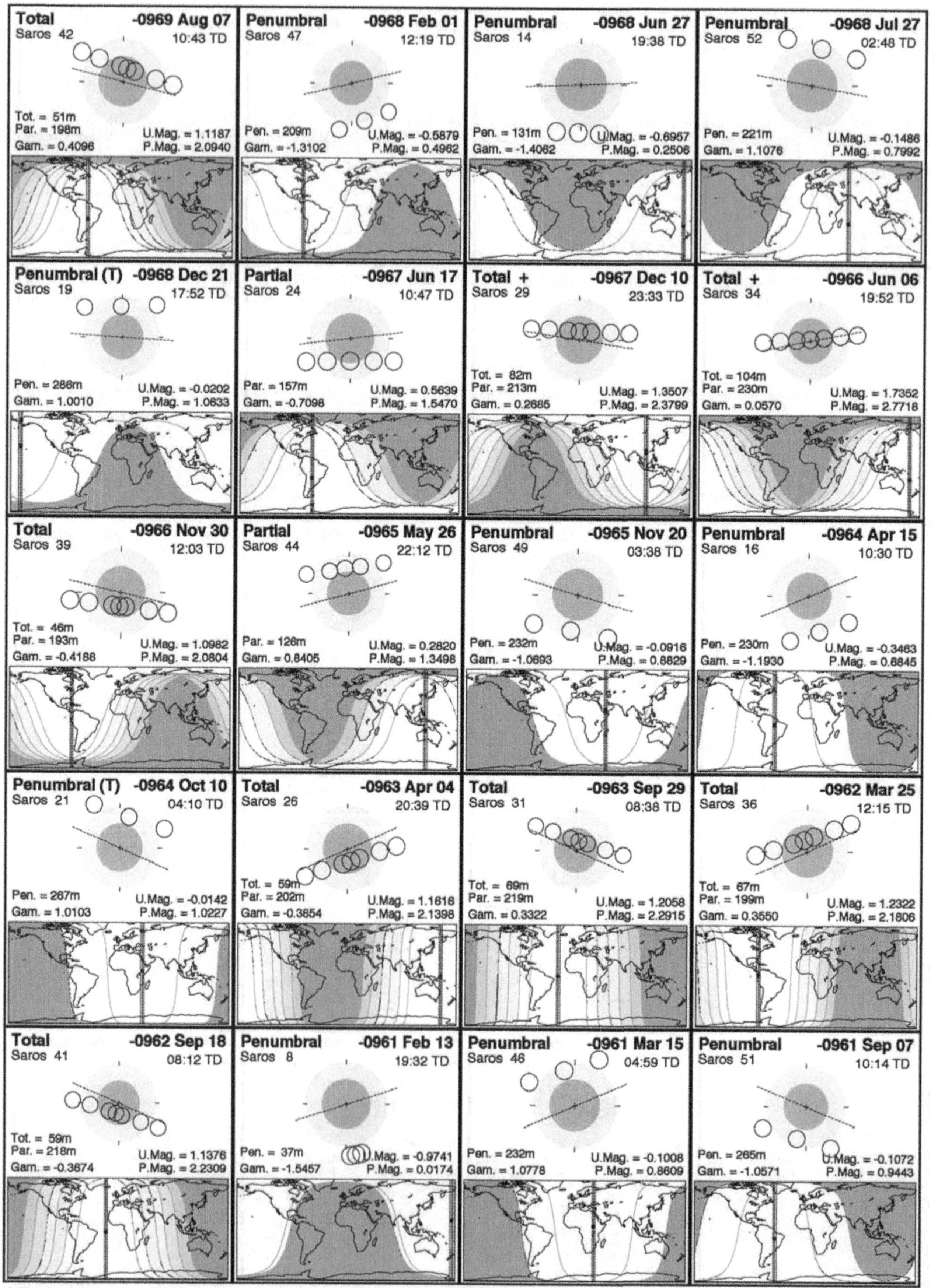

ΔT= 24750 s [= 06h53m] std.err. = ± 610 s [= ± 2.5°] **Plate 126**

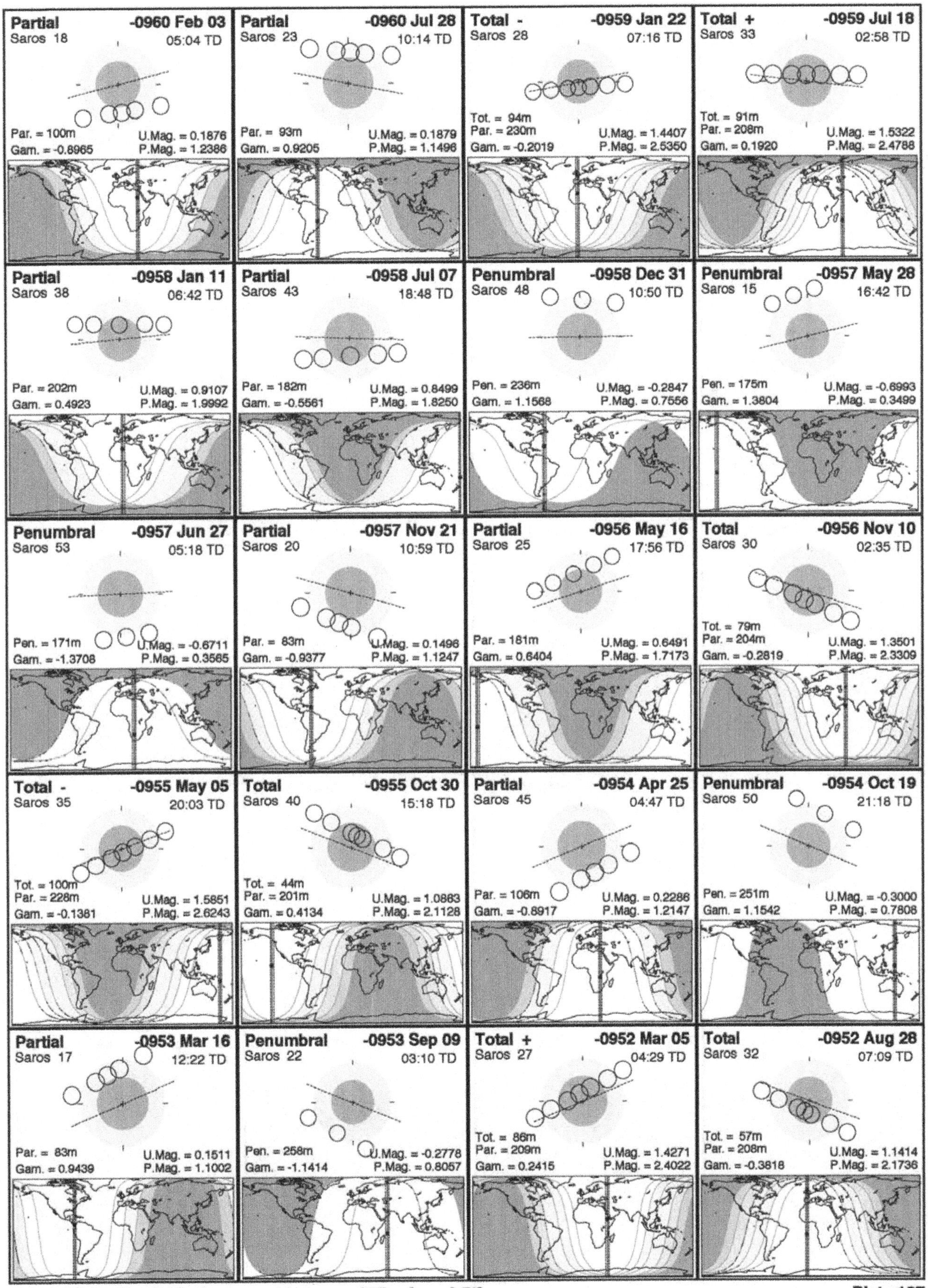

ΔT= 24599 s [= 06h50m] std.err. = ± 607 s [= ± 2.5°] **Plate 127**

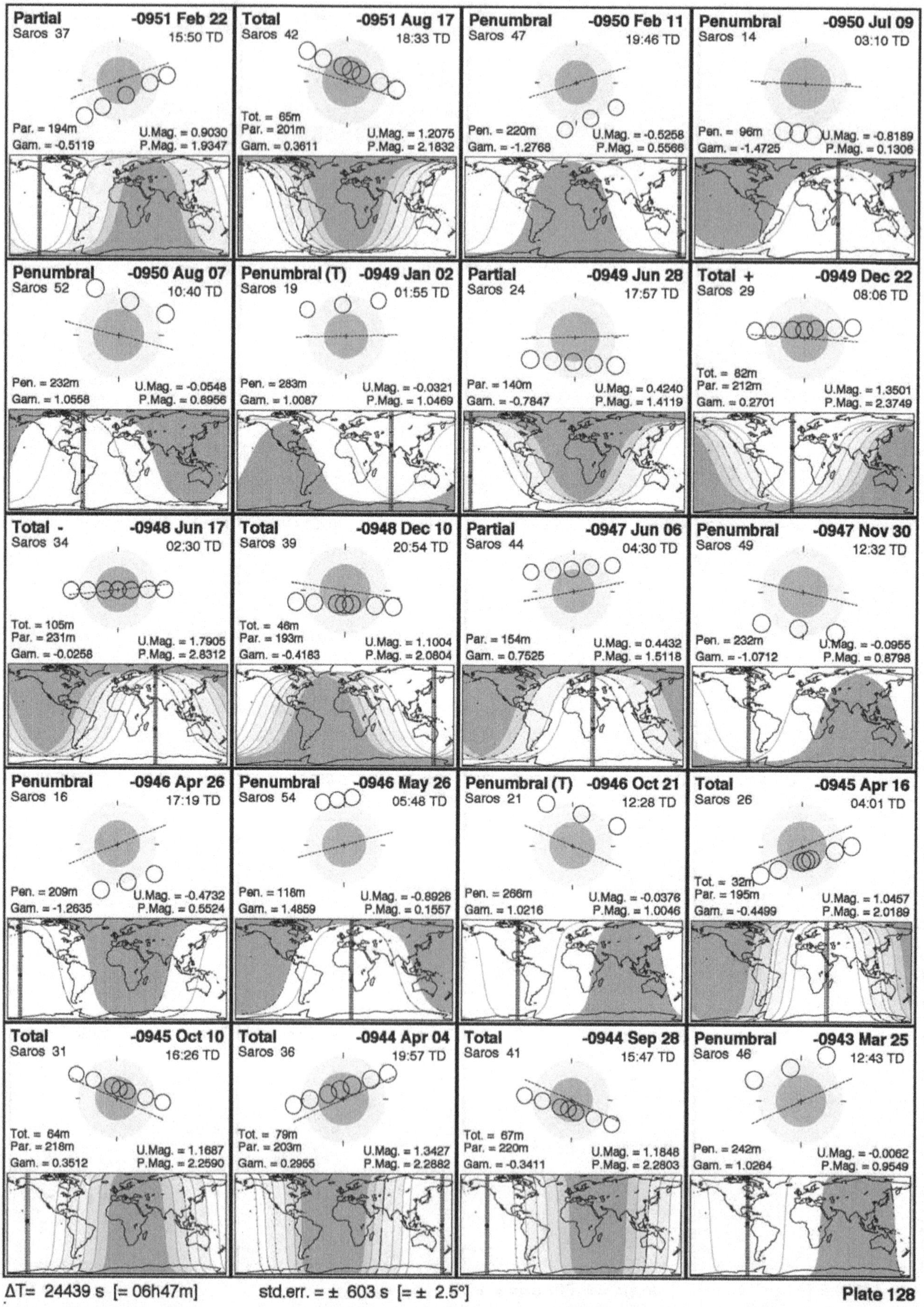

ΔT= 24439 s [= 06h47m] std.err. = ± 603 s [= ± 2.5°] **Plate 128**

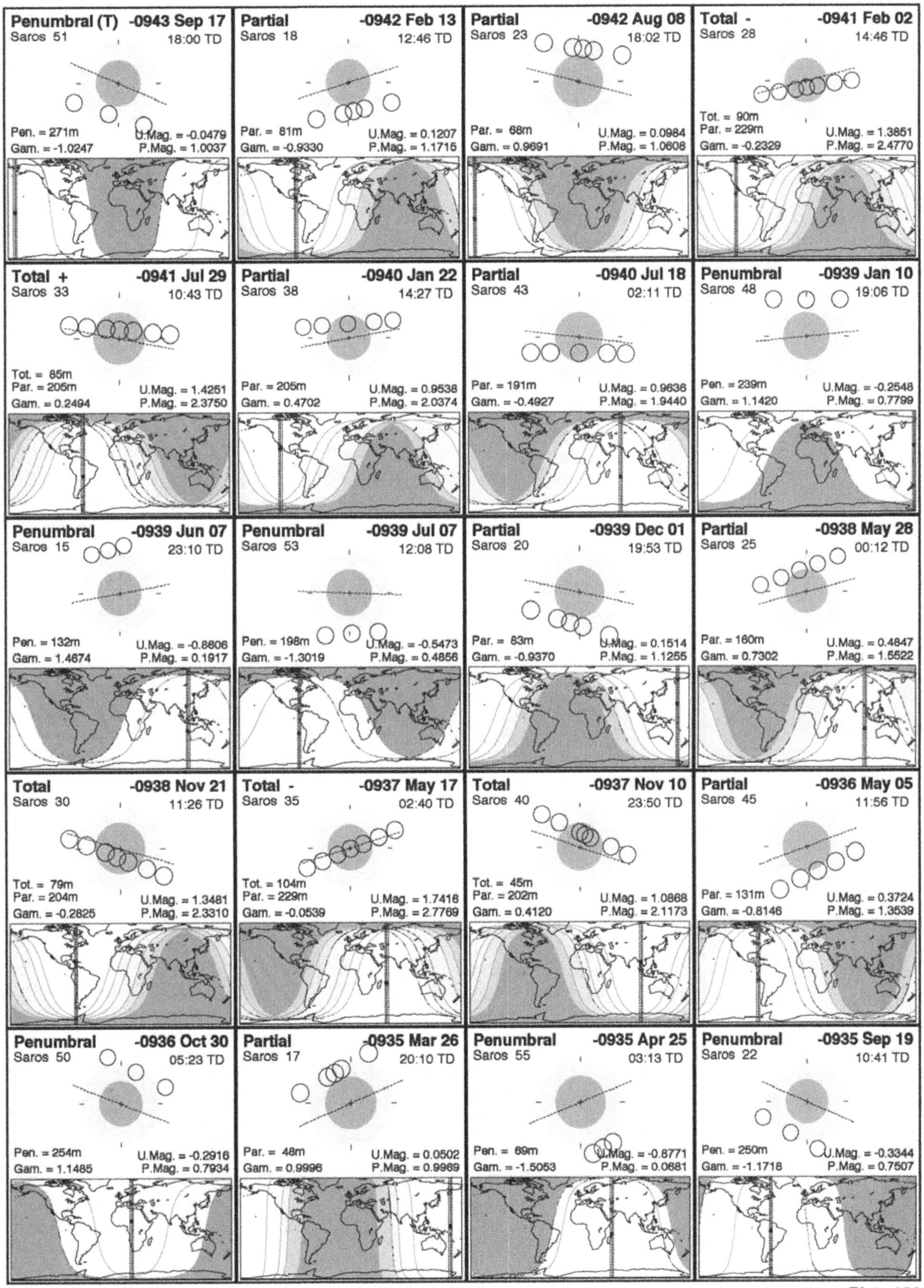

ΔT= 24288 s [= 06h45m]   std.err. = ± 600 s [= ± 2.5°]   Plate 129

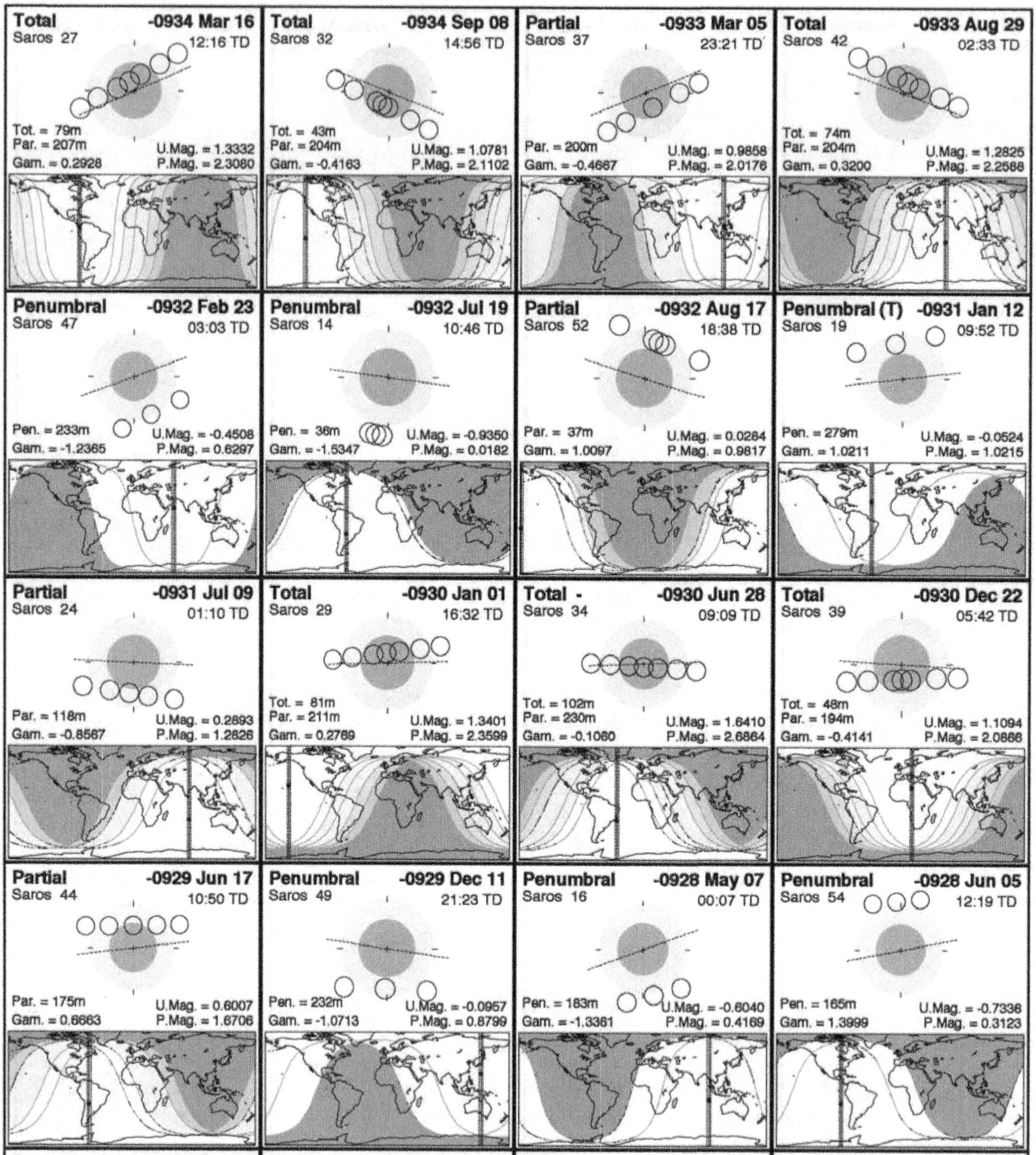

ΔT= 24139 s [= 06h42m] std.err. = ± 597 s [= ± 2.5°] **Plate 130**

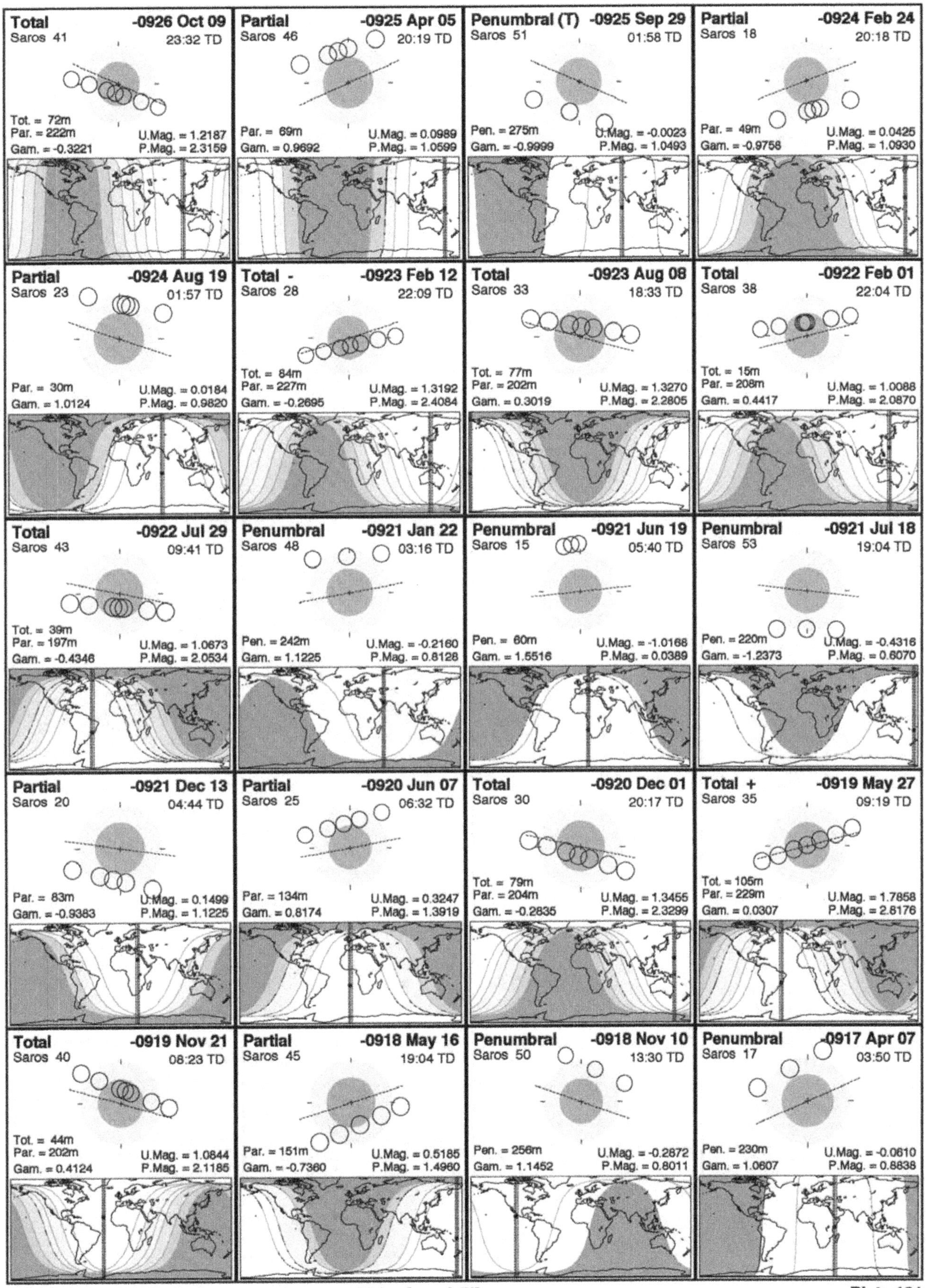

ΔT= 23989 s [= 06h40m] std.err. = ± 594 s [= ± 2.5°] **Plate 131**

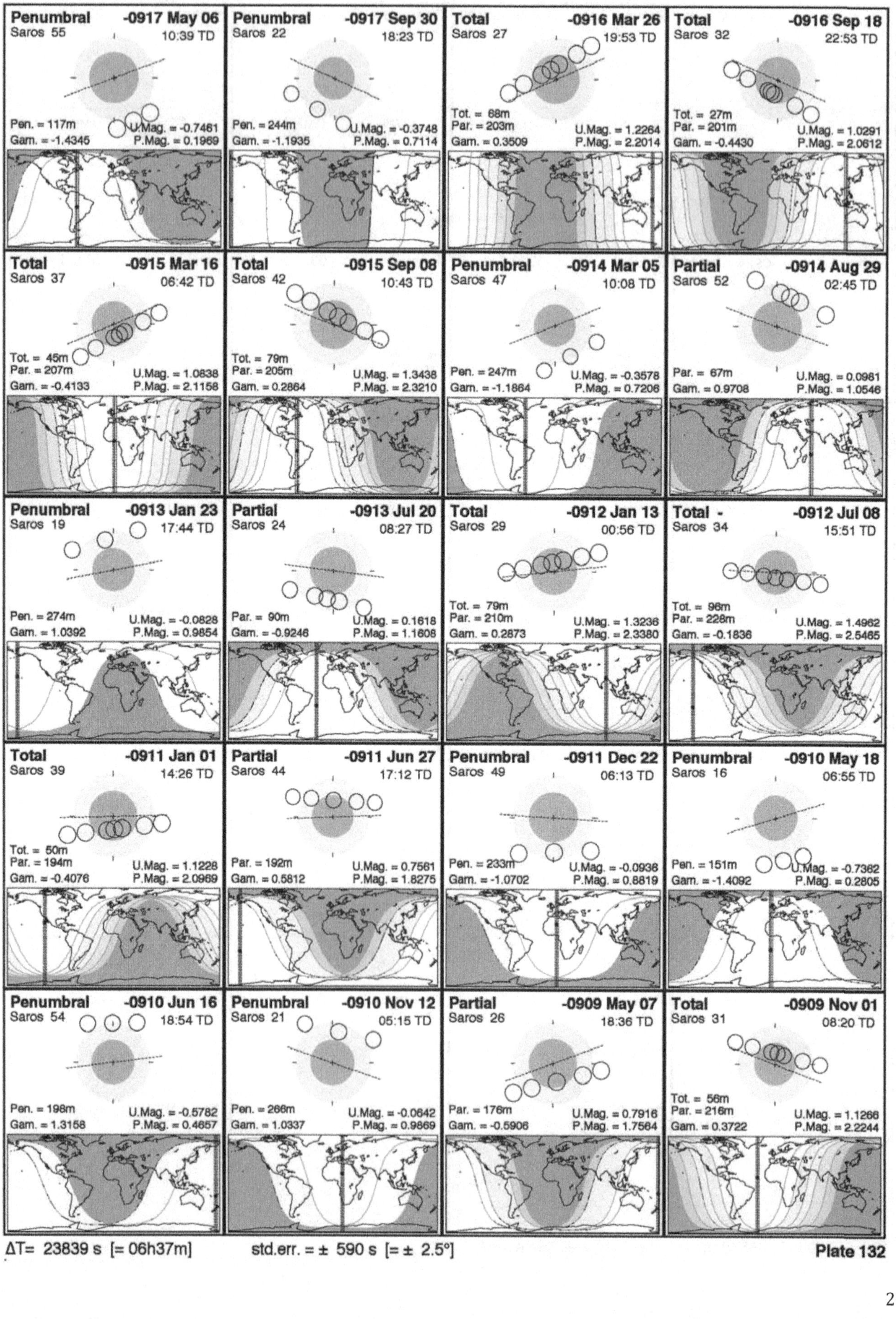

ΔT= 23839 s [= 06h37m] std.err. = ± 590 s [= ± 2.5°] **Plate 132**

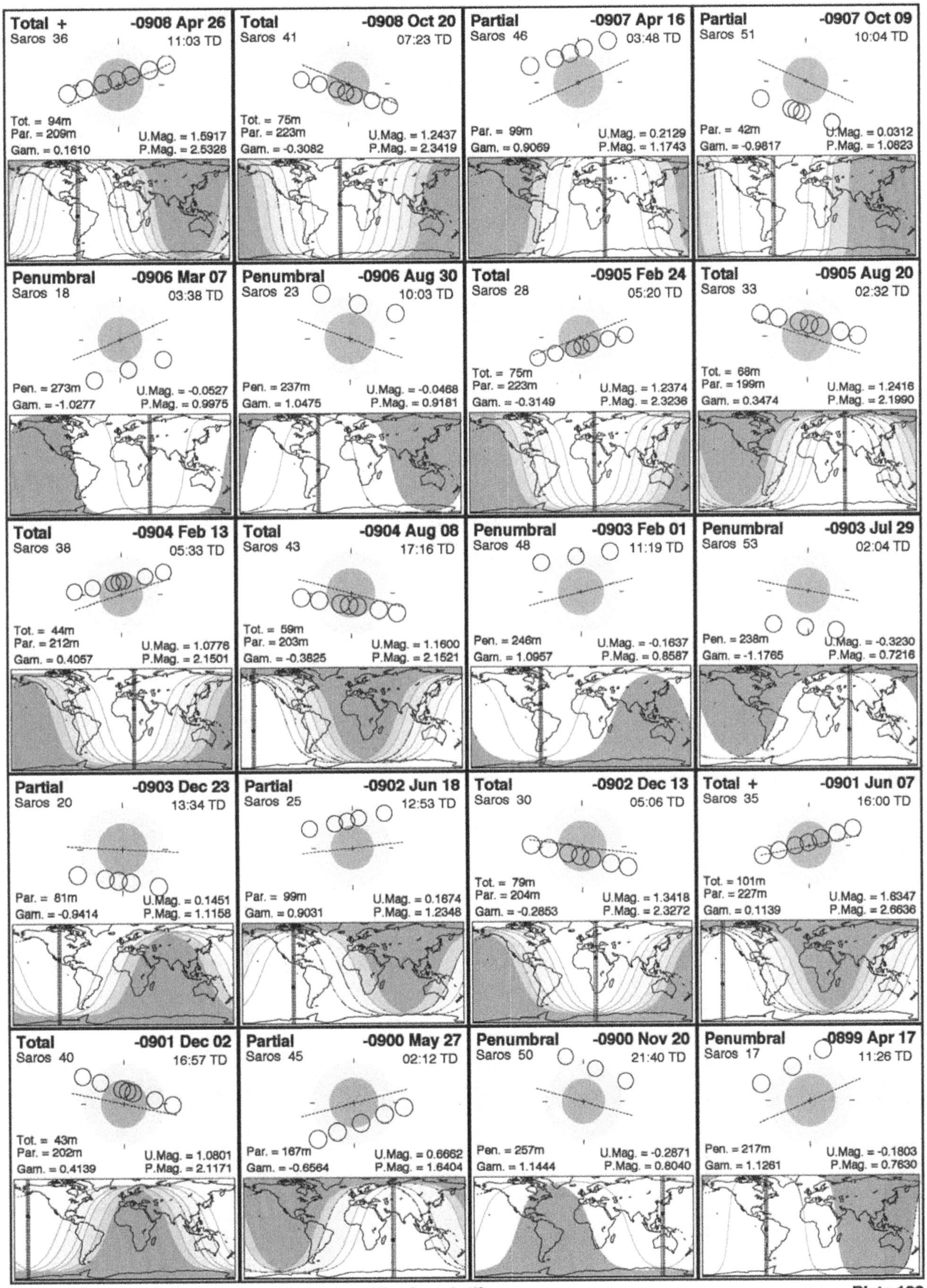

ΔT= 23683 s [= 06h35m] std.err. = ± 587 s [= ± 2.4°] **Plate 133**

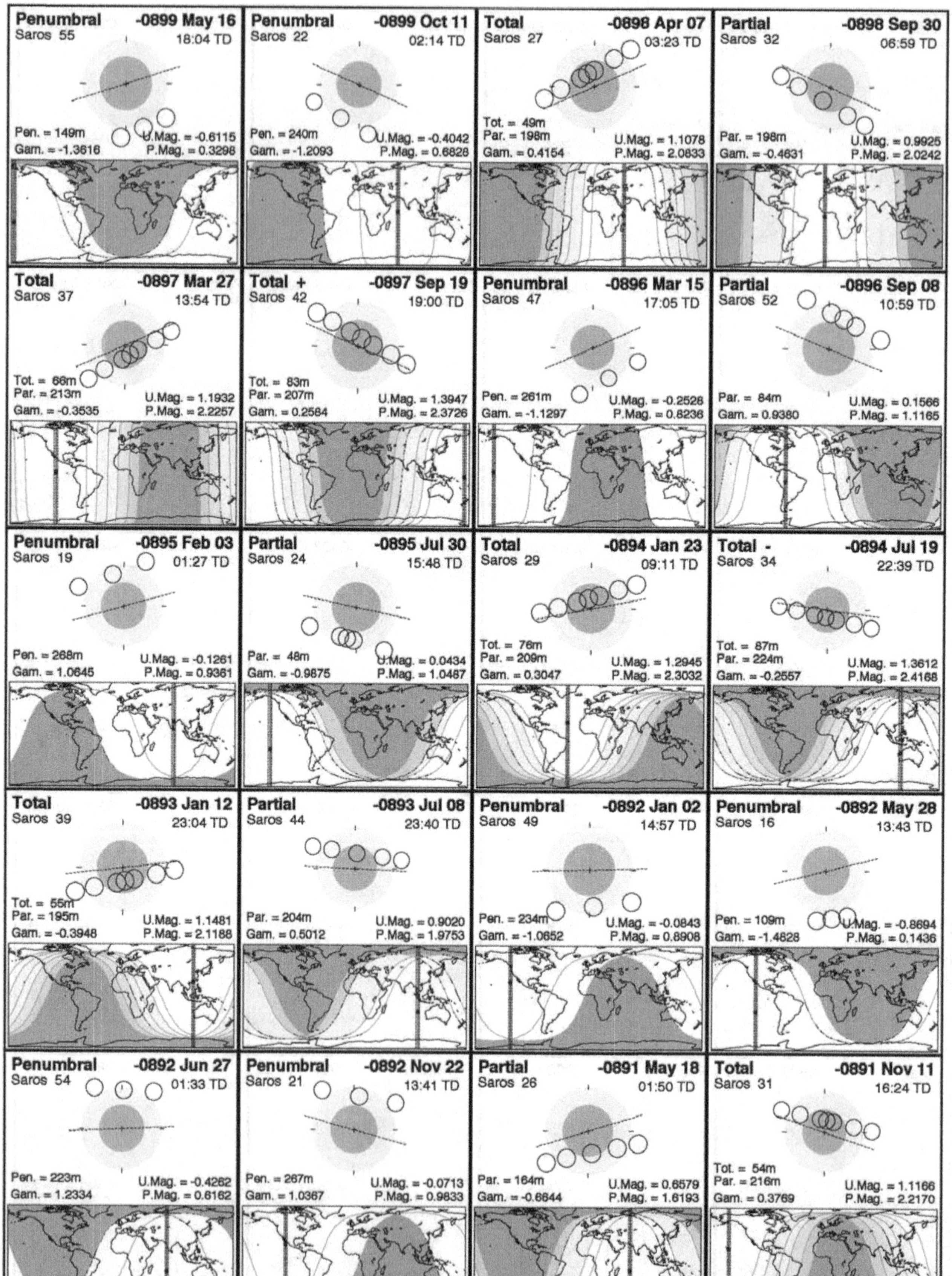

ΔT= 23526 s [= 06h32m] std.err. = ± 583 s [= ± 2.4°] **Plate 134**

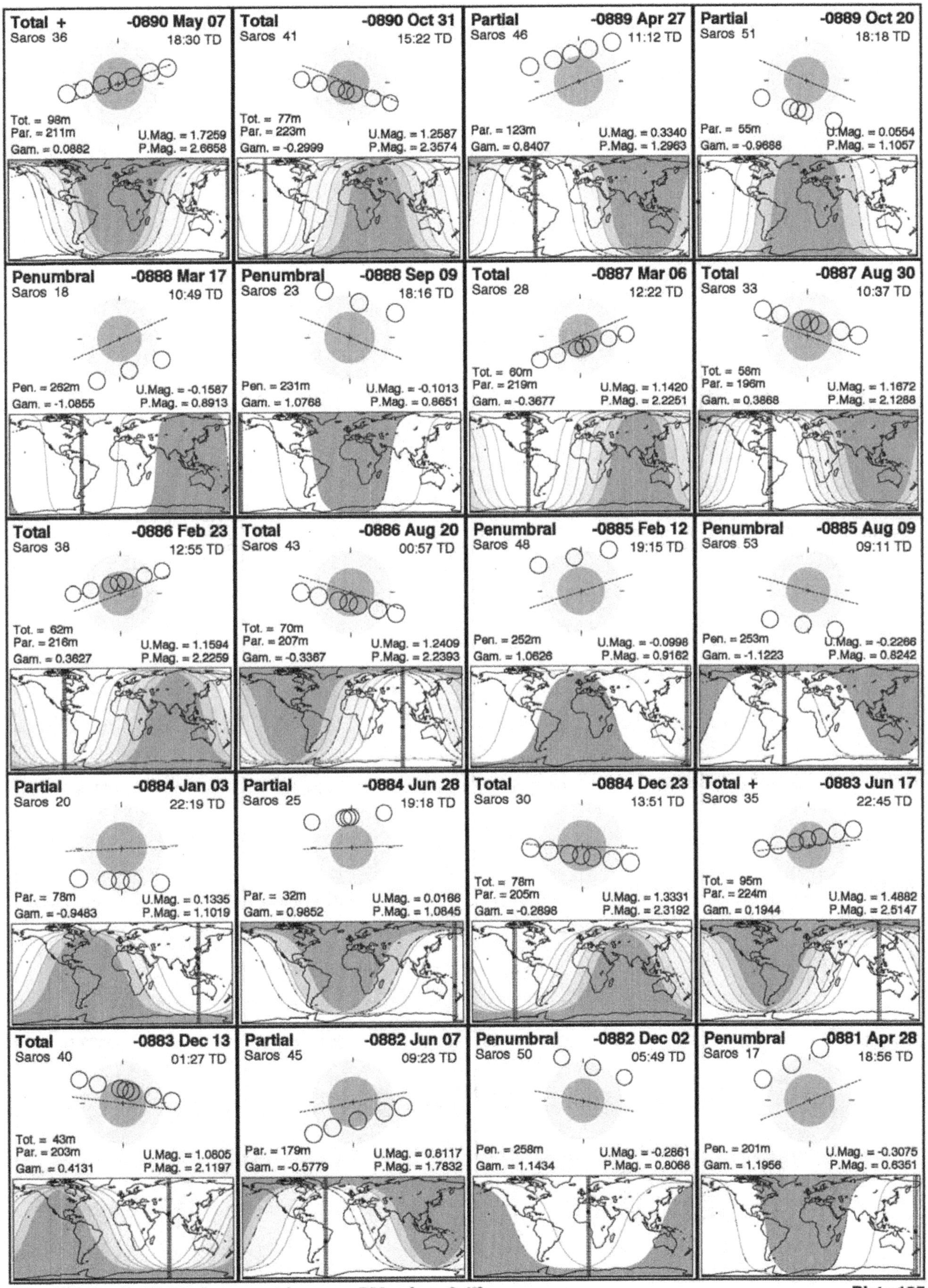

ΔT= 23370 s [= 06h30m] std.err. = ± 580 s [= ± 2.4°] **Plate 135**

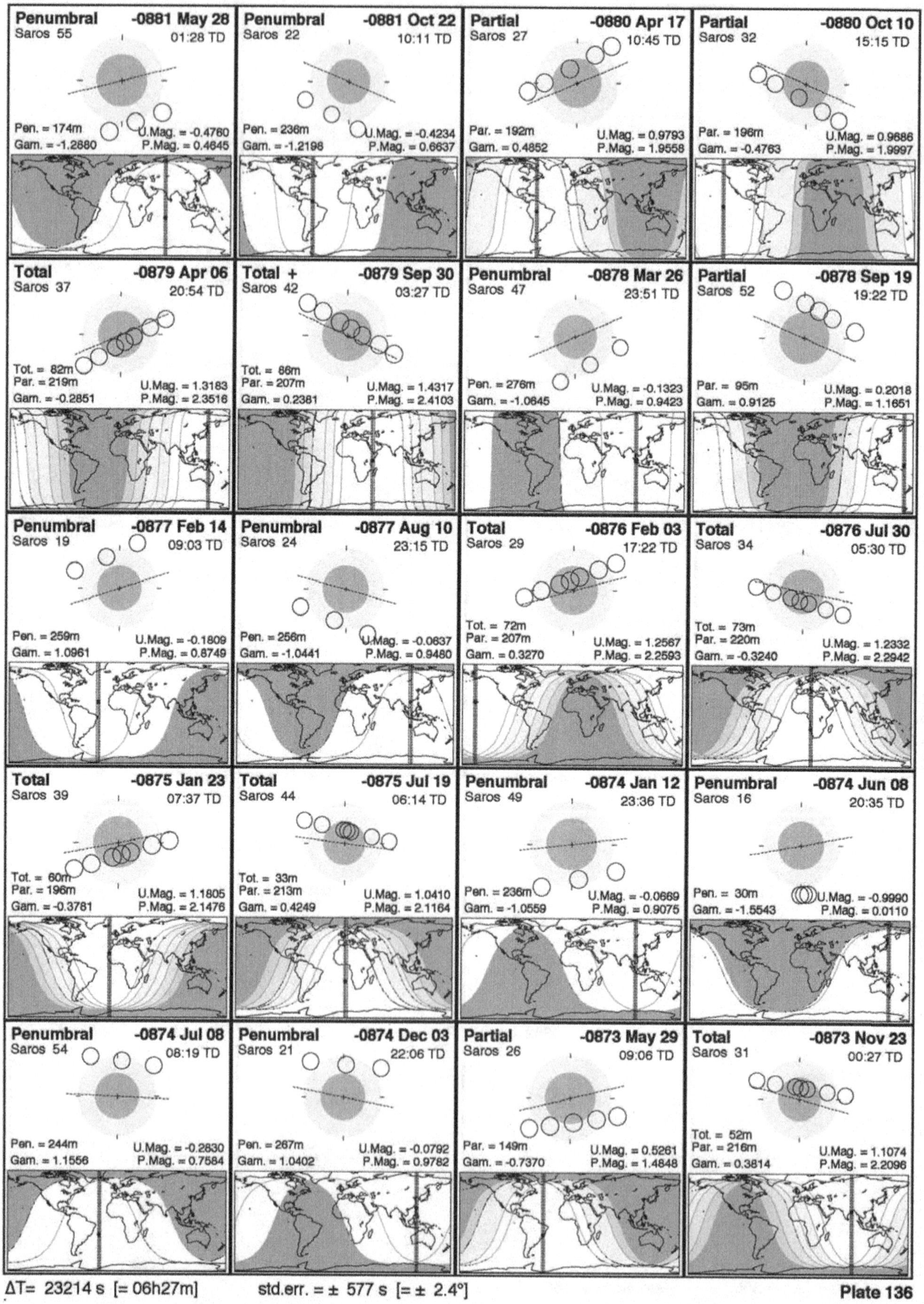

ΔT= 23214 s [= 06h27m] std.err. = ± 577 s [= ± 2.4°] **Plate 136**

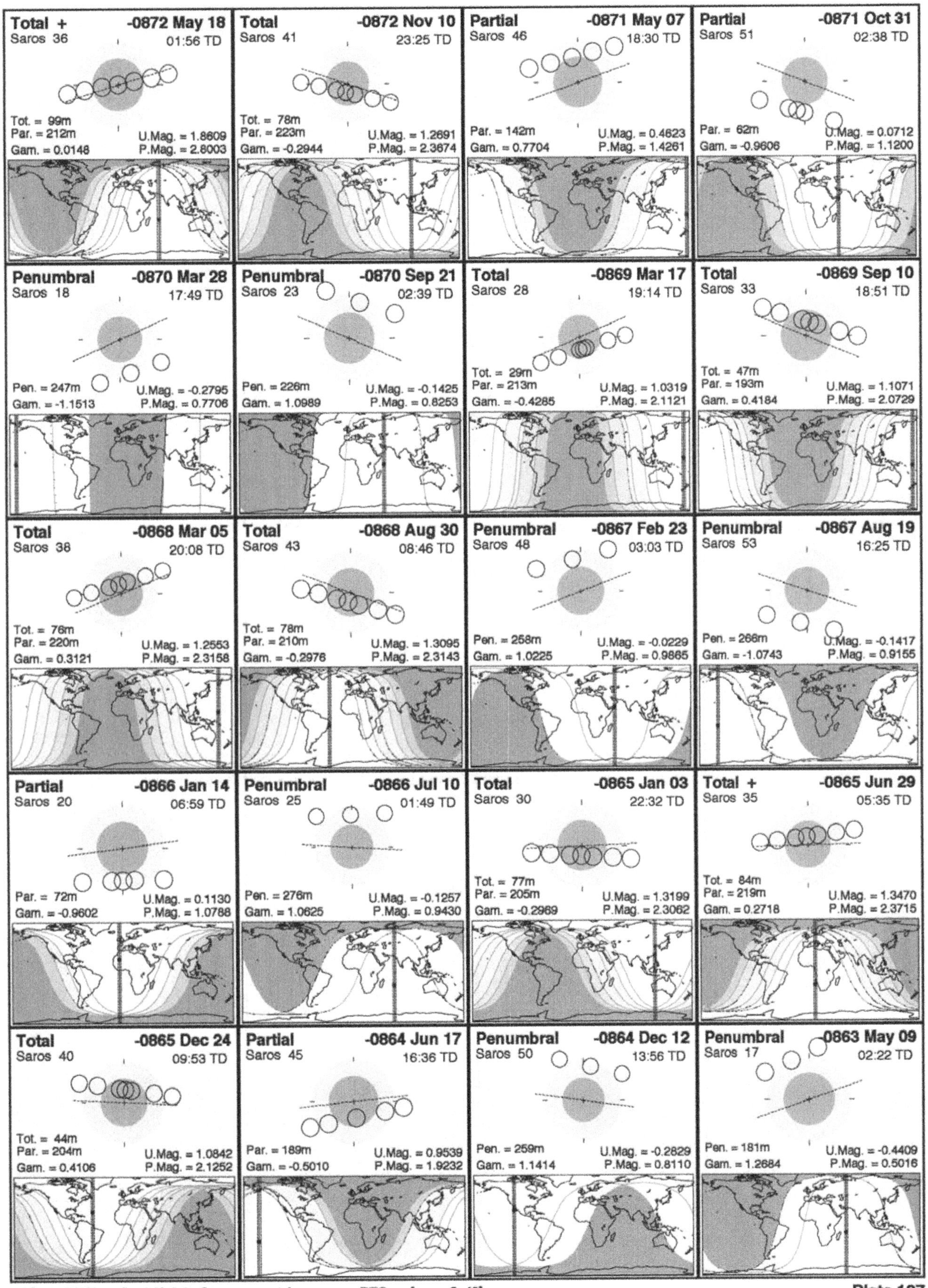

ΔT= 23060 s [= 06h24m] std.err. = ± 573 s [= ± 2.4°] **Plate 137**

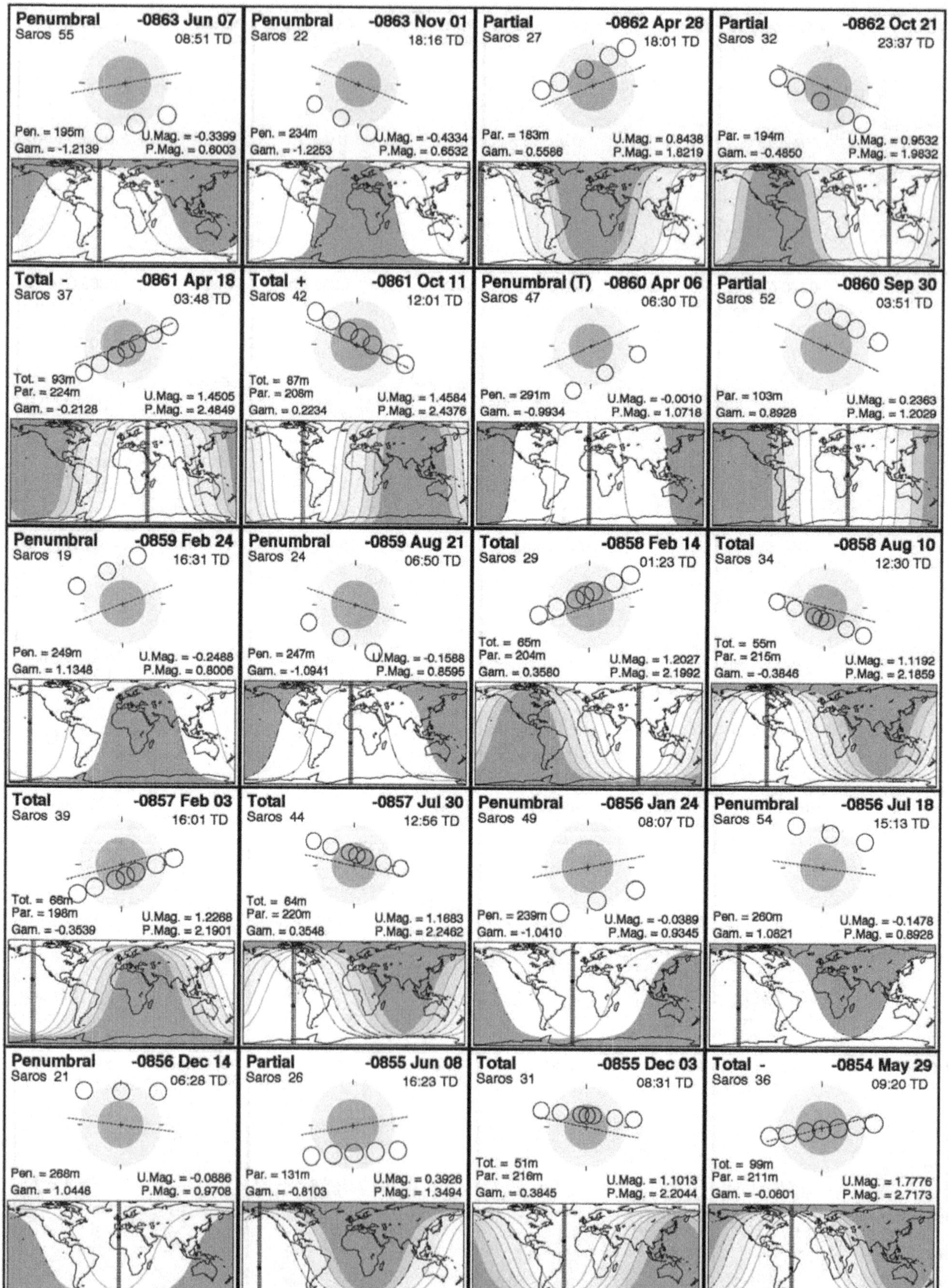

ΔT= 22905 s [= 06h22m] std.err. = ± 570 s [= ± 2.4°] **Plate 138**

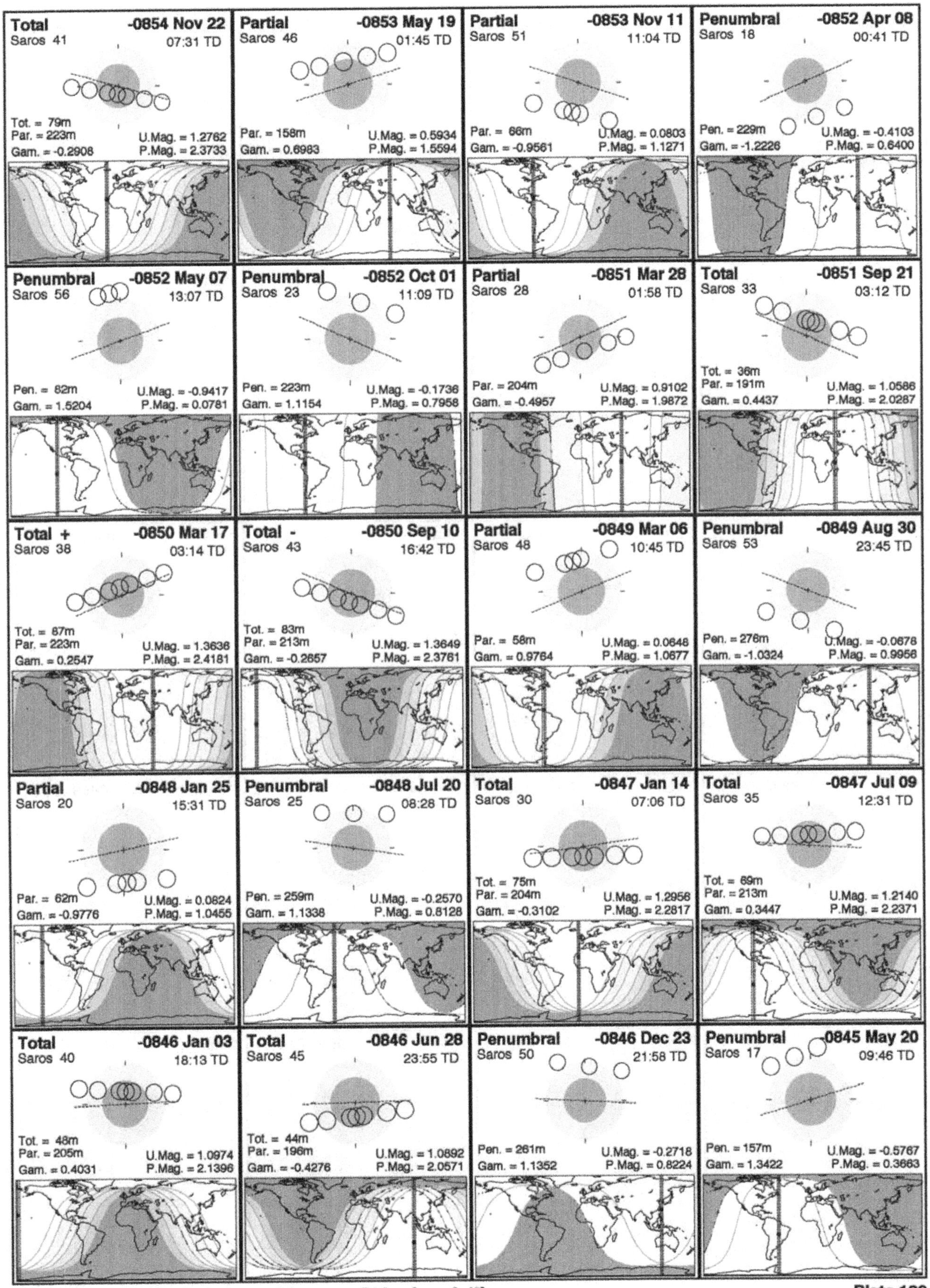

ΔT= 22744 s [= 06h19m] std.err. = ± 566 s [= ± 2.4°] **Plate 139**

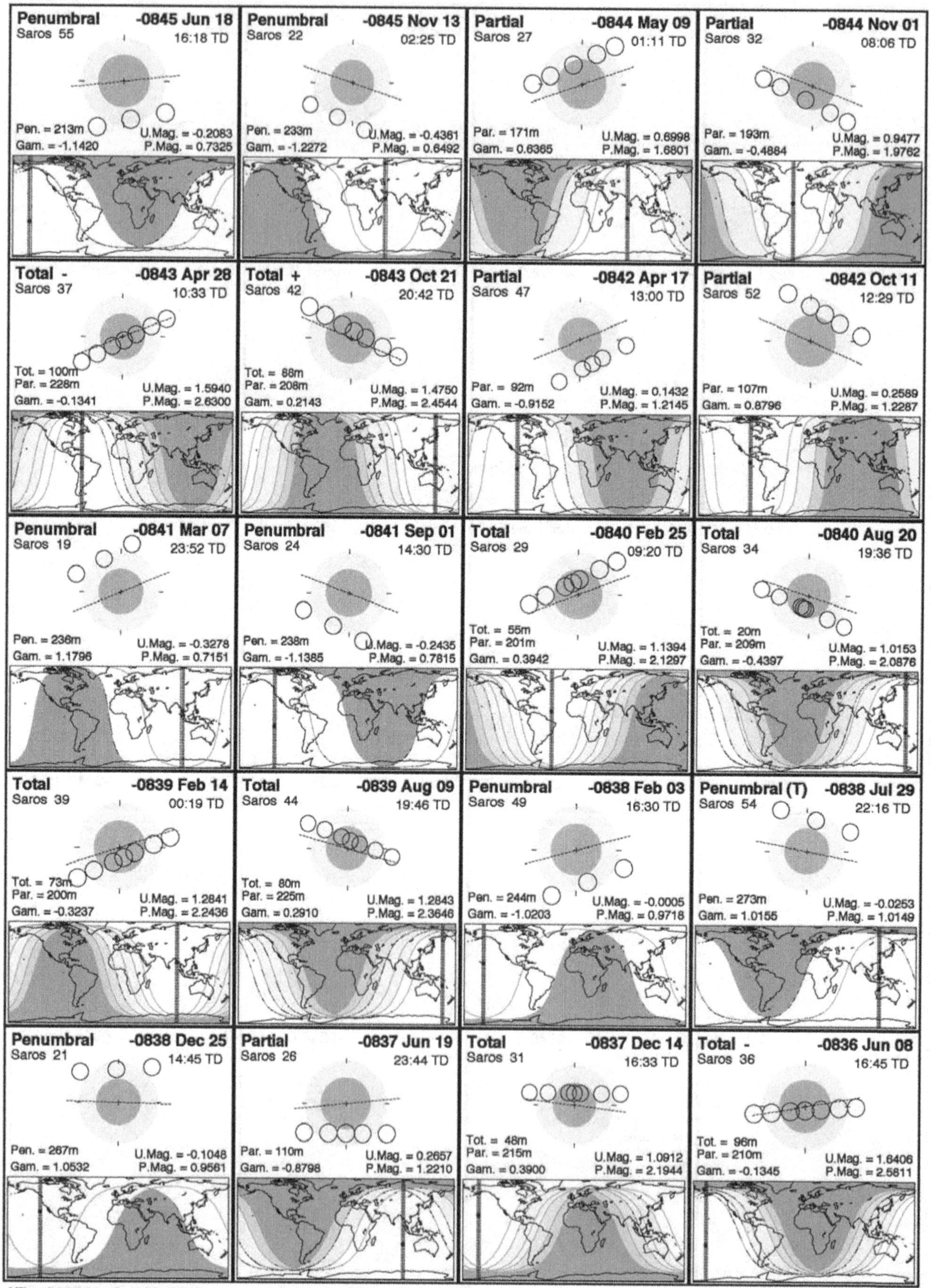

ΔT= 22598 s [= 06h17m] std.err. = ± 563 s [= ± 2.3°] **Plate 140**

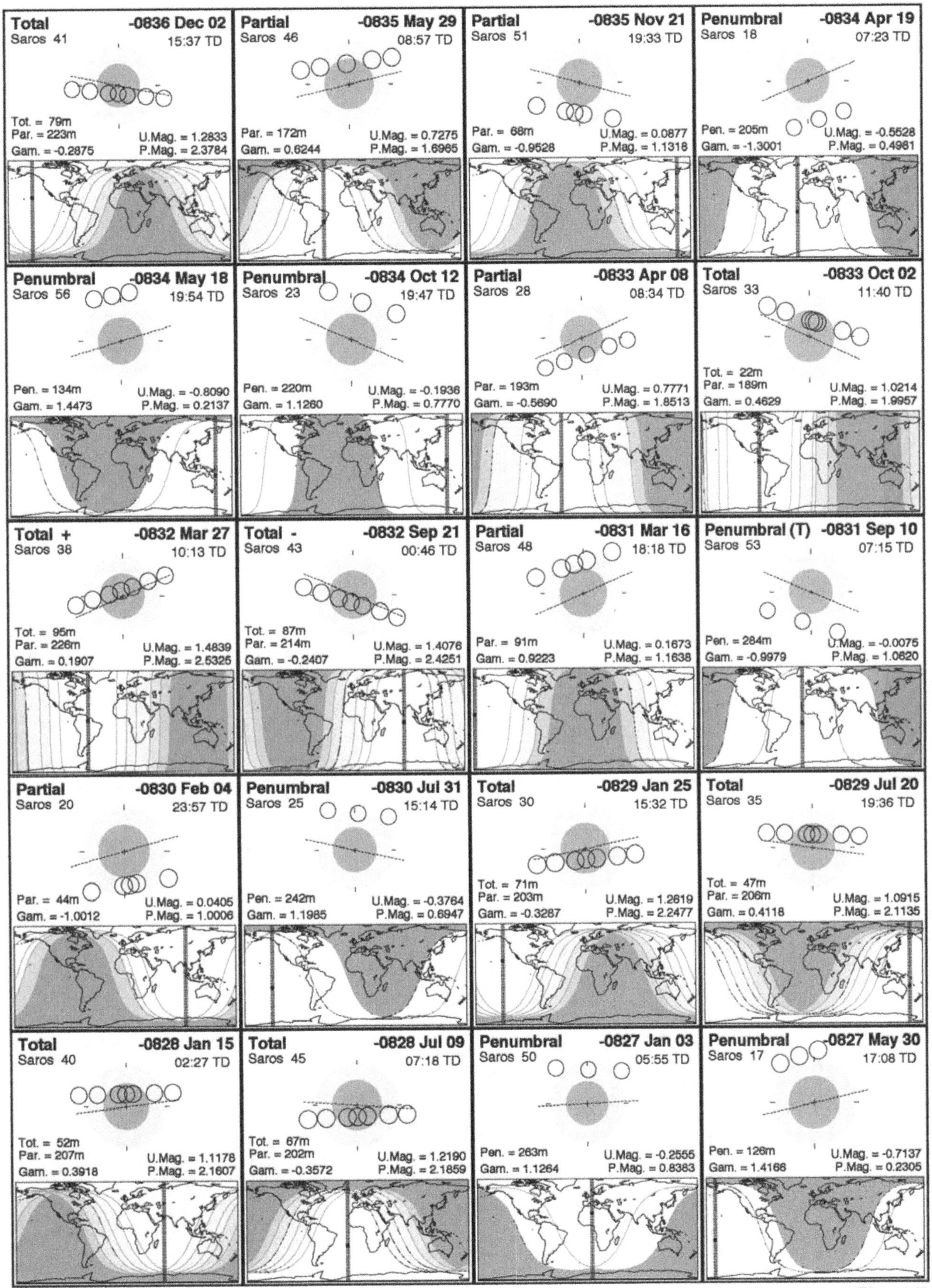

ΔT= 22438 s [= 06h14m] std.err. = ± 559 s [= ± 2.3°] **Plate 141**

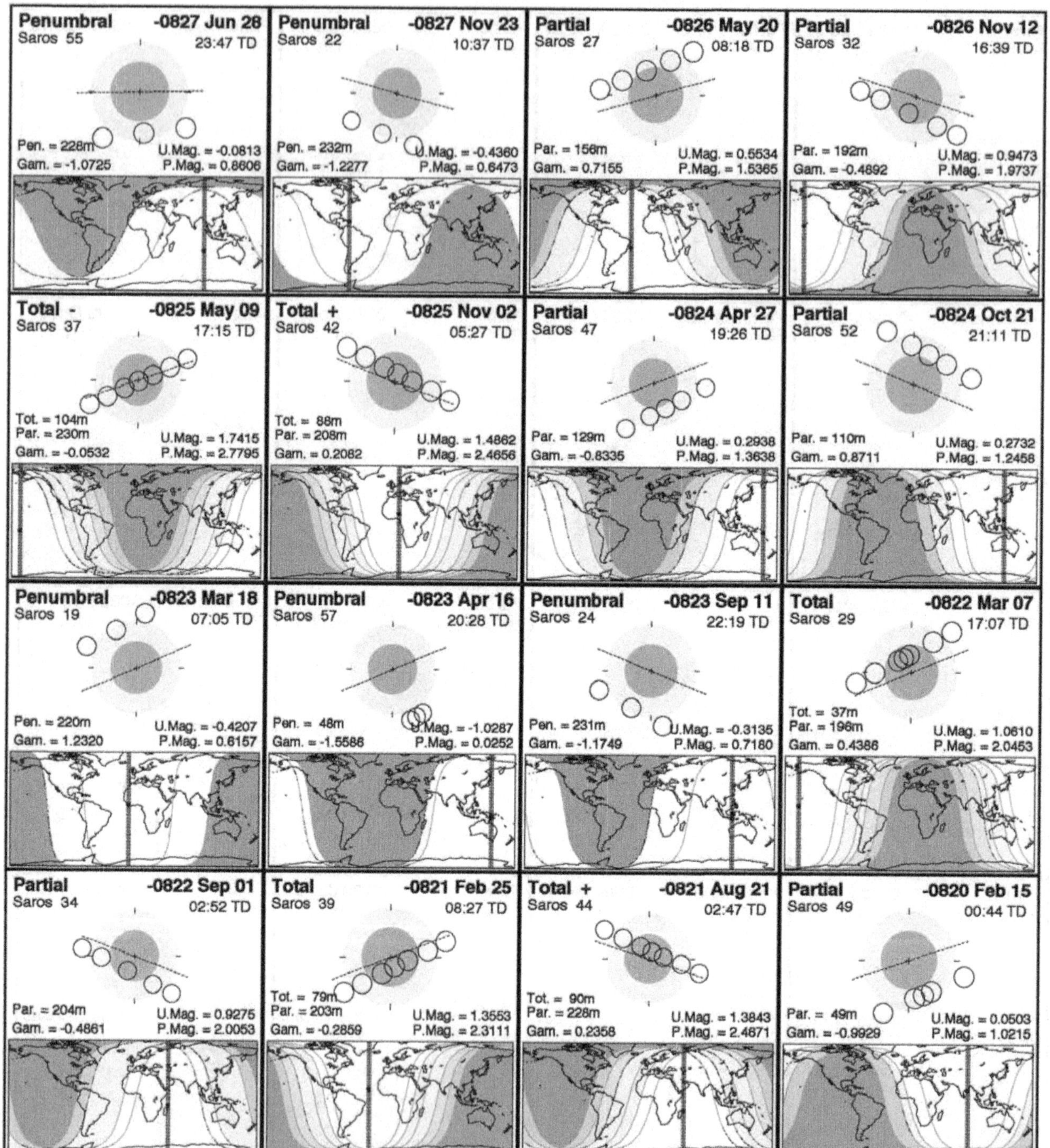

ΔT= 22293 s [= 06h12m]   std.err. = ± 556 s [= ± 2.3°]   **Plate 142**

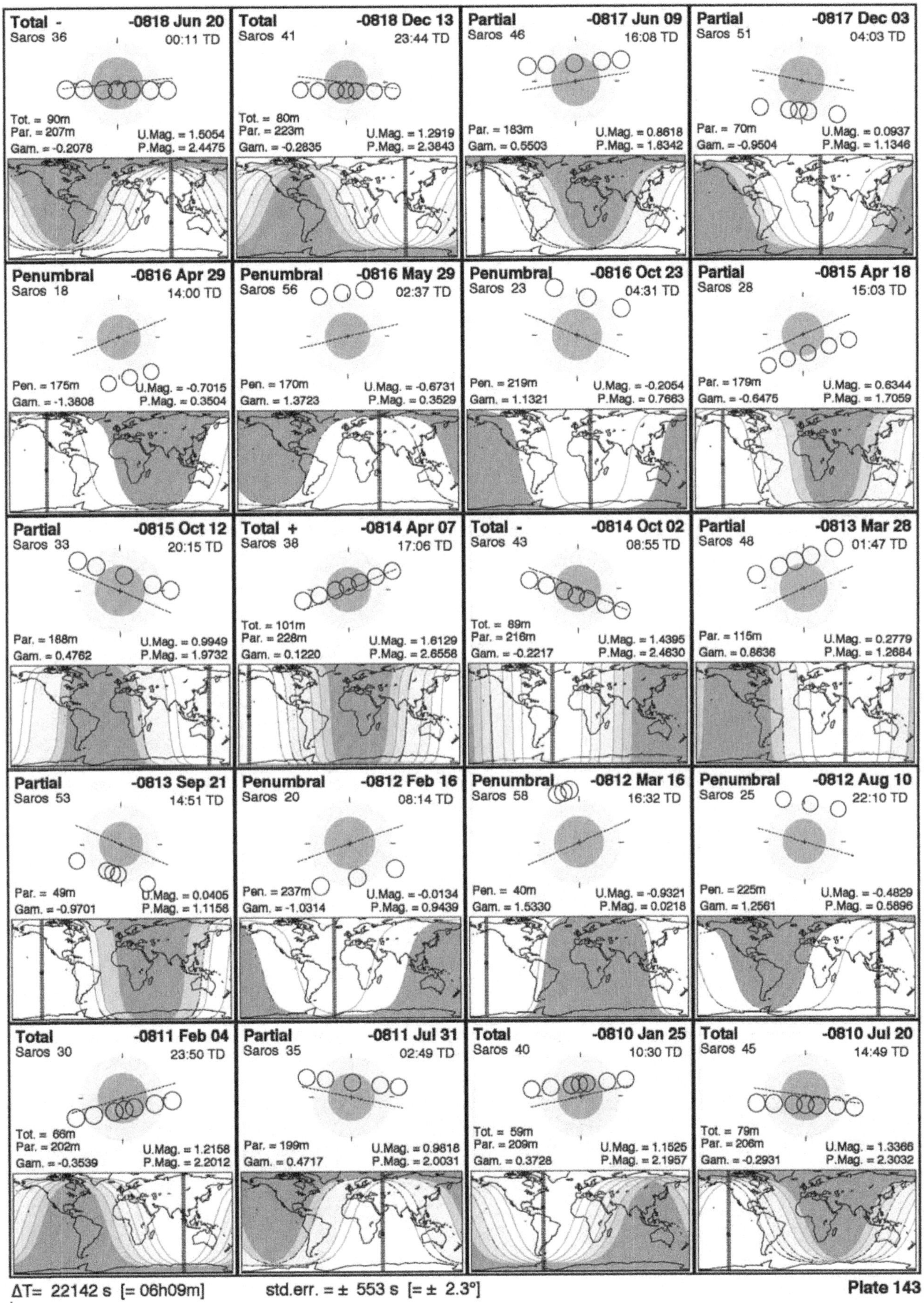

ΔT= 22142 s [= 06h09m] std.err. = ± 553 s [= ± 2.3°] **Plate 143**

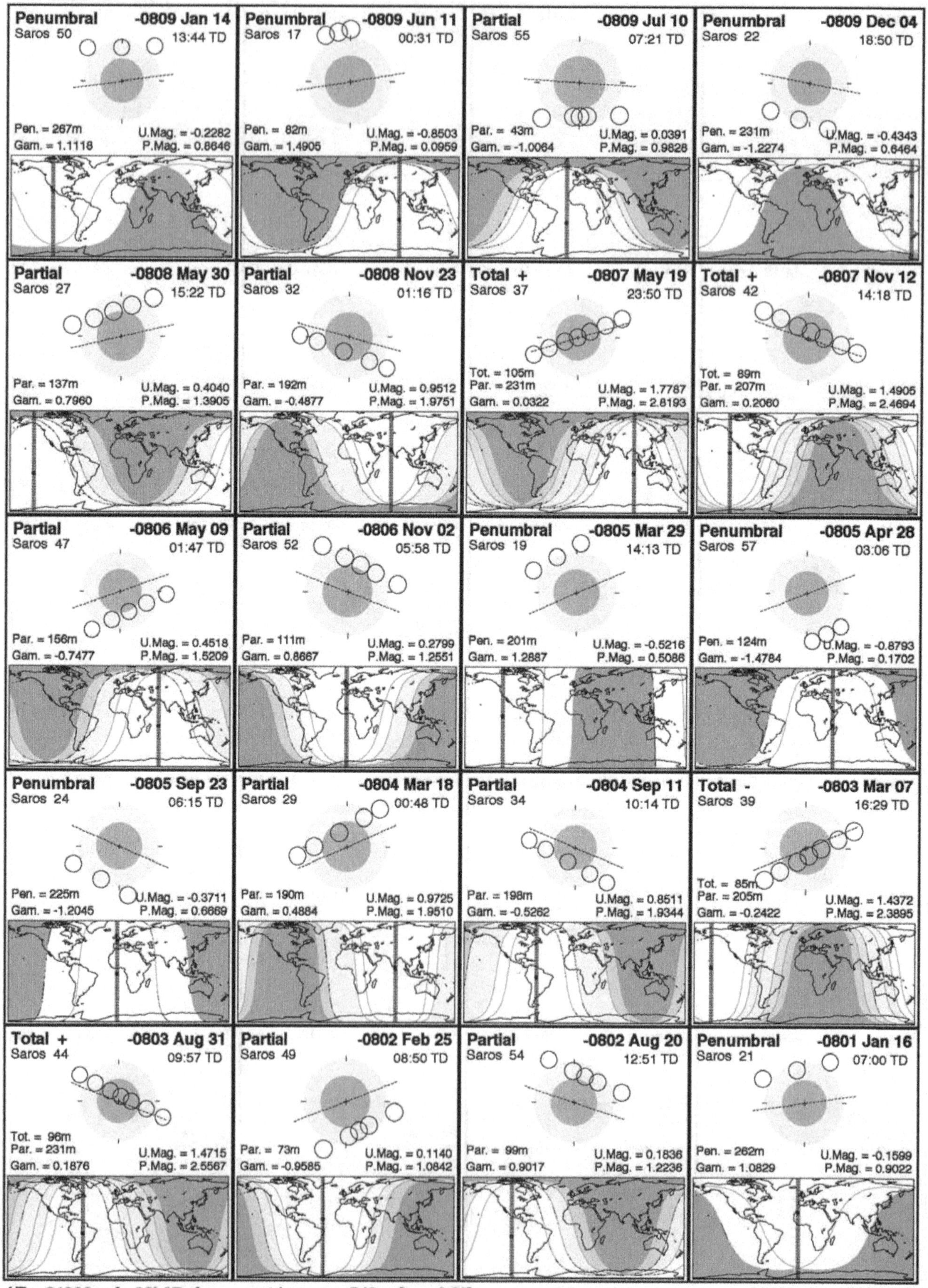

ΔT= 21998 s [= 06h07m] std.err. = ± 549 s [= ± 2.3°] **Plate 144**

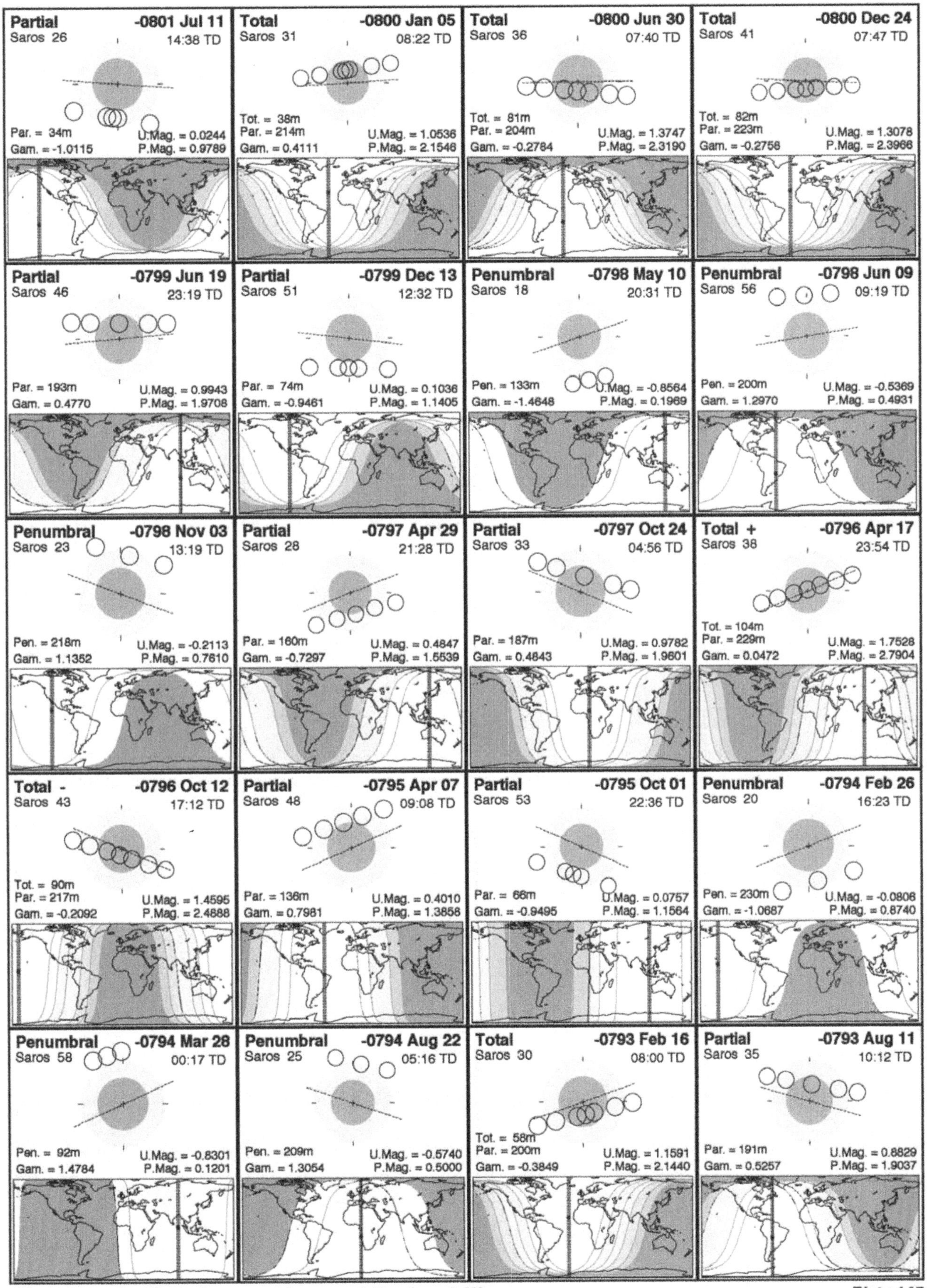

ΔT= 21856 s [= 06h04m] std.err. = ± 546 s [= ± 2.3°] Plate 145

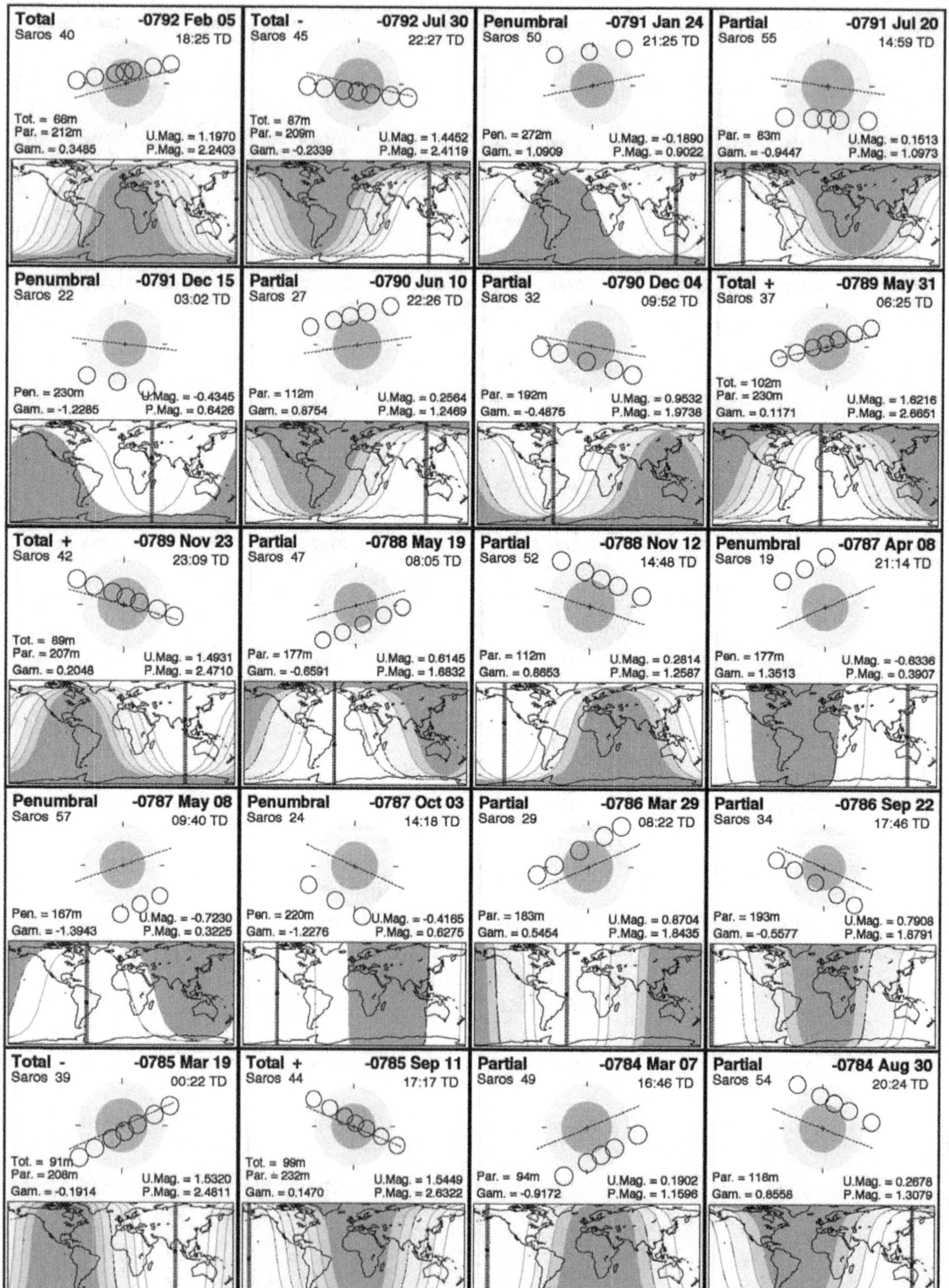

ΔT= 21713 s [= 06h02m] std.err. = ± 543 s [= ± 2.3°] **Plate 146**

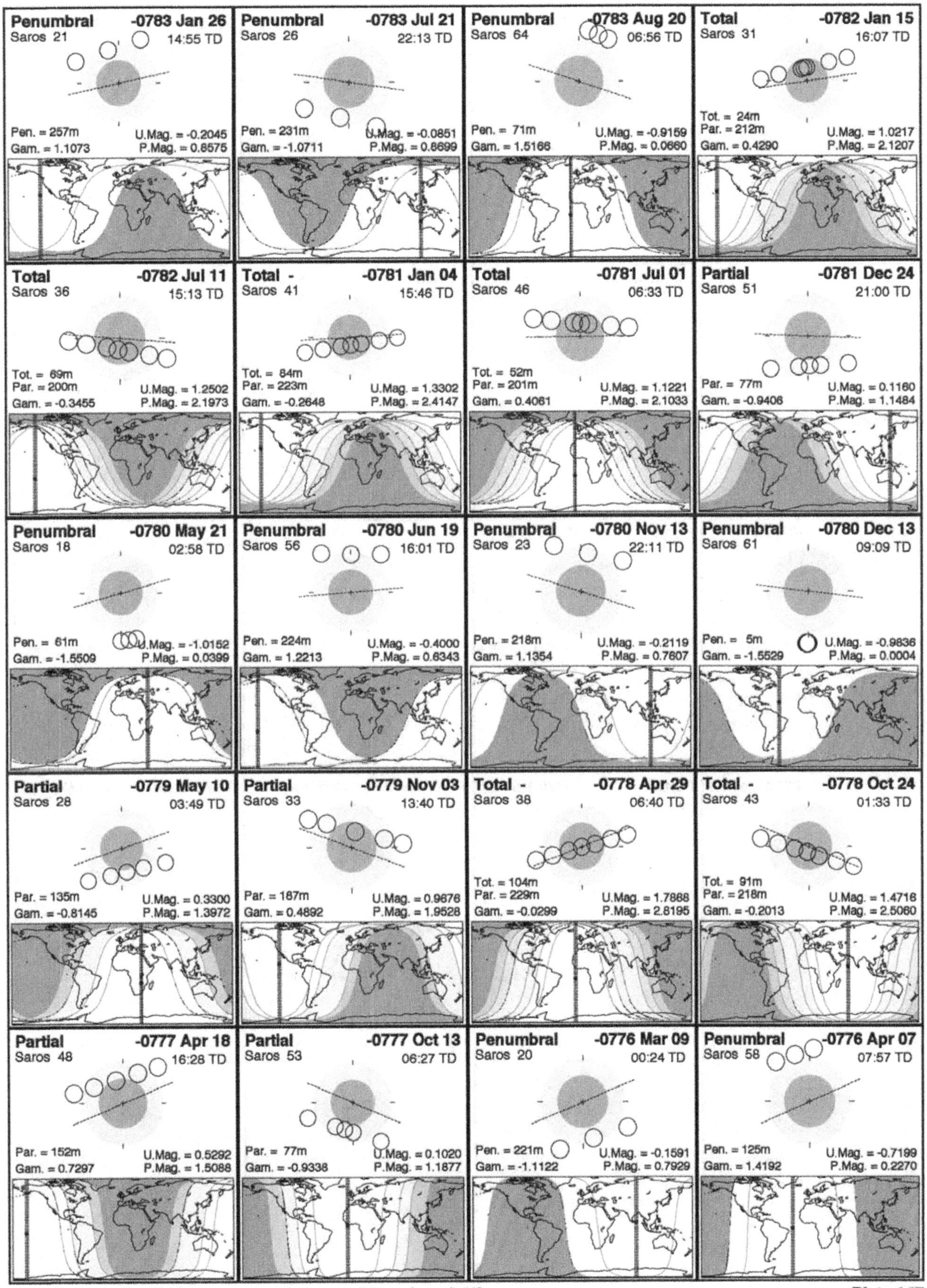

ΔT= 21564 s [= 05h59m] std.err. = ± 539 s [= ± 2.2°] **Plate 147**

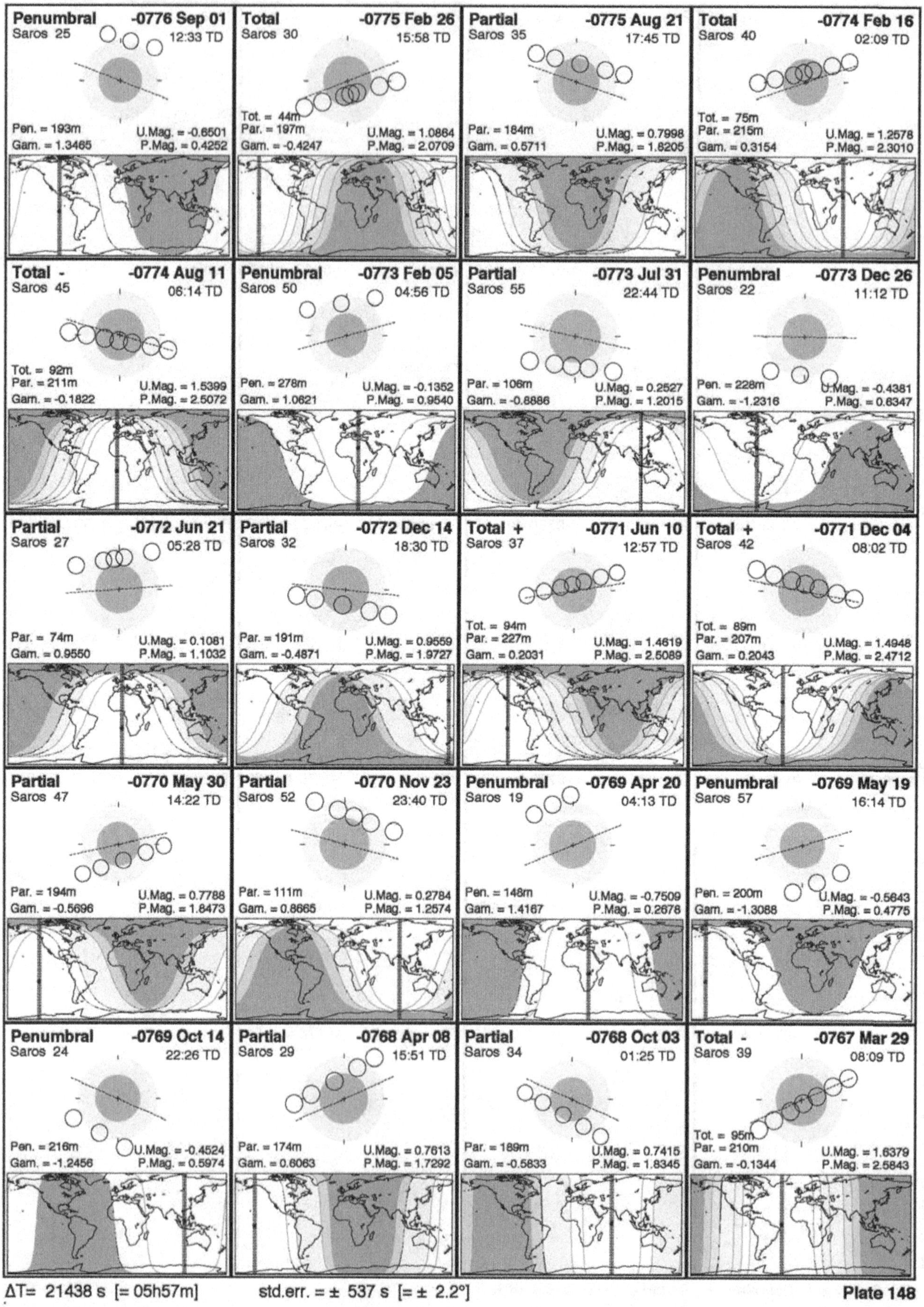

ΔT= 21438 s [= 05h57m] std.err. = ± 537 s [= ± 2.2°] **Plate 148**

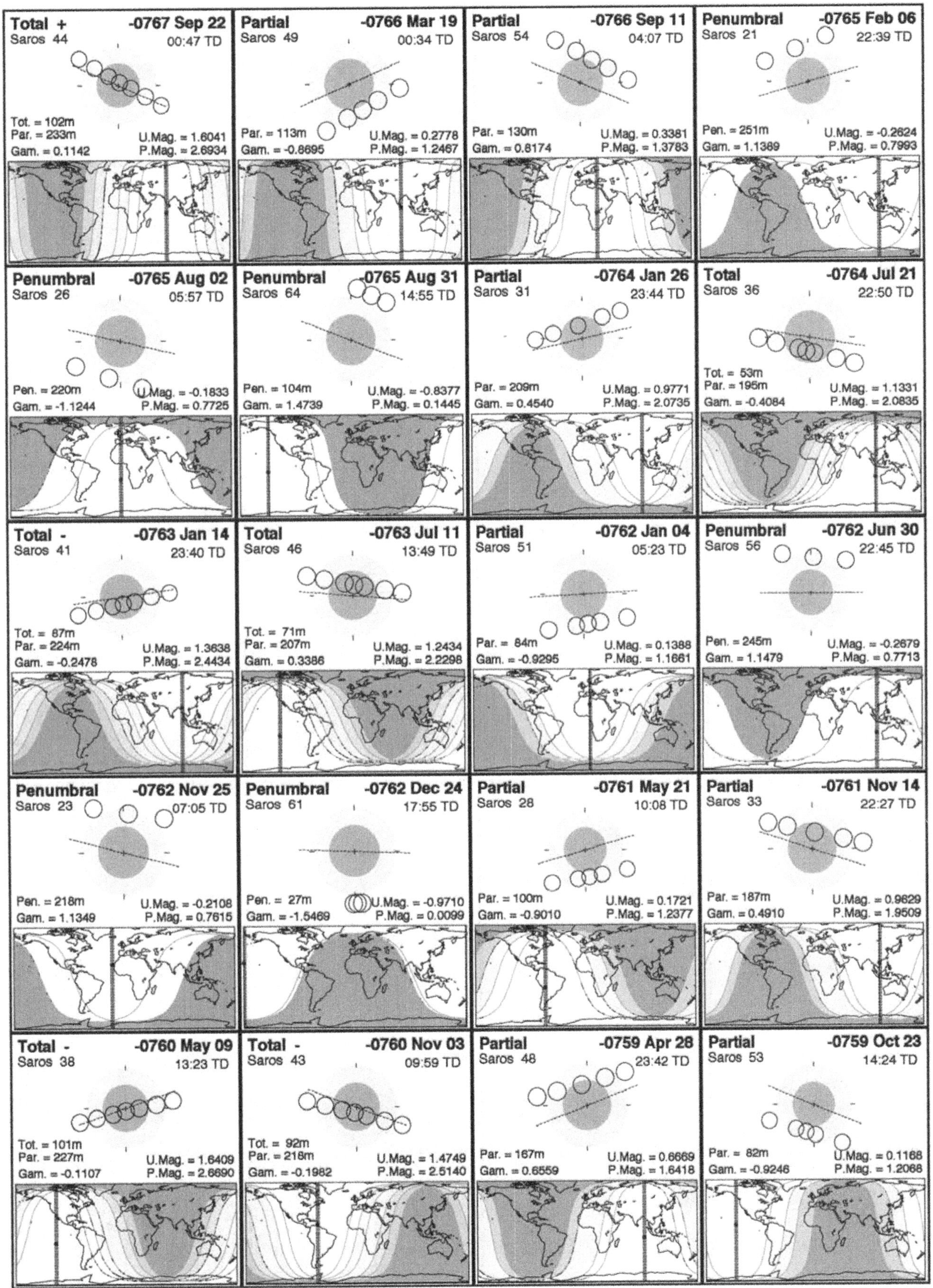

ΔT= 21288 s [= 05h55m] std.err. = ± 533 s [= ± 2.2°] **Plate 149**

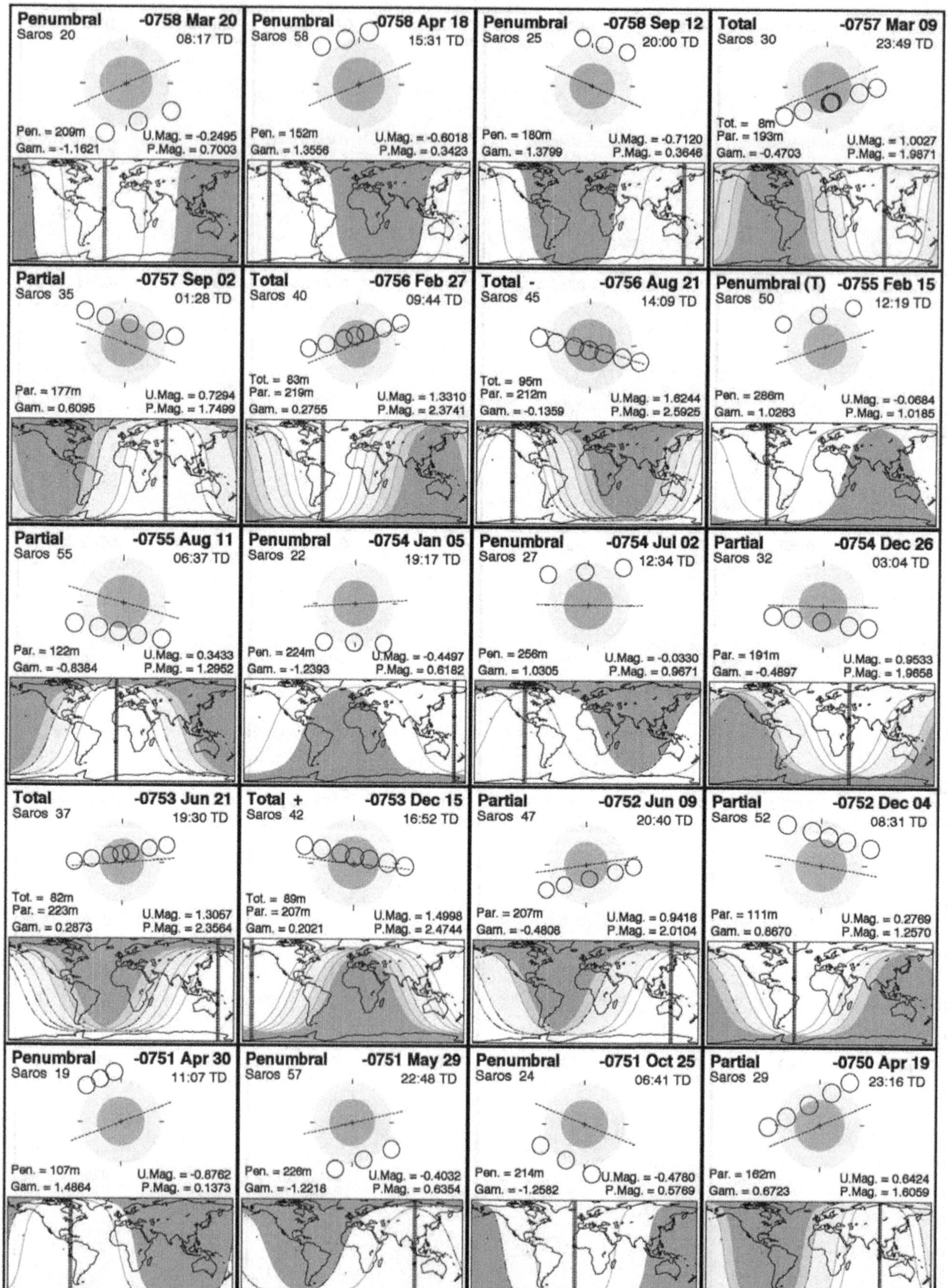

ΔT= 21149 s [= 05h52m] std.err. = ± 530 s [= ± 2.2°] **Plate 150**

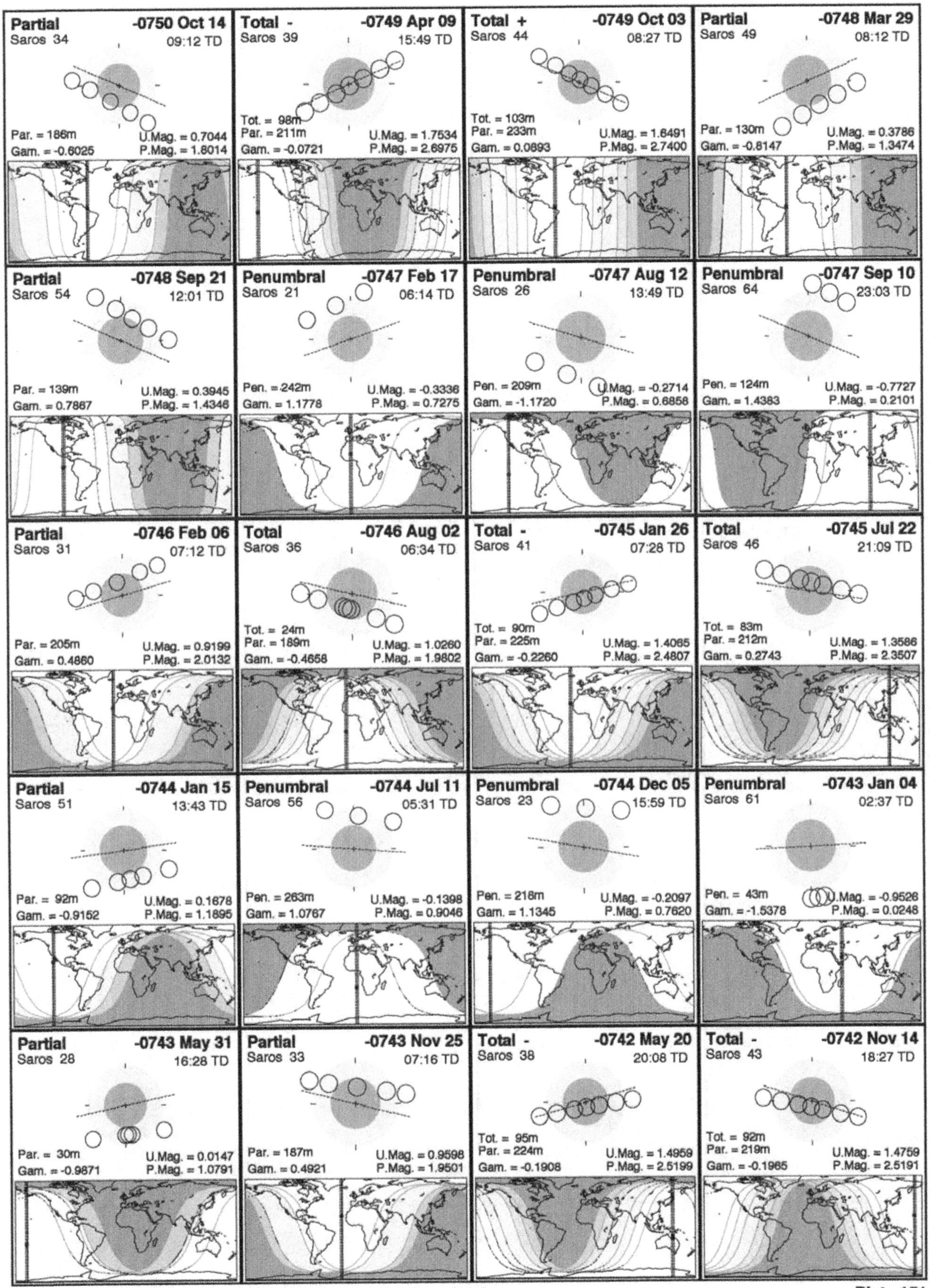

ΔT= 21008 s [= 05h50m] std.err. = ± 527 s [= ± 2.2°] **Plate 151**

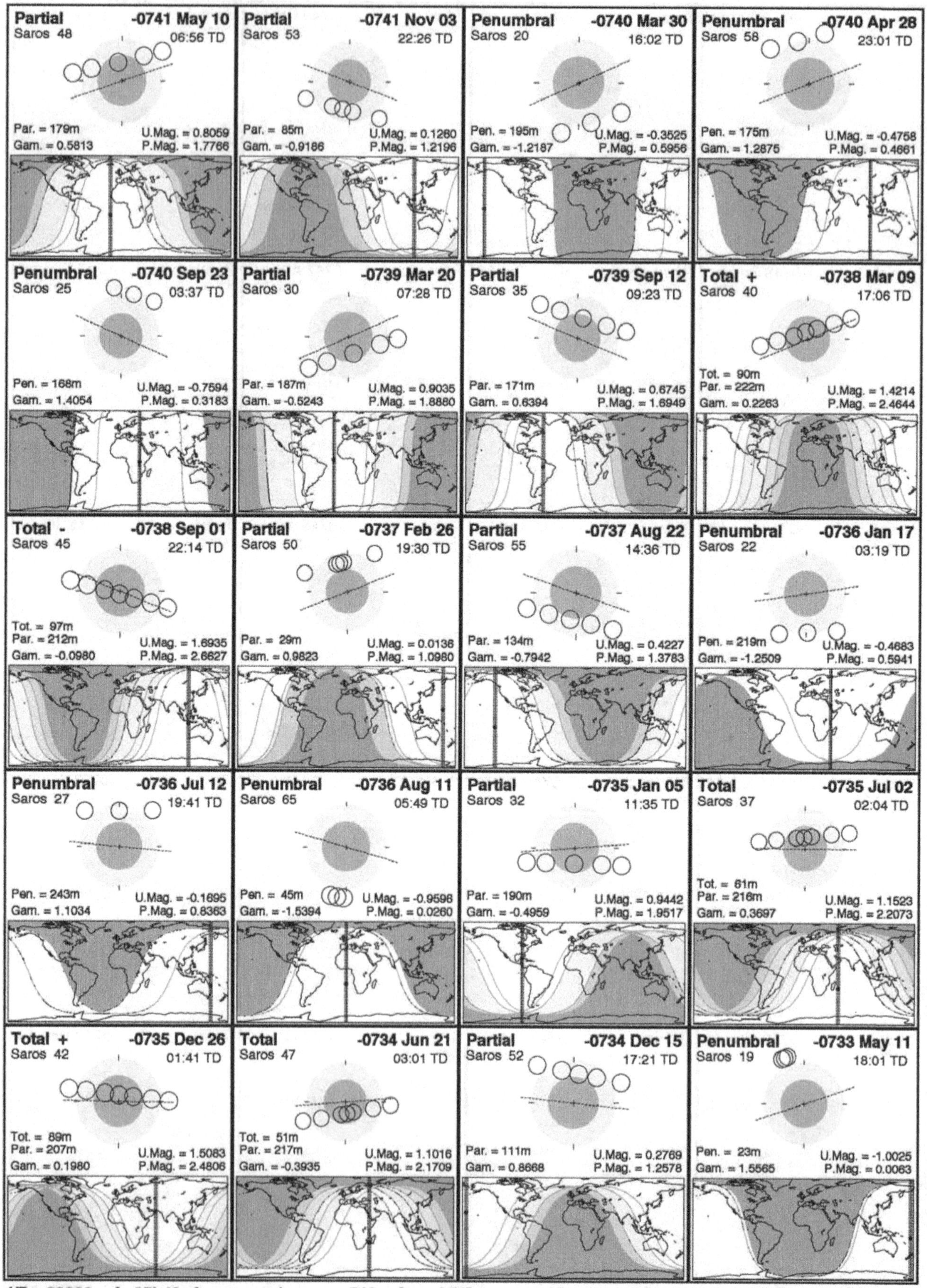

ΔT= 20868 s [= 05h48m] std.err. = ± 523 s [= ± 2.2°] **Plate 152**

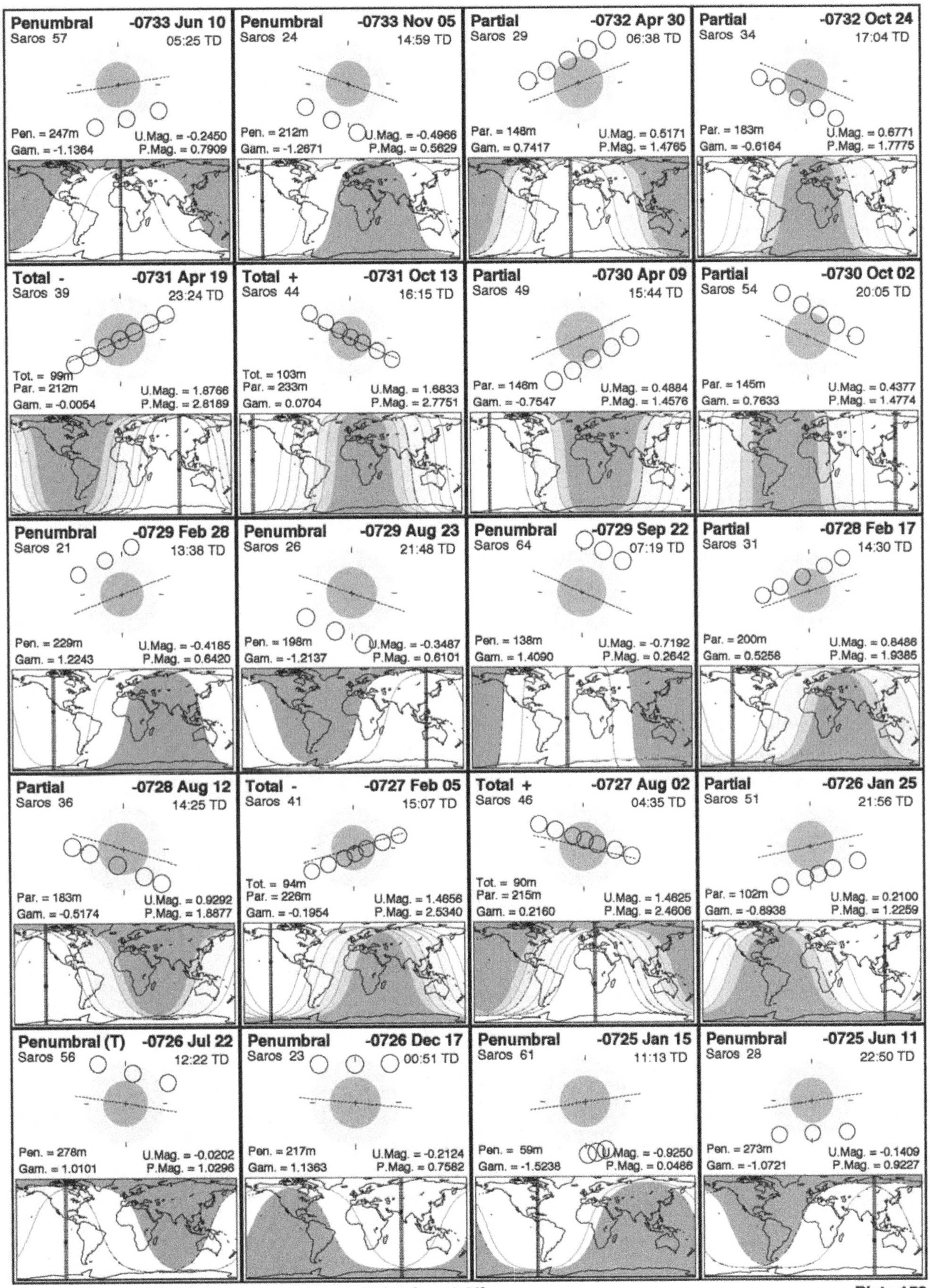

ΔT= 20736 s [= 05h46m] std.err. = ± 520 s [= ± 2.2°] **Plate 153**

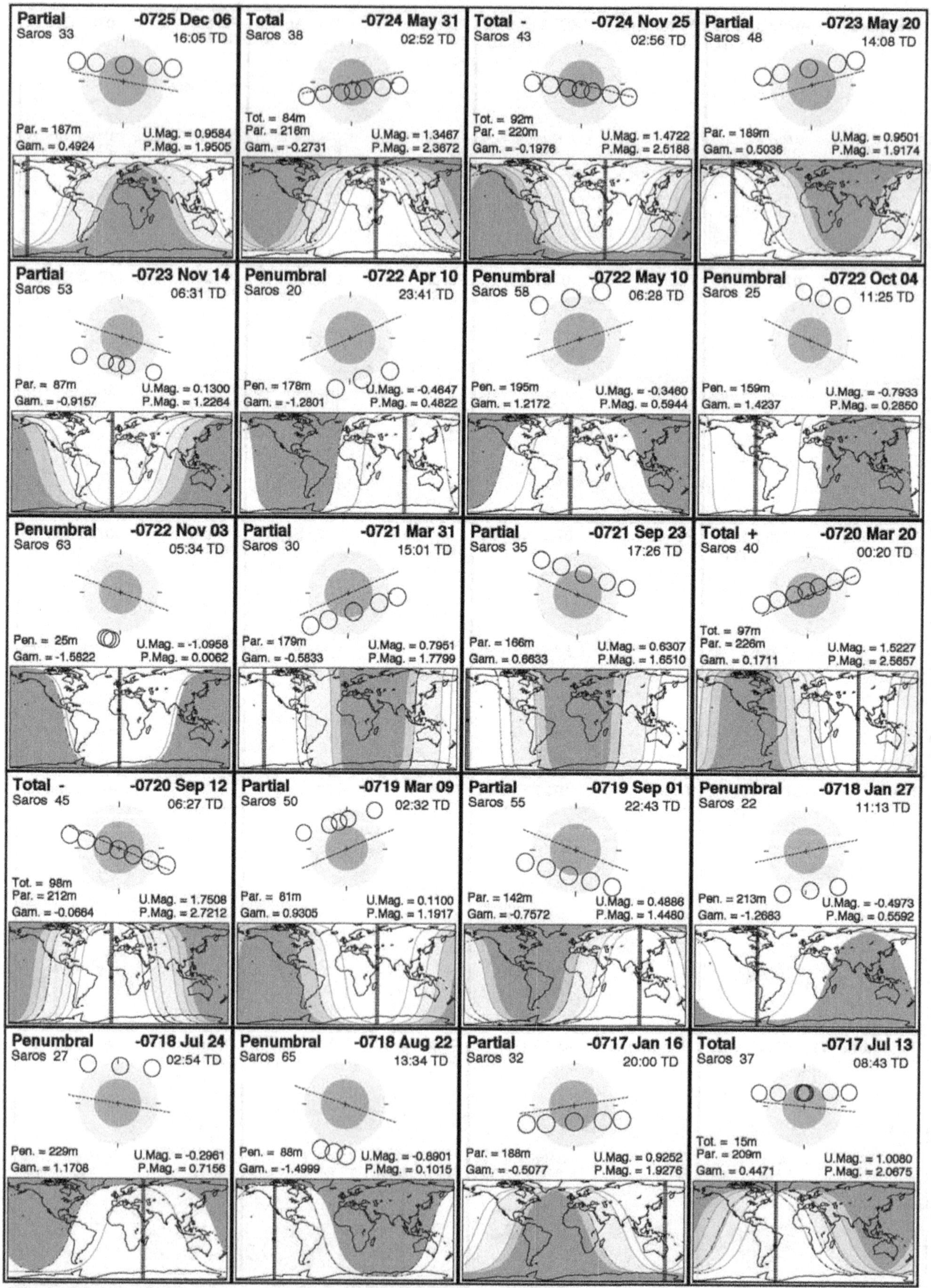

ΔT= 20598 s [= 05h43m] std.err. = ± 517 s [= ± 2.2°] **Plate 154**

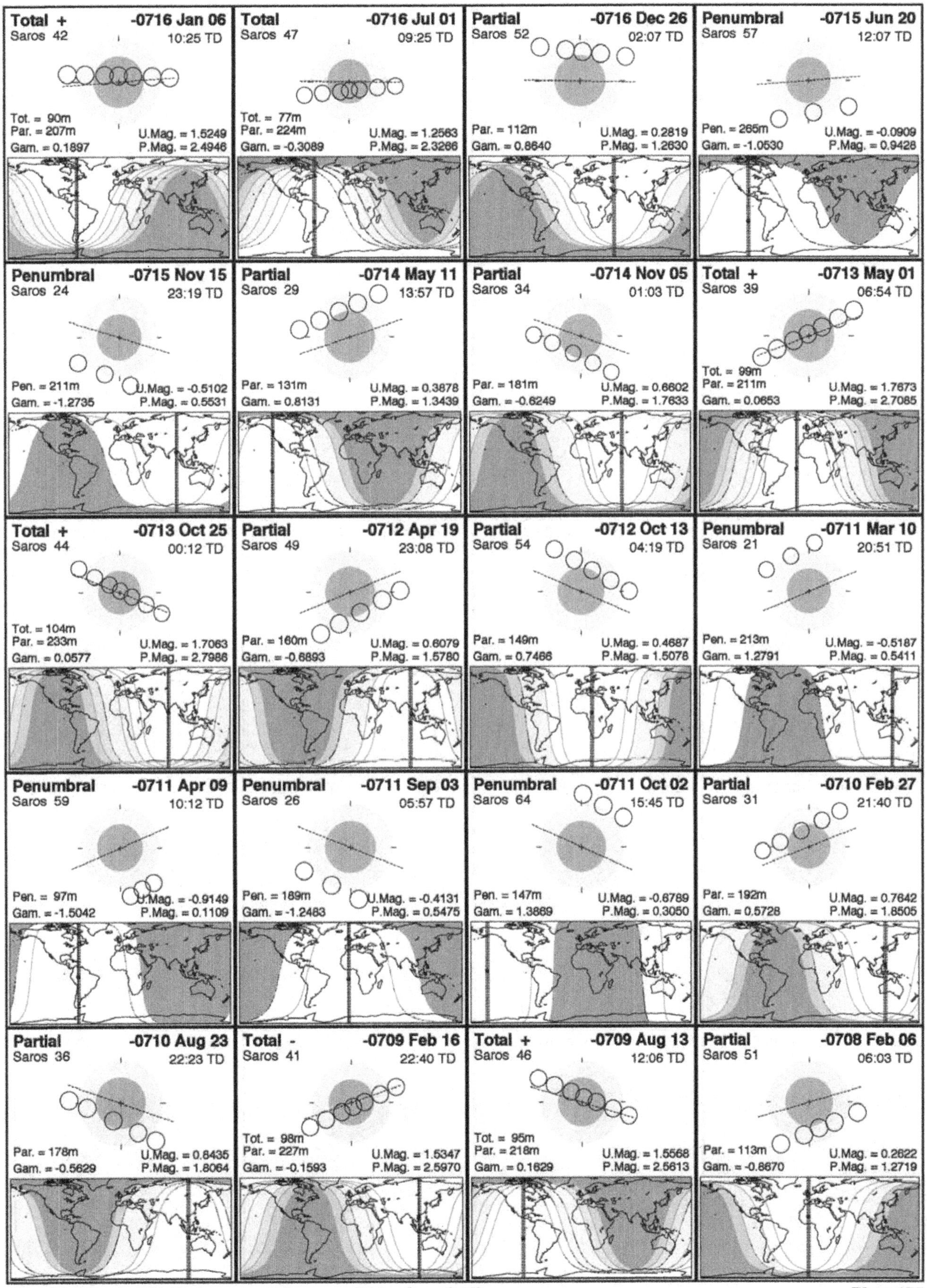

ΔT= 20468 s [= 05h41m] std.err. = ± 514 s [= ± 2.1°] **Plate 155**

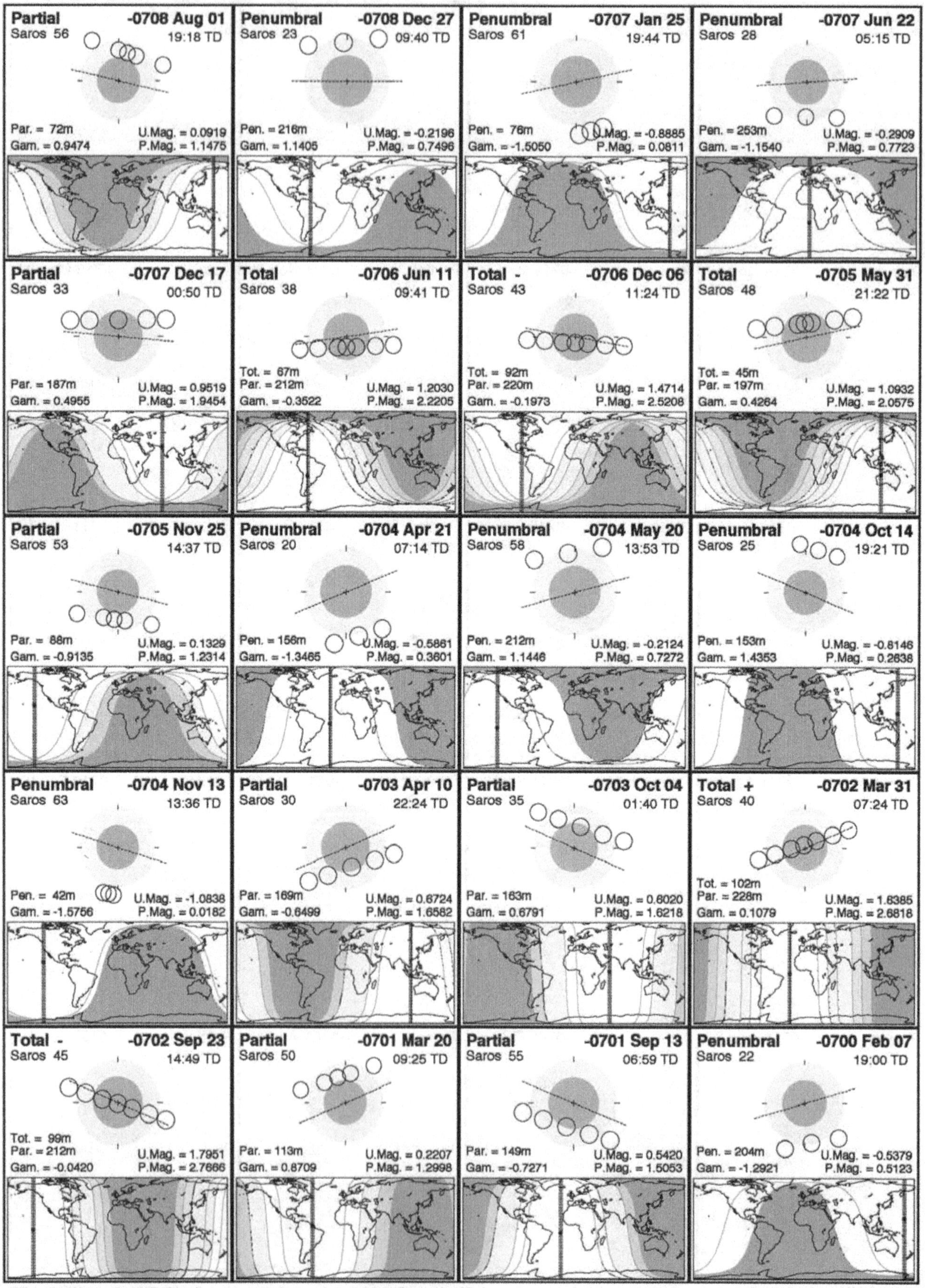

ΔT= 20329 s [= 05h39m] std.err. = ± 511 s [= ± 2.1°] **Plate 156**

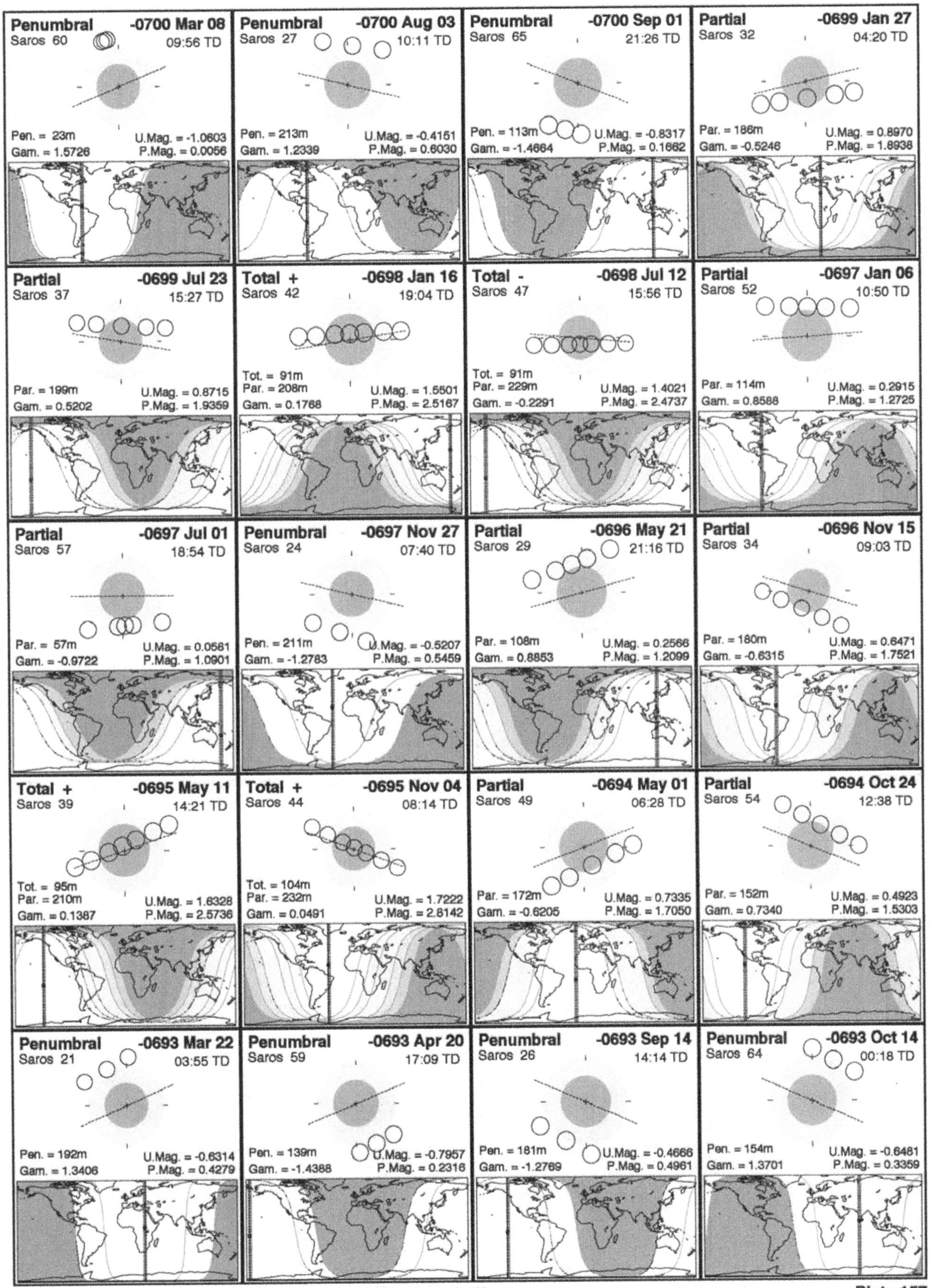

ΔT= 20207 s [= 05h37m] std.err. = ± 508 s [= ± 2.1°] **Plate 157**

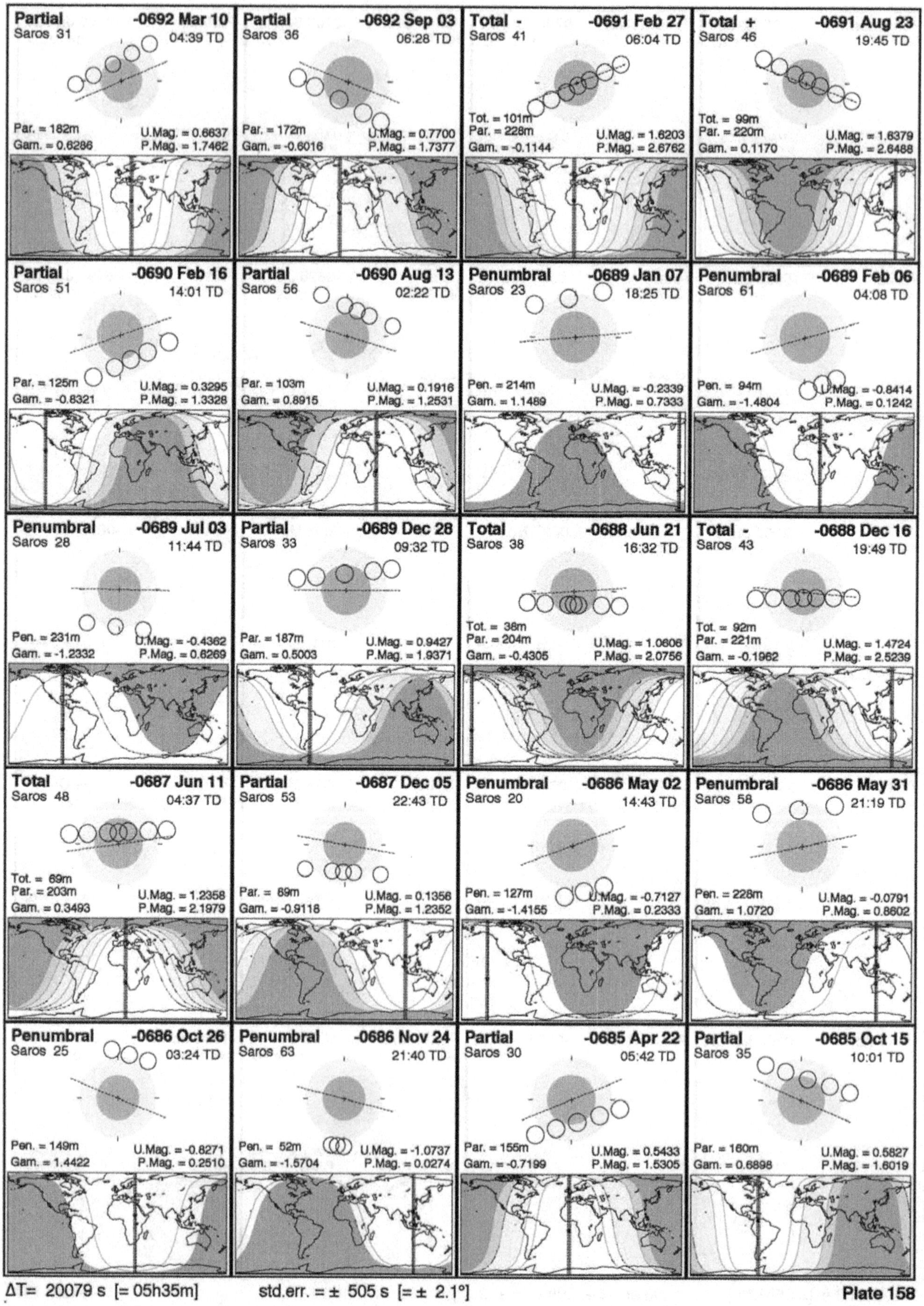

ΔT= 20079 s [= 05h35m] std.err. = ± 505 s [= ± 2.1°] **Plate 158**

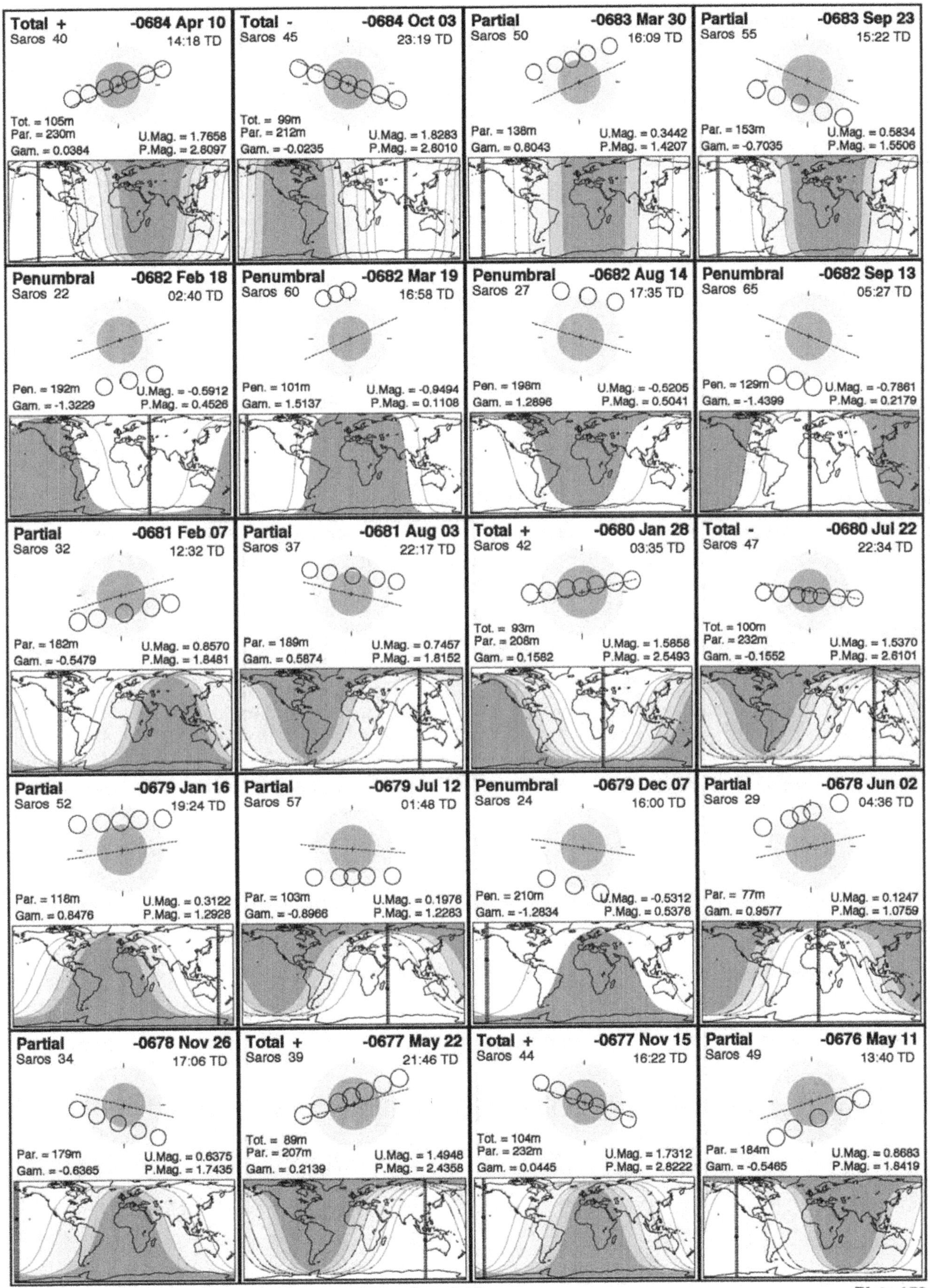

ΔT= 19950 s [= 05h32m]   std.err. = ± 501 s [= ± 2.1°]   Plate 159

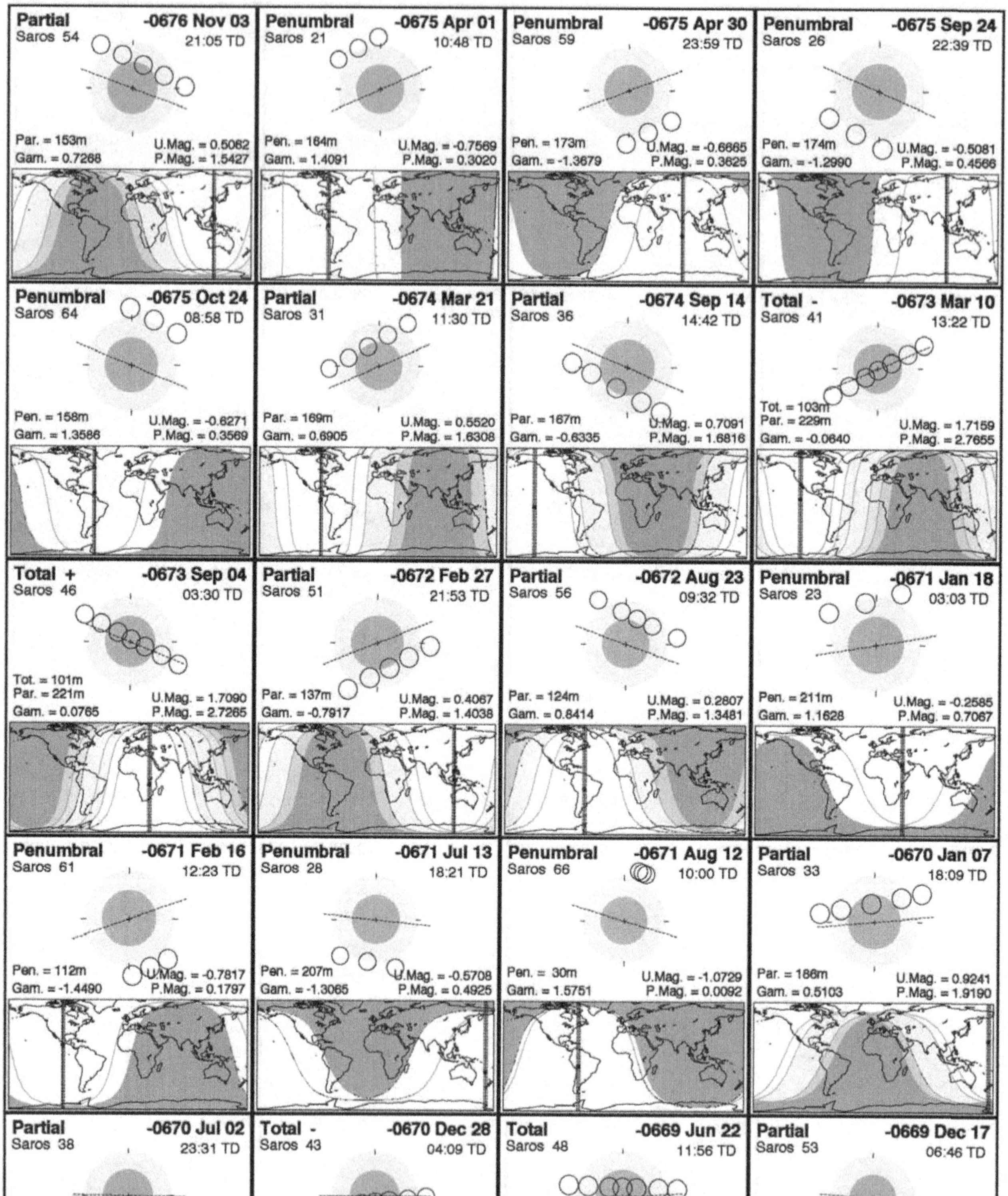

ΔT= 19813 s [= 05h30m] std.err. = ± 498 s [= ± 2.1°] **Plate 160**

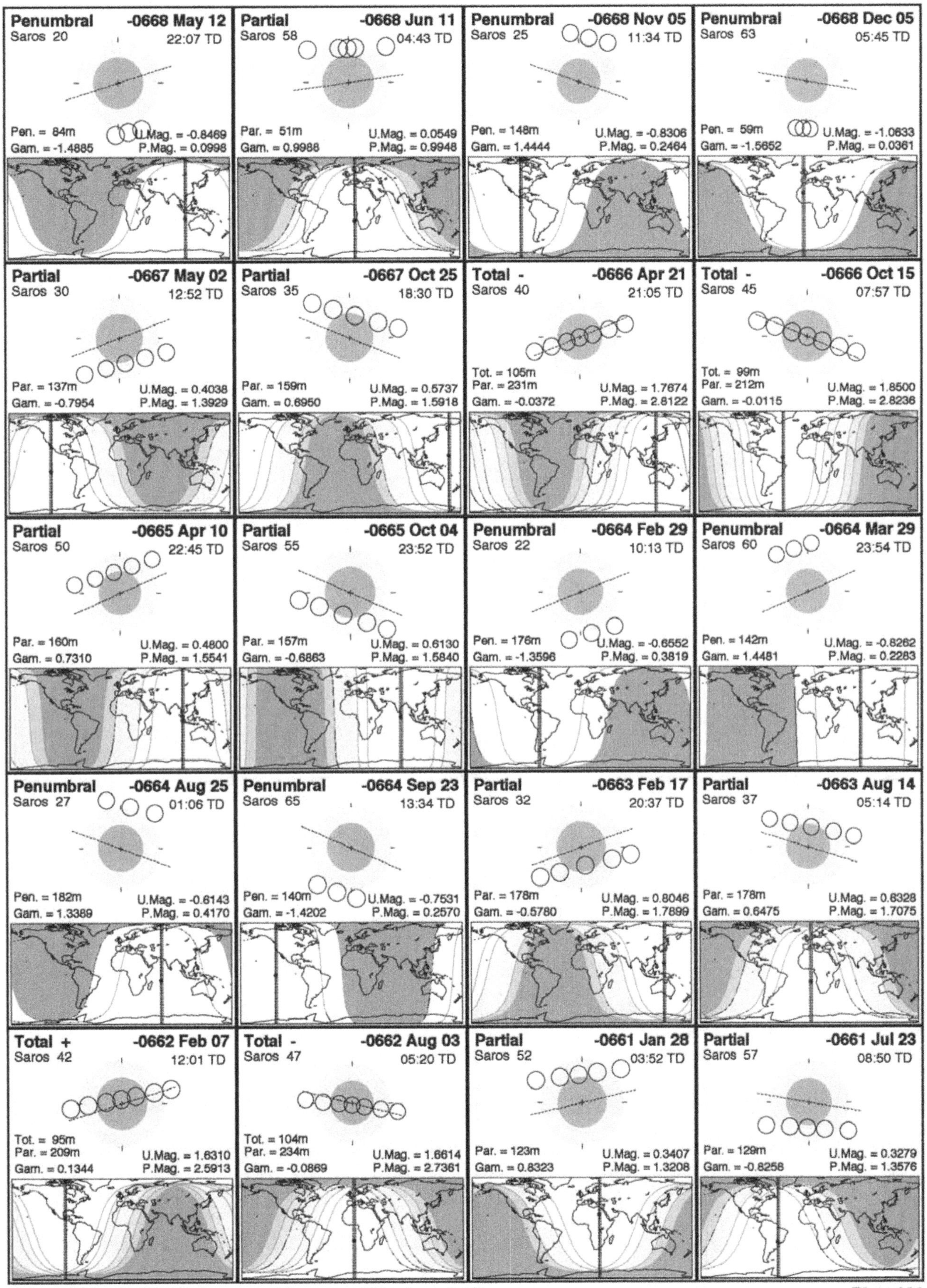

ΔT= 19694 s [= 05h28m] std.err. = ± 495 s [= ± 2.1°] **Plate 161**

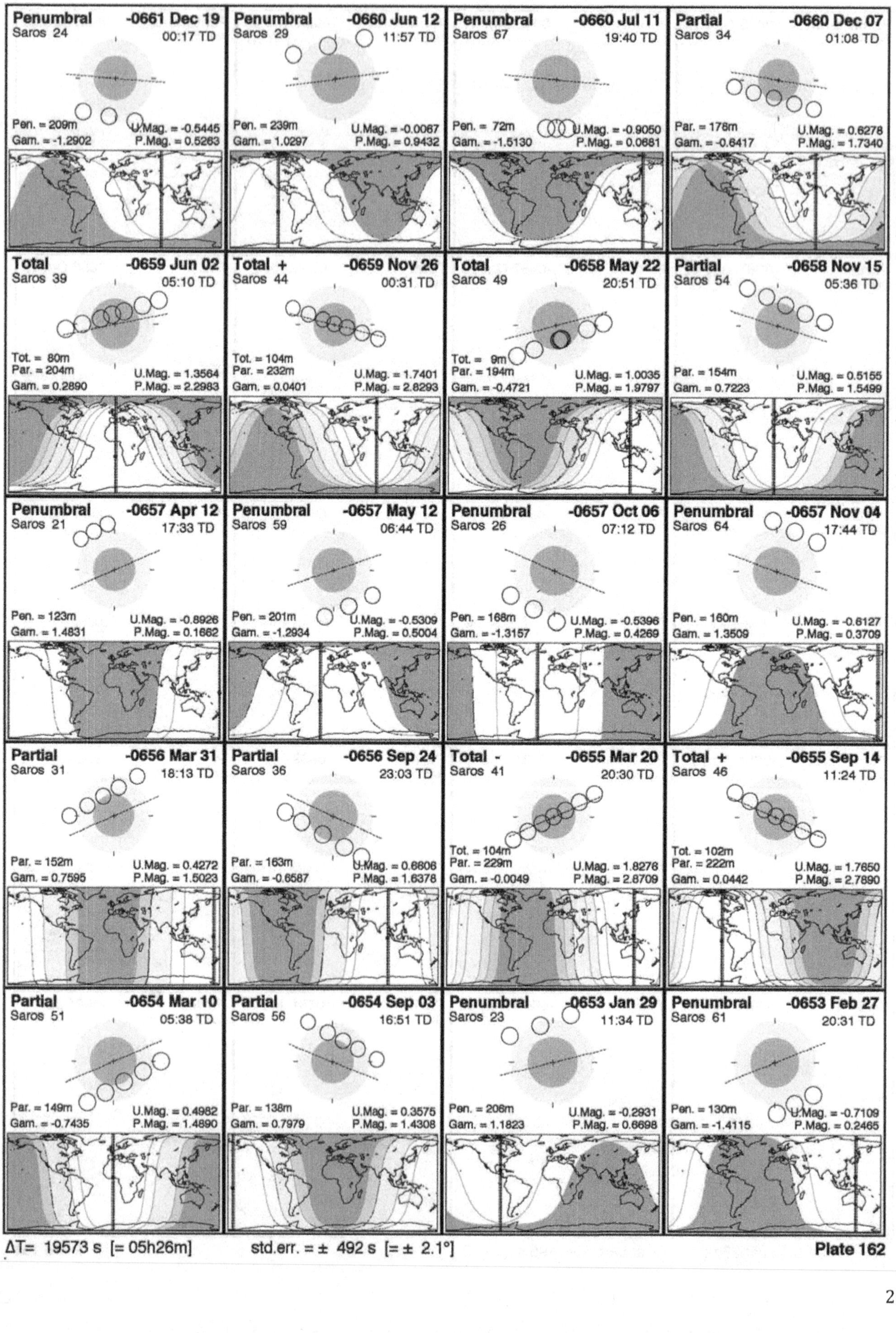

ΔT= 19573 s [= 05h26m] std.err. = ± 492 s [= ± 2.1°] **Plate 162**

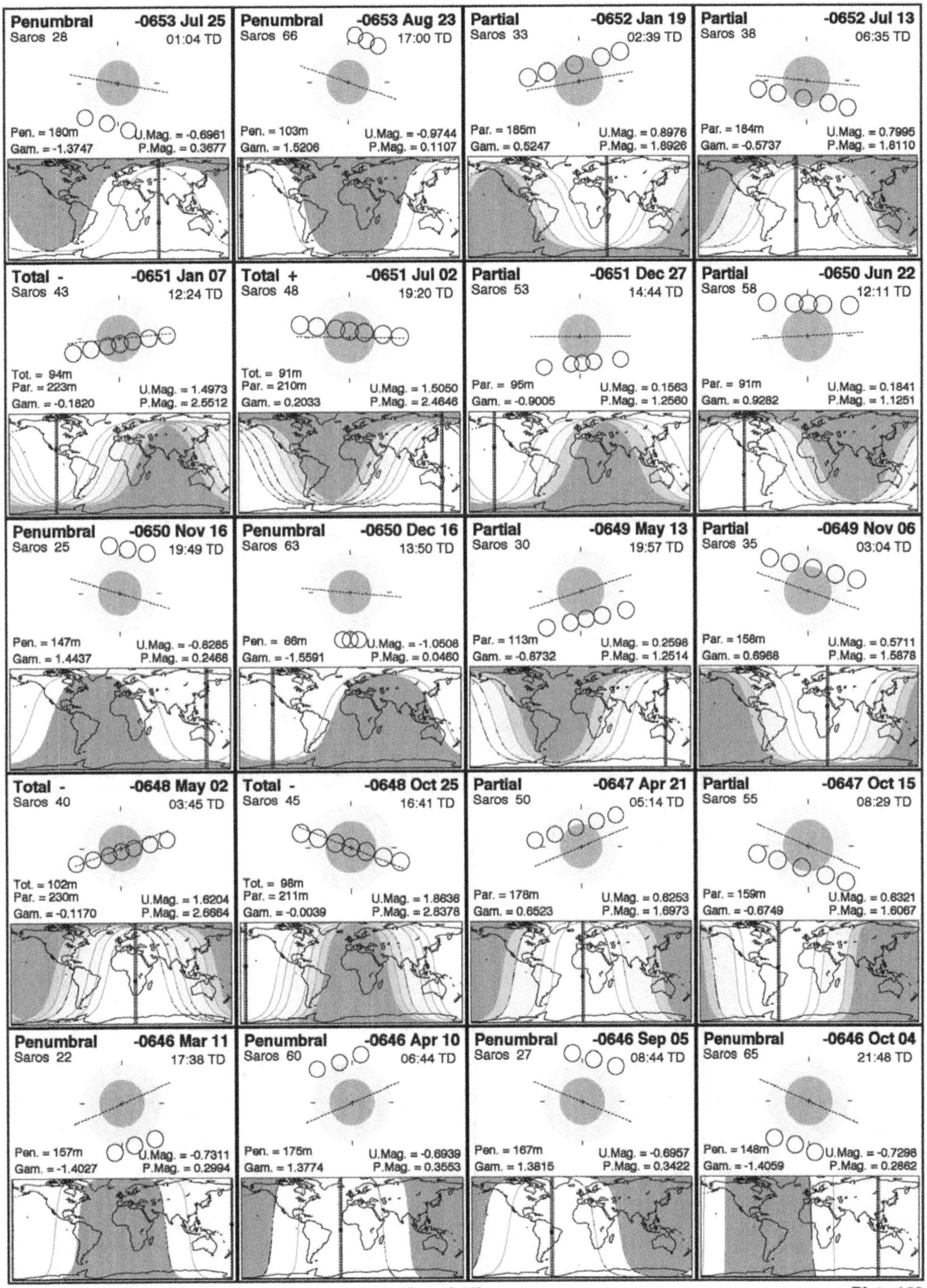

ΔT= 19453 s [= 05h24m] std.err. = ± 489 s [= ± 2.0°] Plate 163

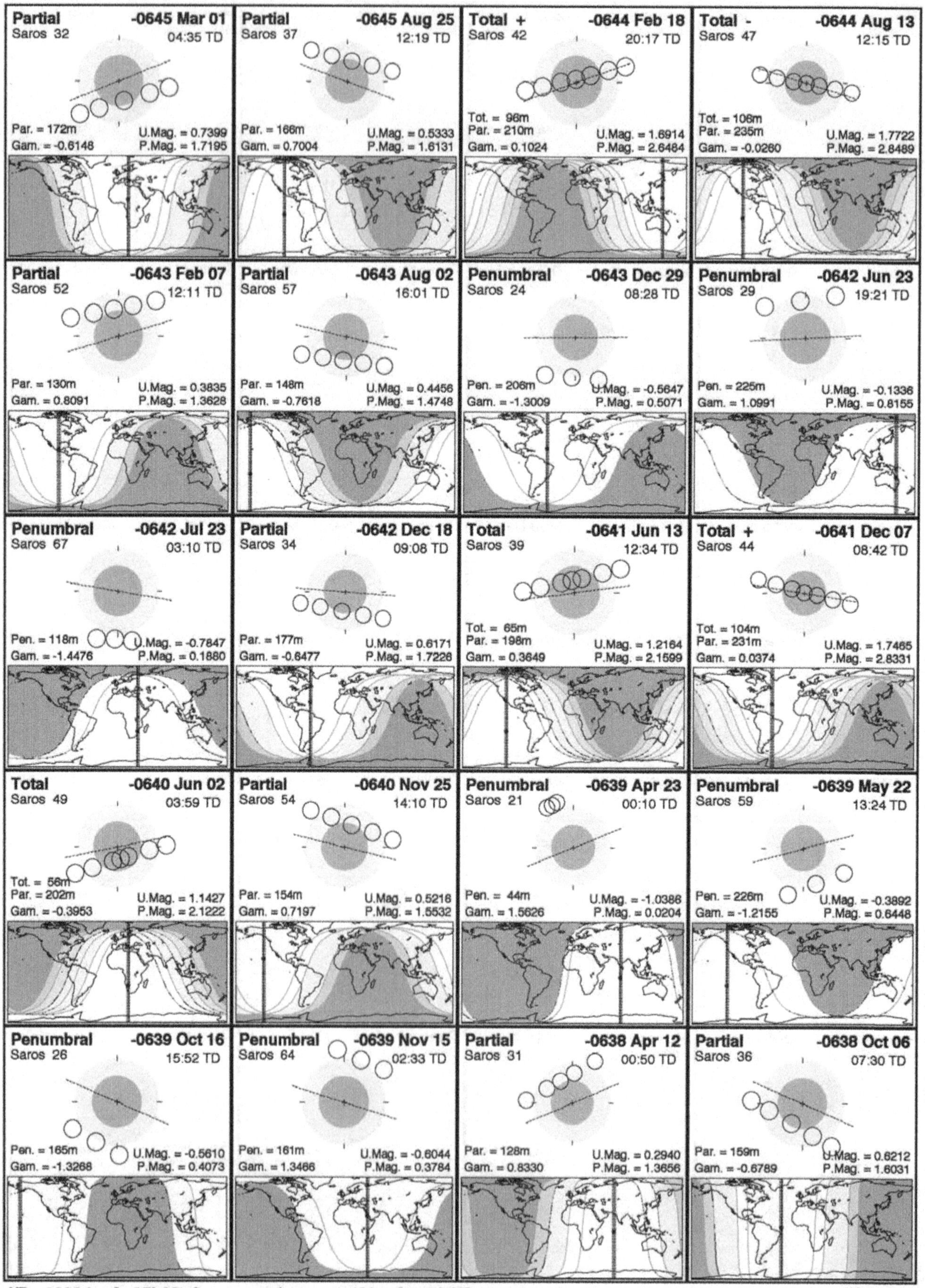

ΔT= 19334 s [= 05h22m] std.err. = ± 486 s [= ± 2.0°] Plate 164

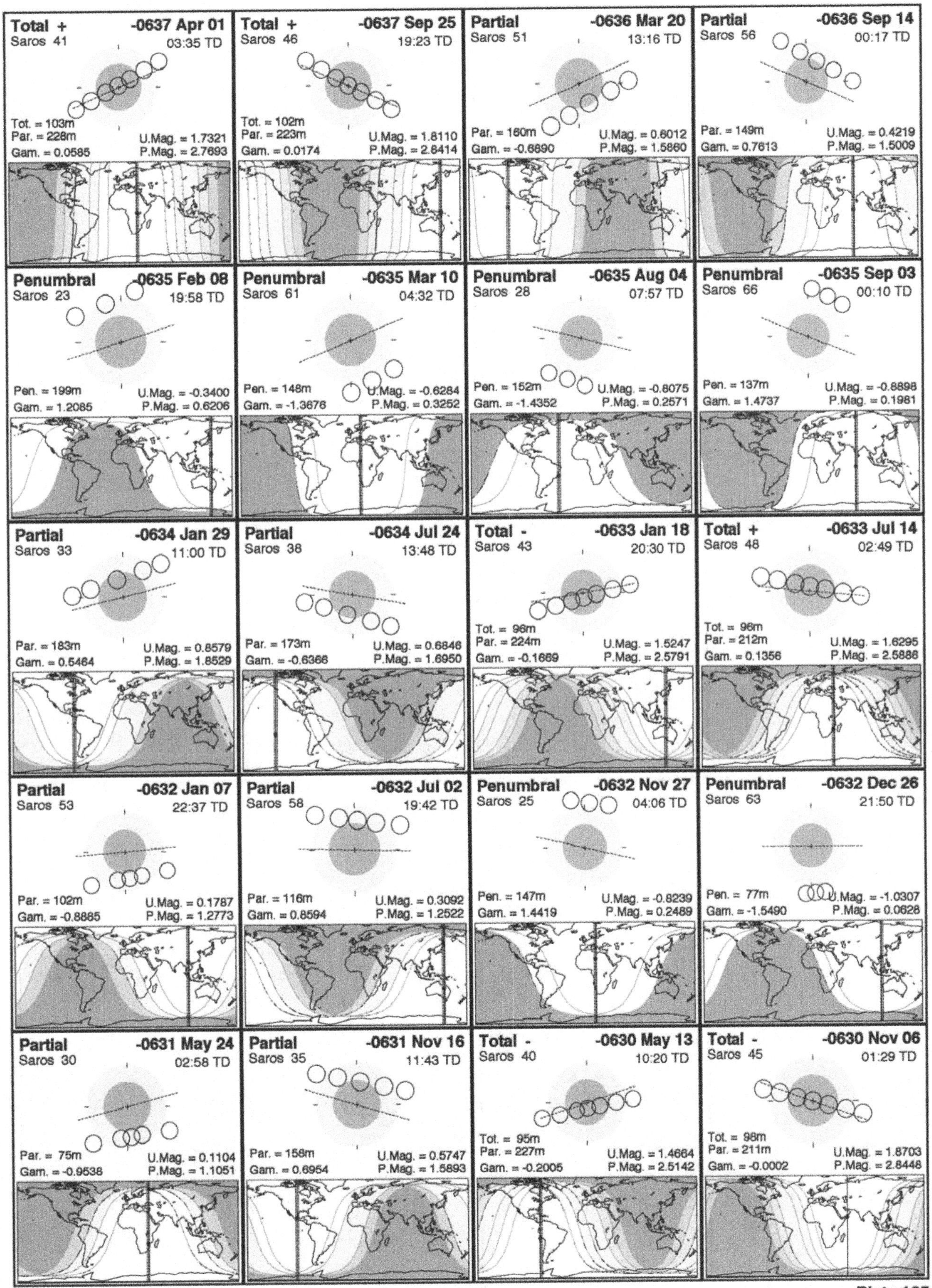

ΔT= 19207 s [= 05h20m] std.err. = ± 483 s [= ± 2.0°] **Plate 165**

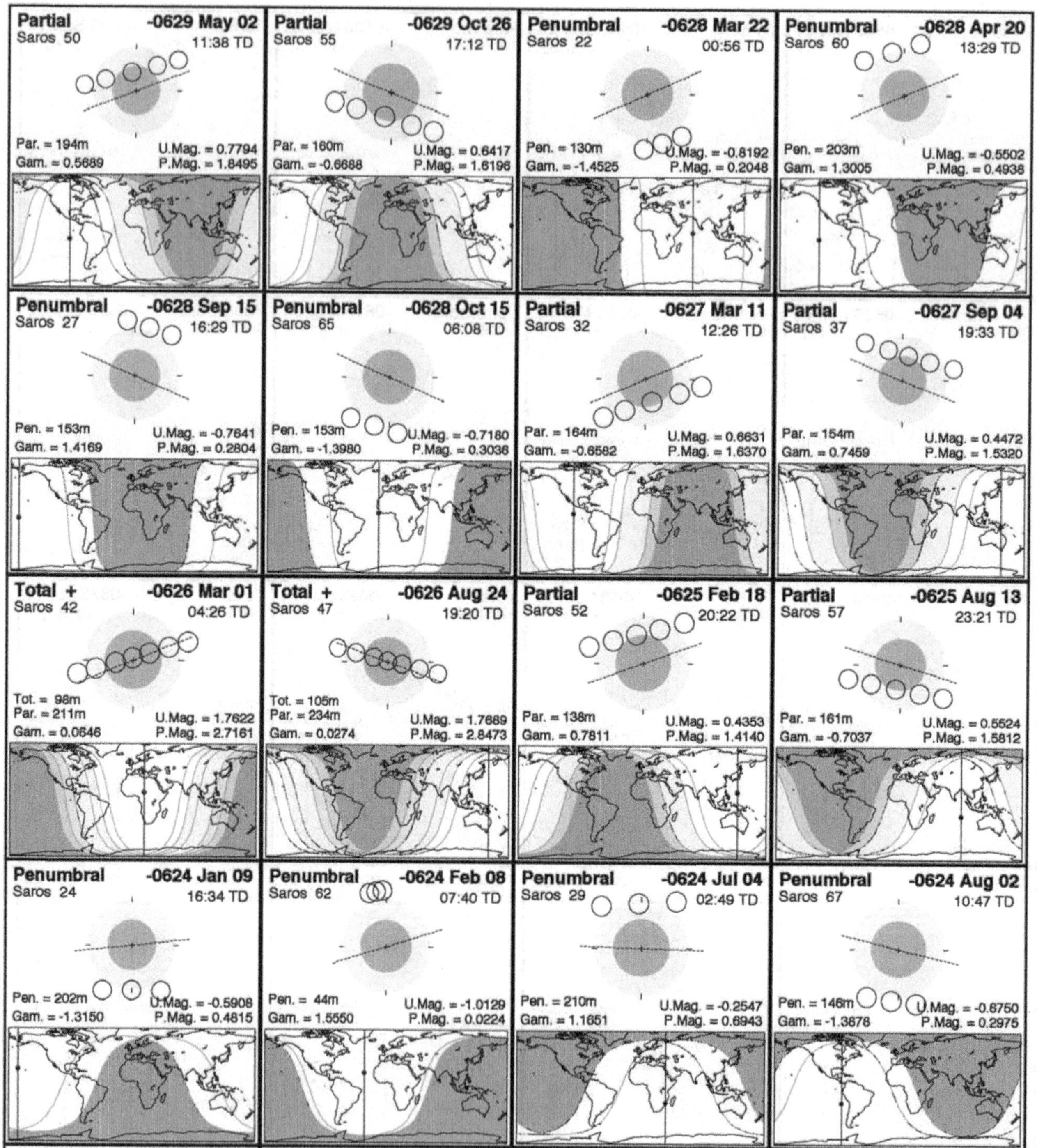

ΔT= 19081 s [= 05h18m] std.err. = ± 480 s [= ± 2.0°] **Plate 166**

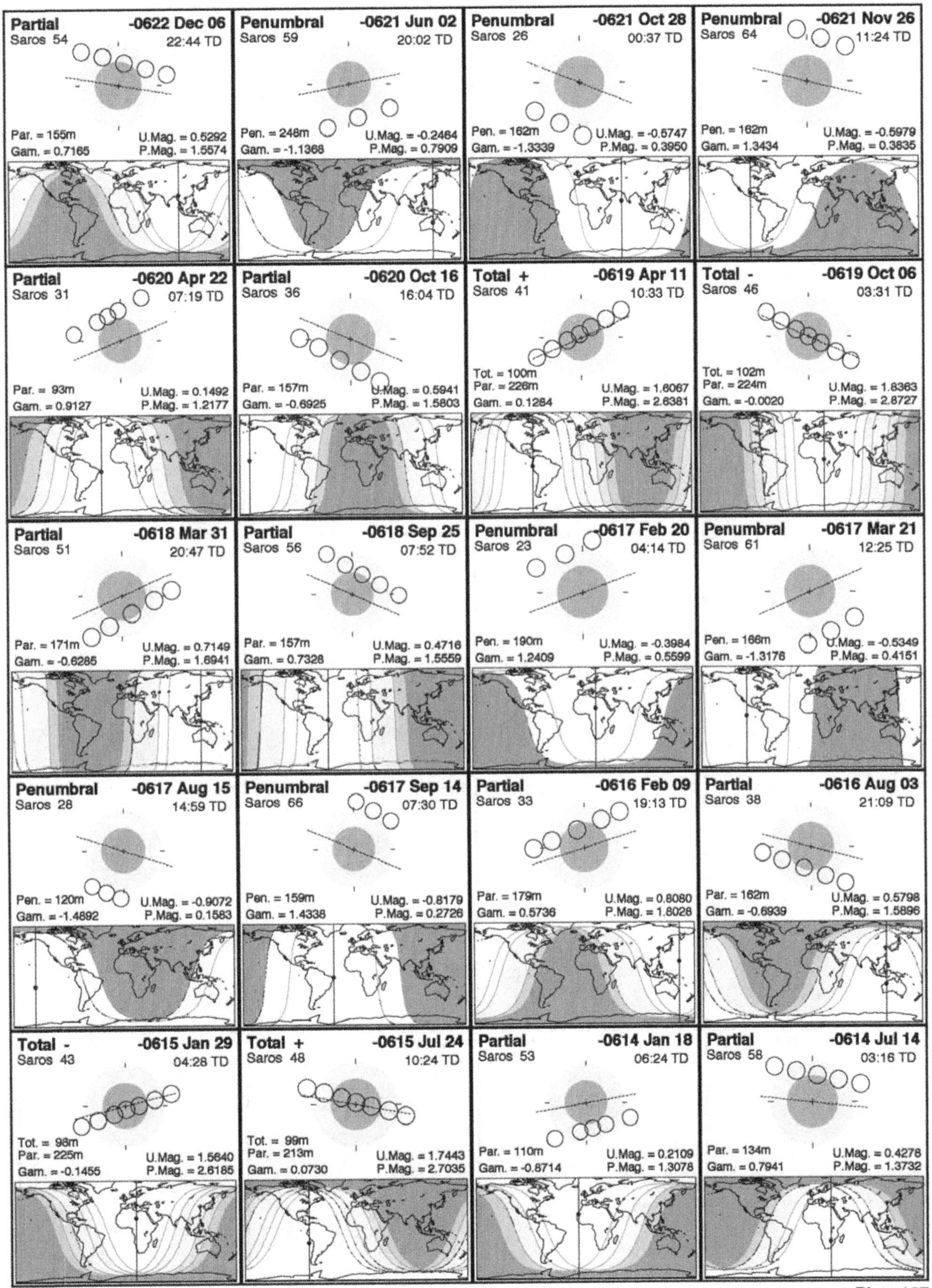

ΔT= 18962 s [= 05h16m] std.err. = ± 477 s [= ± 2.0°] Plate 167

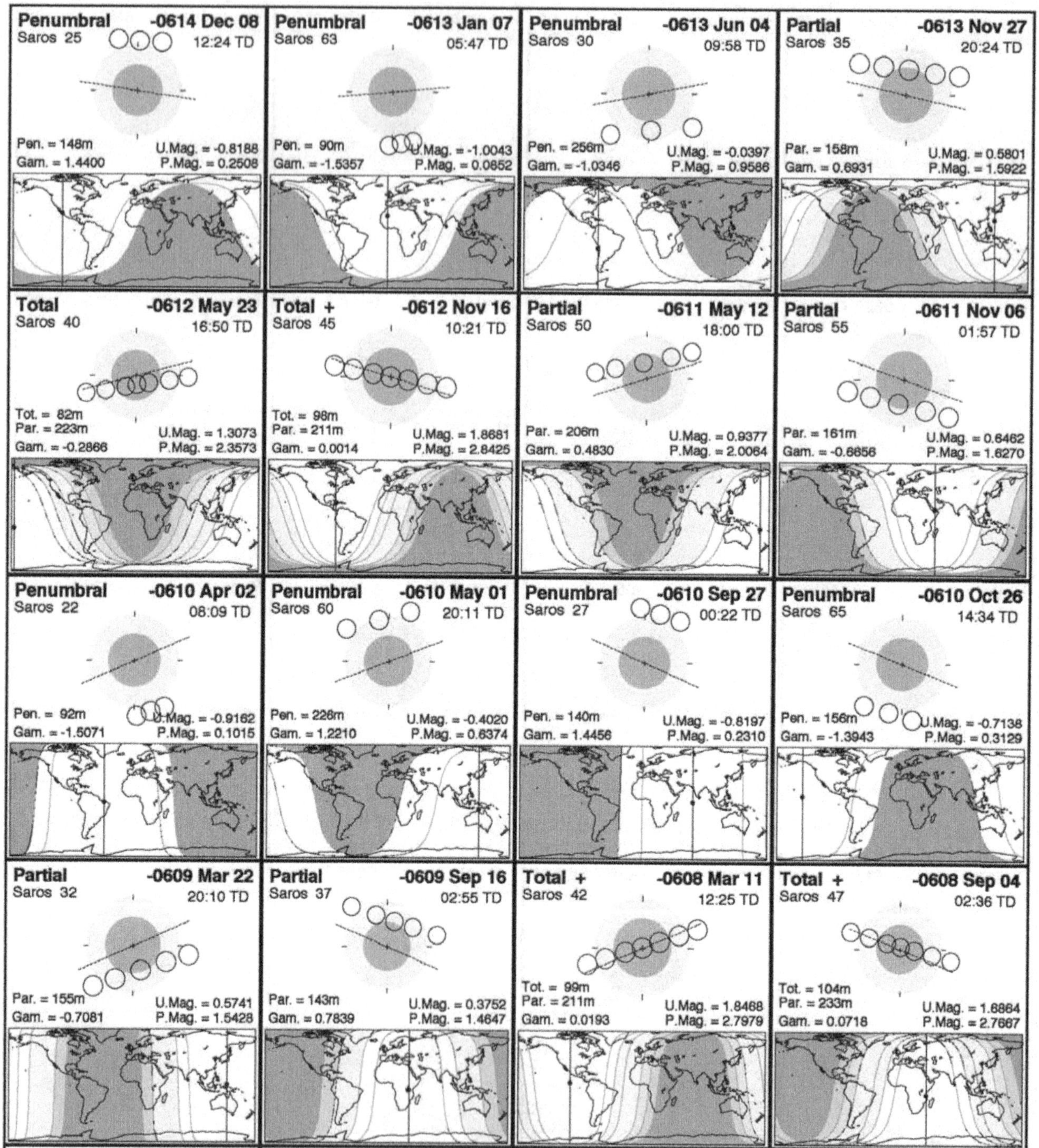

ΔT= 18838 s [= 05h14m] std.err. = ± 474 s [= ± 2.0°] **Plate 168**

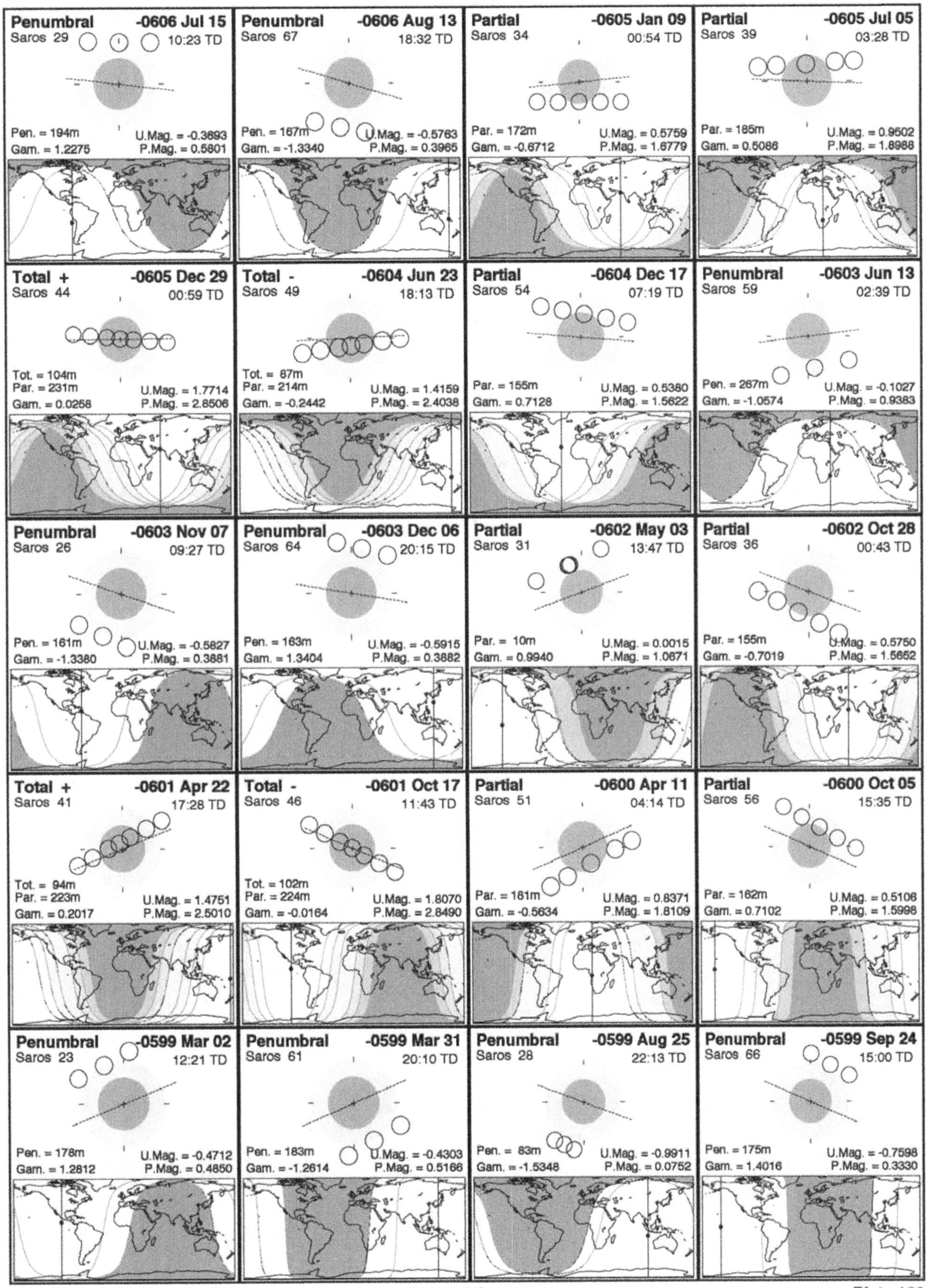

ΔT= 18720 s [= 05h12m] std.err. = ± 471 s [= ± 2.0°] **Plate 169**

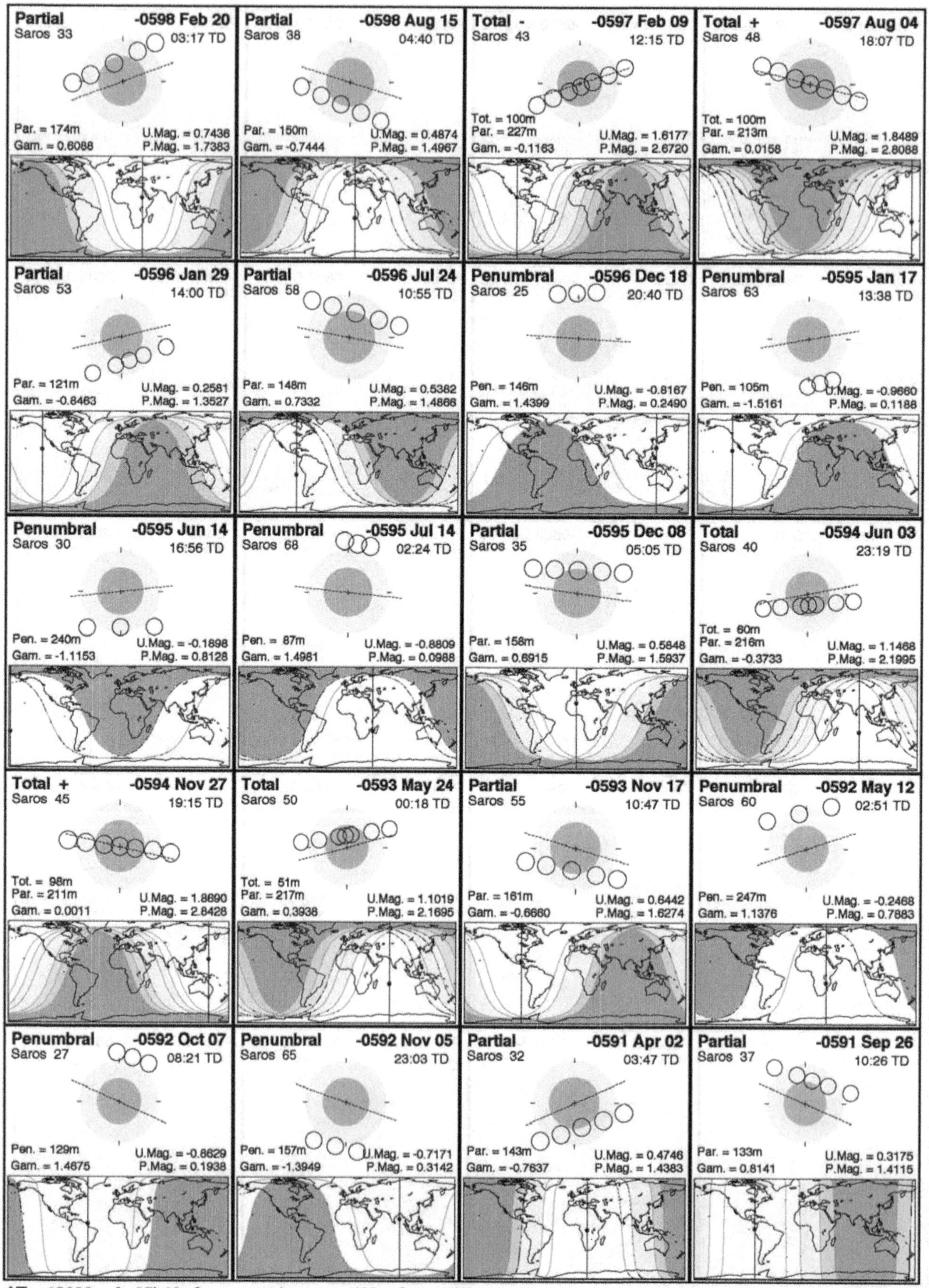

ΔT= 18603 s [= 05h10m] std.err. = ± 468 s [= ± 1.9°] **Plate 170**

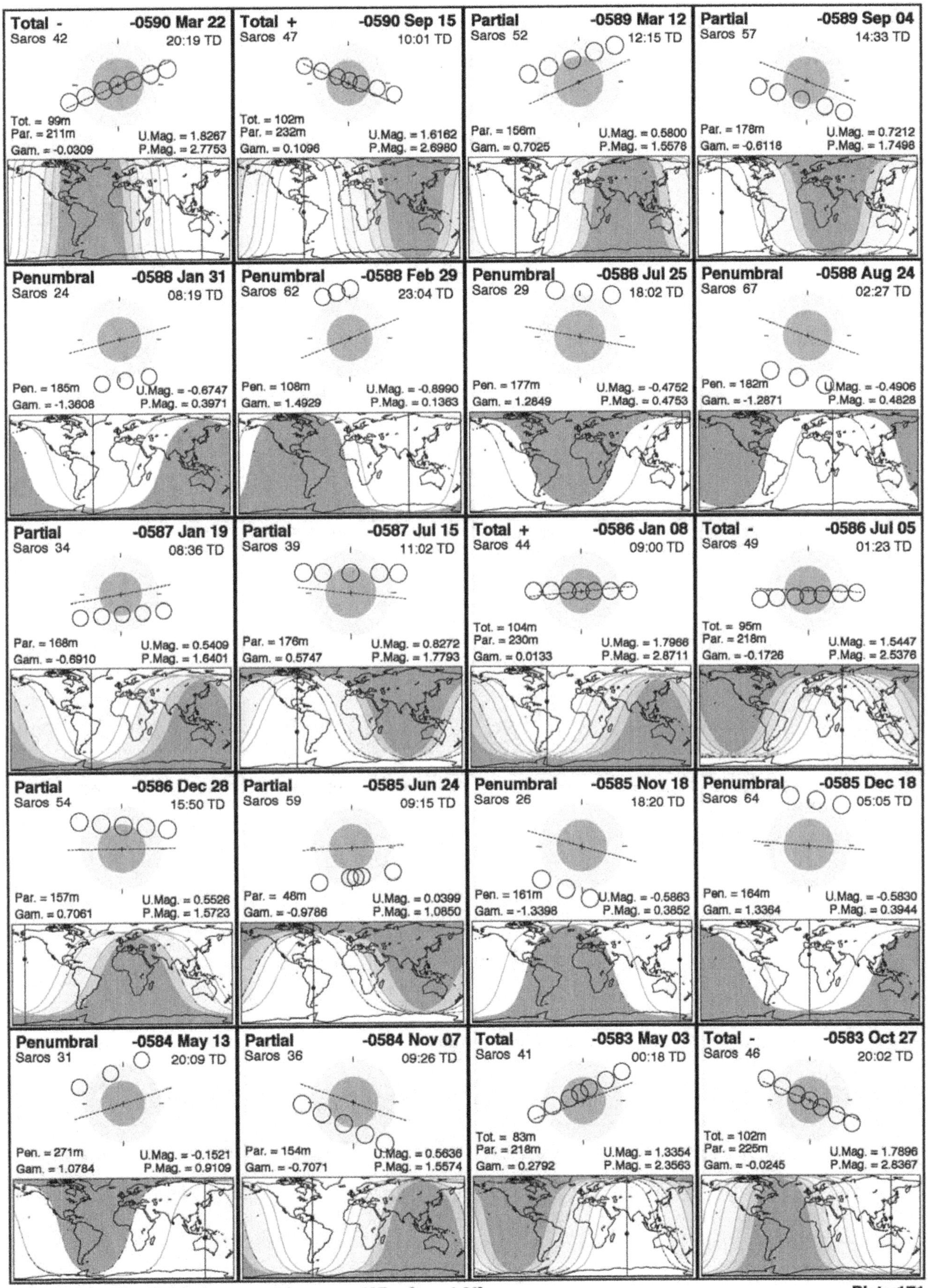

ΔT= 18479 s [= 05h08m] std.err. = ± 465 s [= ± 1.9°] **Plate 171**

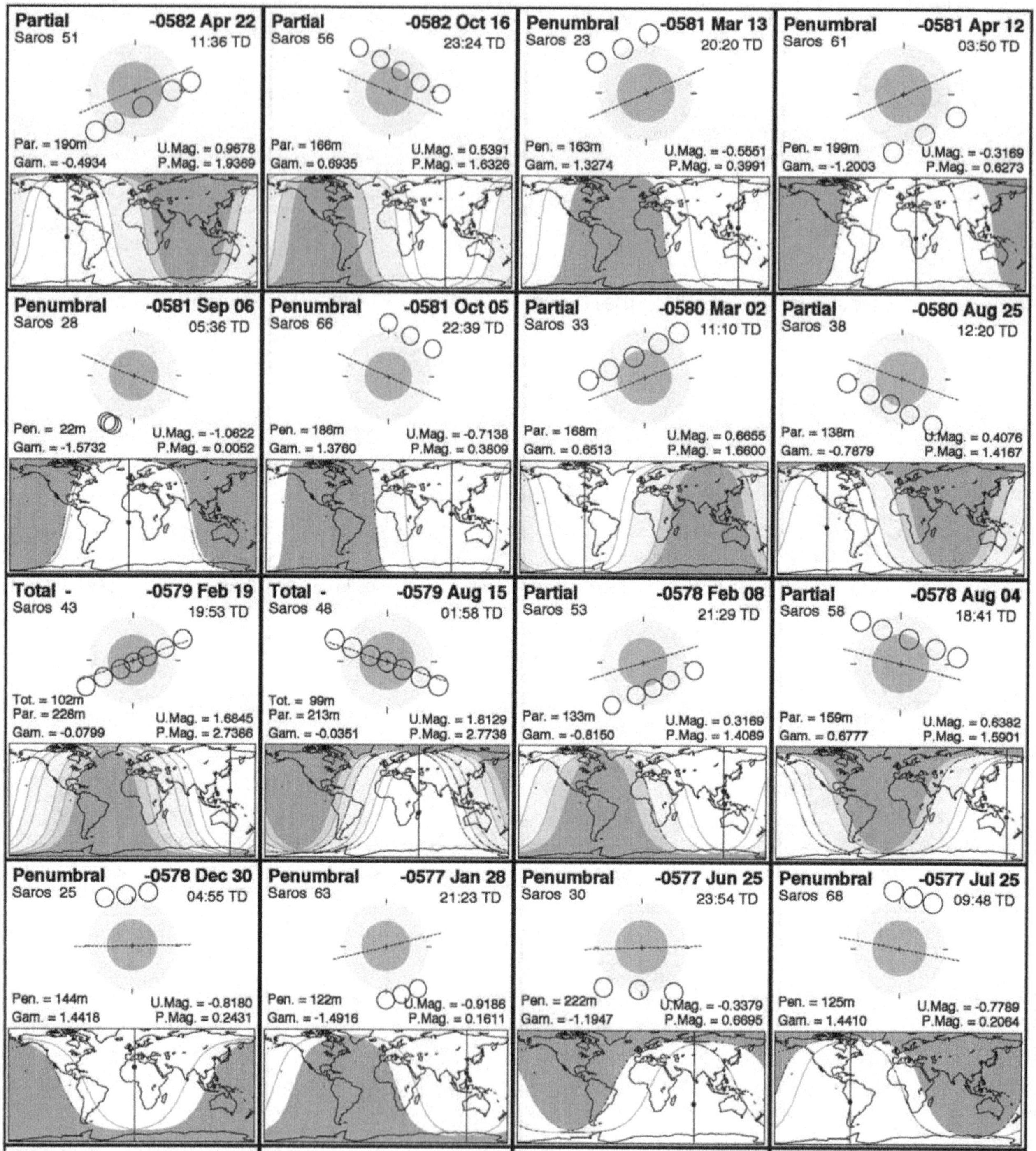

ΔT= 18355 s [= 05h06m] std.err. = ± 461 s [= ± 1.9°] **Plate 172**

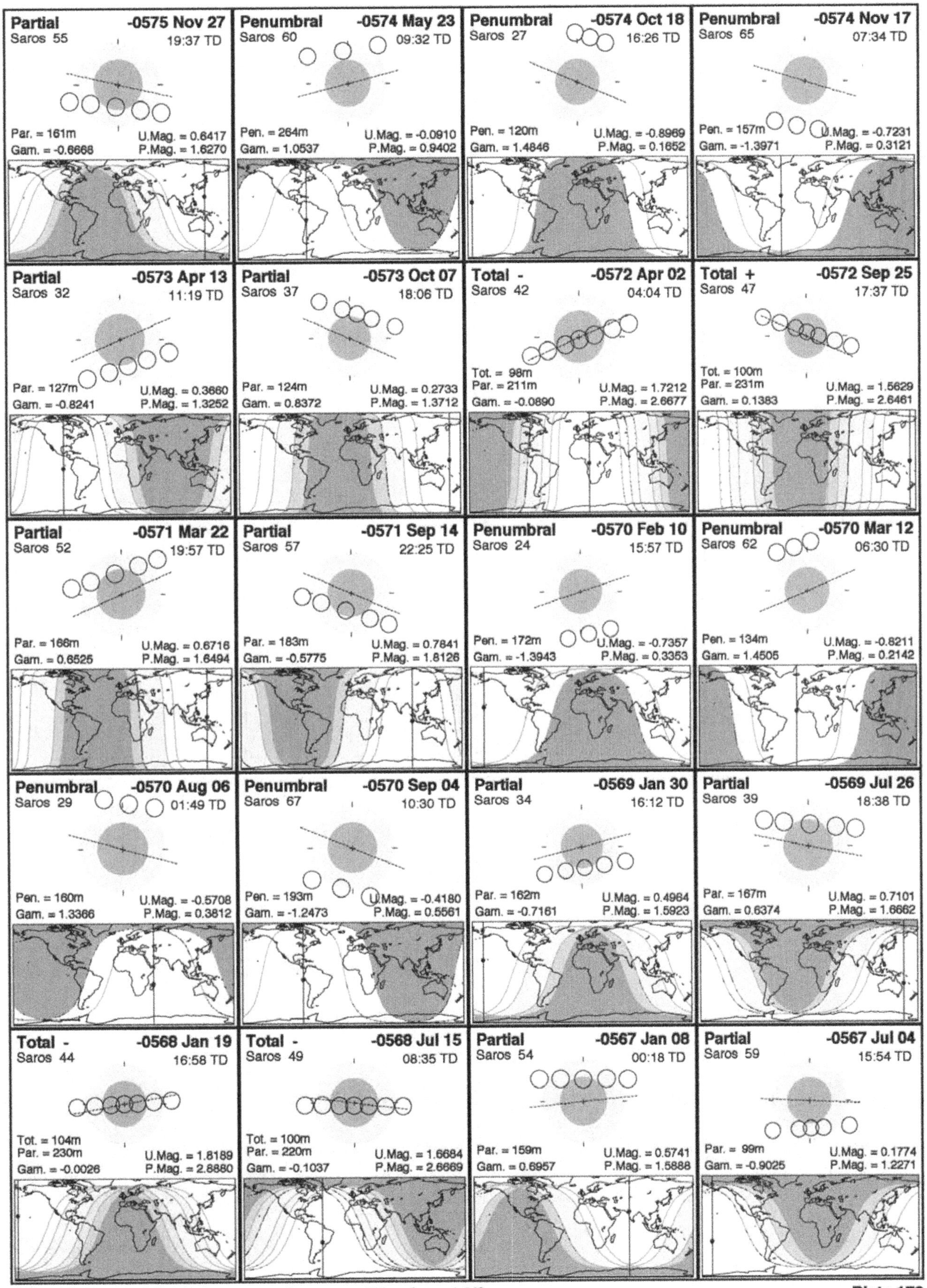

ΔT= 18239 s [= 05h04m] std.err. = ± 459 s [= ± 1.9°] **Plate 173**

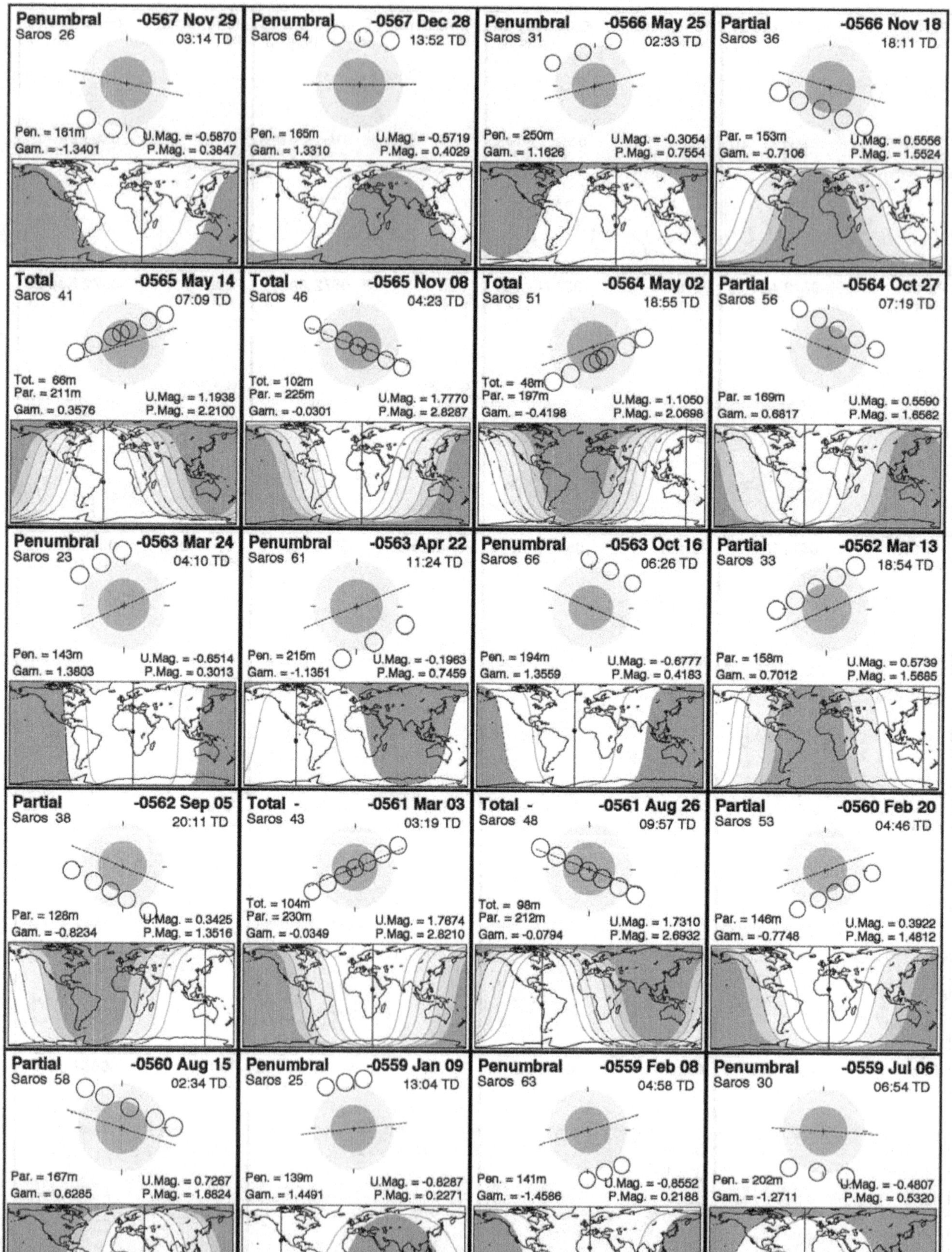

ΔT= 18117 s [= 05h02m] std.err. = ± 455 s [= ± 1.9°] **Plate 174**

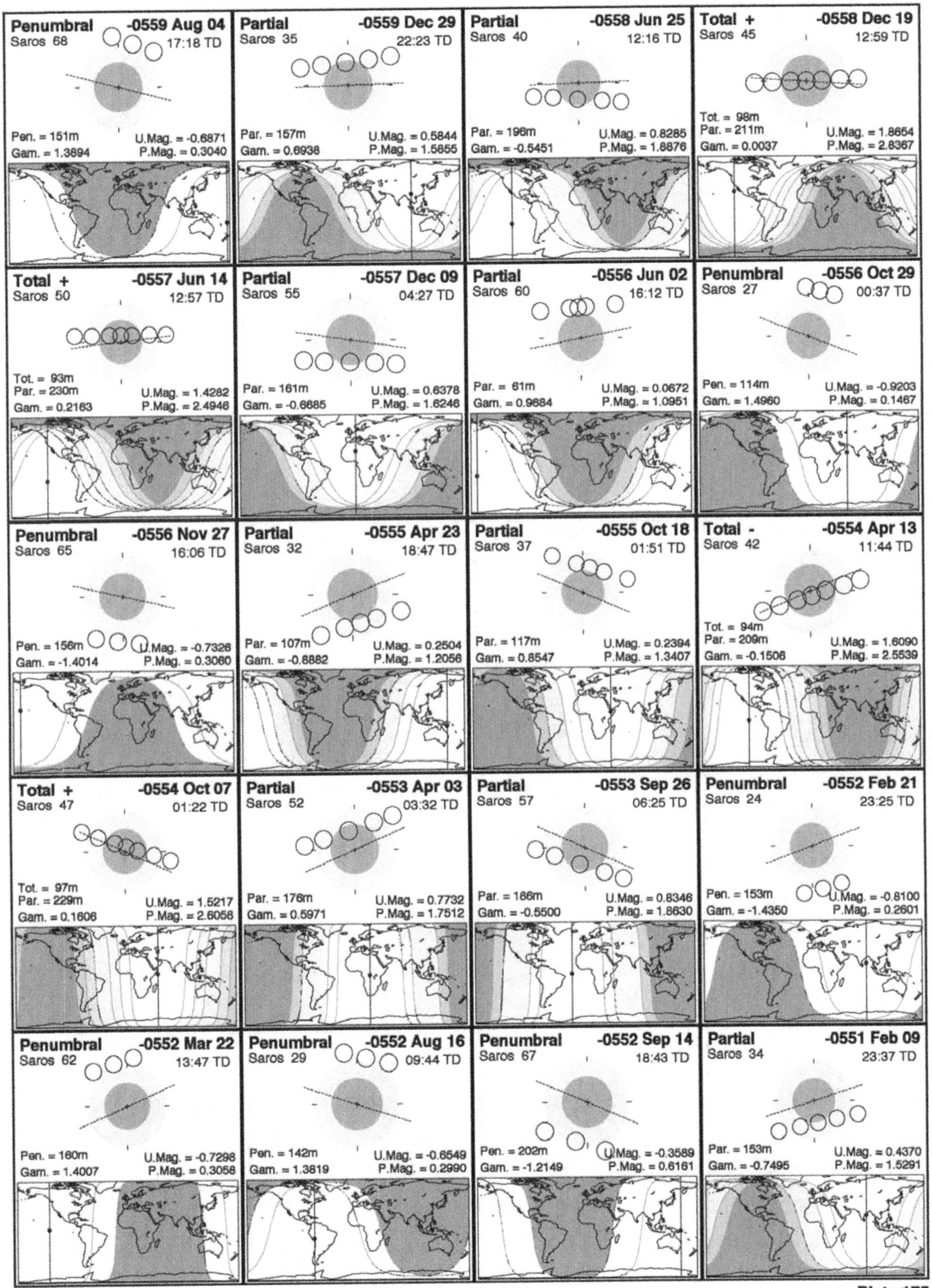

ΔT= 18000 s [= 05h00m] std.err. = ± 453 s [= ± 1.9°] **Plate 175**

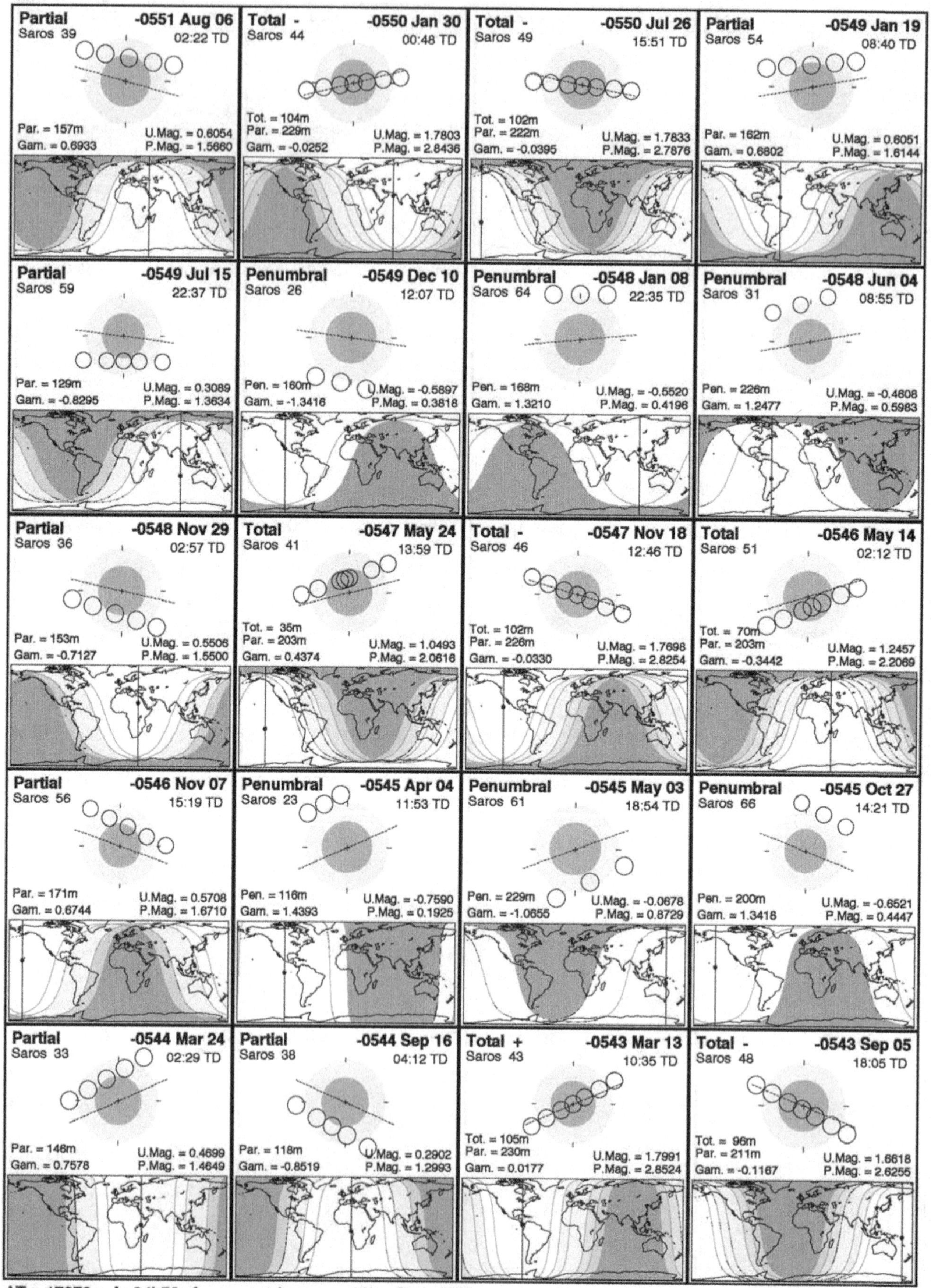

ΔT= 17879 s [= 04h58m] std.err. = ± 449 s [= ± 1.9°] **Plate 176**

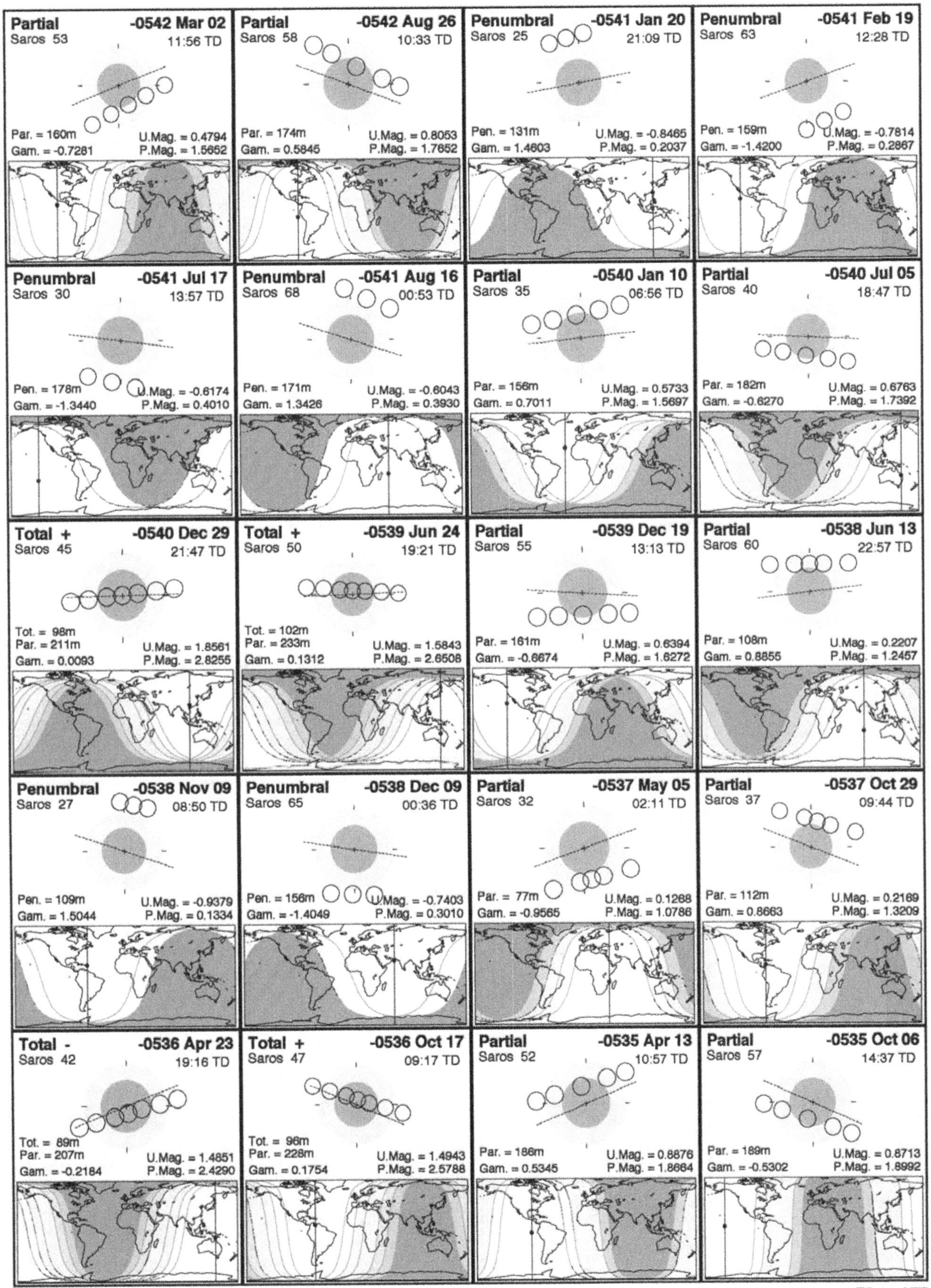

ΔT= 17750 s [= 04h56m] std.err. = ± 446 s [= ± 1.9°] Plate 177

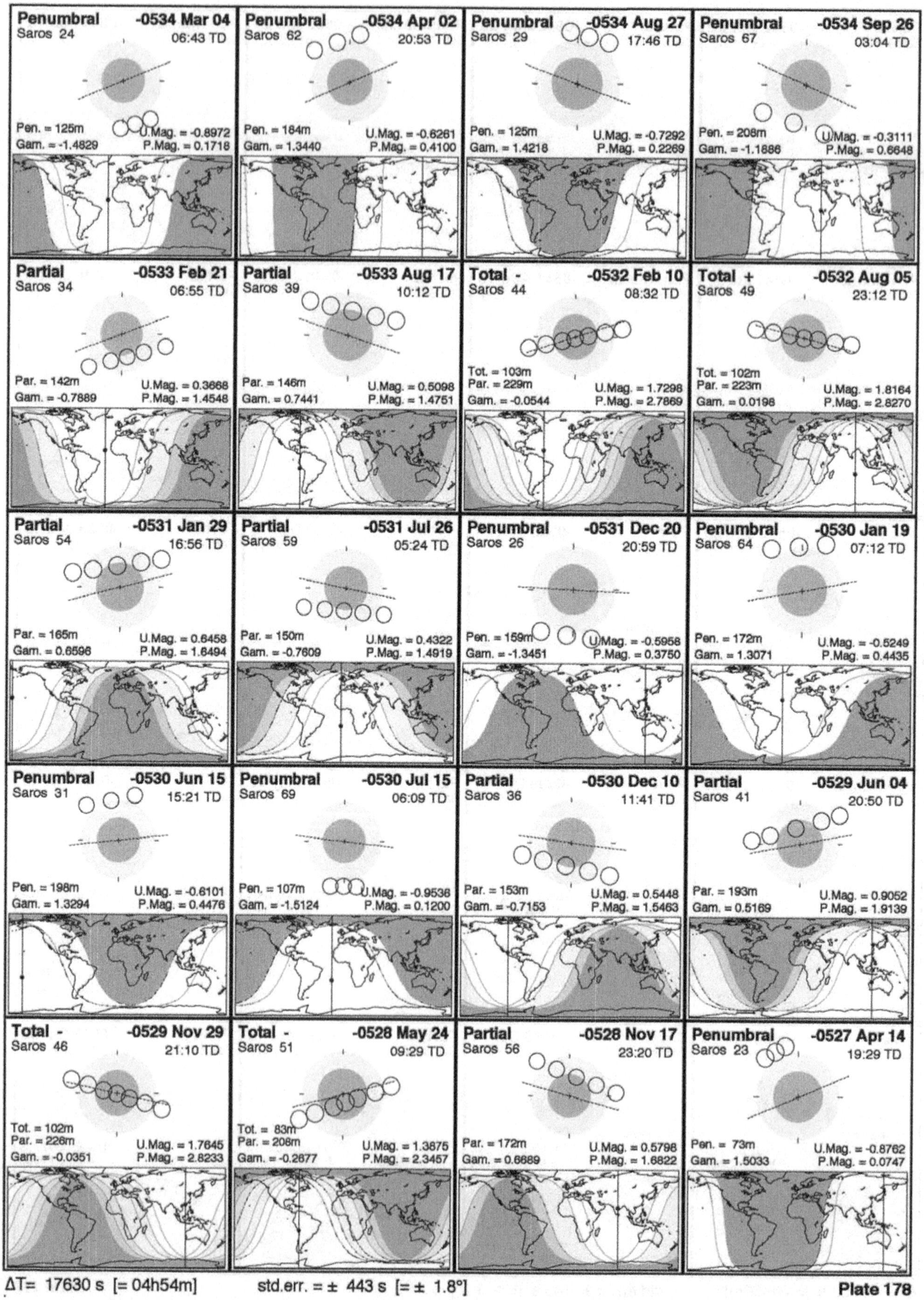

ΔT= 17630 s [= 04h54m] std.err. = ± 443 s [= ± 1.8°] **Plate 178**

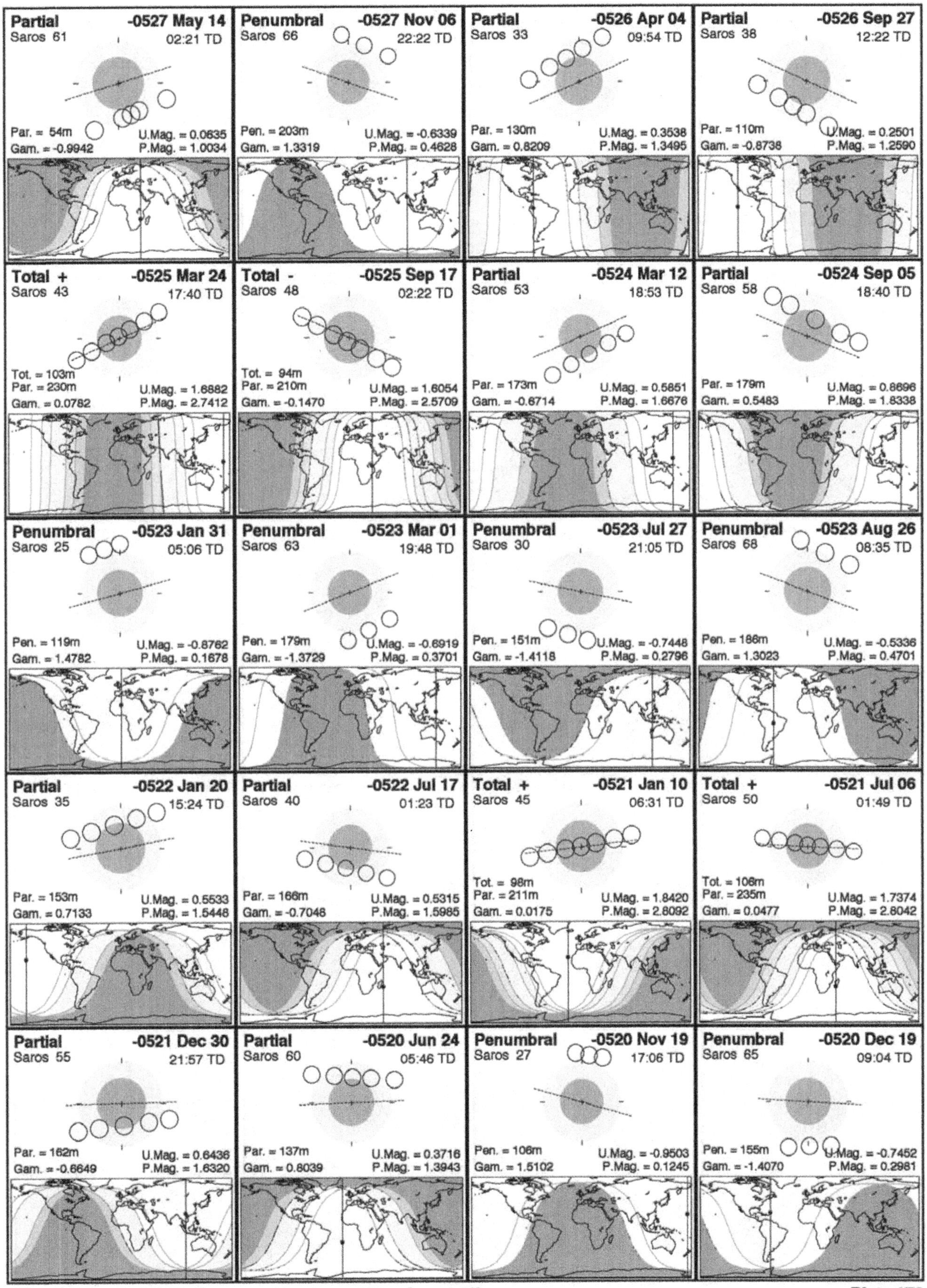

ΔT= 17522 s [= 04h52m] std.err. = ± 441 s [= ± 1.8°] **Plate 179**

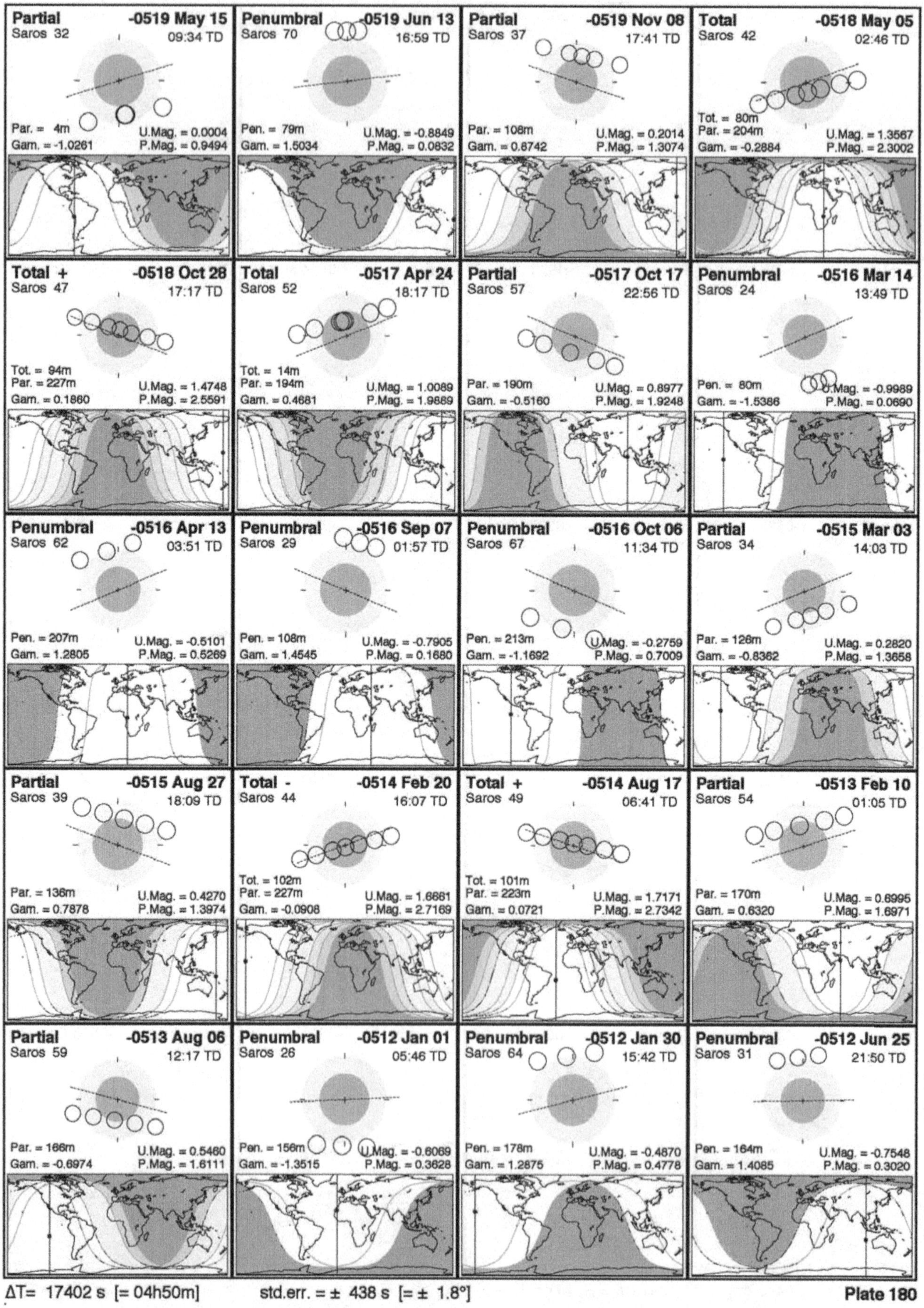

ΔT= 17402 s [= 04h50m] std.err. = ± 438 s [= ± 1.8°] **Plate 180**

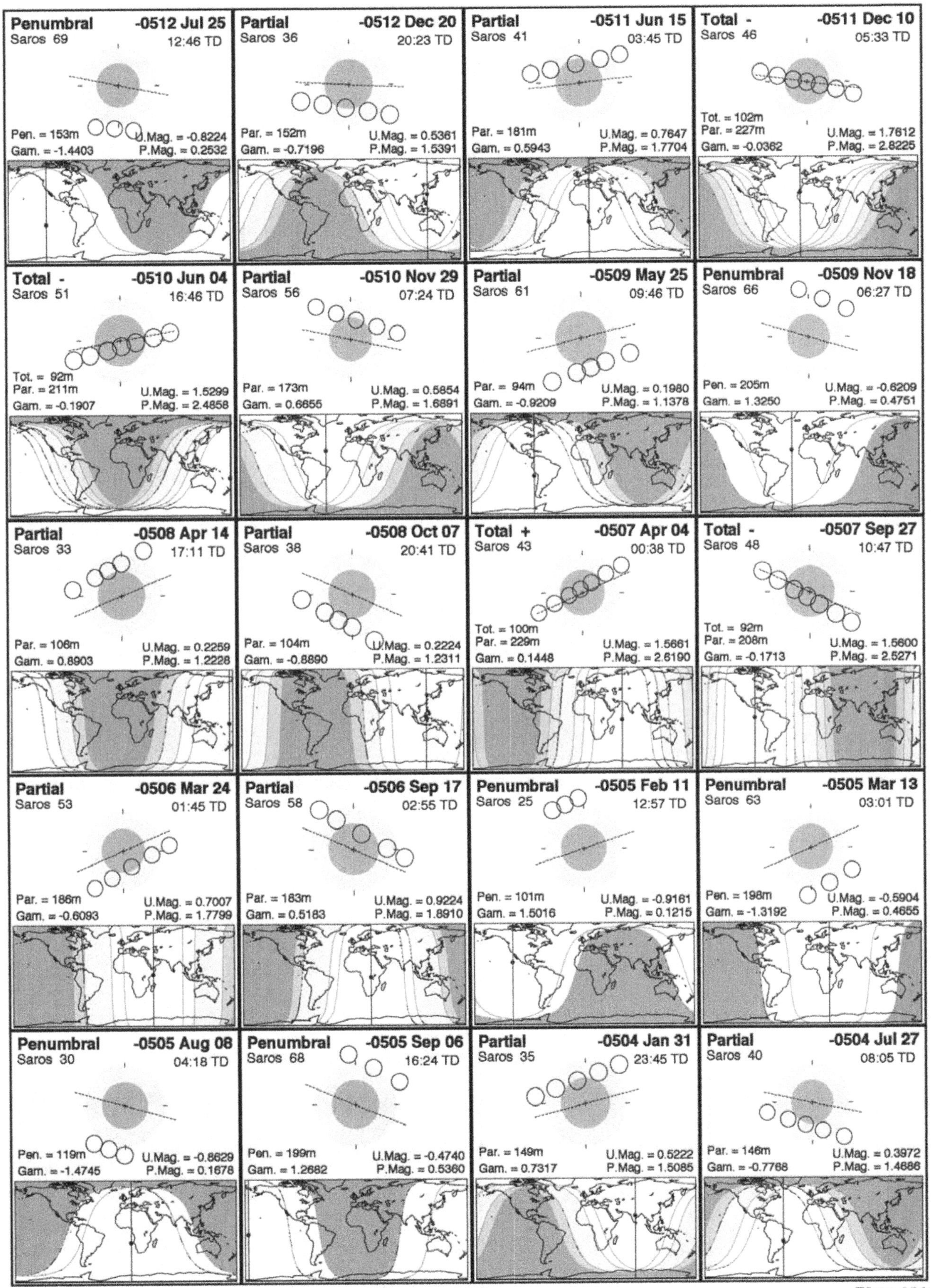

ΔT= 17295 s [= 04h48m] std.err. = ± 435 s [= ± 1.8°] **Plate 181**

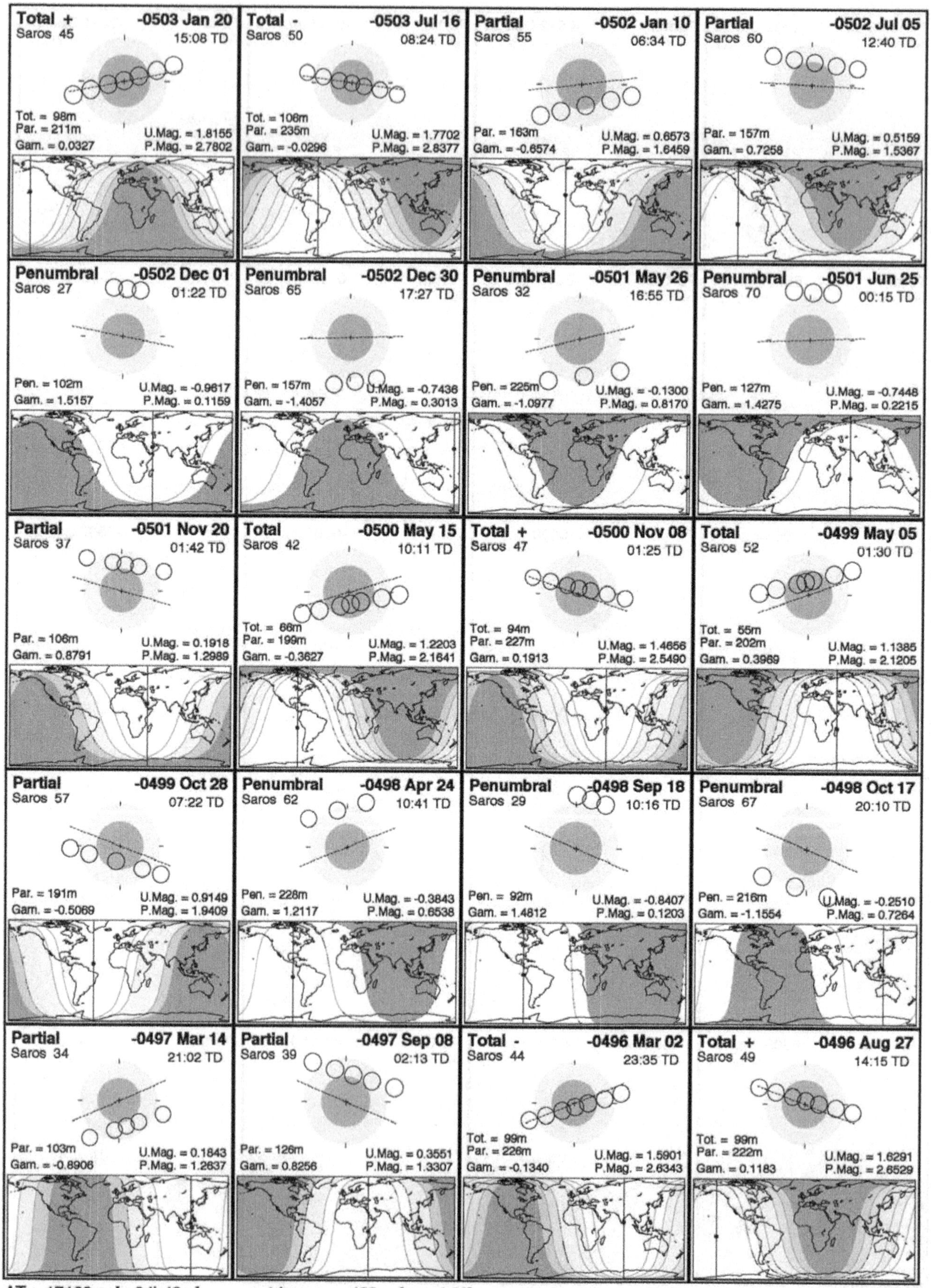

ΔT= 17169 s [= 04h46m] std.err. = ± 432 s [= ± 1.8°] **Plate 182**

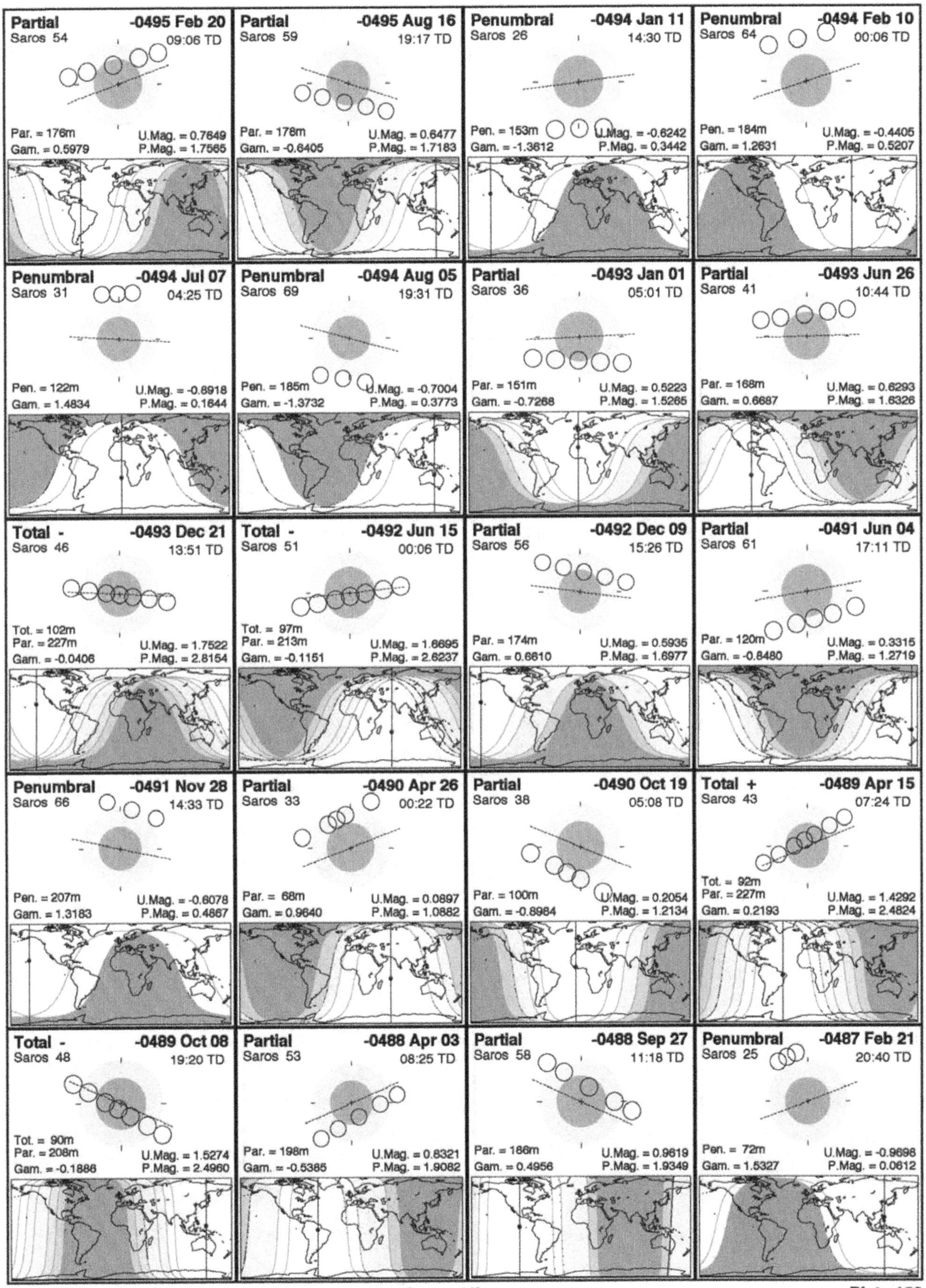

ΔT= 17033 s [= 04h44m] std.err. = ± 429 s [= ± 1.8°] **Plate 183**

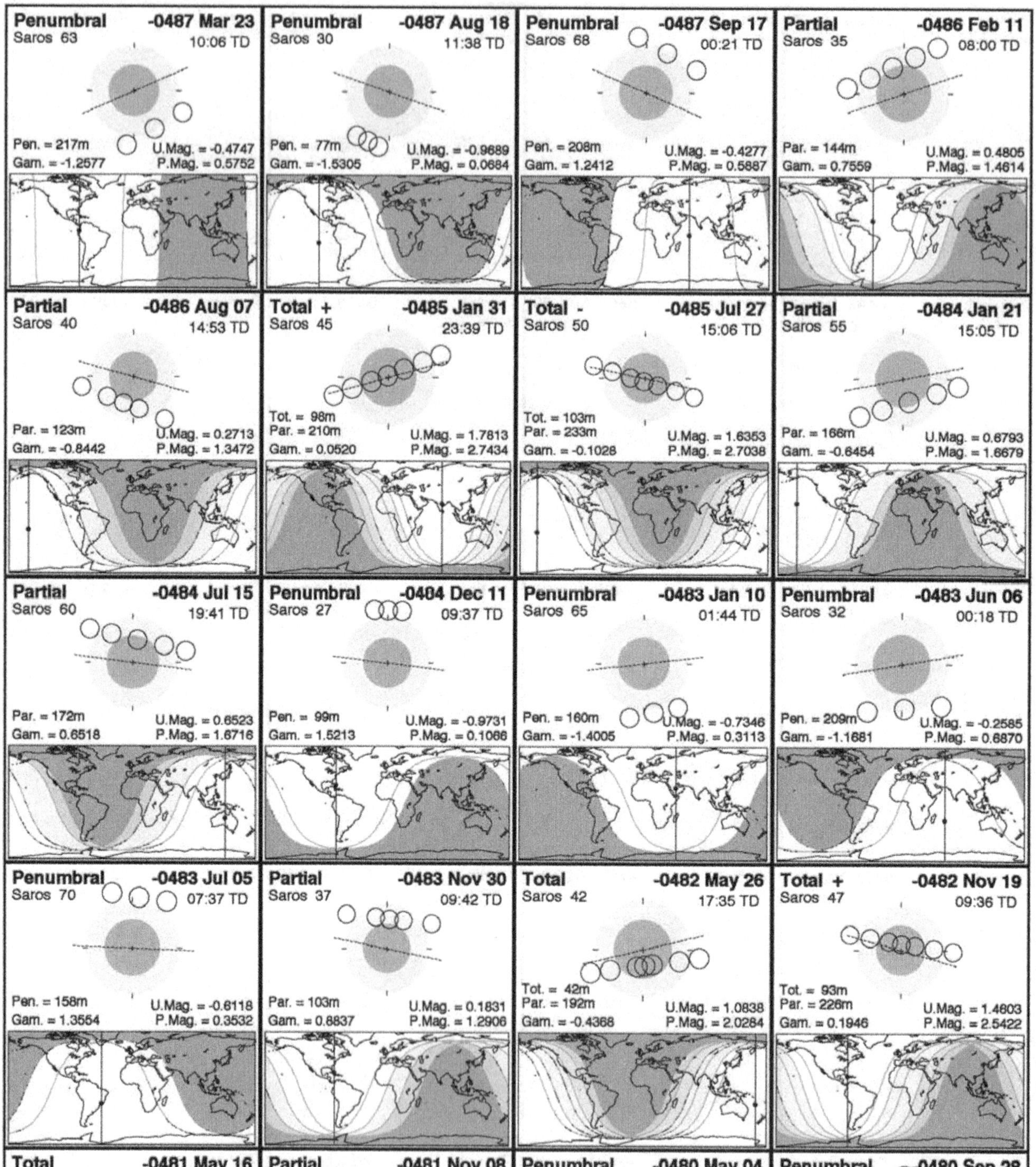

ΔT= 16890 s [= 04h41m] std.err. = ± 426 s [= ± 1.8°] **Plate 184**

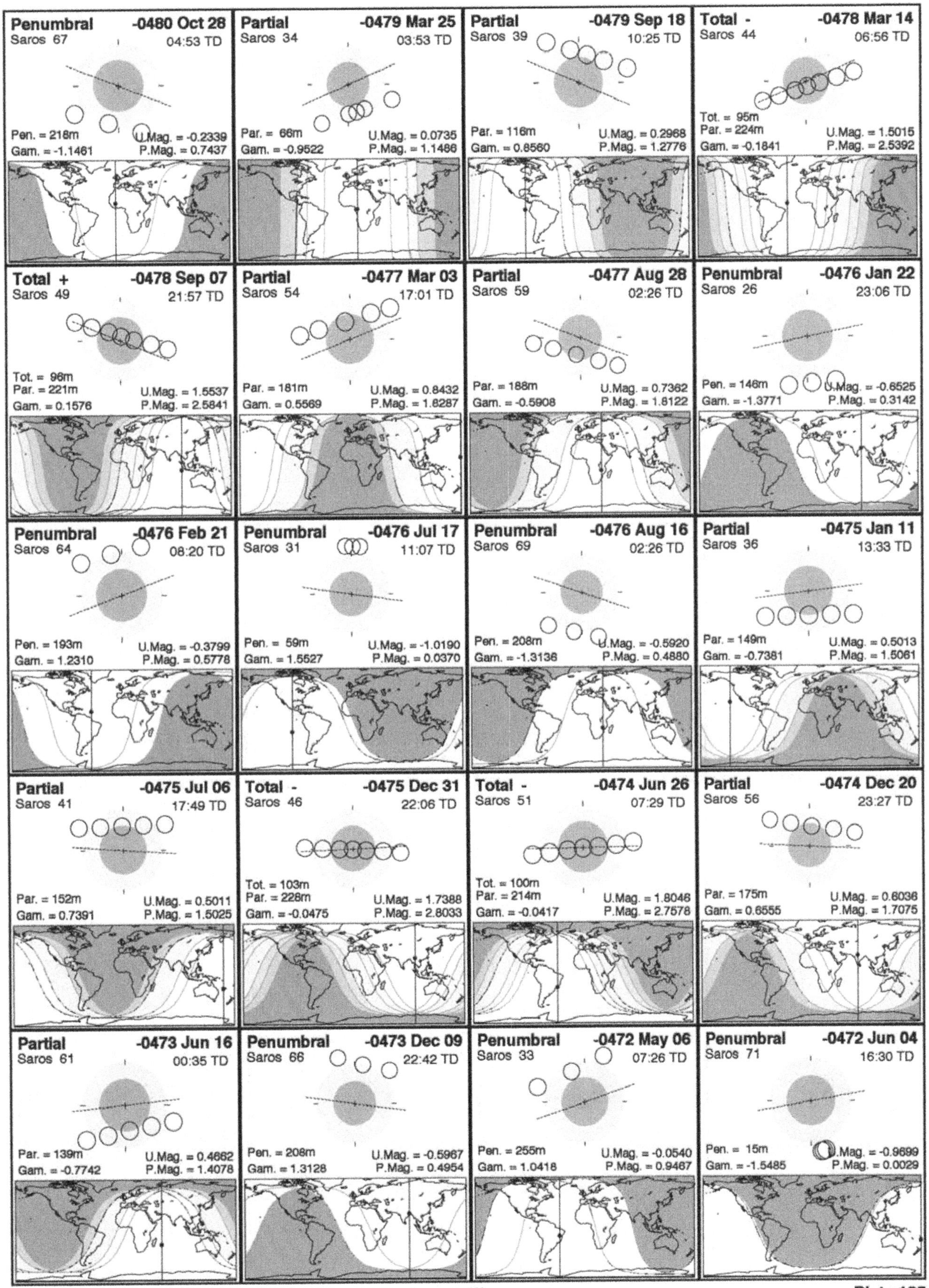

ΔT= 16757 s [= 04h39m] std.err. = ± 423 s [= ± 1.8°] Plate 185

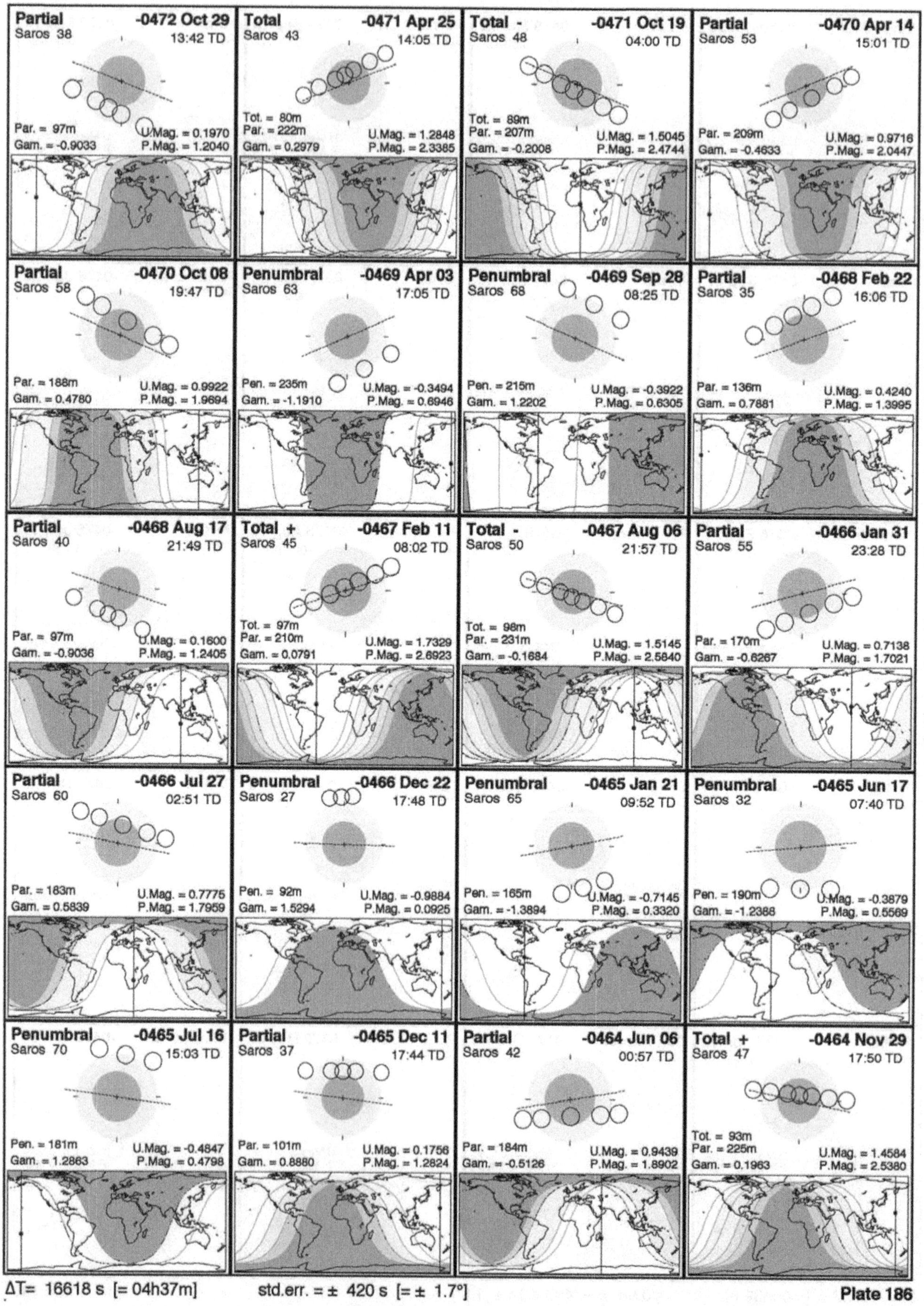

ΔT= 16618 s [= 04h37m] std.err. = ± 420 s [= ± 1.7°] **Plate 186**

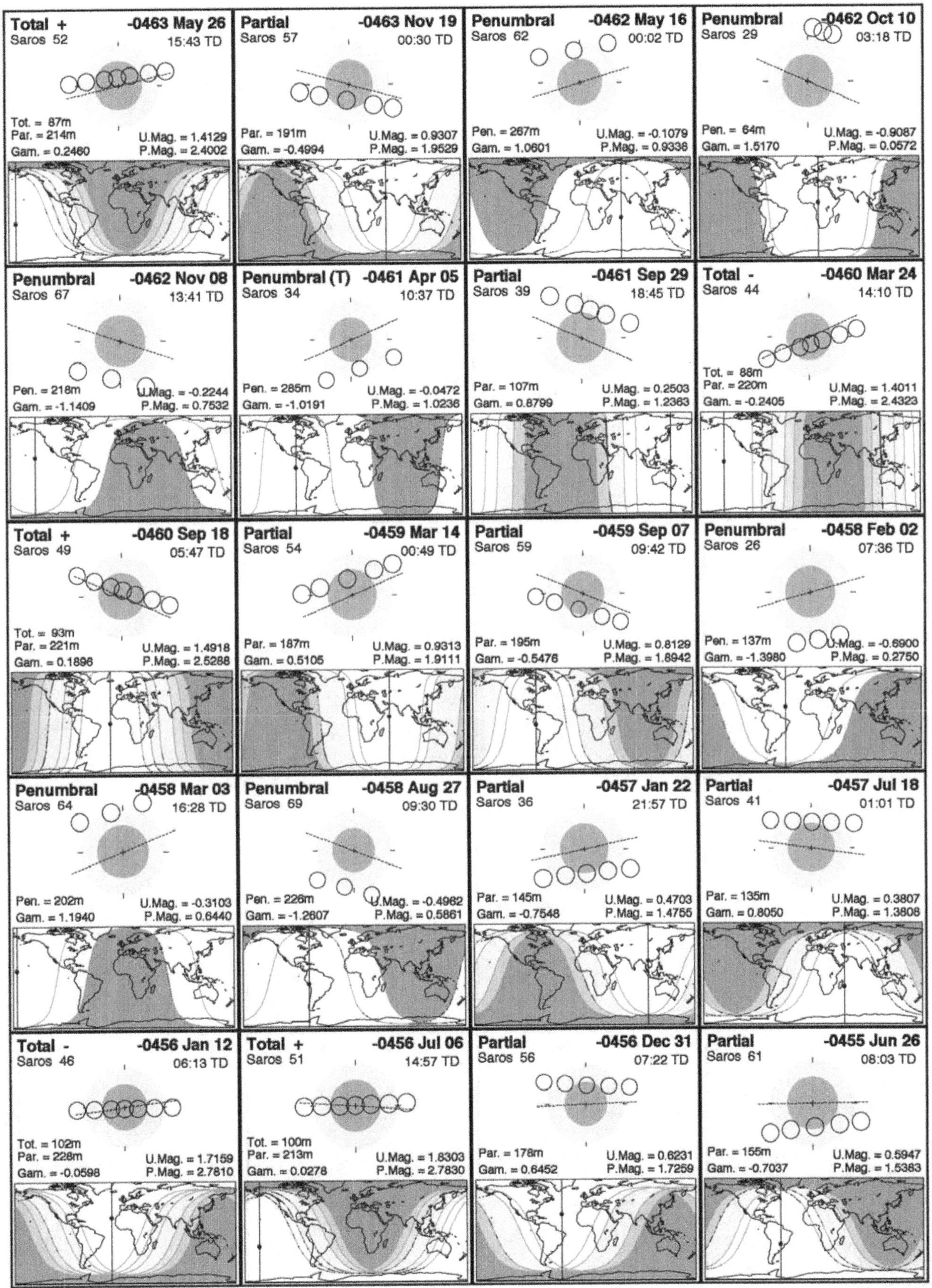

ΔT= 16472 s [= 04h35m] std.err. = ± 417 s [= ± 1.7°] Plate 187

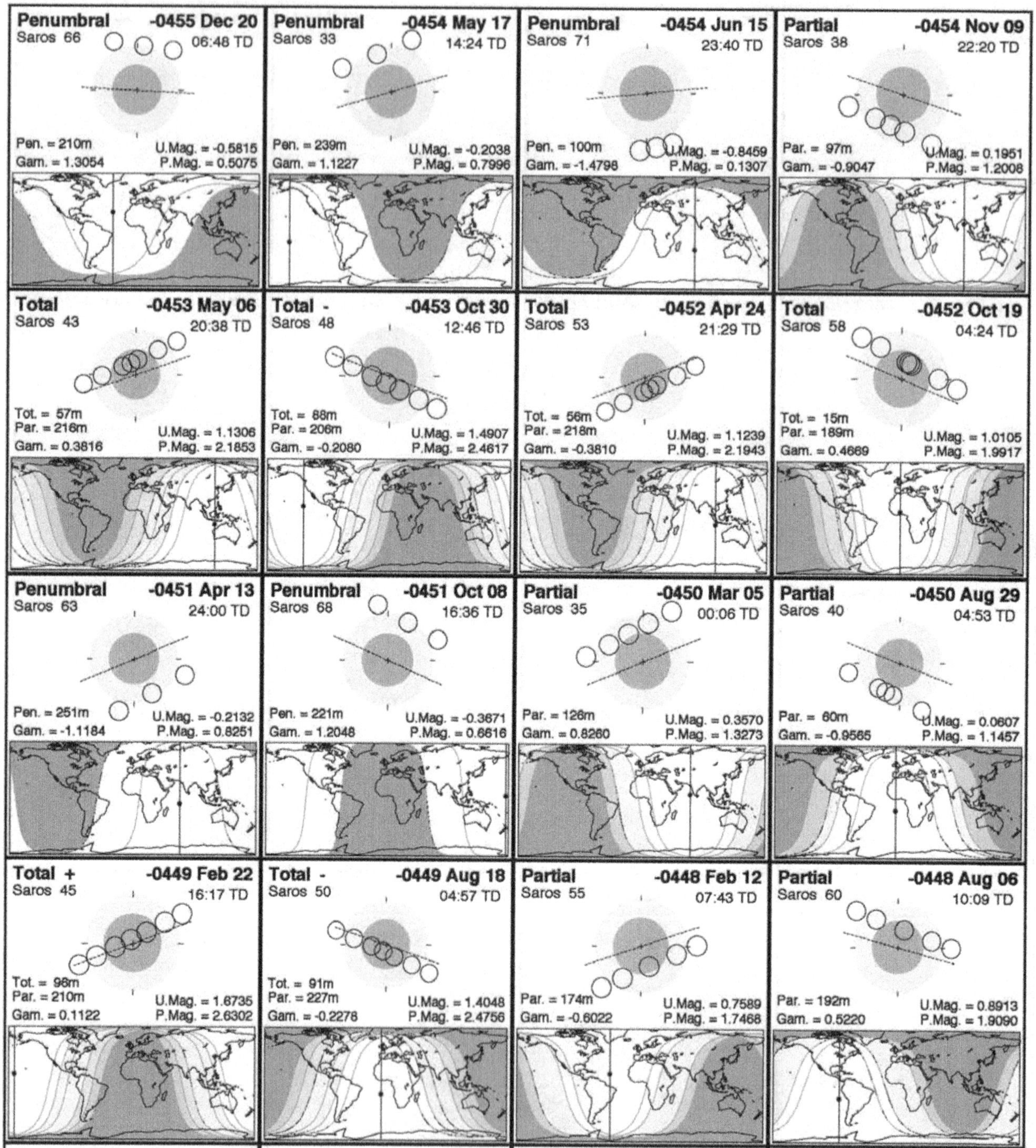

ΔT= 16328 s [= 04h32m] std.err. = ± 414 s [= ± 1.7°] **Plate 188**

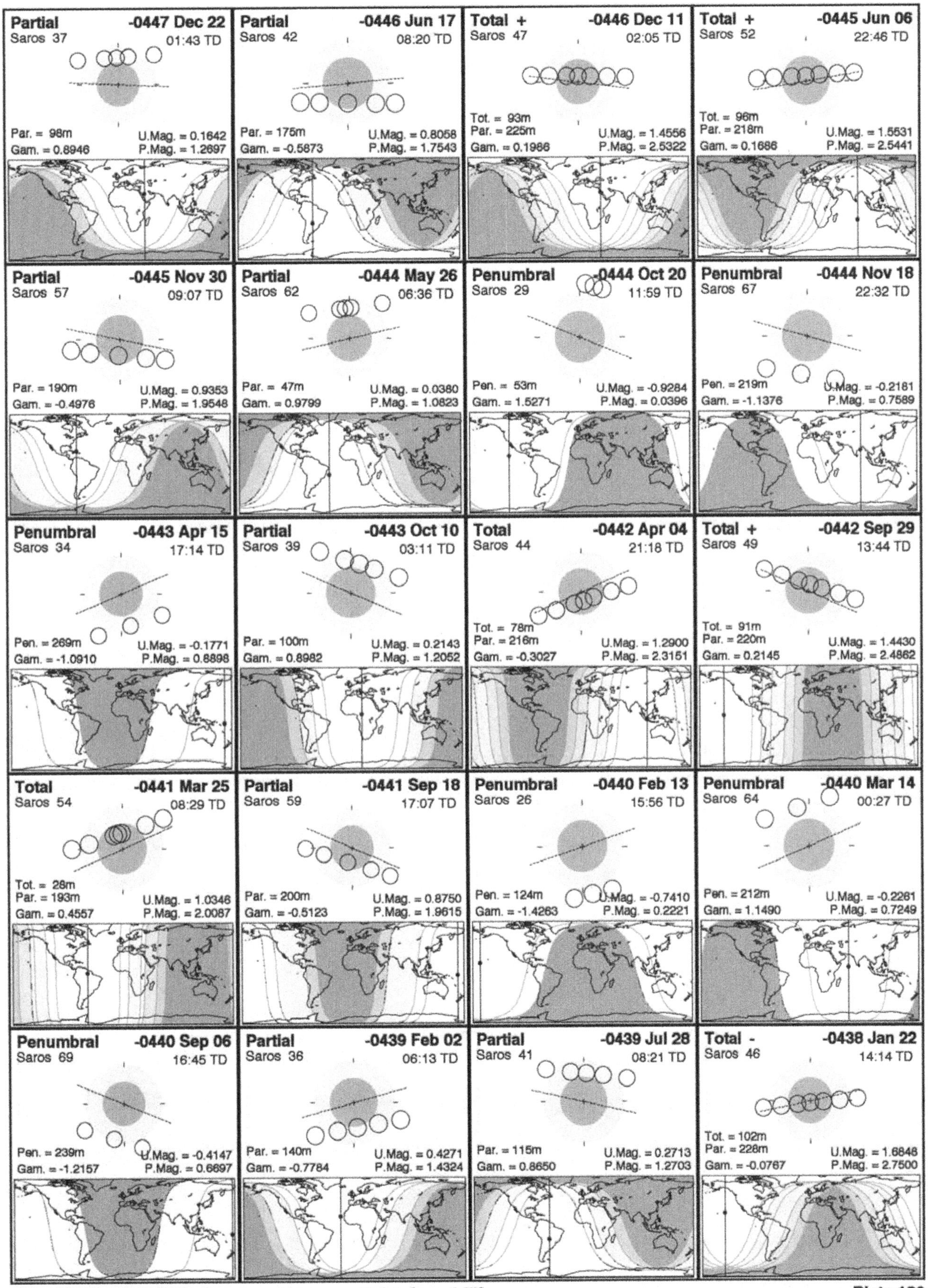

ΔT= 16195 s [= 04h30m] std.err. = ± 411 s [= ± 1.7°] **Plate 189**

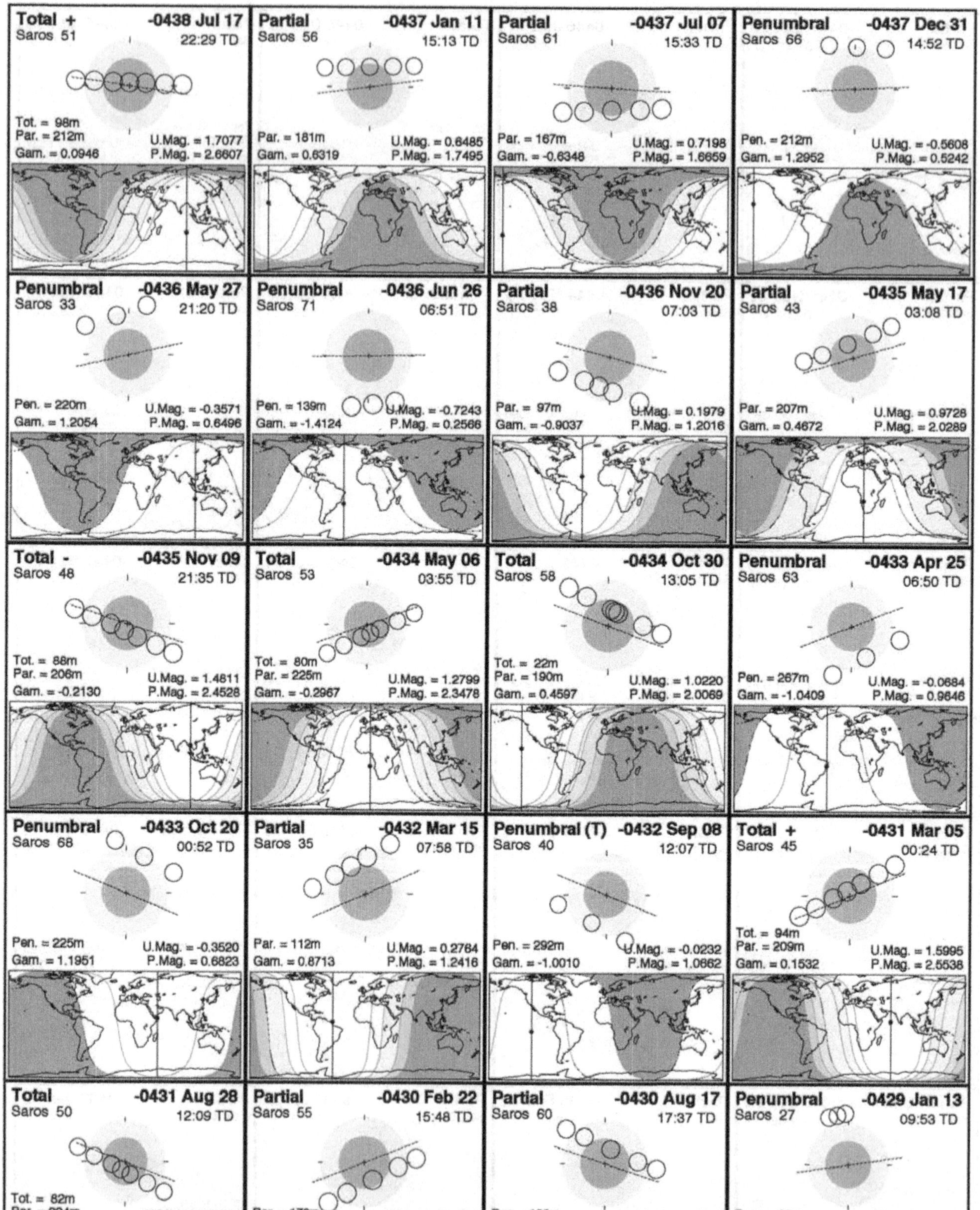

ΔT= 16054 s [= 04h28m] std.err. = ± 408 s [= ± 1.7°] **Plate 190**

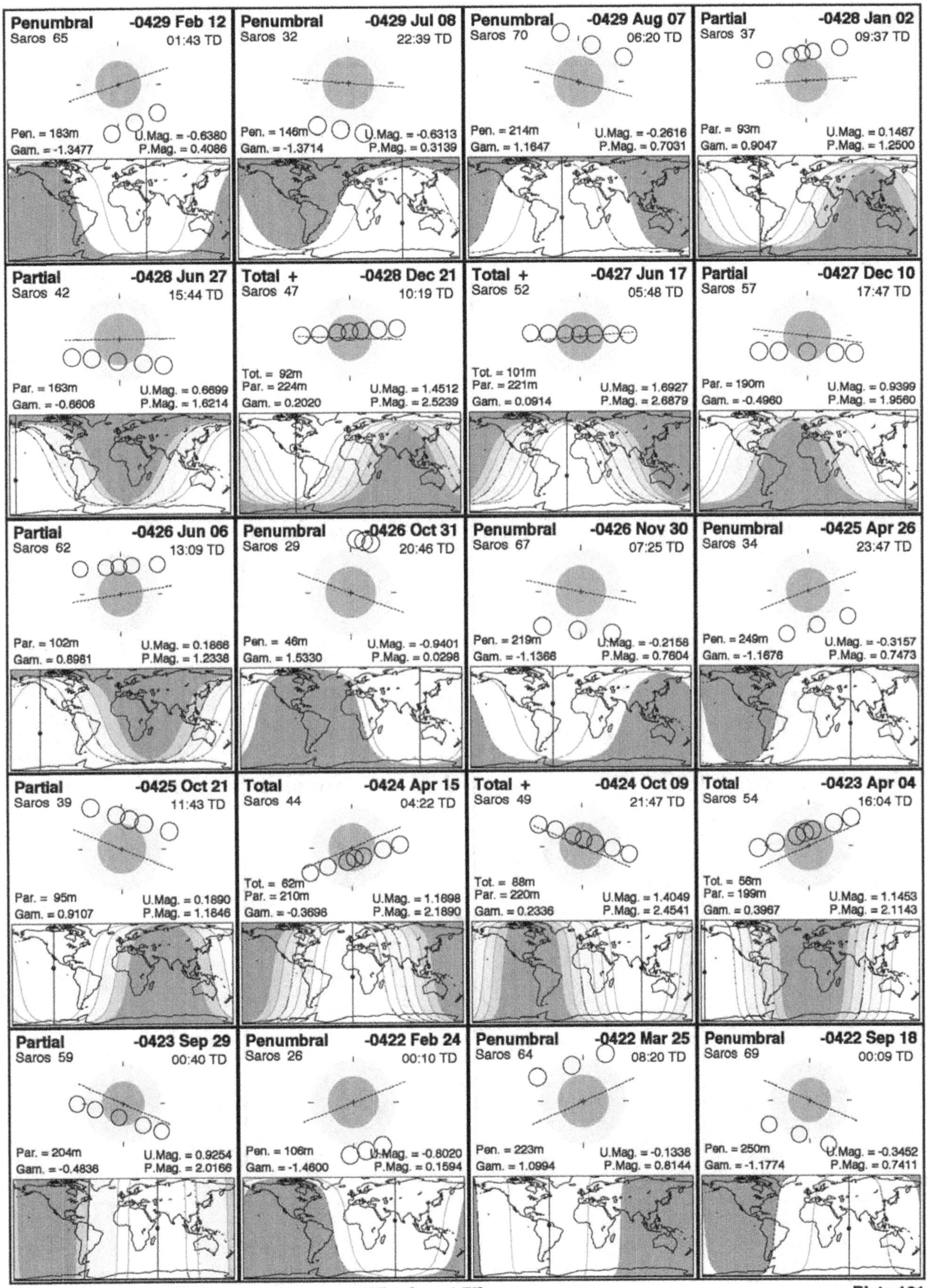

ΔT= 15915 s [= 04h25m] std.err. = ± 405 s [= ± 1.7°] **Plate 191**

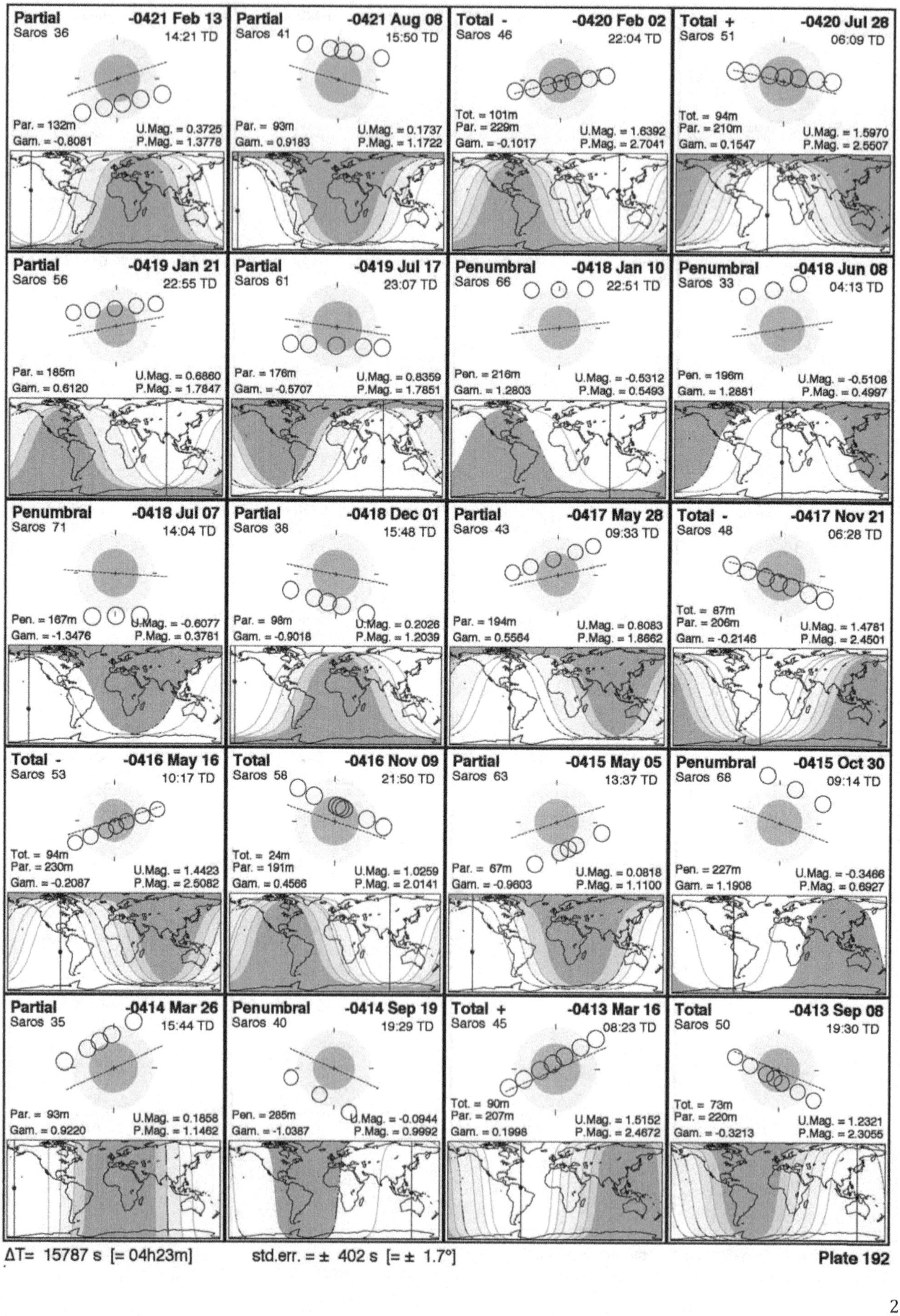

ΔT= 15787 s [= 04h23m] std.err. = ± 402 s [= ± 1.7°] **Plate 192**

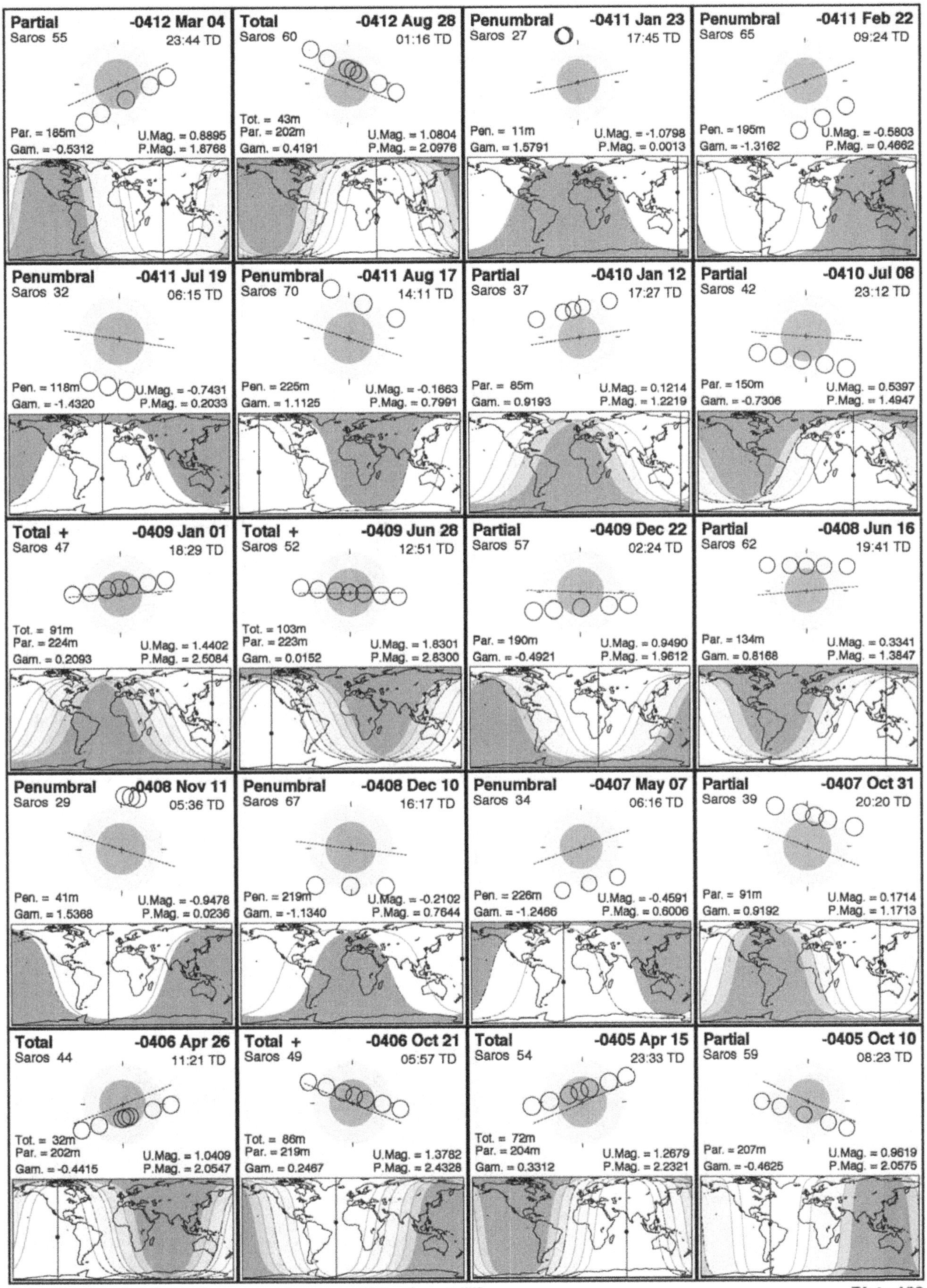

ΔT= 15643 s [= 04h21m] std.err. = ± 398 s [= ± 1.7°] **Plate 193**

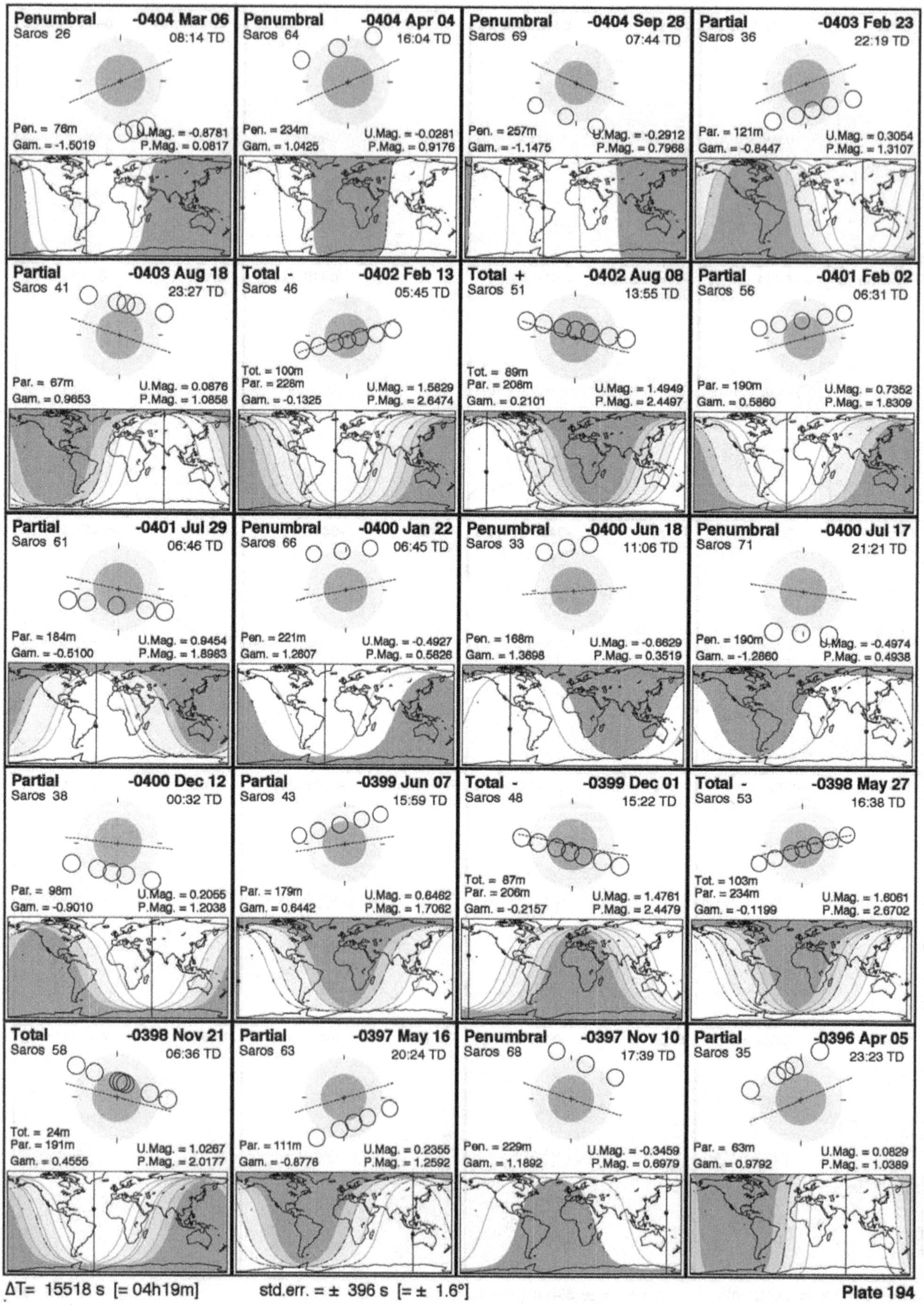

ΔT= 15518 s [= 04h19m] std.err. = ± 396 s [= ± 1.6°] **Plate 194**

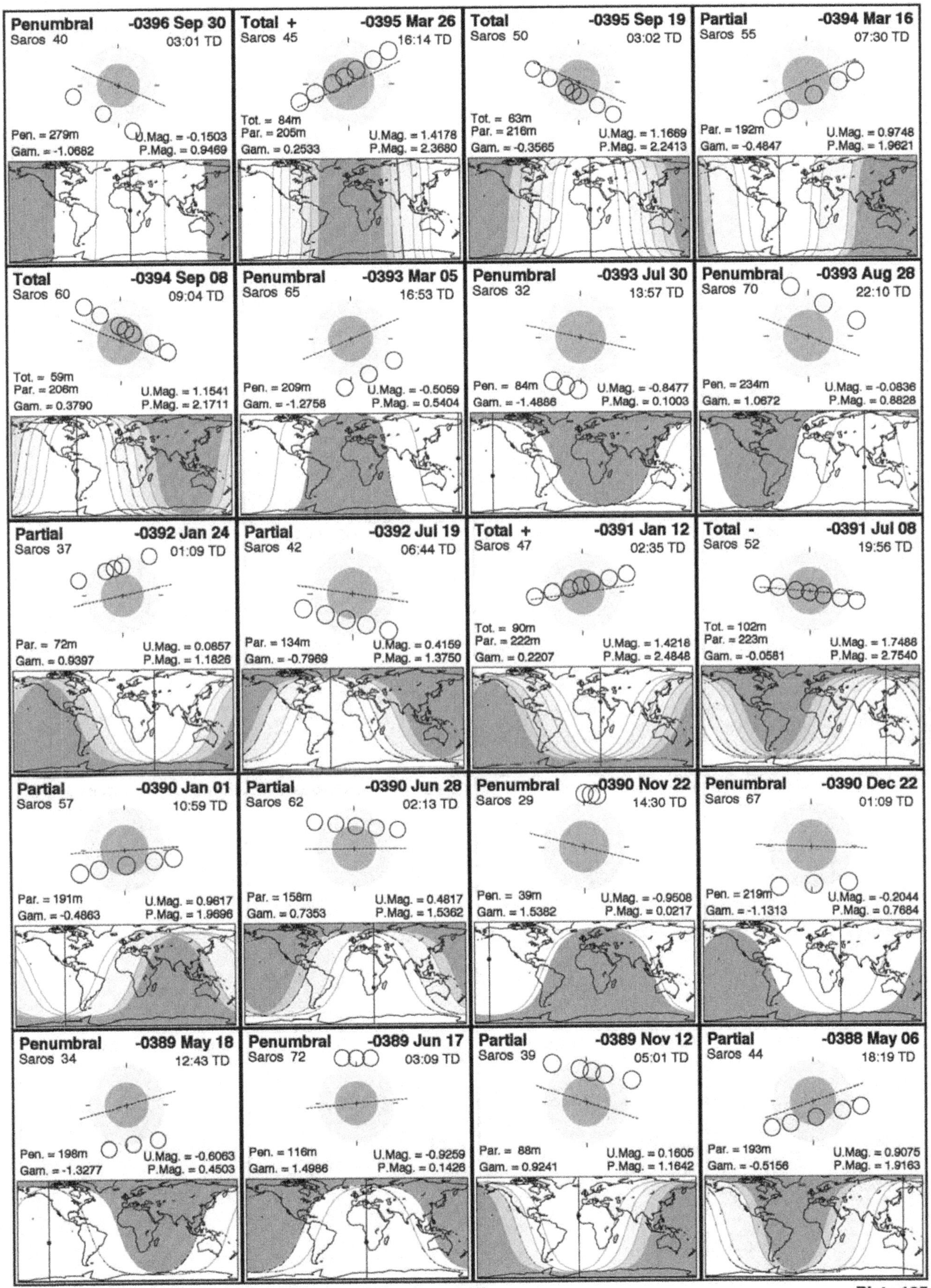

ΔT= 15386 s [= 04h16m] std.err. = ± 393 s [= ± 1.6°] **Plate 195**

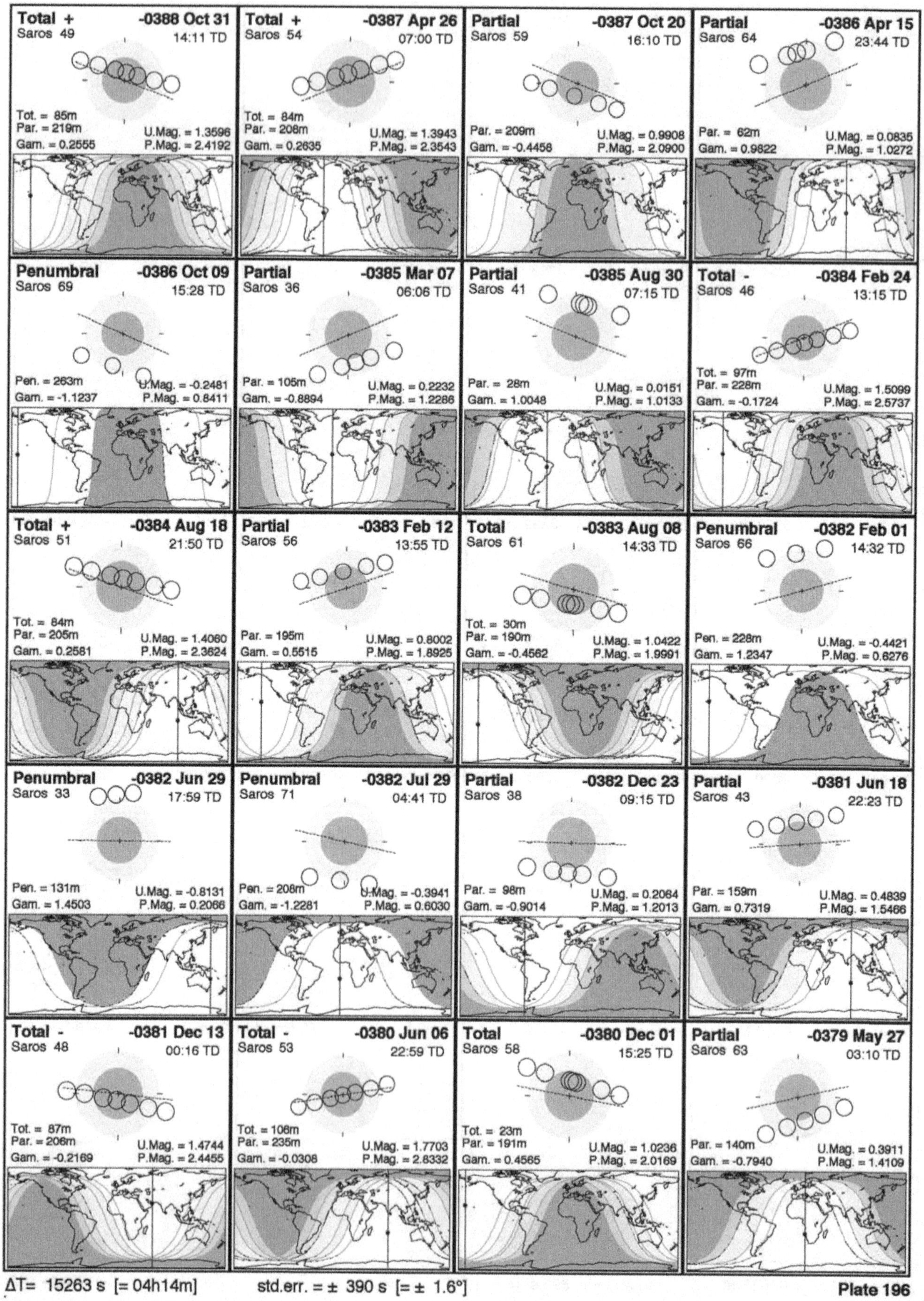

ΔT= 15263 s [= 04h14m] std.err. = ± 390 s [= ± 1.6°] **Plate 196**

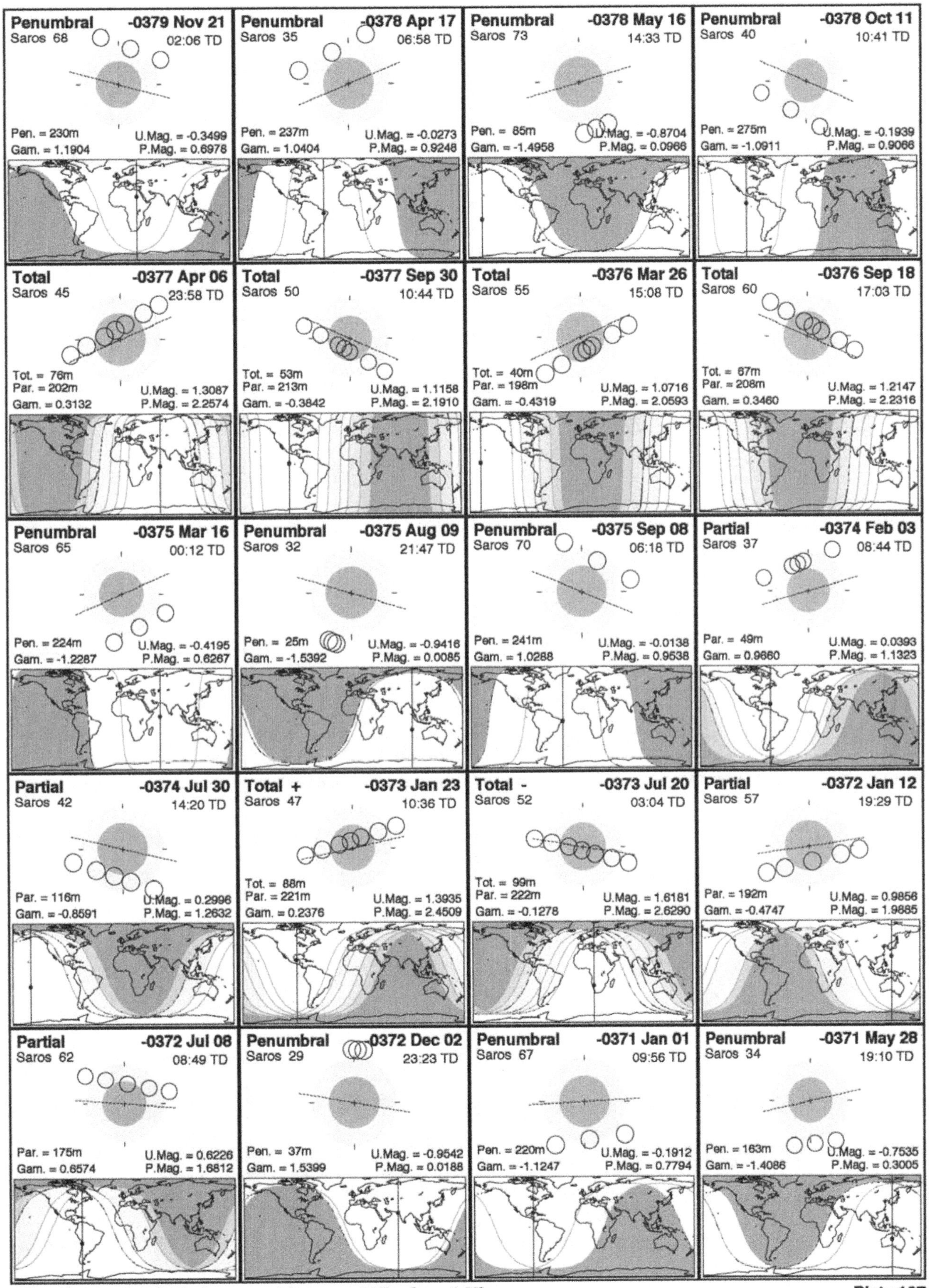

ΔT= 15126 s [= 04h12m] std.err. = ± 387 s [= ± 1.6°] **Plate 197**

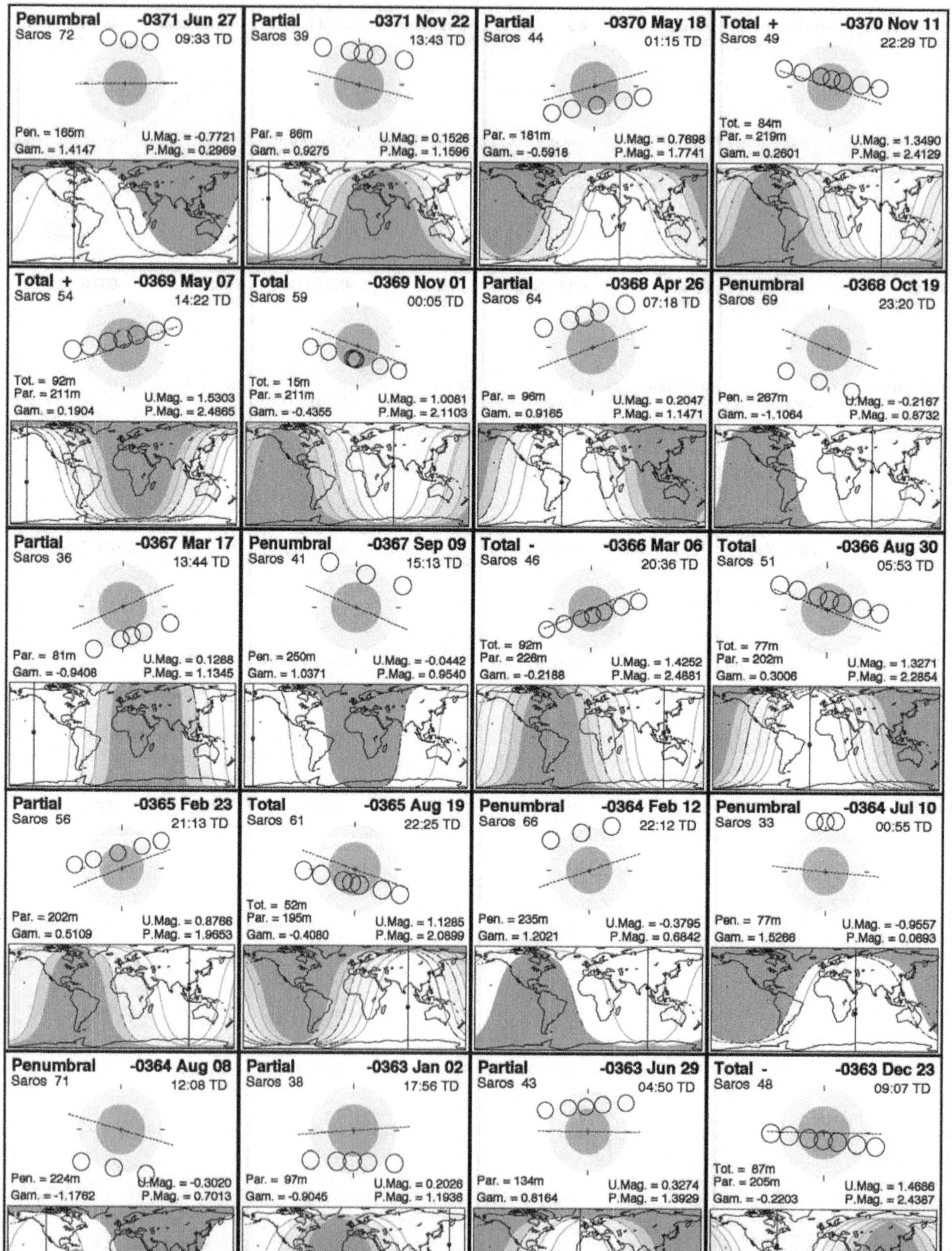

ΔT= 15013 s [= 04h10m] std.err. = ± 384 s [= ± 1.6°] **Plate 198**

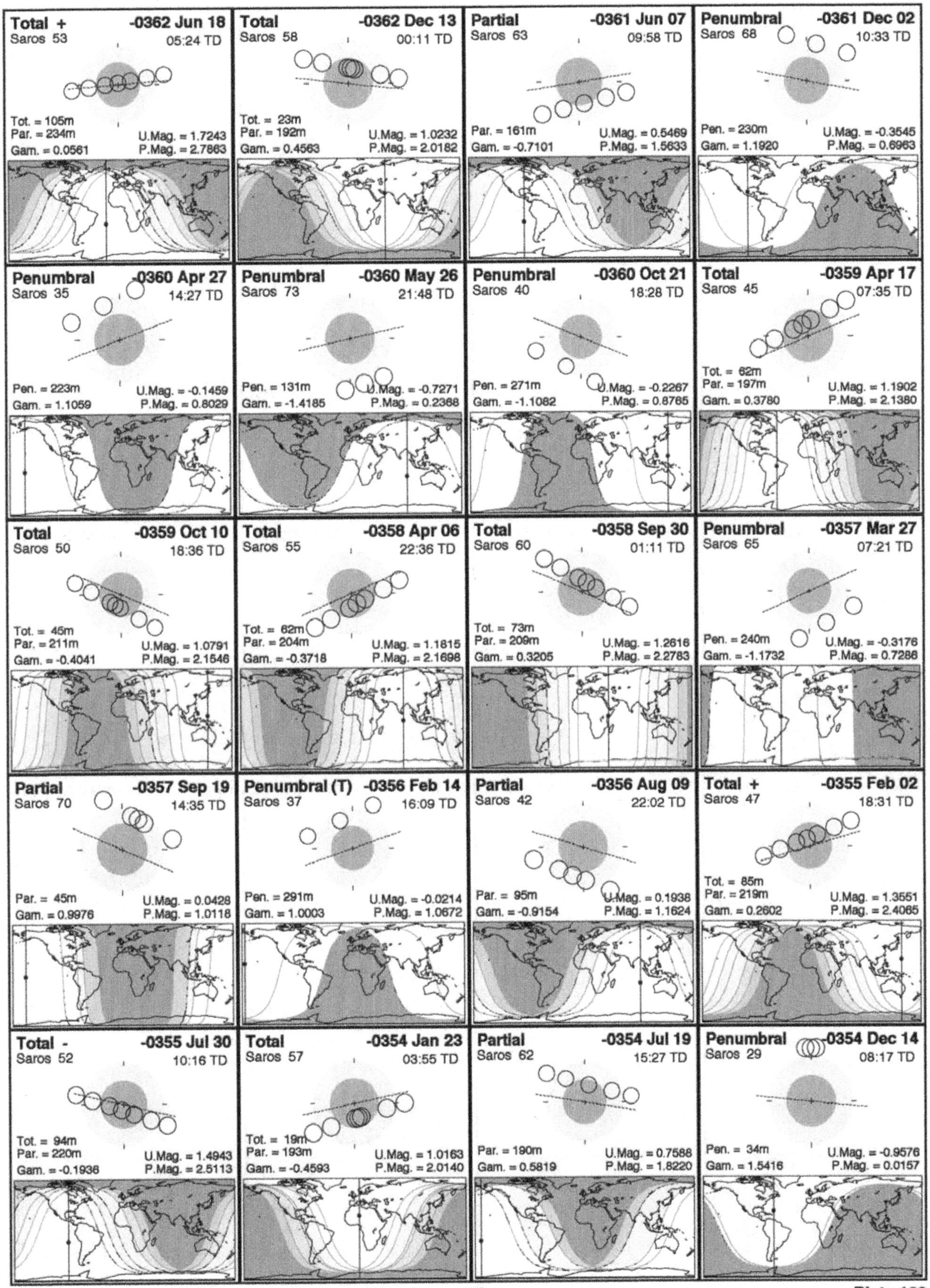

ΔT= 14880 s [= 04h08m] std.err. = ± 381 s [= ± 1.6°] **Plate 199**

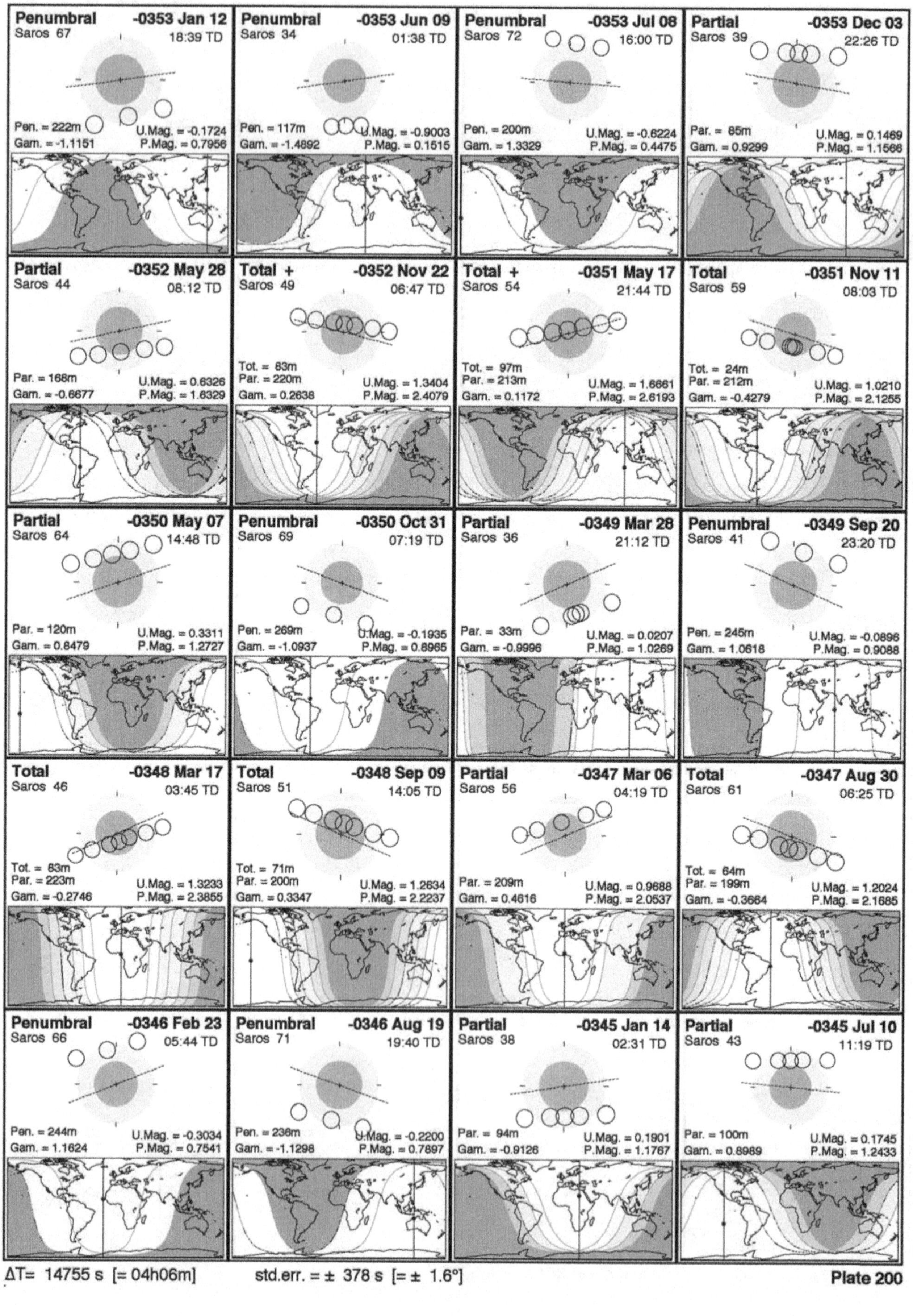

ΔT= 14755 s [= 04h06m] std.err. = ± 378 s [= ± 1.6°] **Plate 200**

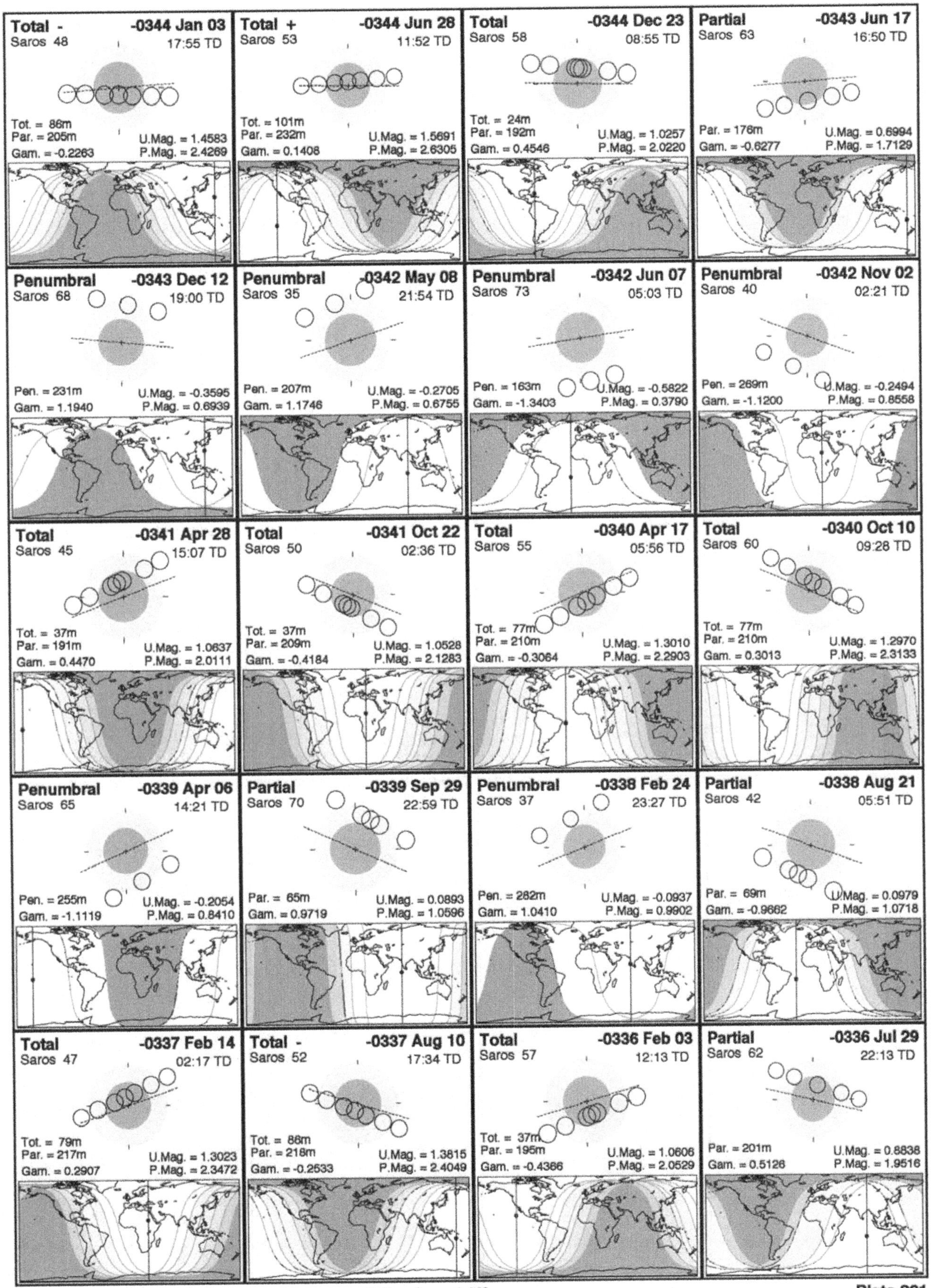

ΔT= 14626 s [= 04h04m] std.err. = ± 375 s [= ± 1.6°] **Plate 201**

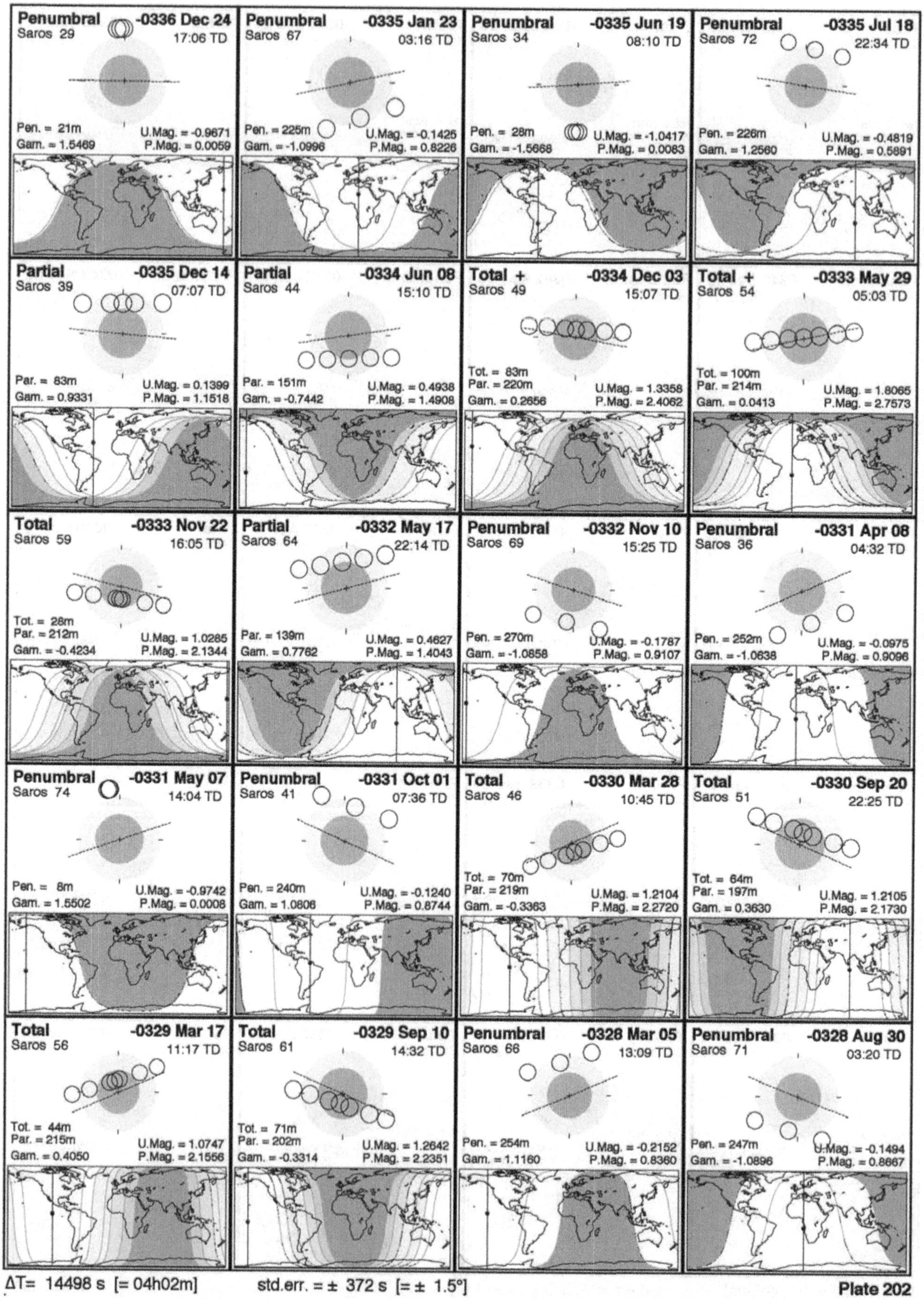

ΔT= 14498 s [= 04h02m] std.err. = ± 372 s [= ± 1.5°] **Plate 202**

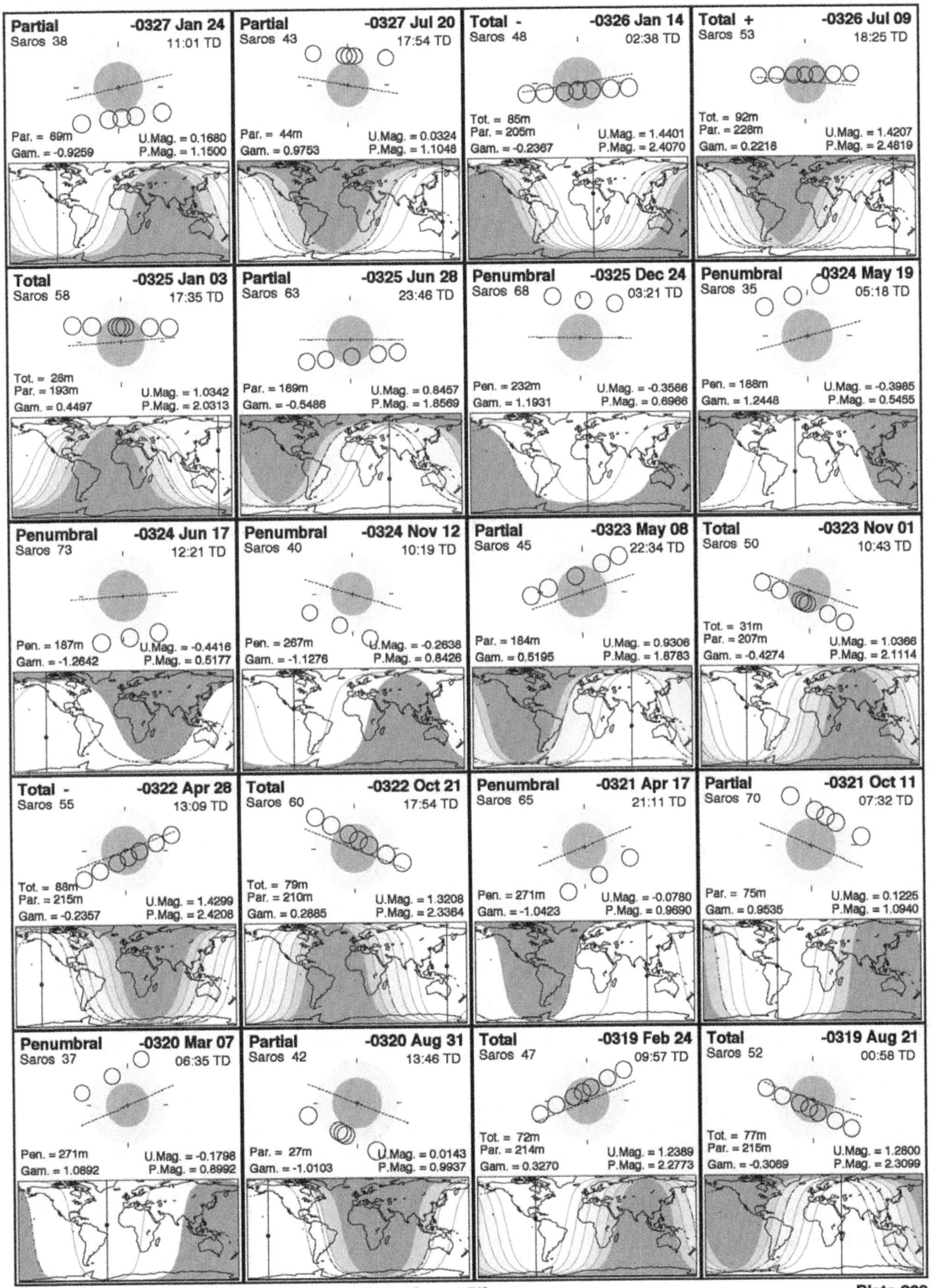

ΔT= 14384 s [= 04h00m] std.err. = ± 369 s [= ± 1.5°] **Plate 203**

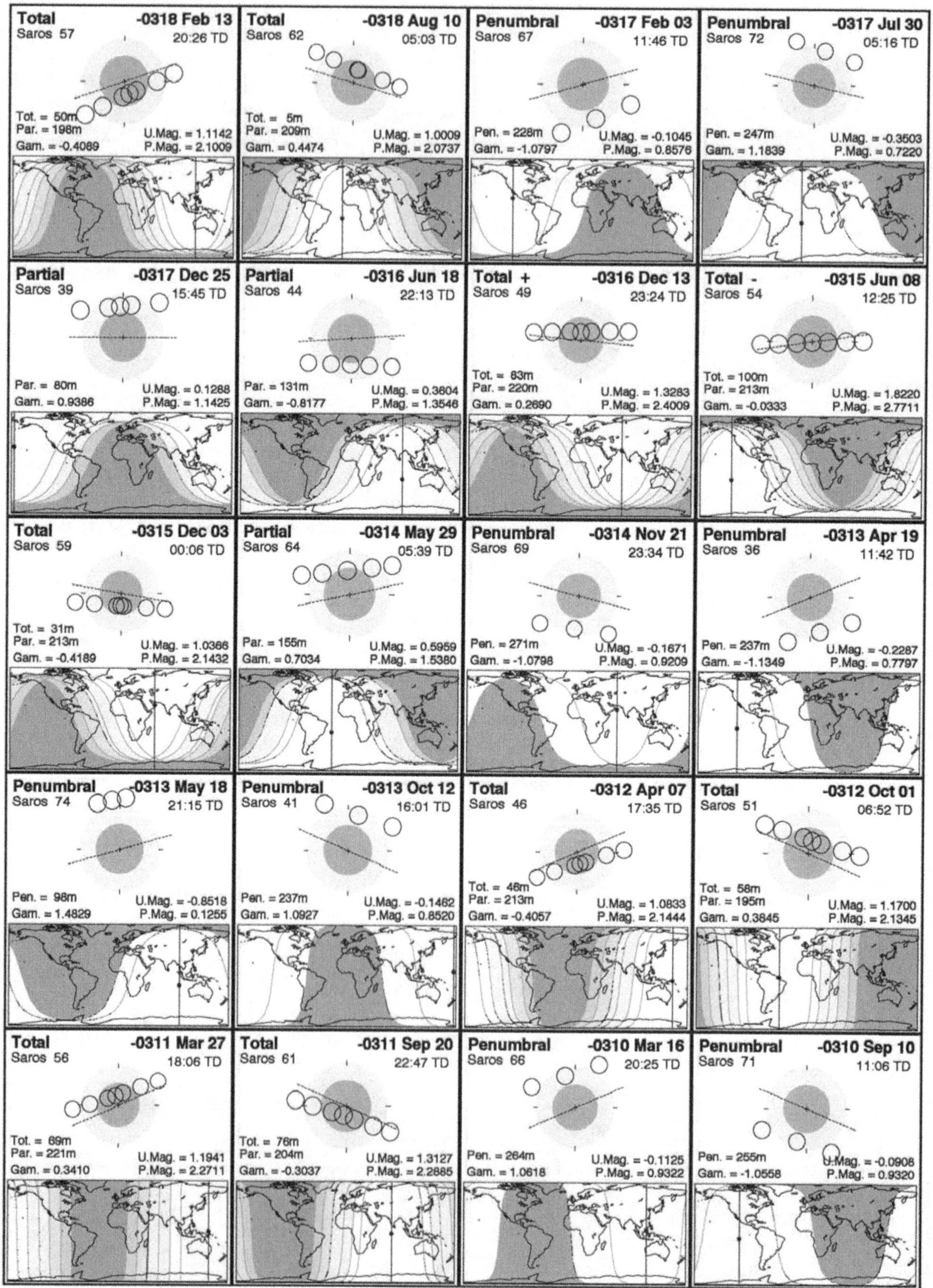

ΔT= 14258 s [= 03h58m] std.err. = ± 366 s [= ± 1.5°] **Plate 204**

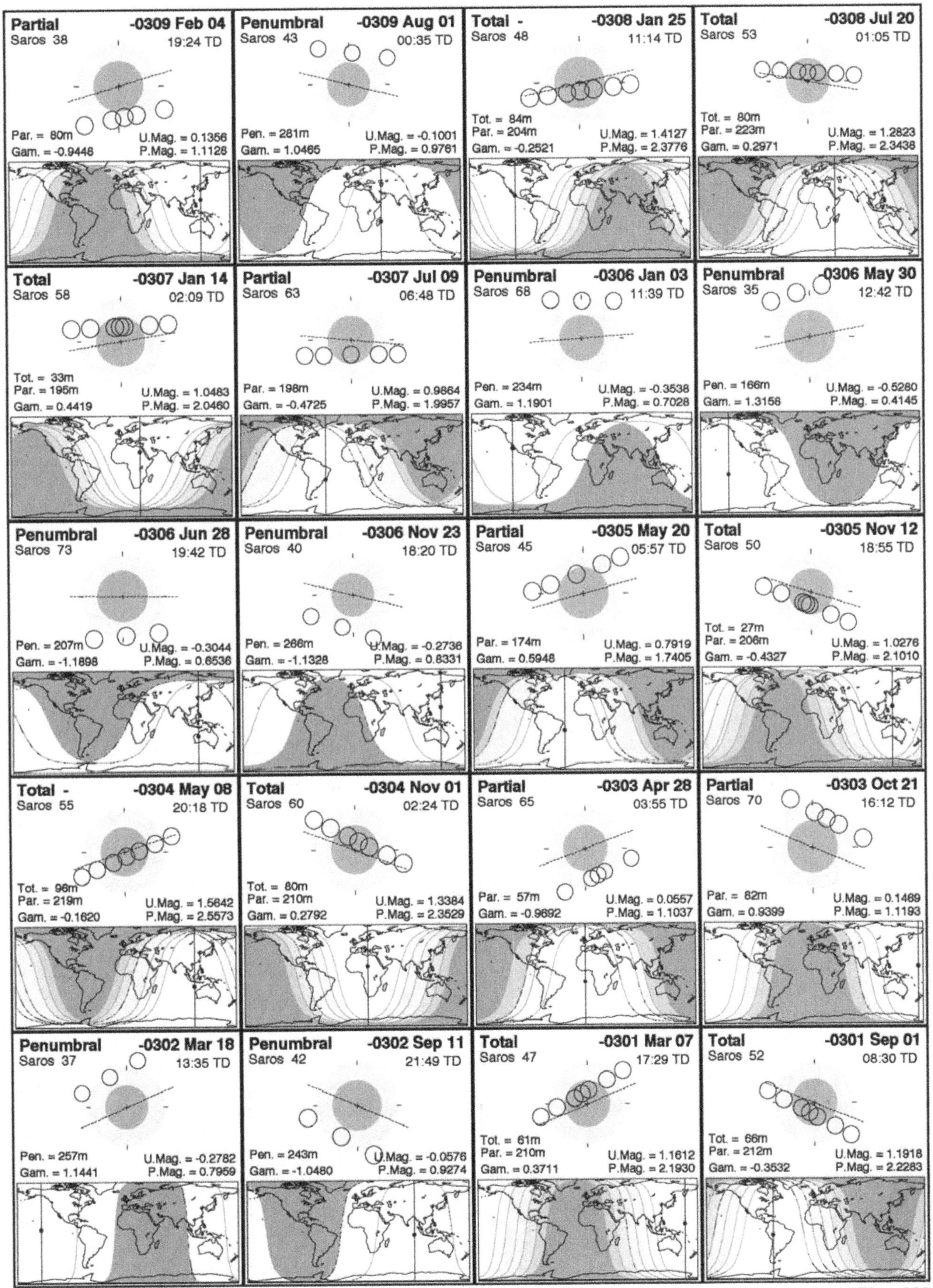

ΔT= 14134 s [= 03h56m] std.err. = ± 363 s [= ± 1.5°] **Plate 205**

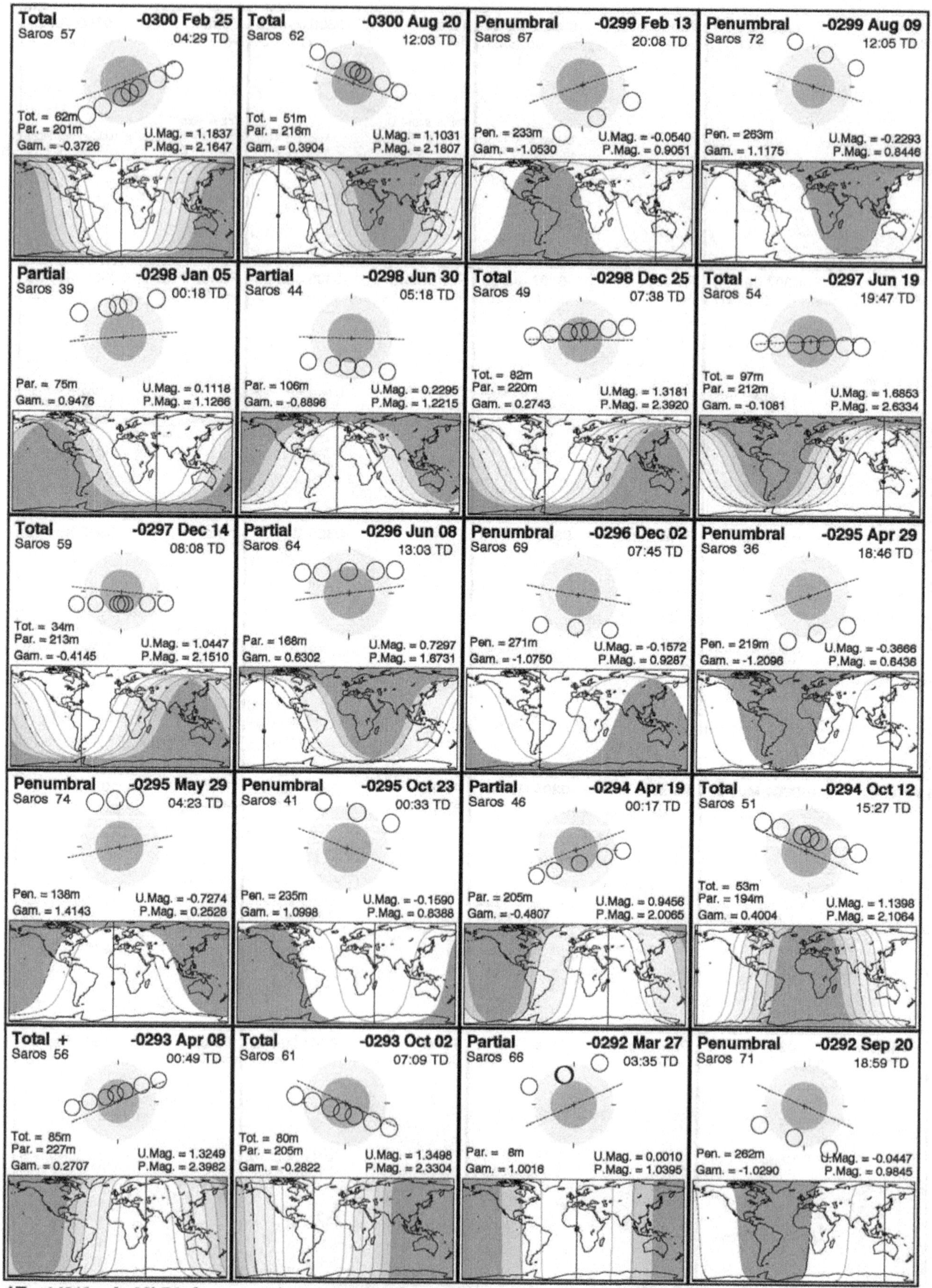

ΔT= 14010 s [= 03h54m] std.err. = ± 359 s [= ± 1.5°] **Plate 206**

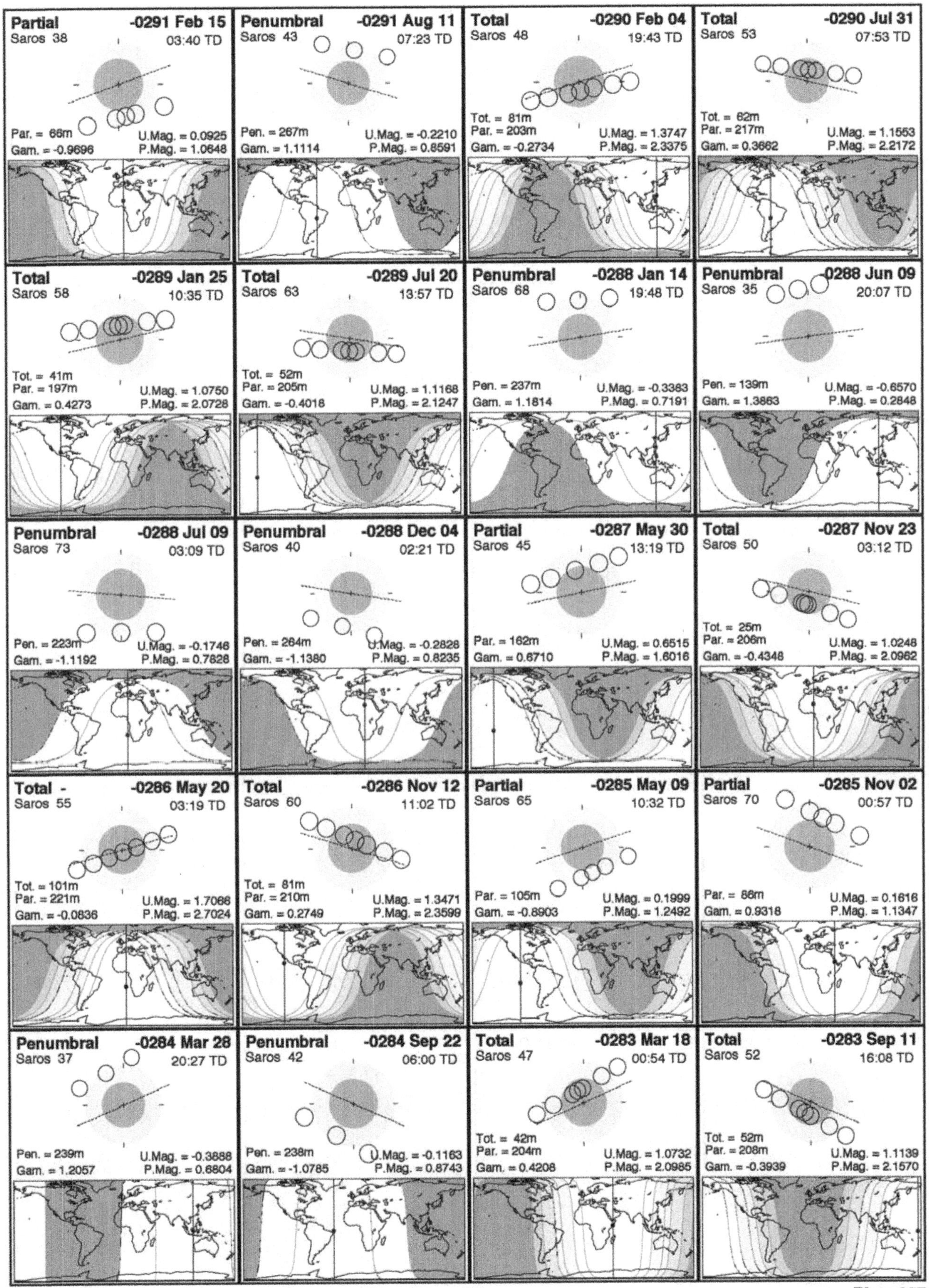

ΔT= 13889 s [= 03h51m] std.err. = ± 356 s [= ± 1.5°] **Plate 207**

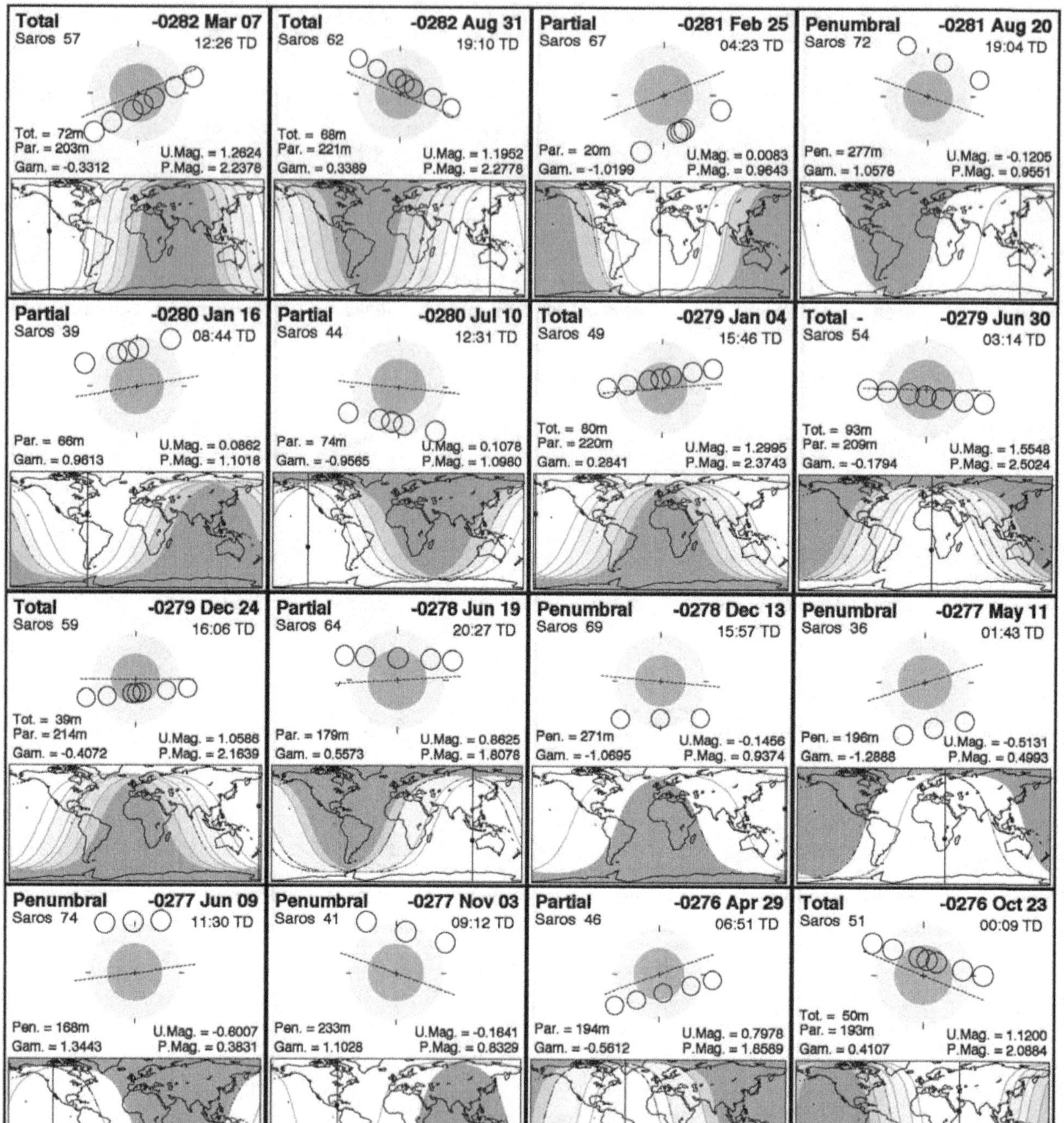

ΔT= 13768 s [= 03h49m] std.err. = ± 353 s [= ± 1.5°] **Plate 208**

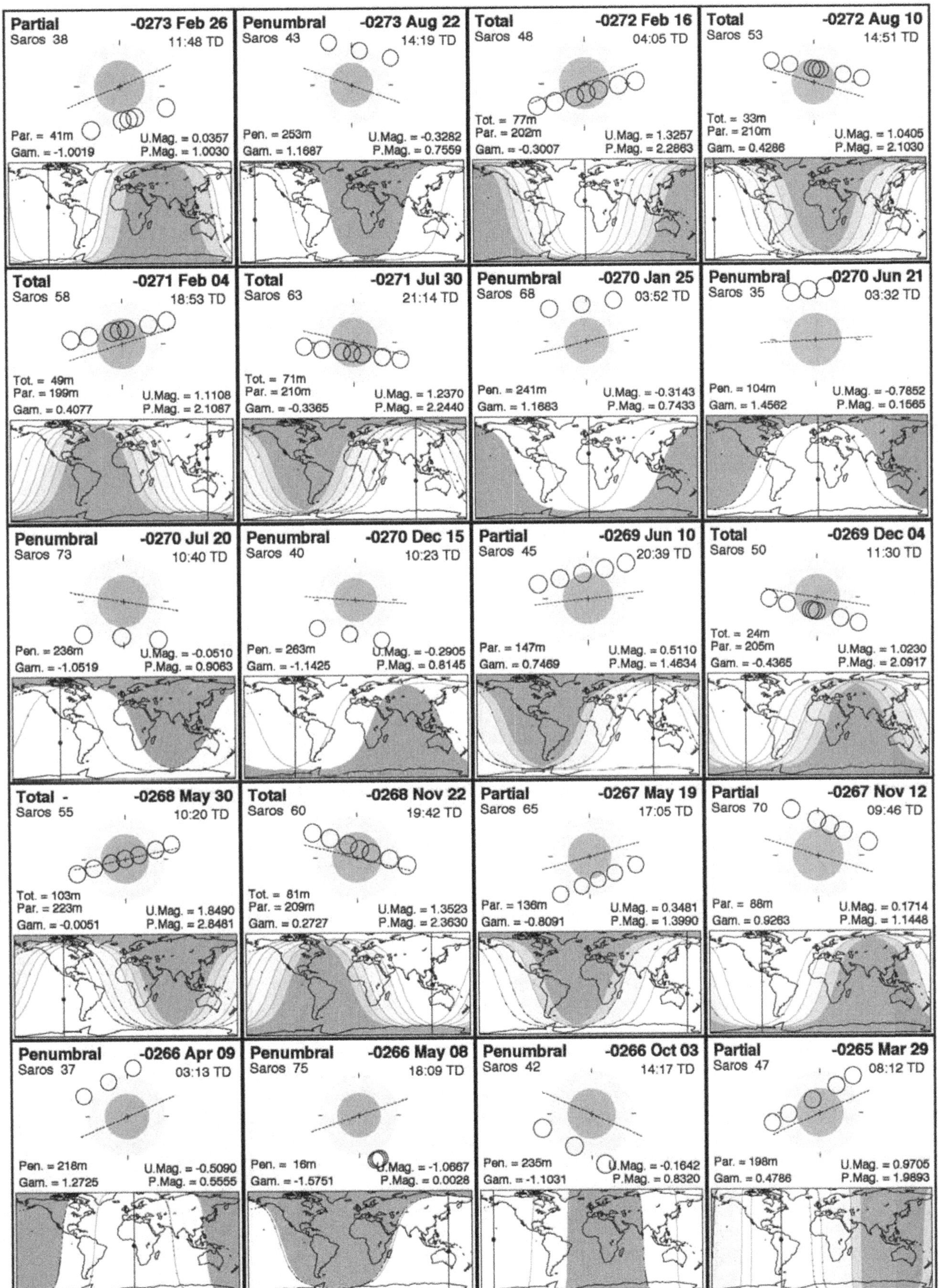

ΔT= 13650 s [= 03h47m] std.err. = ± 350 s [= ± 1.5°] **Plate 209**

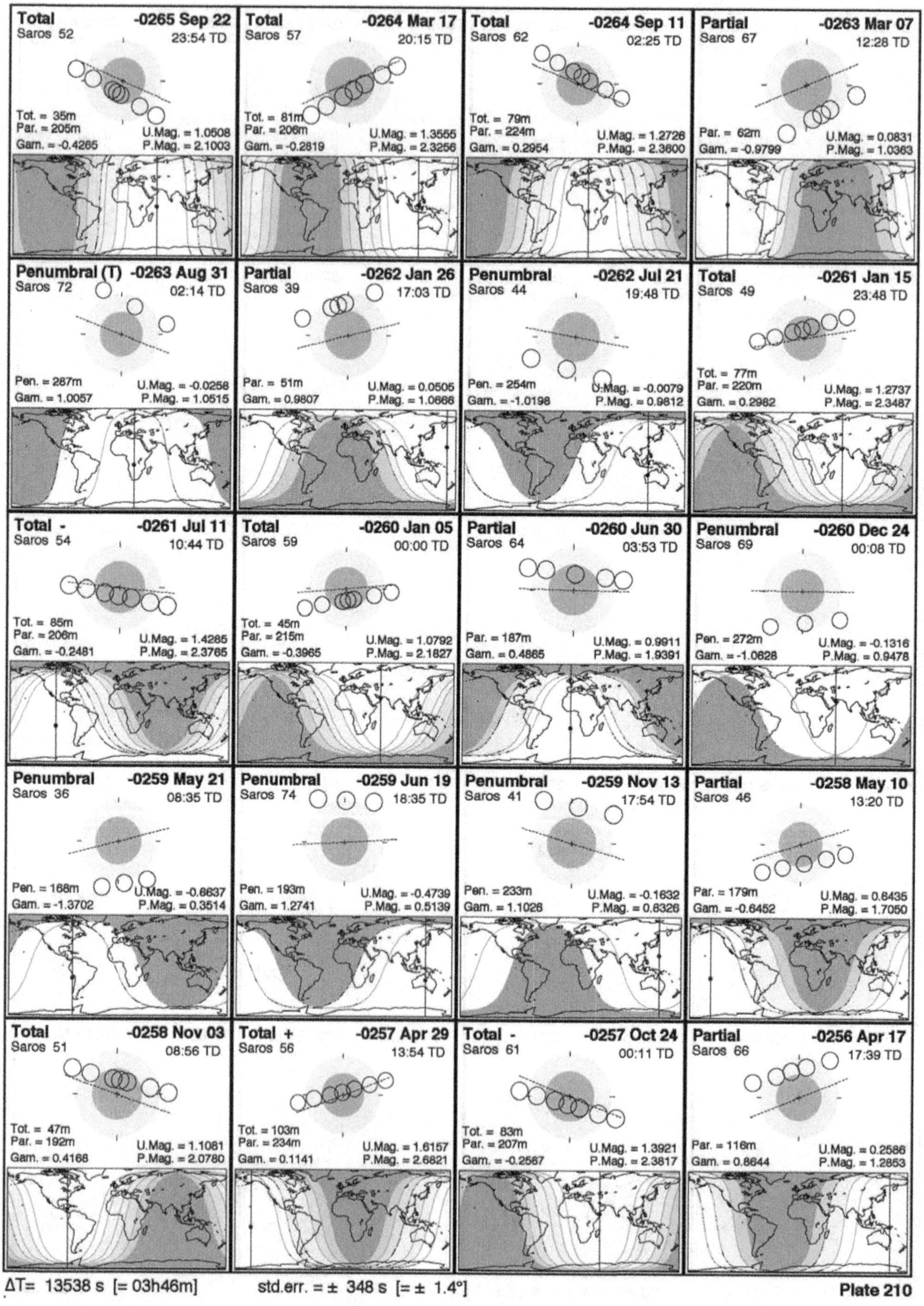

ΔT= 13538 s [= 03h46m] std.err. = ± 348 s [= ± 1.4°] **Plate 210**

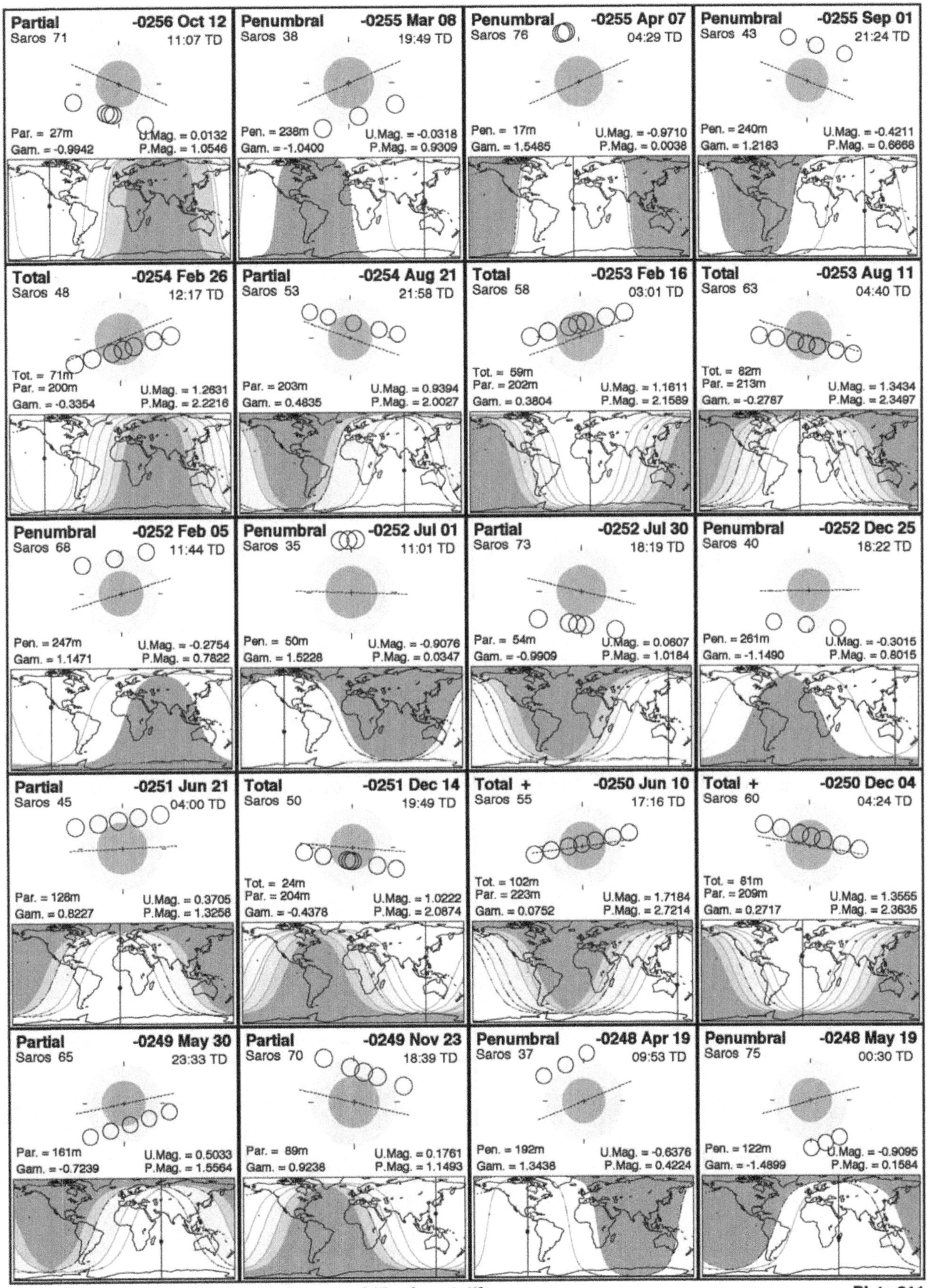

ΔT= 13421 s [= 03h44m] std.err. = ± 345 s [= ± 1.4°] Plate 211

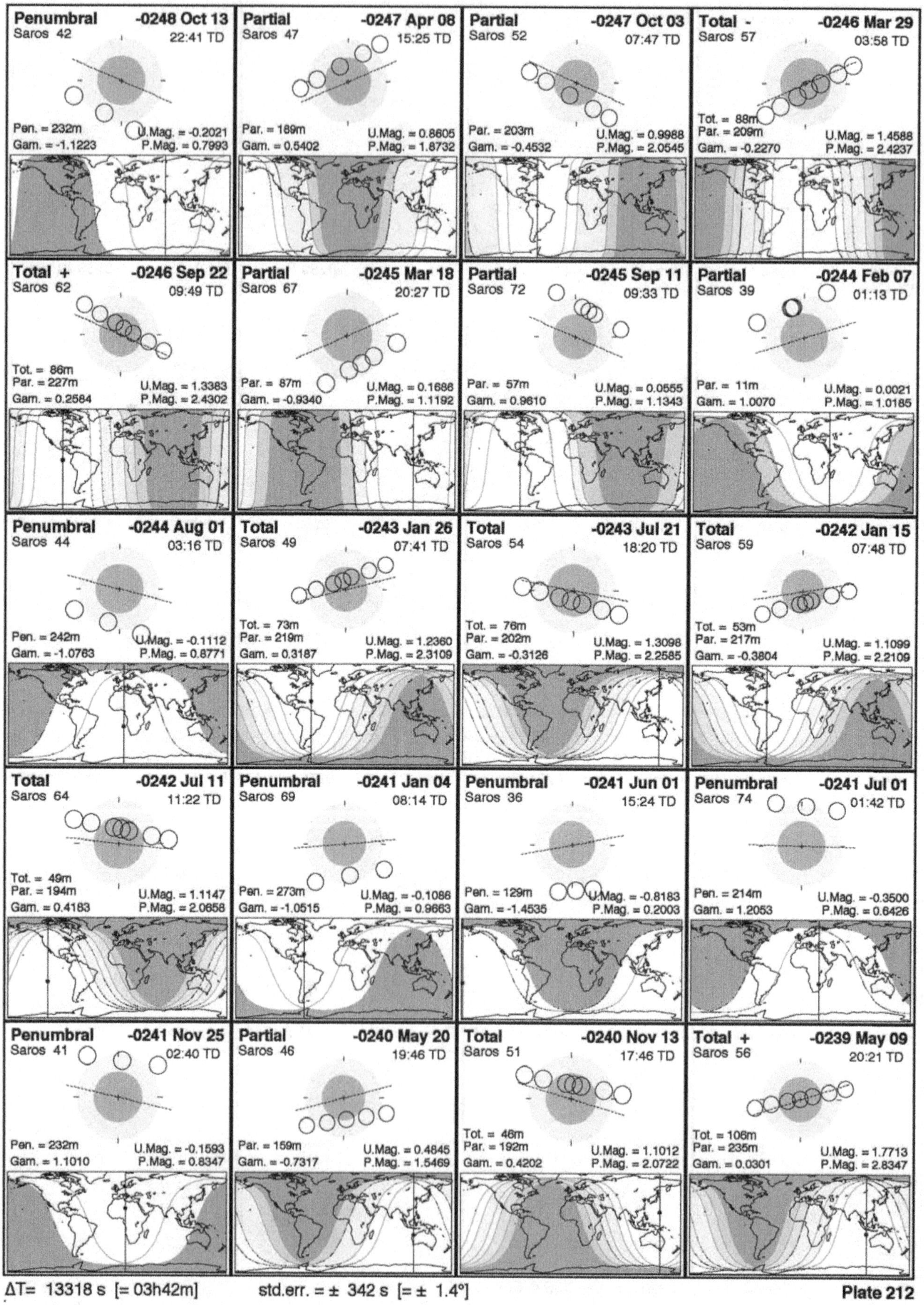

ΔT= 13318 s [= 03h42m] std.err. = ± 342 s [= ± 1.4°] **Plate 212**

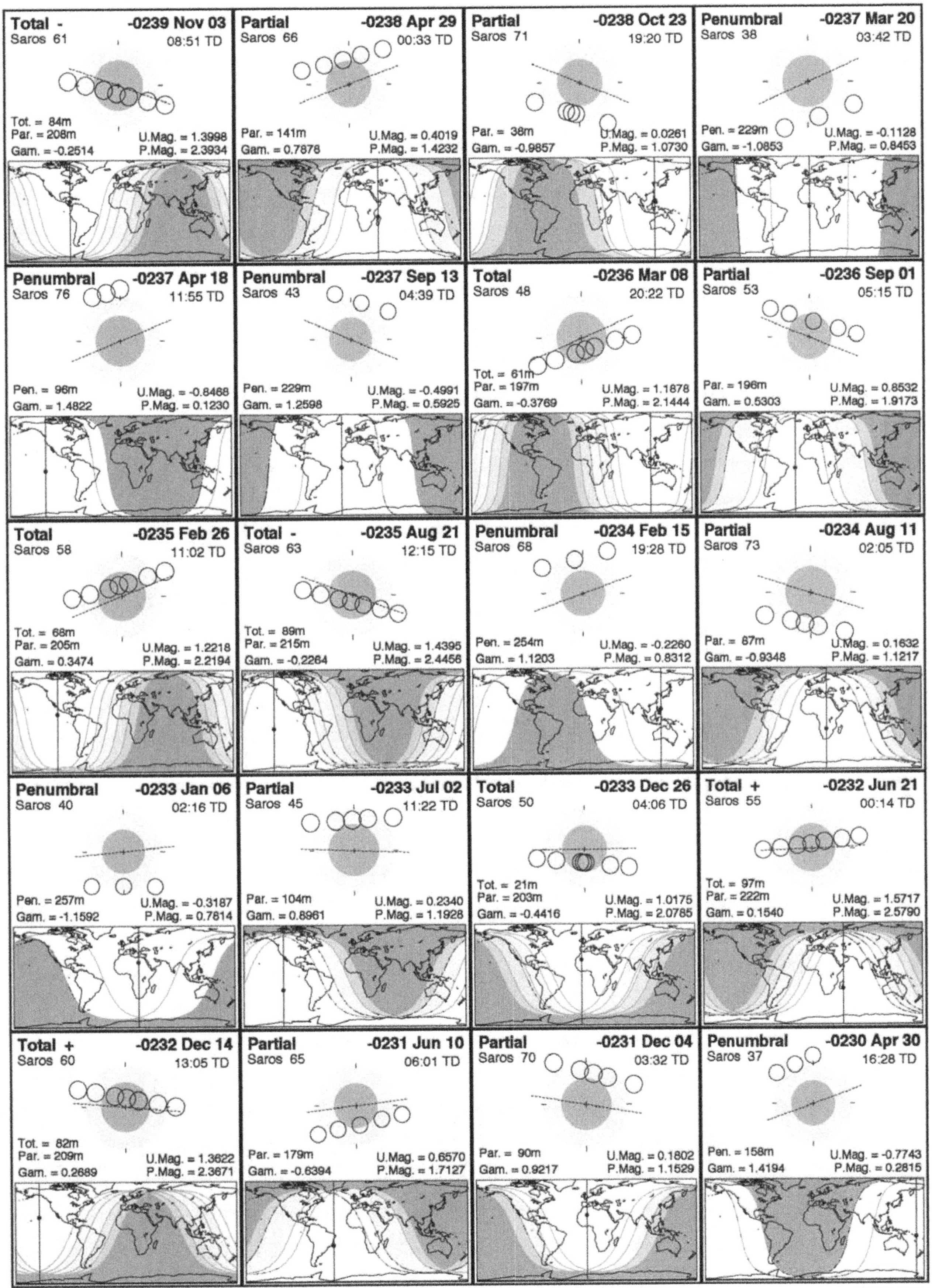

ΔT= 13203 s [= 03h40m] std.err. = ± 339 s [= ± 1.4°] **Plate 213**

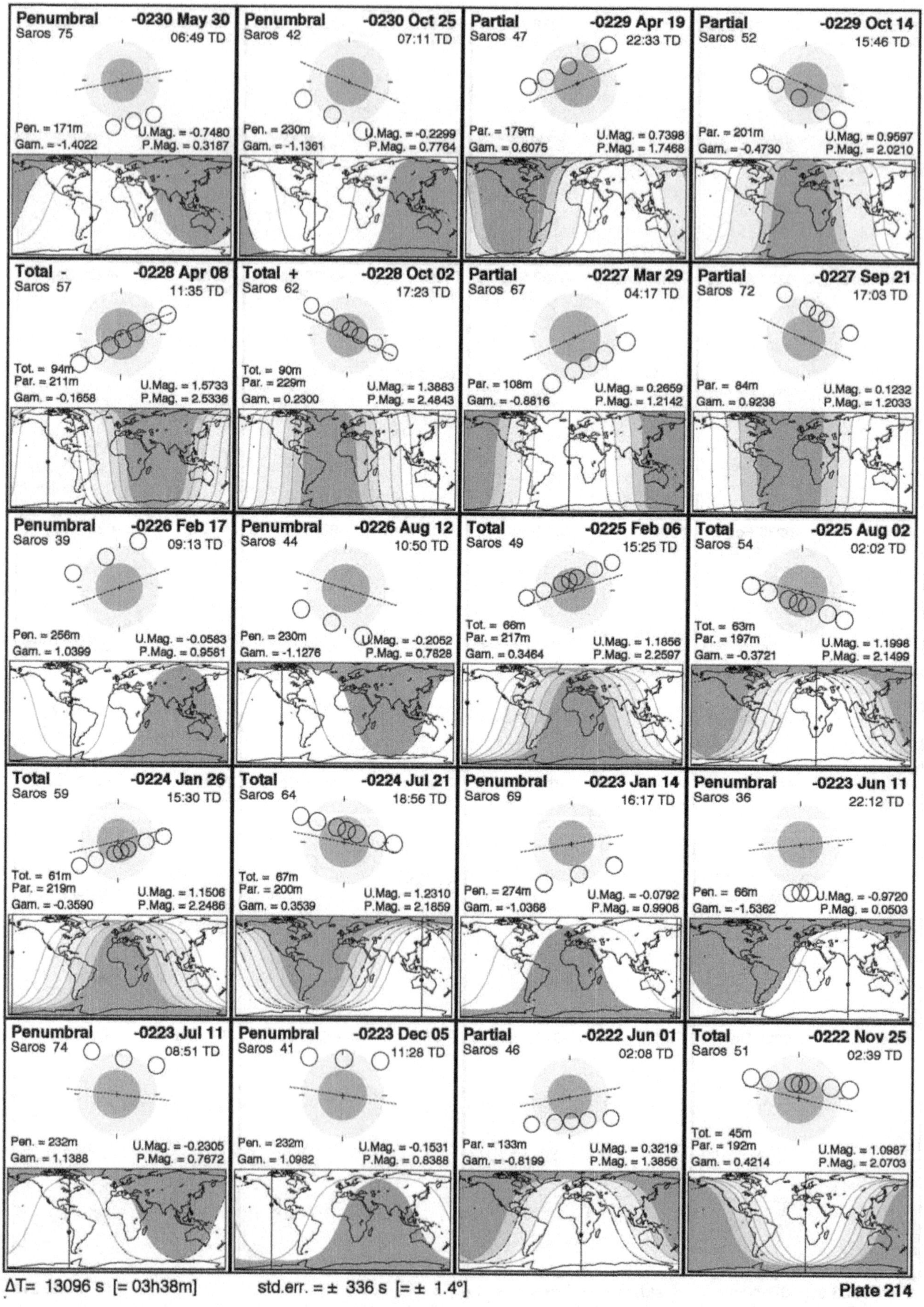

ΔT= 13096 s [= 03h38m] std.err. = ± 336 s [= ± 1.4°] **Plate 214**

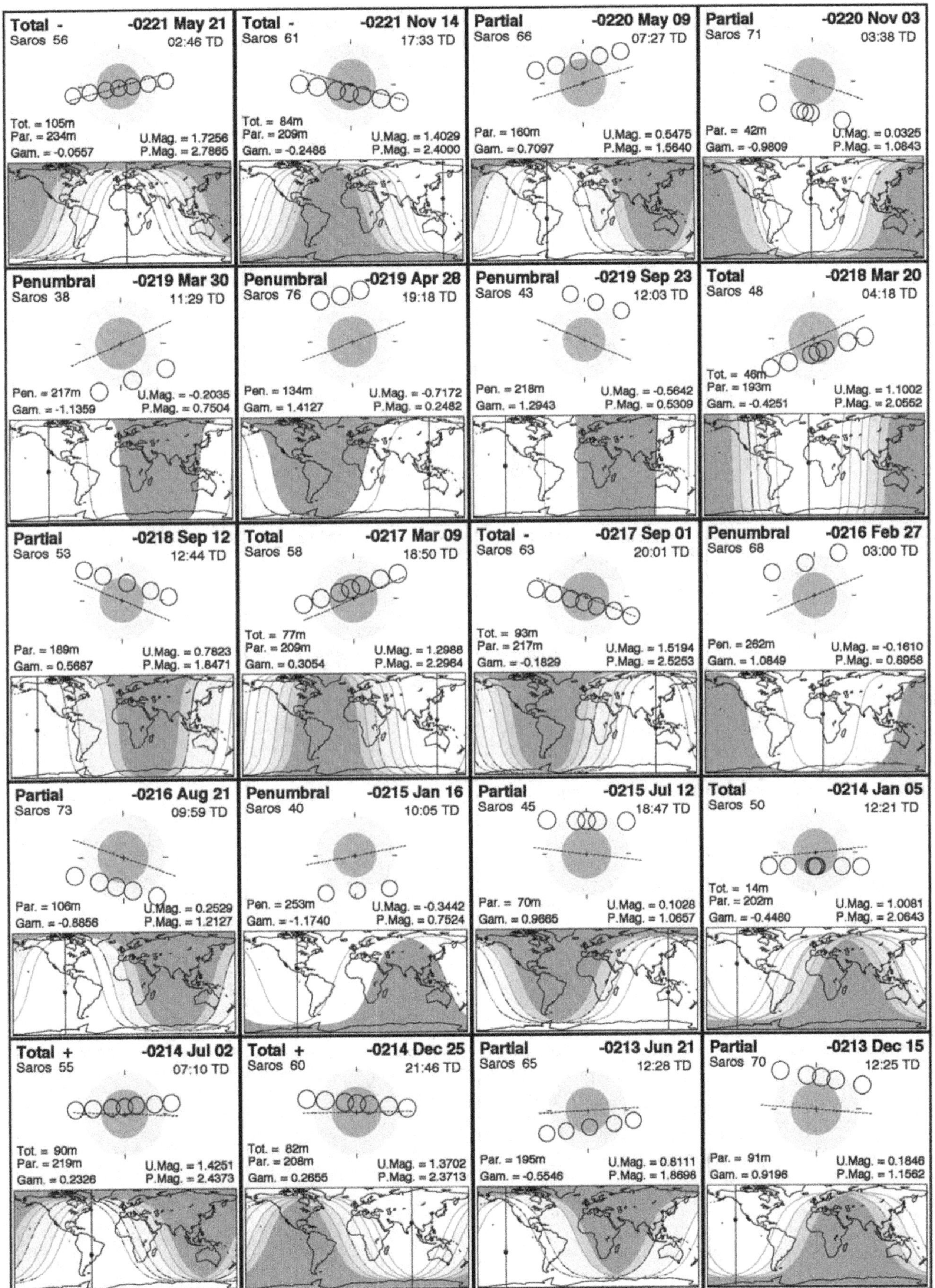

ΔT= 12984 s [= 03h36m] std.err. = ± 333 s [= ± 1.4°] **Plate 215**

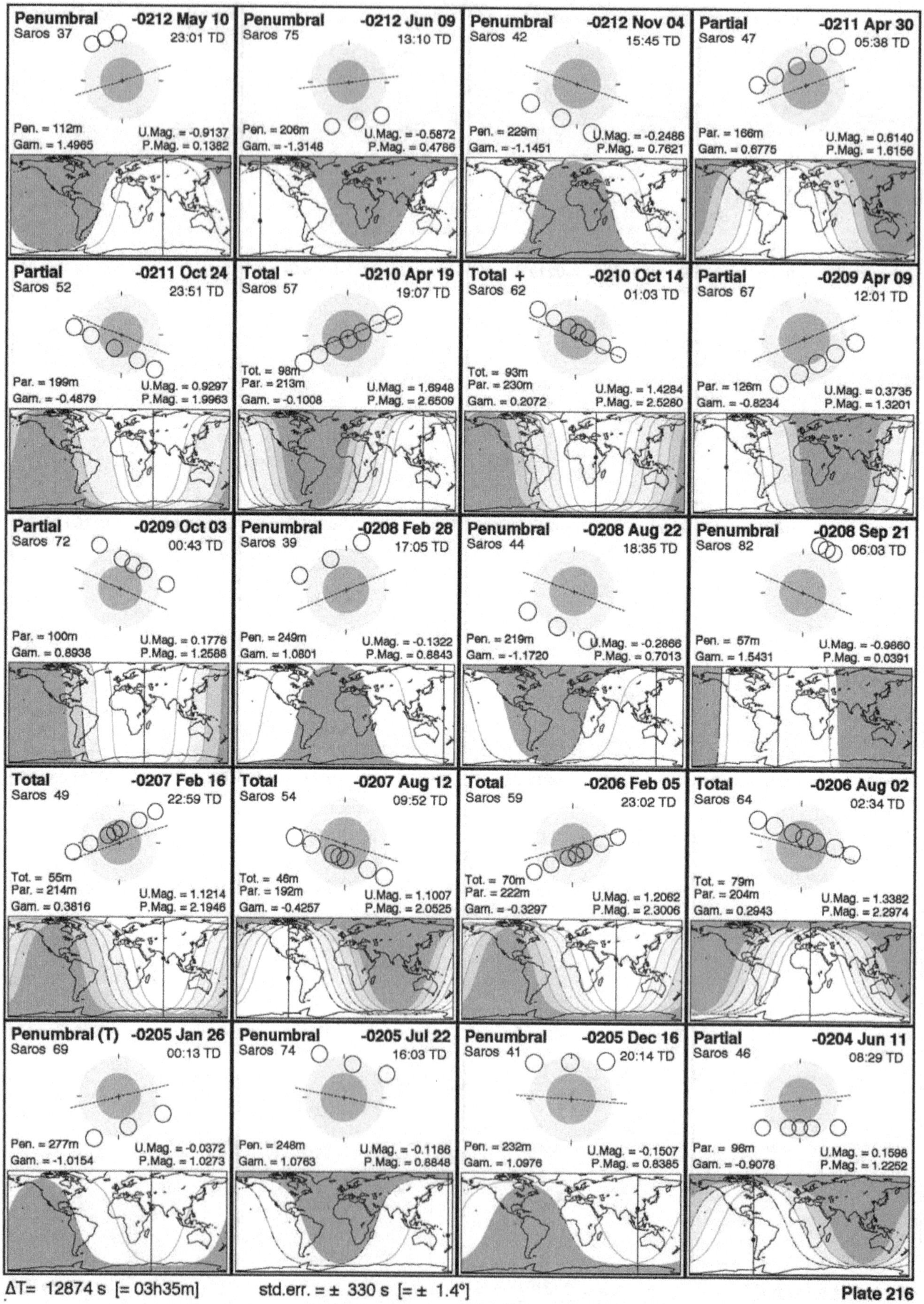

ΔT= 12874 s [= 03h35m] std.err. = ± 330 s [= ± 1.4°] **Plate 216**

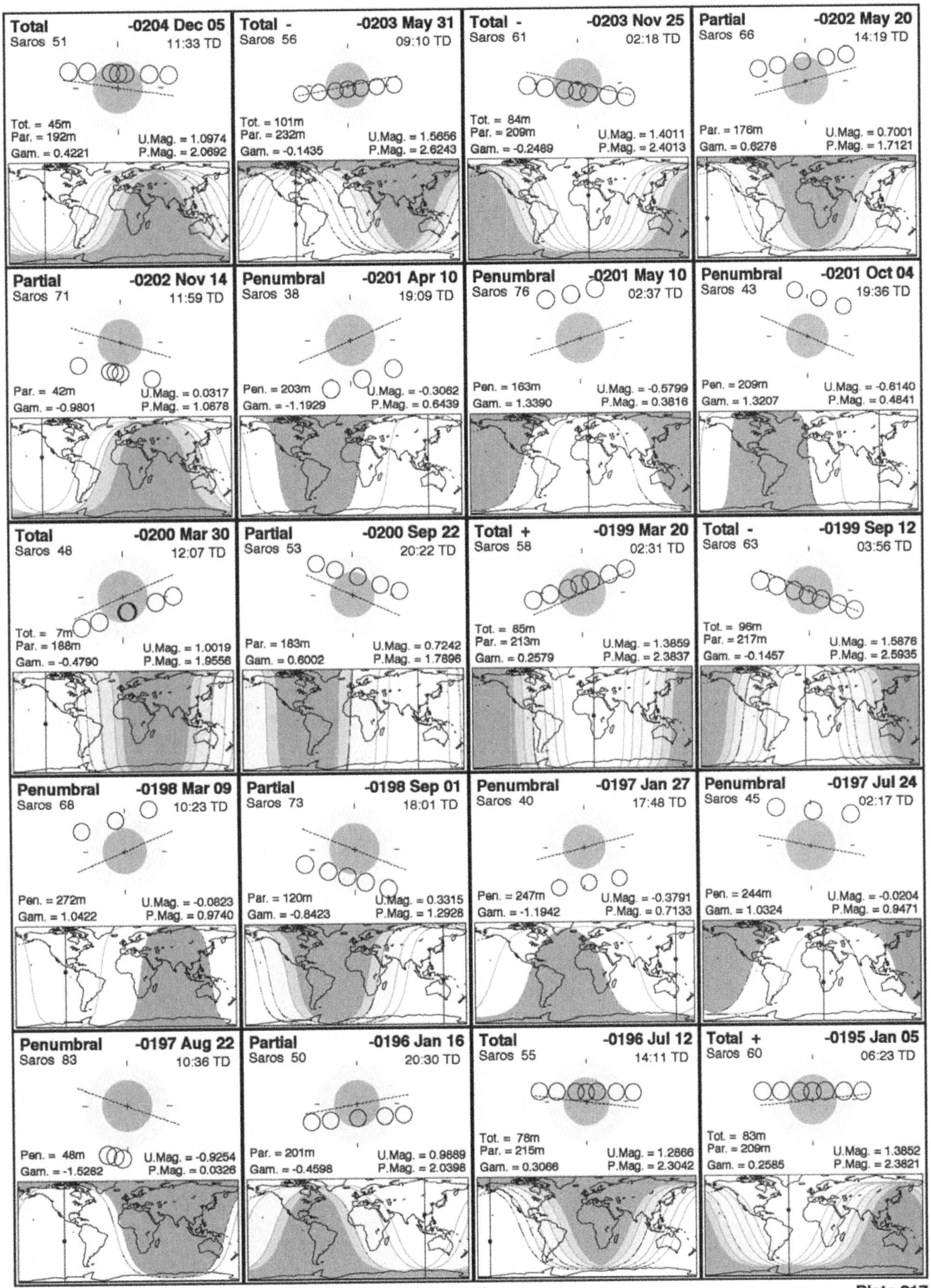

ΔT= 12769 s [= 03h33m] std.err. = ± 327 s [= ± 1.4°] Plate 217

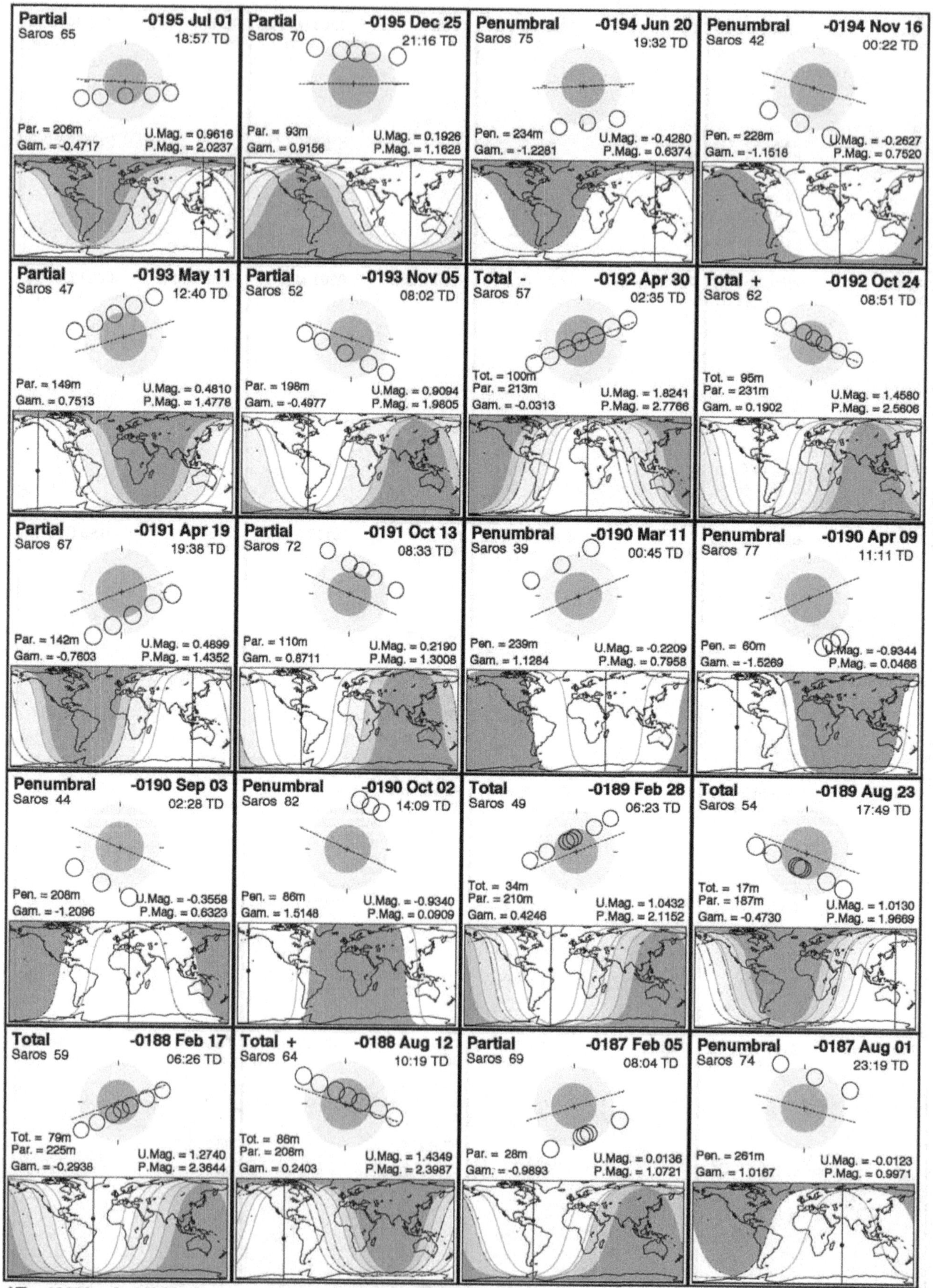

ΔT= 12666 s [= 03h31m] std.err. = ± 325 s [= ± 1.4°] **Plate 218**

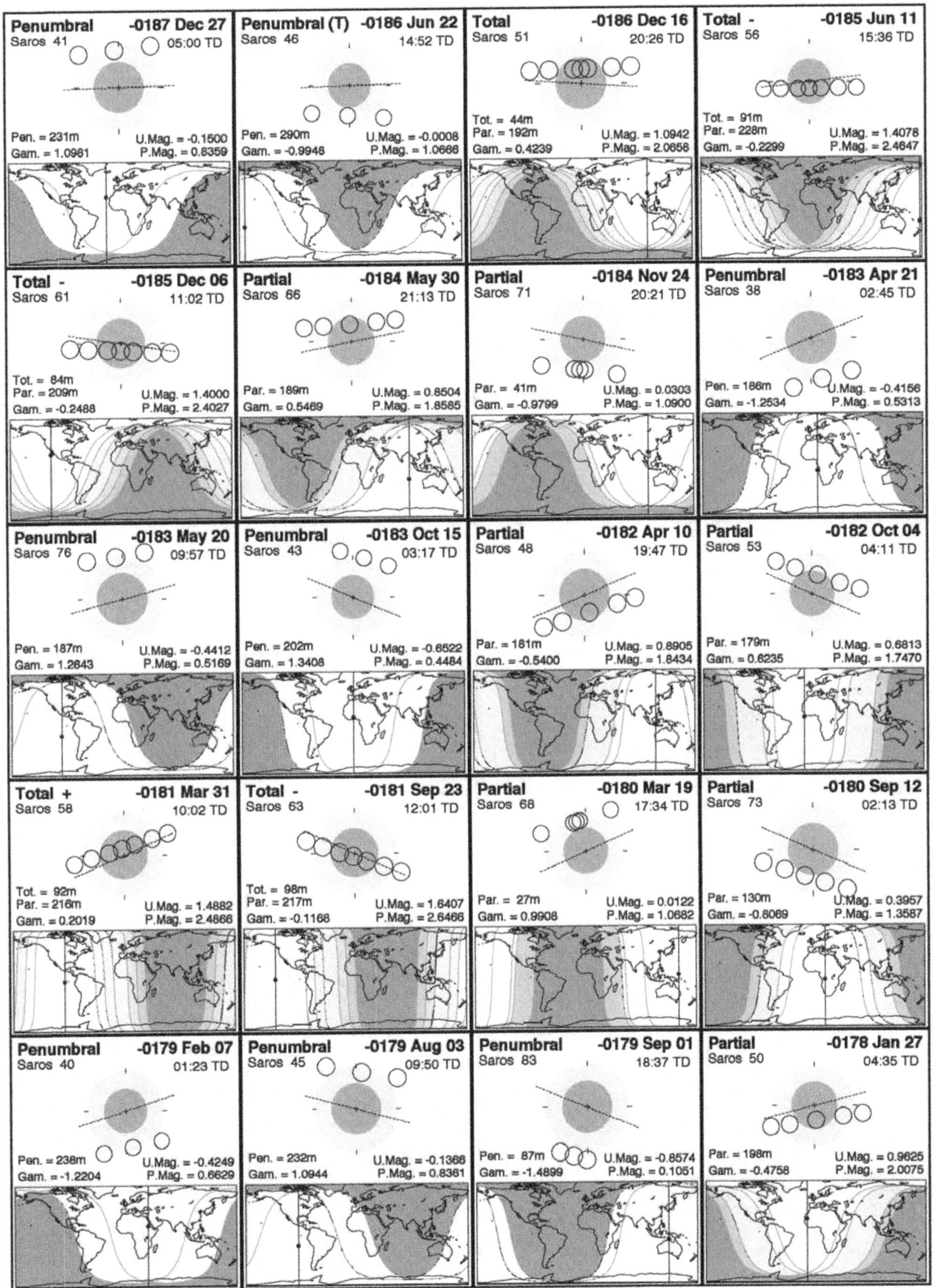

ΔT= 12564 s [= 03h29m] std.err. = ± 322 s [= ± 1.3°] **Plate 219**

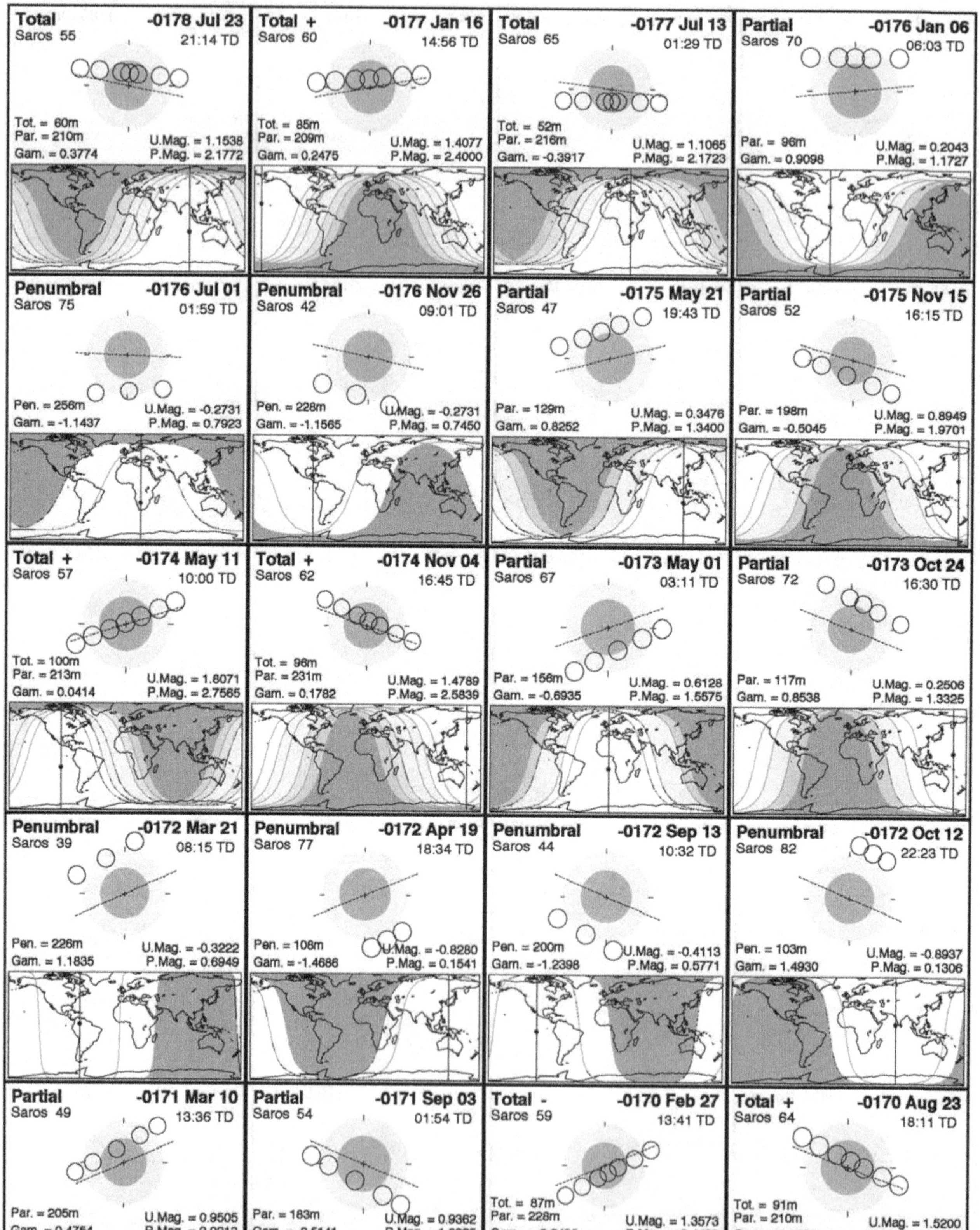

ΔT= 12463 s [= 03h28m] std.err. = ± 319 s [= ± 1.3°] **Plate 220**

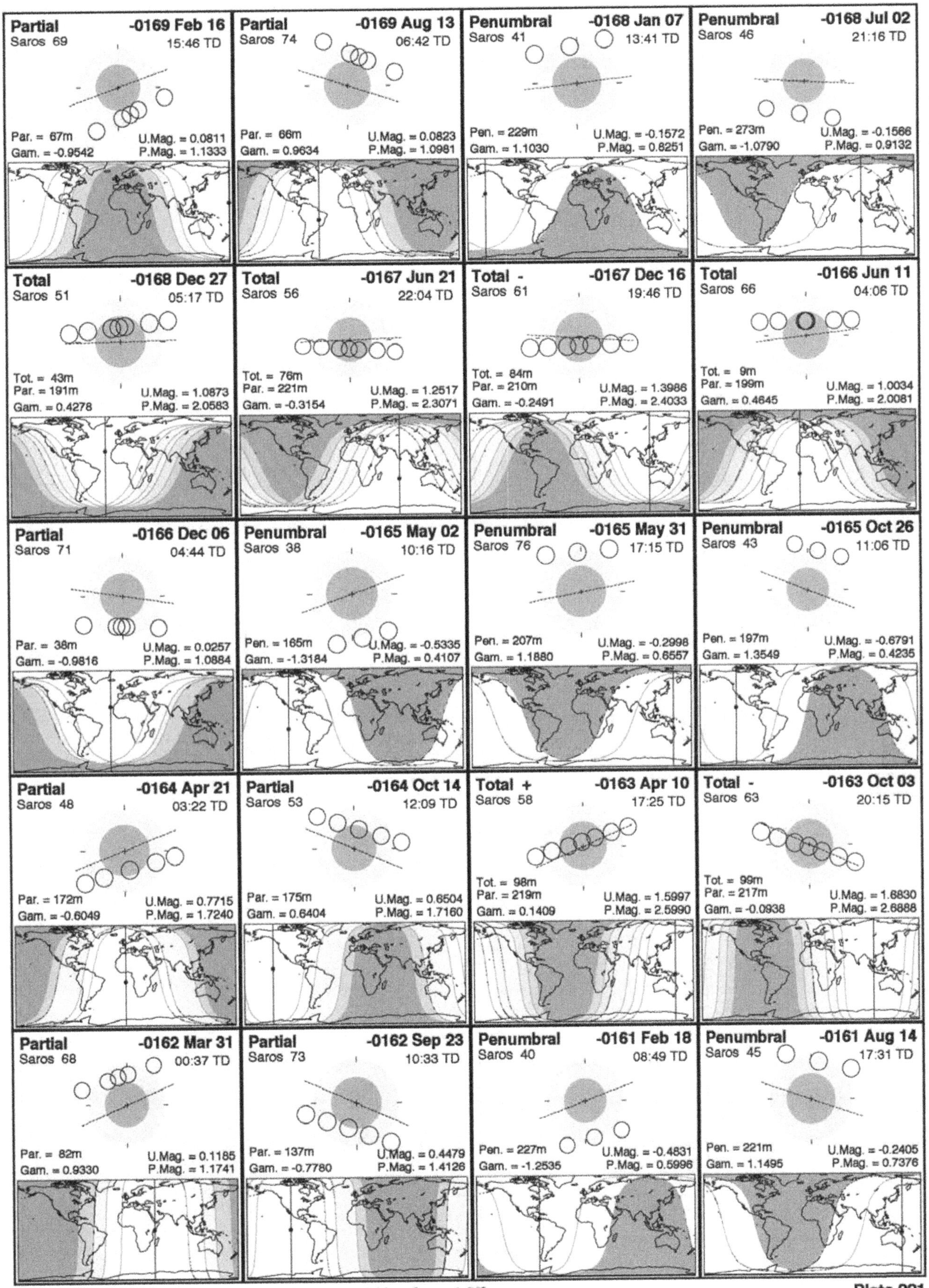

ΔT= 12362 s [= 03h26m] std.err. = ± 316 s [= ± 1.3°] **Plate 221**

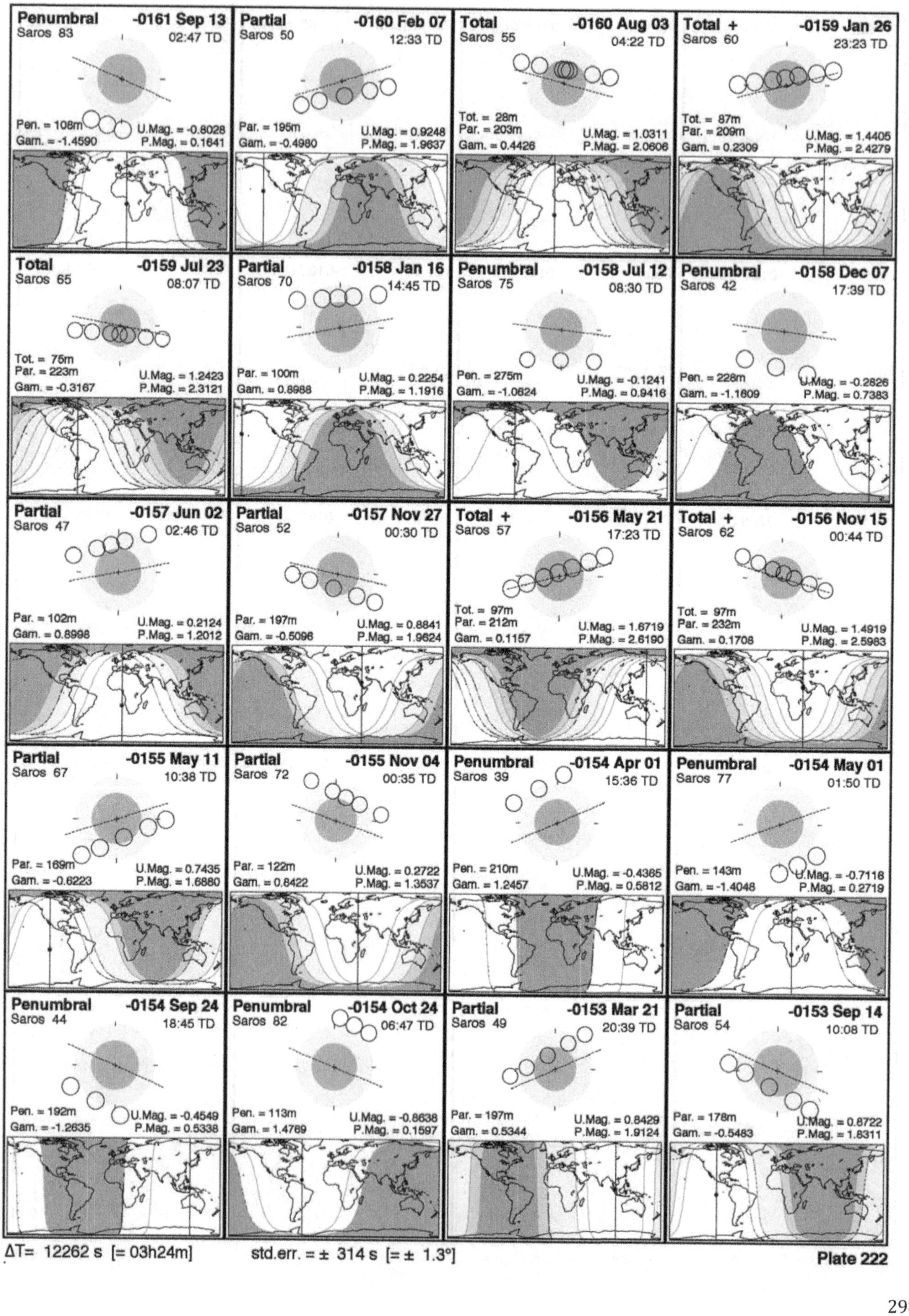

ΔT= 12262 s [= 03h24m] std.err. = ± 314 s [= ± 1.3°] **Plate 222**

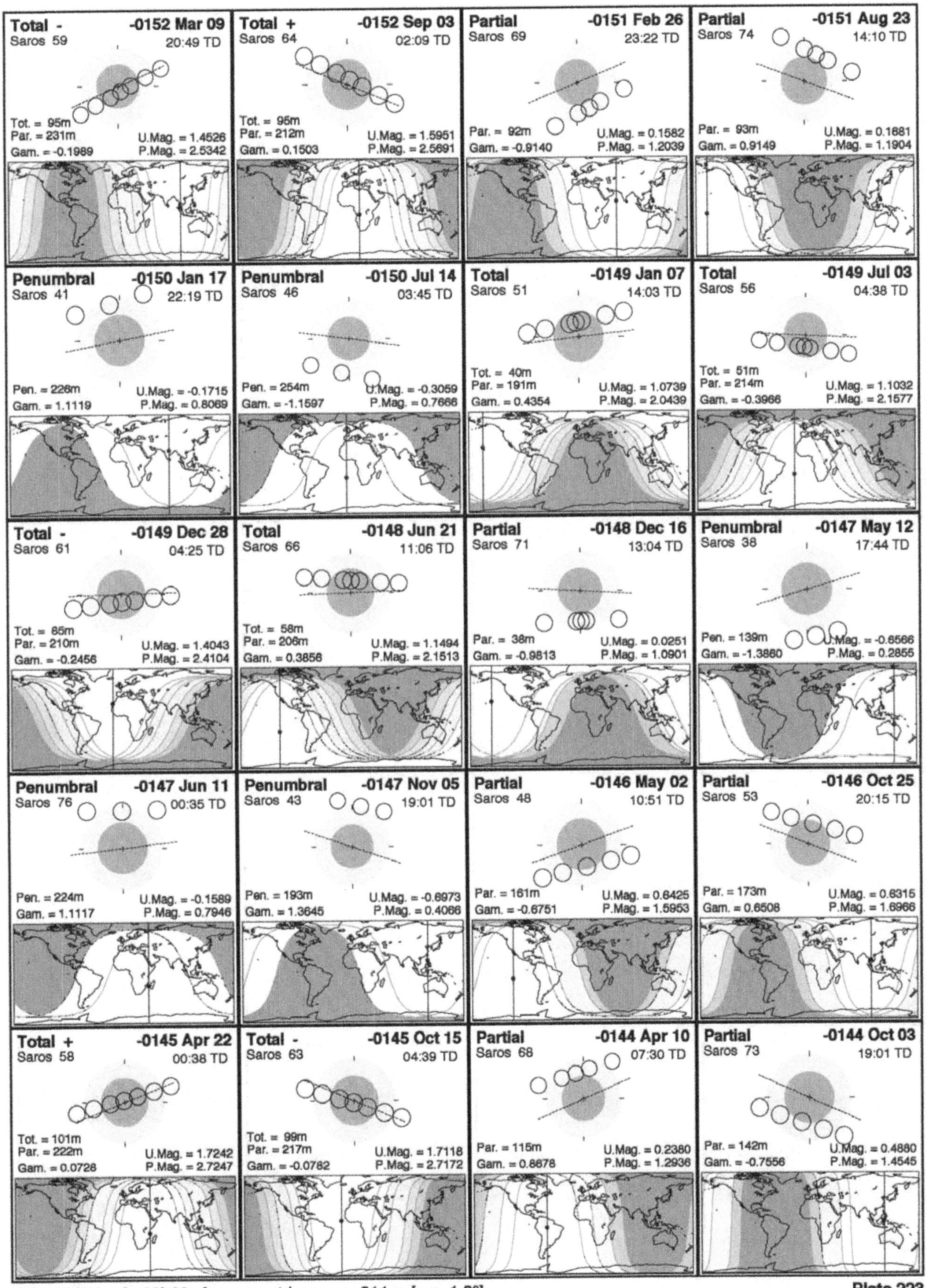

ΔT= 12164 s [= 03h23m] std.err. = ± 311 s [= ± 1.3°] **Plate 223**

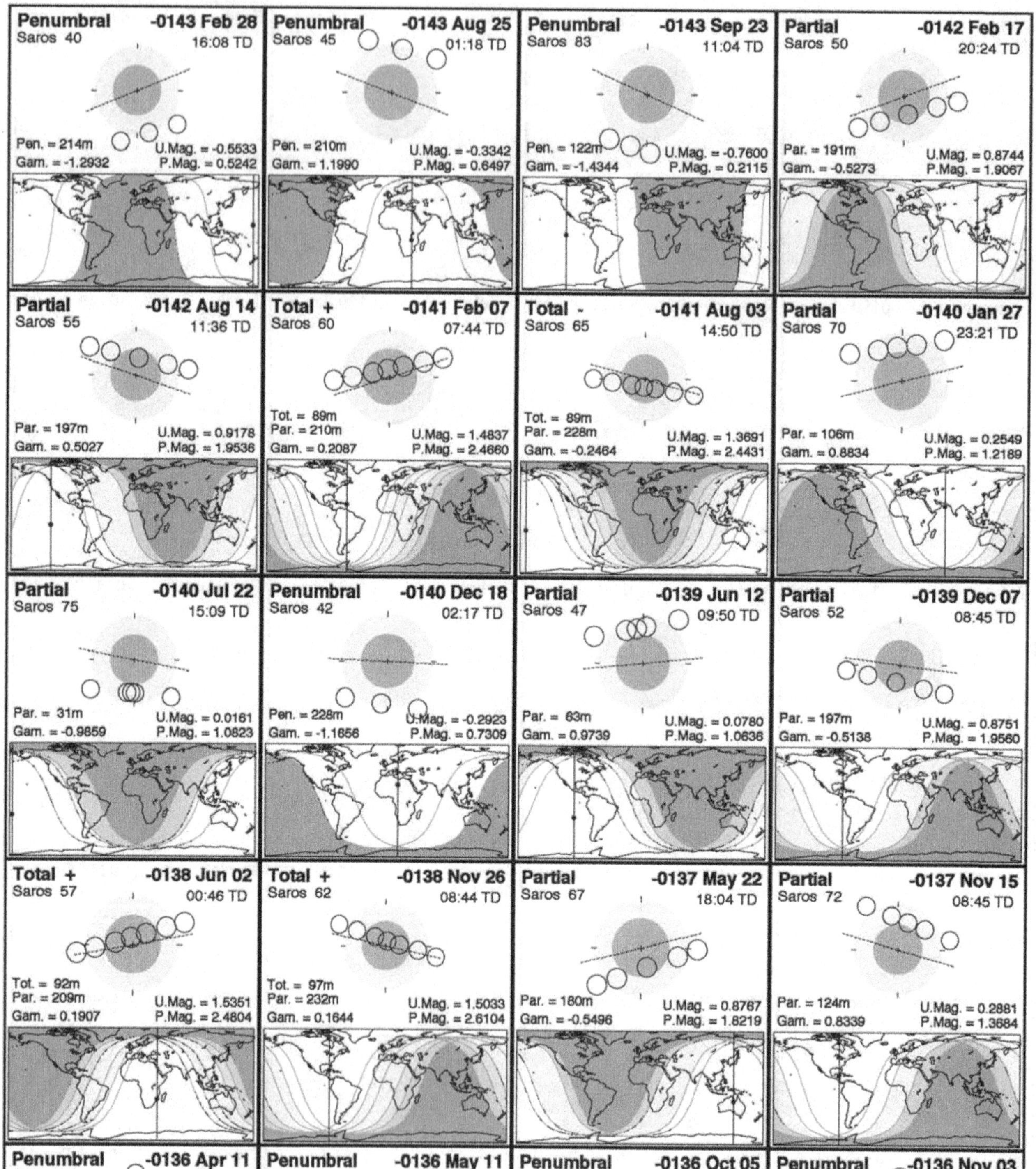

ΔT= 12061 s [= 03h21m] std.err. = ± 308 s [= ± 1.3°] **Plate 224**

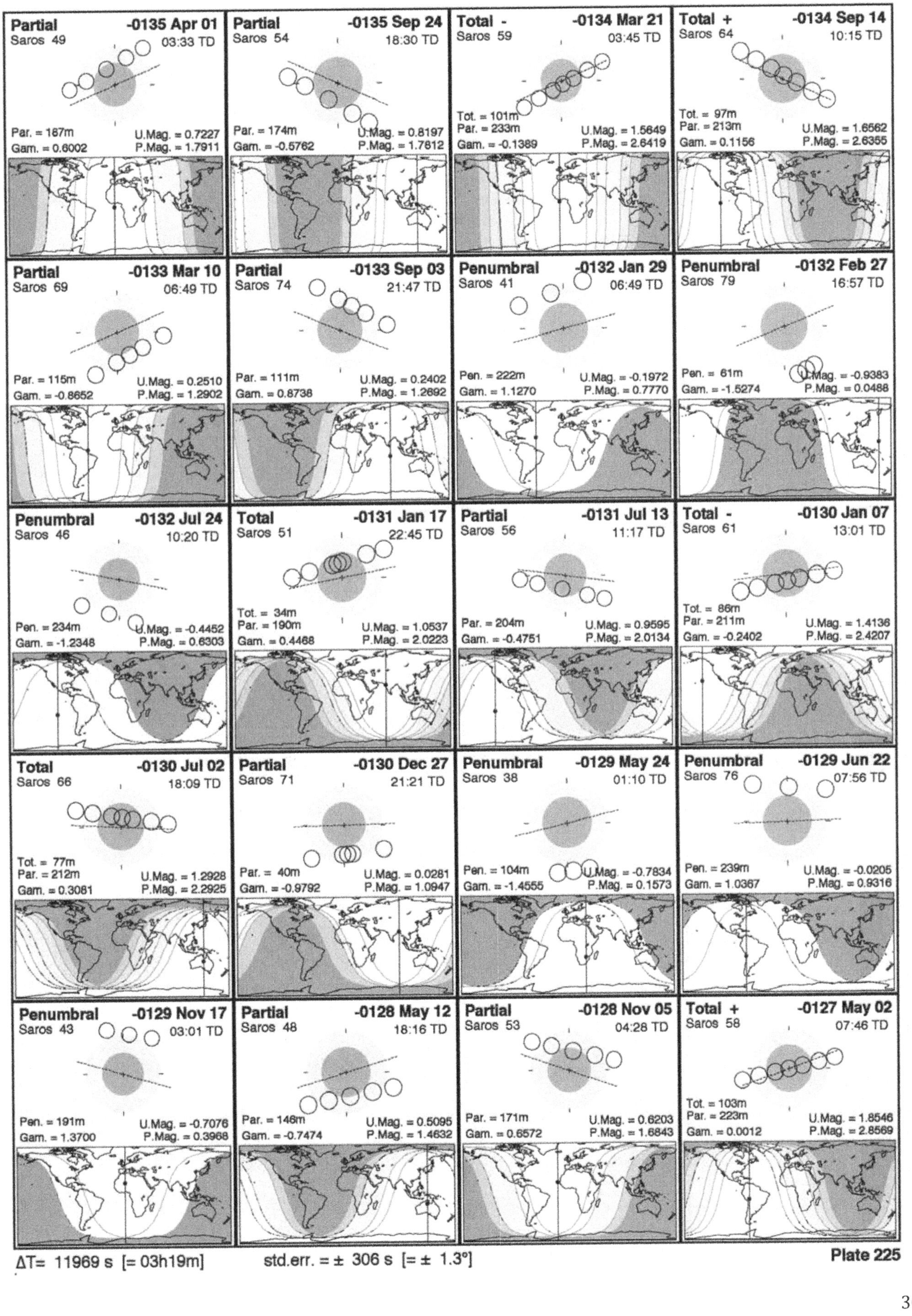

ΔT= 11969 s [= 03h19m] std.err. = ± 306 s [= ± 1.3°] **Plate 225**

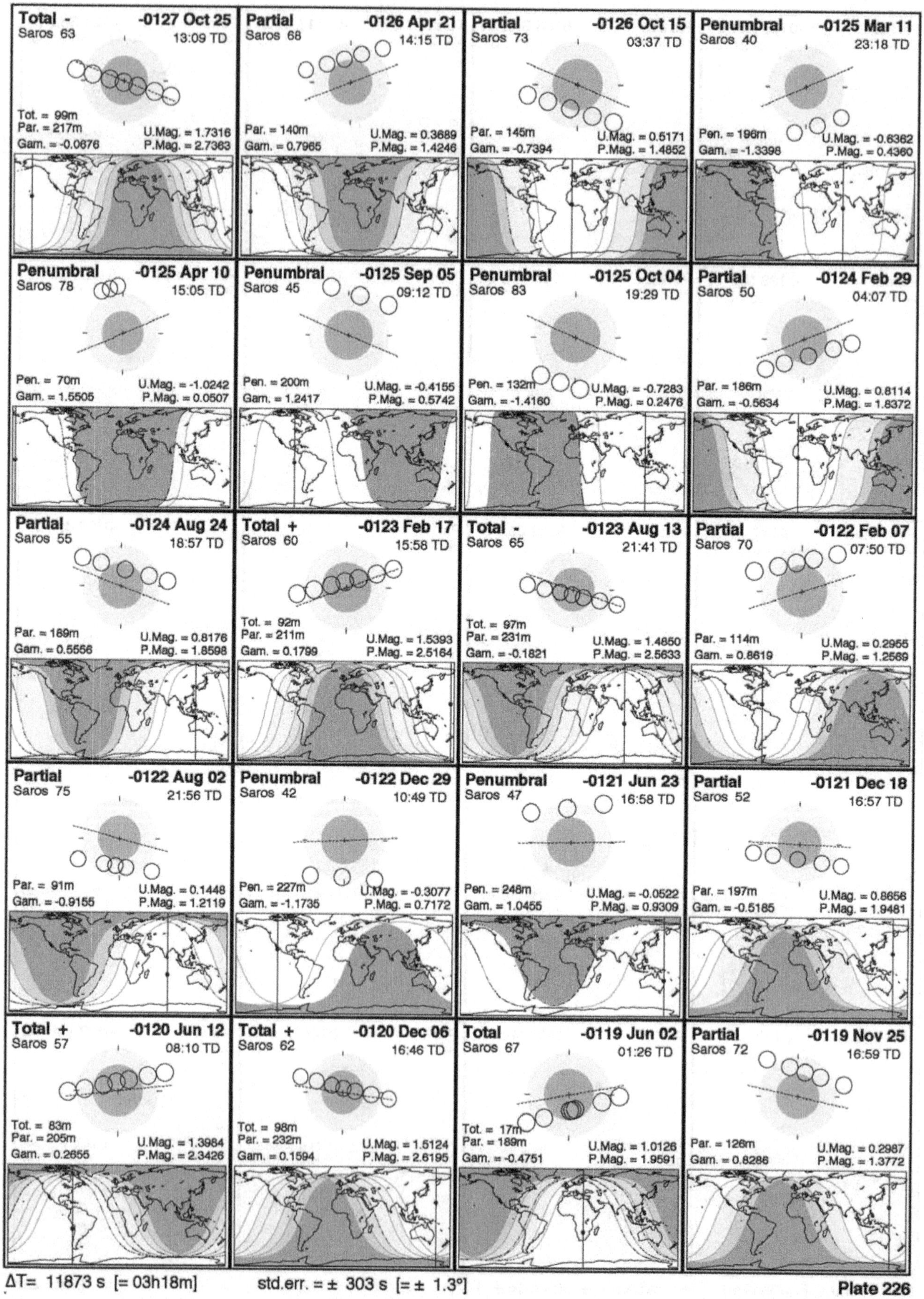

ΔT= 11873 s [= 03h18m] std.err. = ± 303 s [= ± 1.3°] **Plate 226**

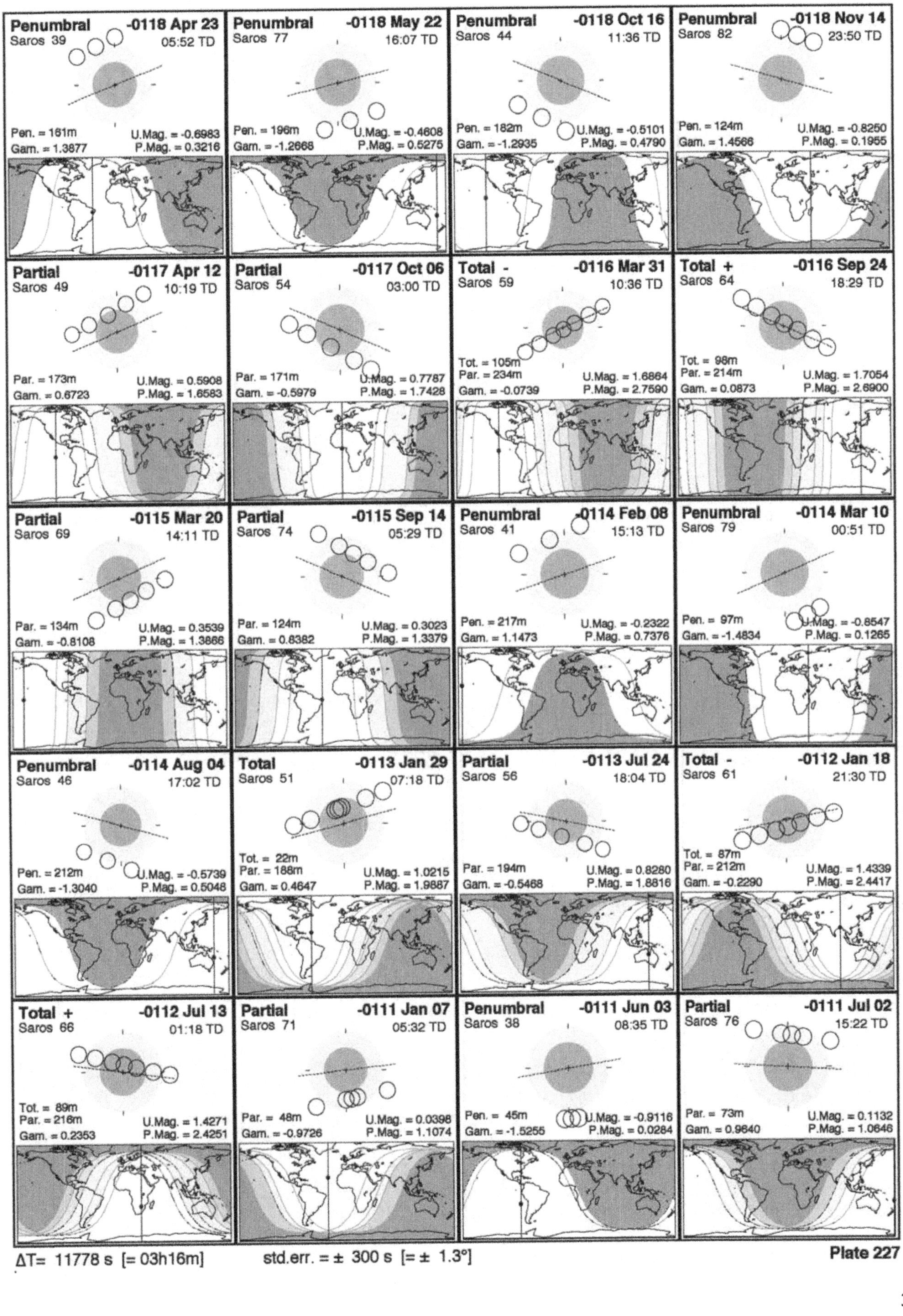

ΔT= 11778 s [= 03h16m] std.err. = ± 300 s [= ± 1.3°] **Plate 227**

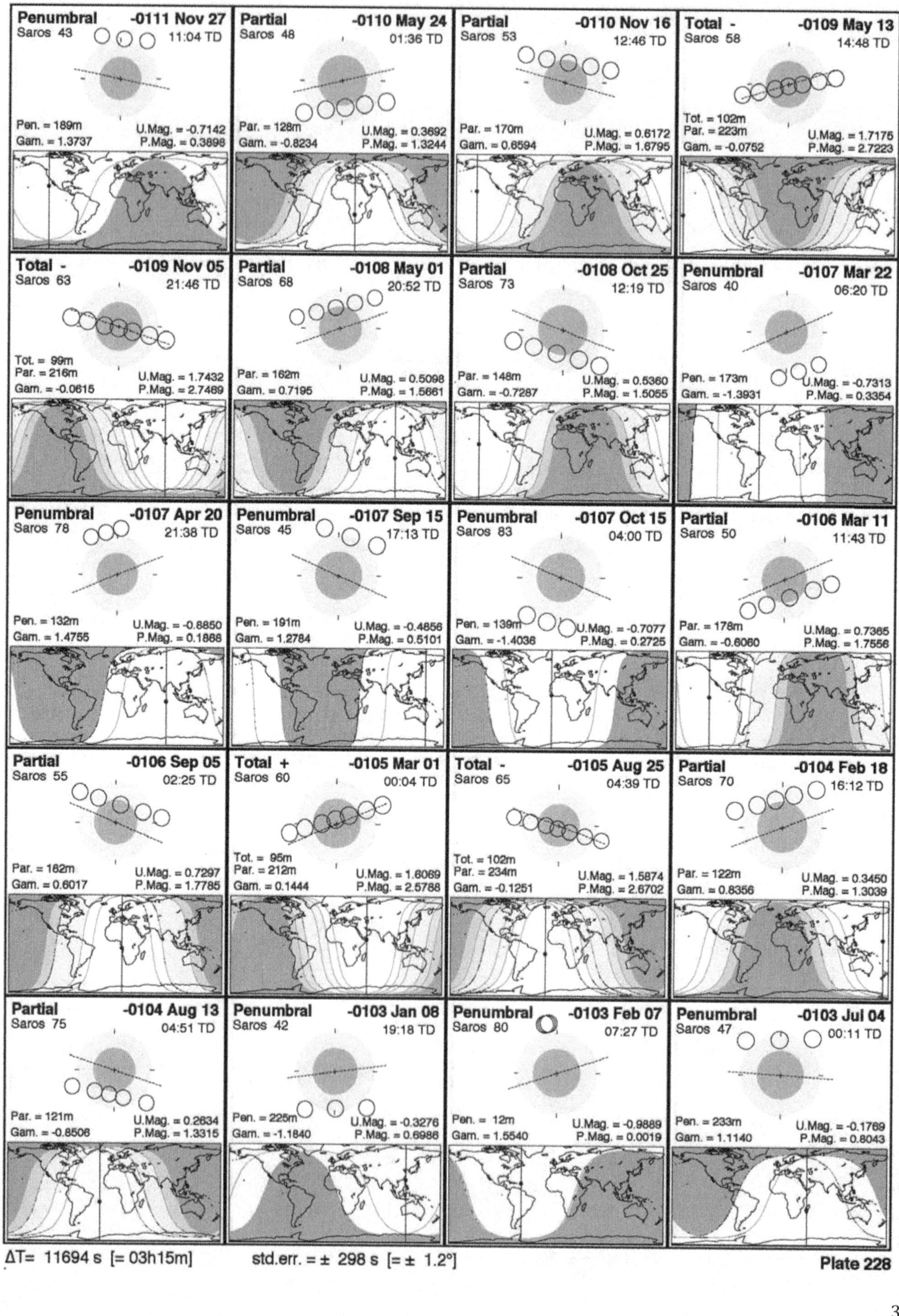

ΔT= 11694 s [= 03h15m] std.err. = ± 298 s [= ± 1.2°] **Plate 228**

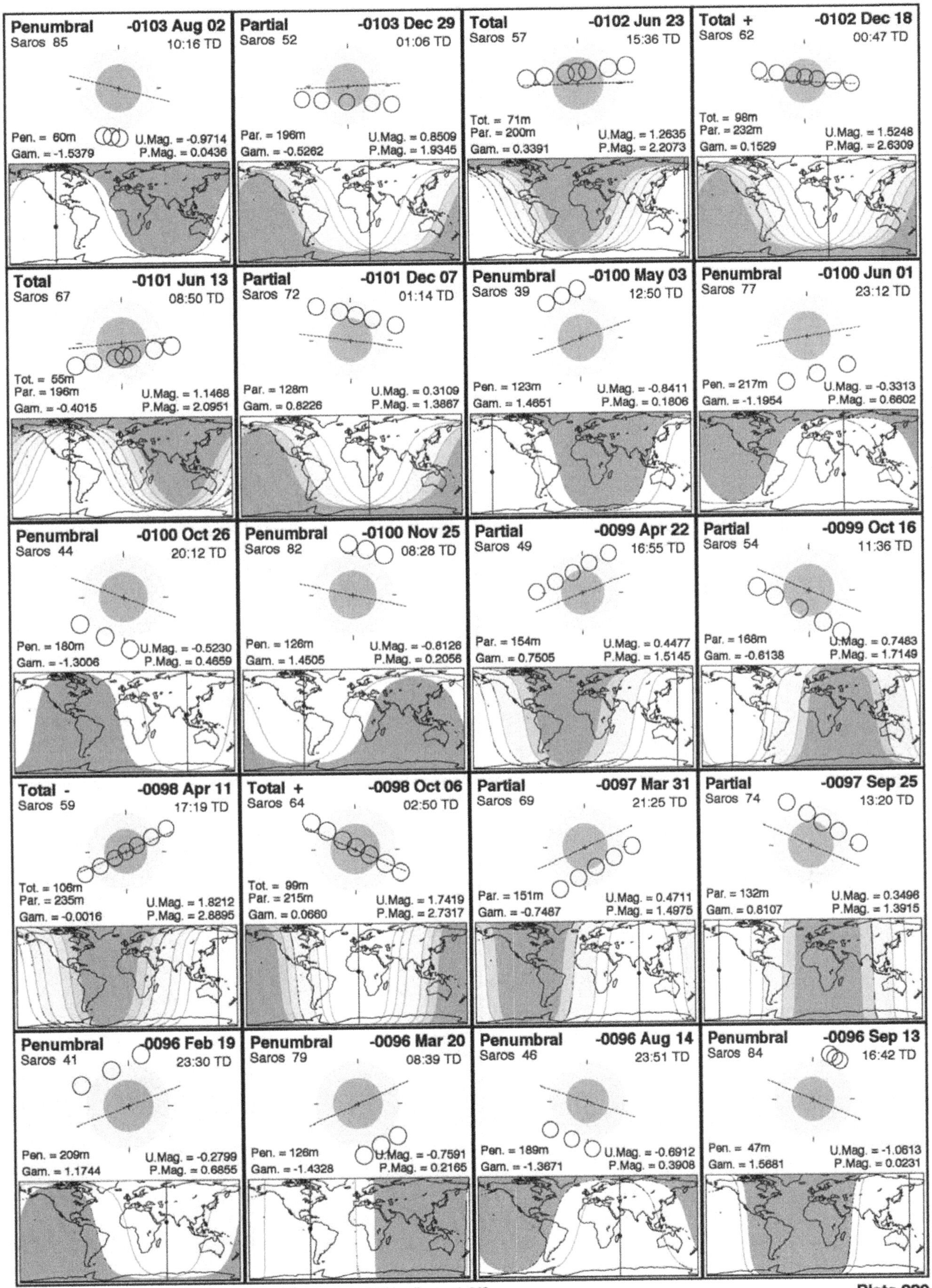

ΔT= 11609 s [= 03h13m] std.err. = ± 296 s [= ± 1.2°] **Plate 229**

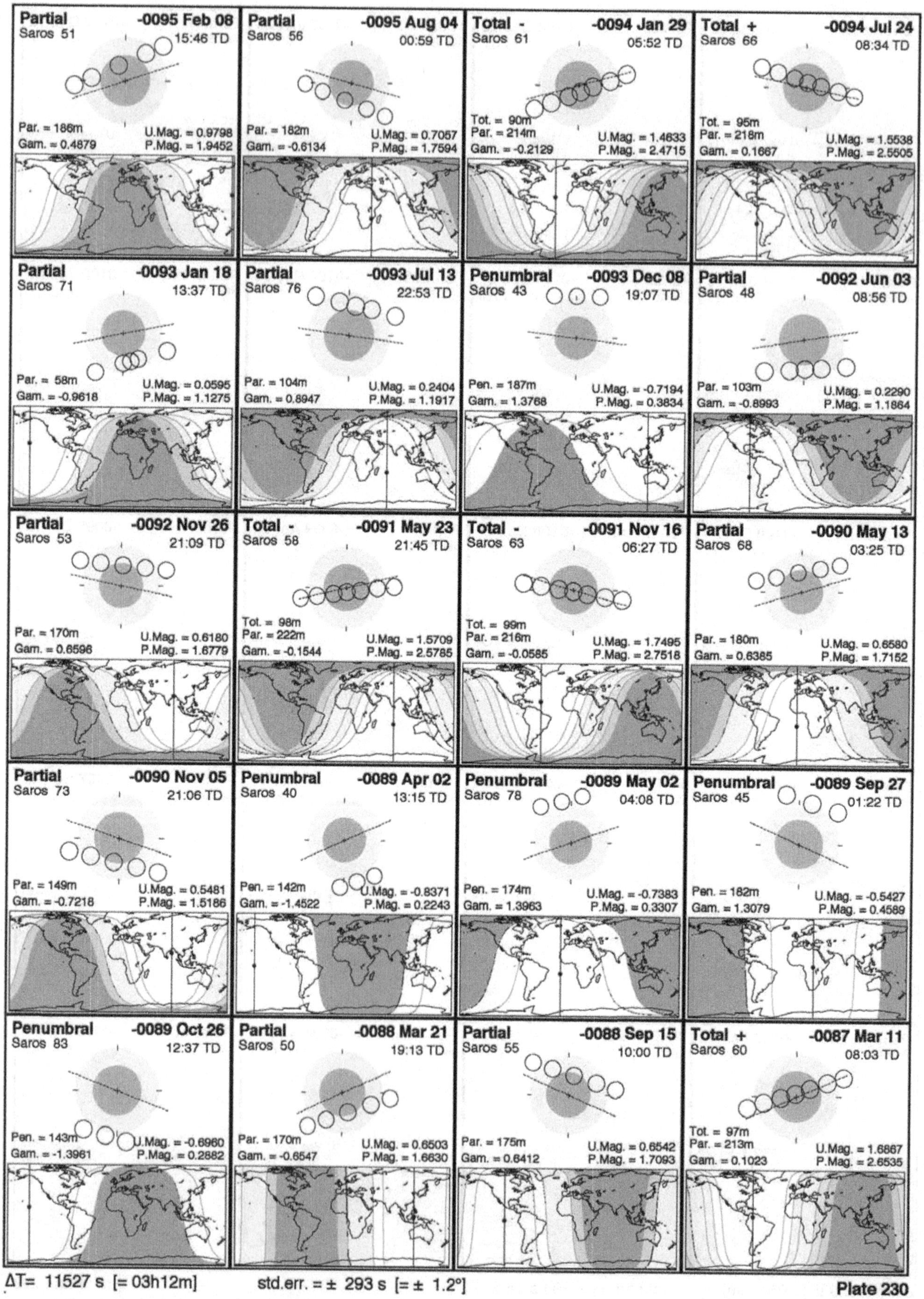

ΔT= 11527 s [= 03h12m] std.err. = ± 293 s [= ± 1.2°] **Plate 230**

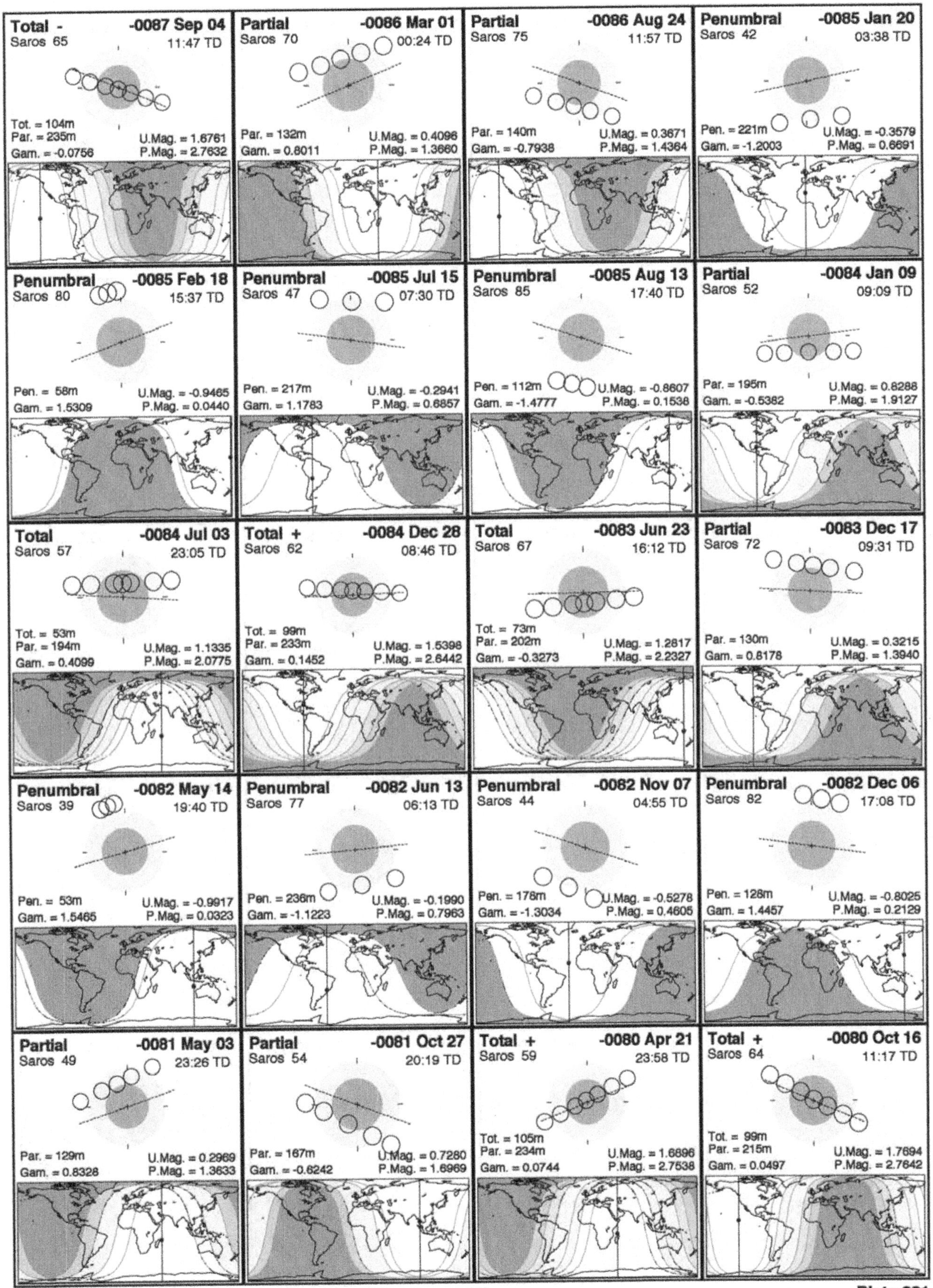

ΔT= 11434 s [= 03h11m] std.err. = ± 291 s [= ± 1.2°] Plate 231

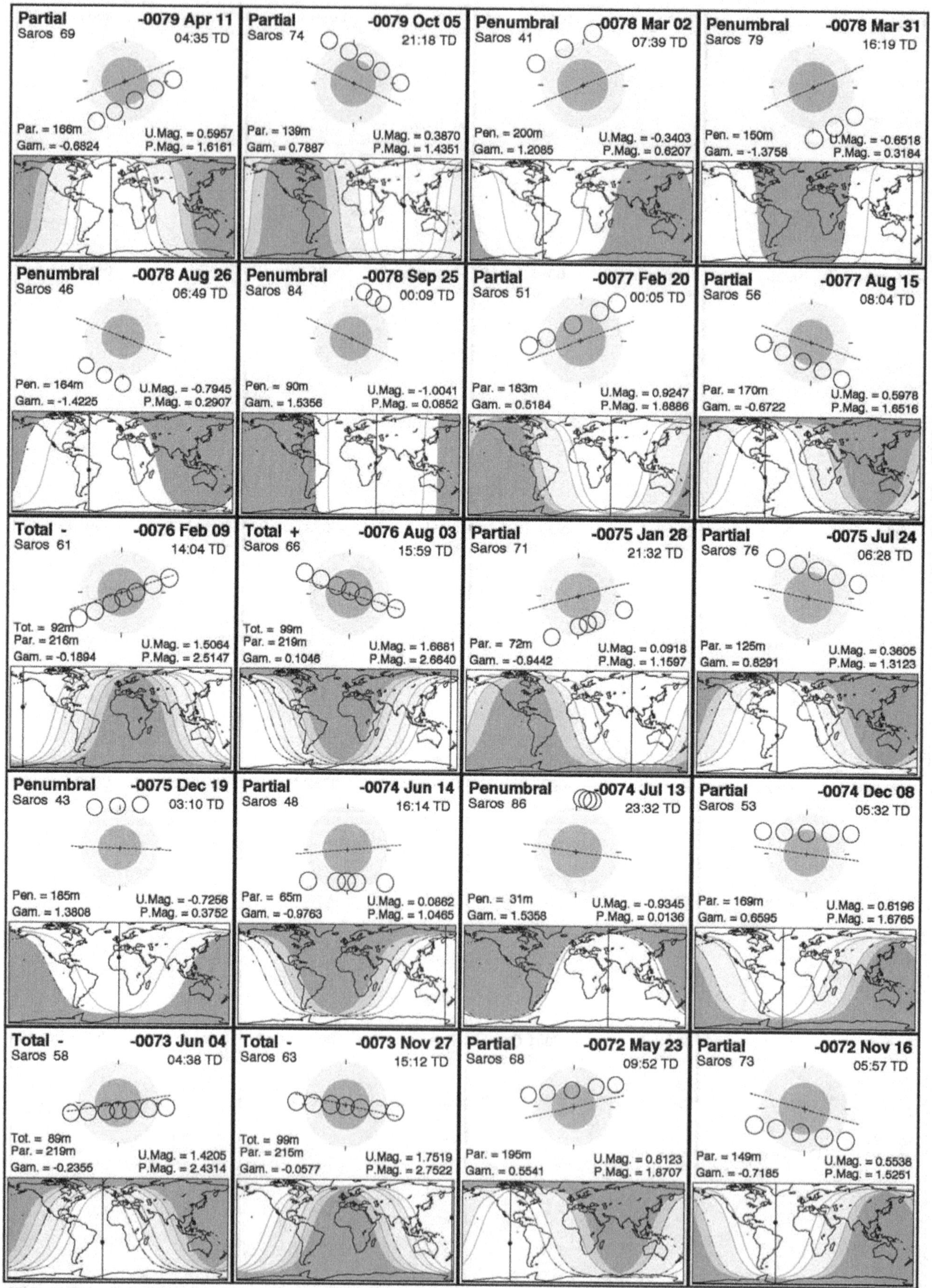

ΔT= 11352 s [= 03h09m] std.err. = ± 288 s [= ± 1.2°]

Plate 232

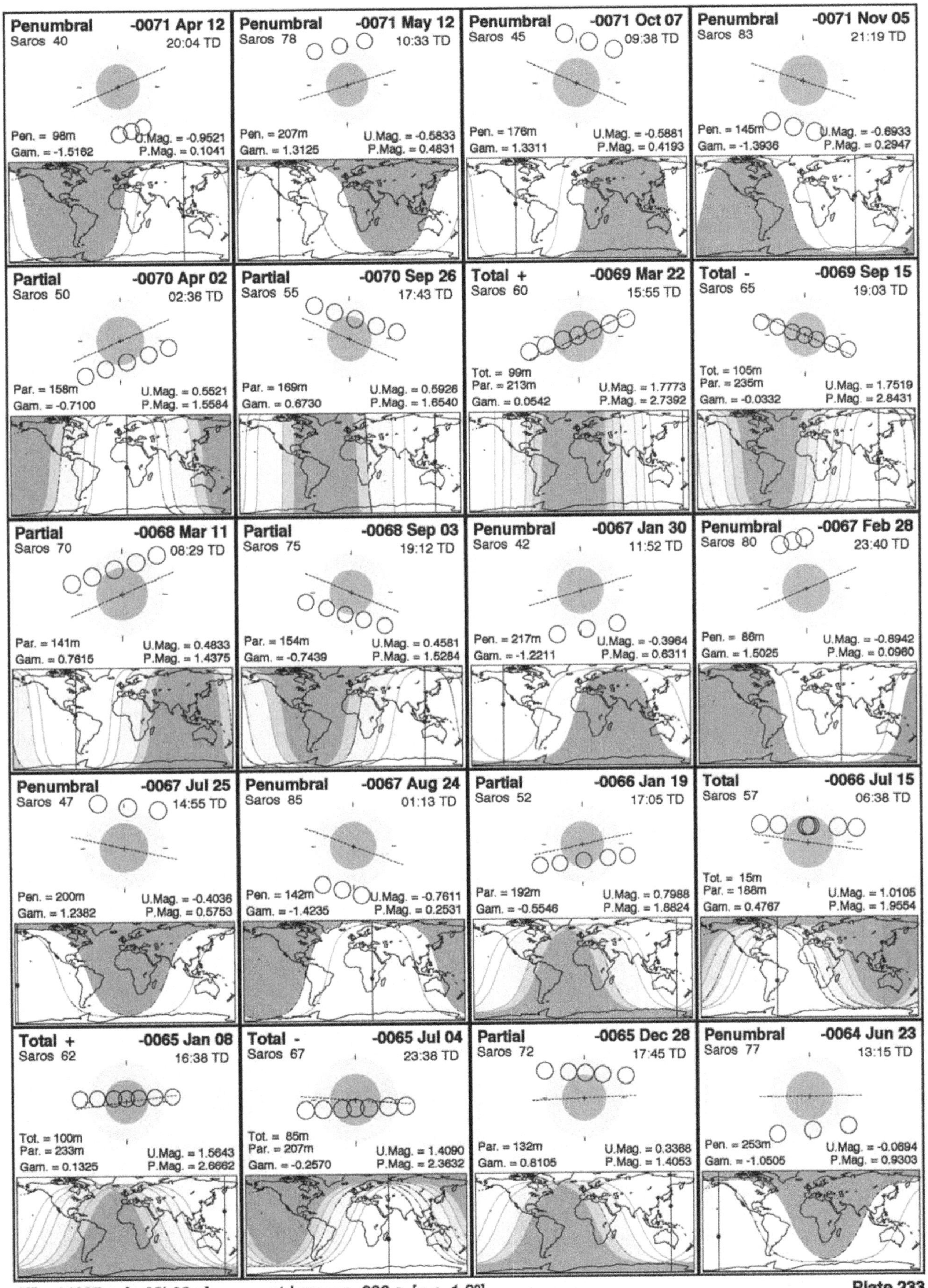

ΔT= 11267 s [= 03h08m] std.err. = ± 286 s [= ± 1.2°] **Plate 233**

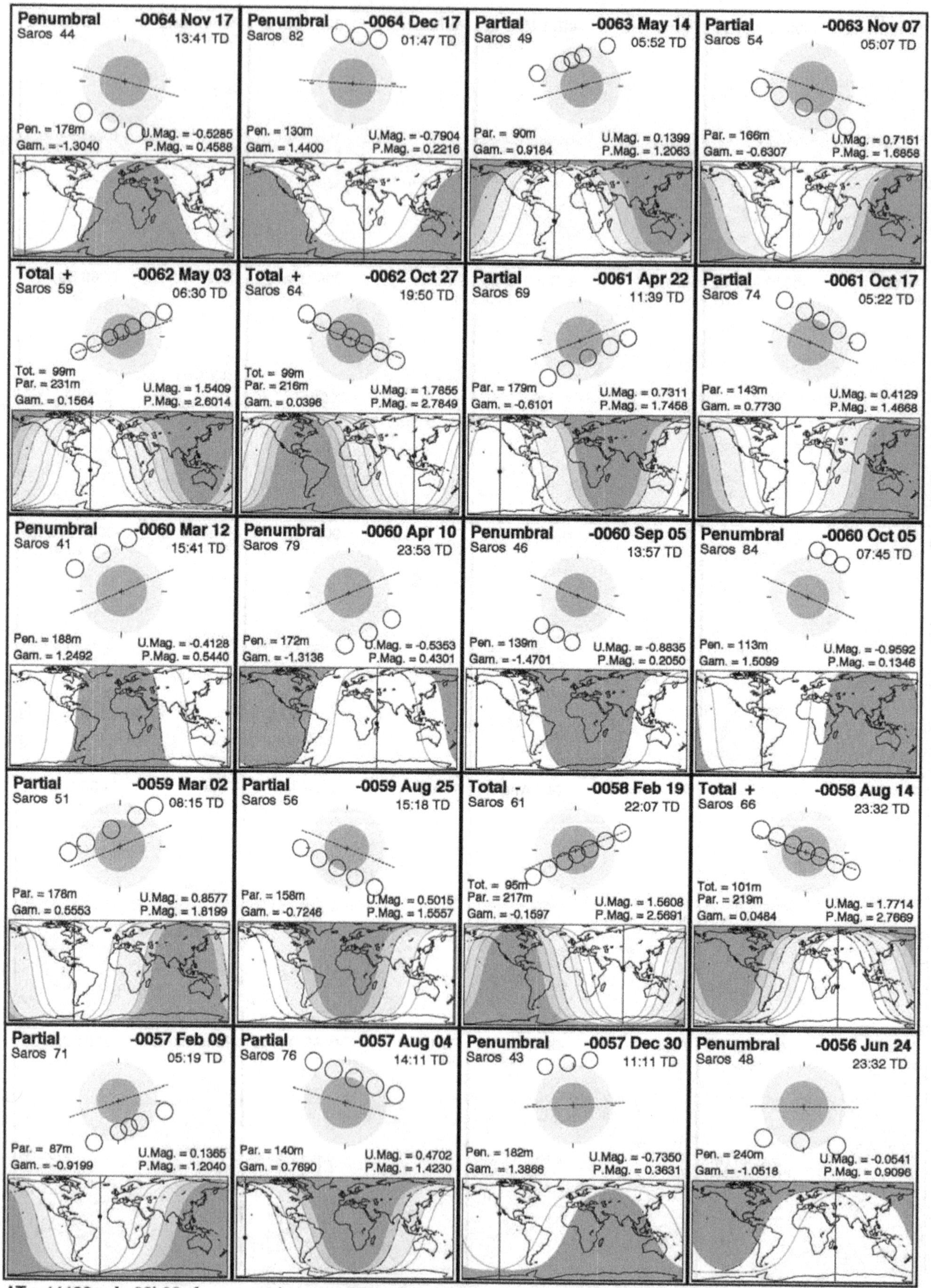

ΔT= 11186 s [= 03h06m] std.err. = ± 284 s [= ± 1.2°] **Plate 234**

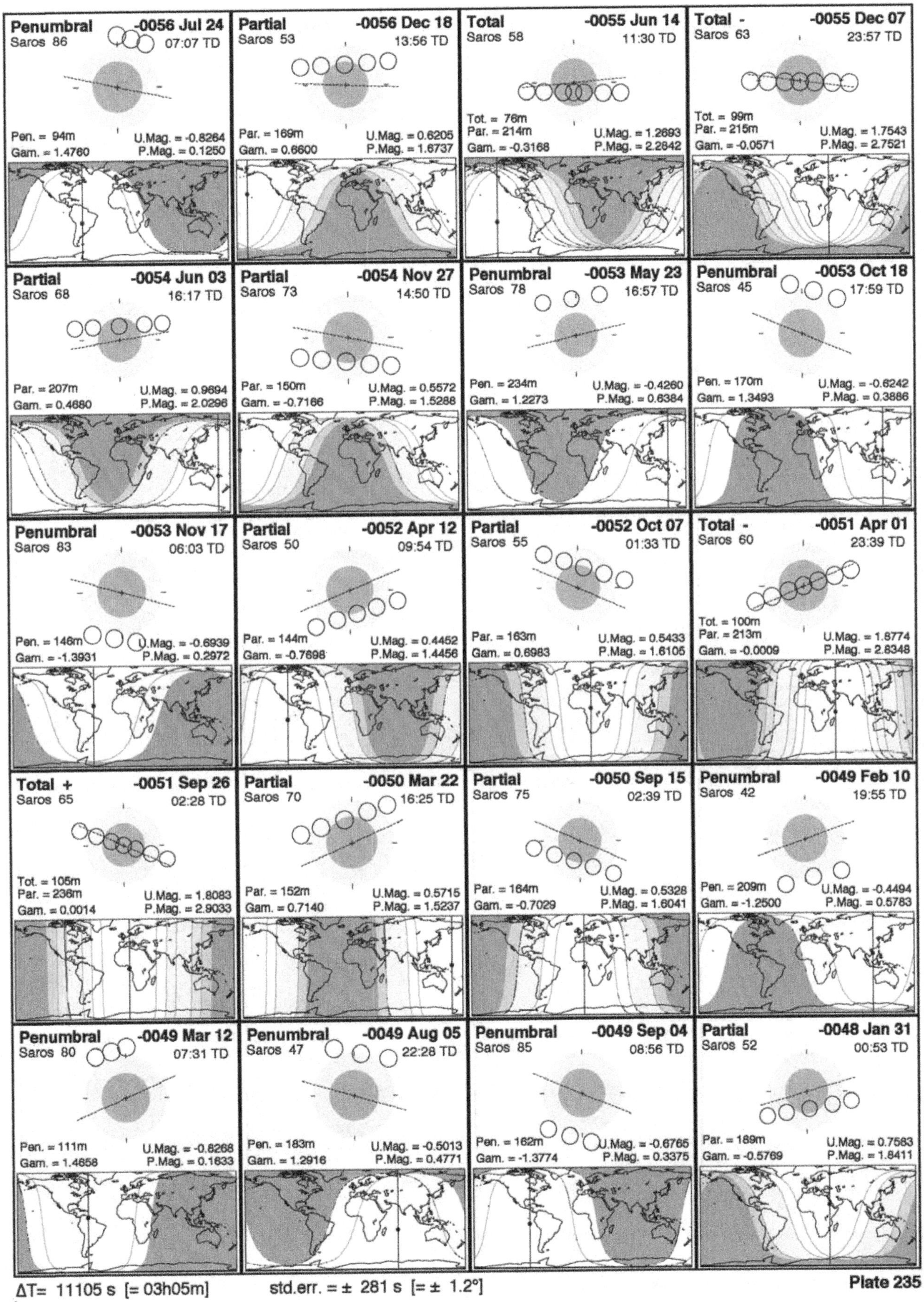

ΔT= 11105 s [= 03h05m] std.err. = ± 281 s [= ± 1.2°] Plate 235

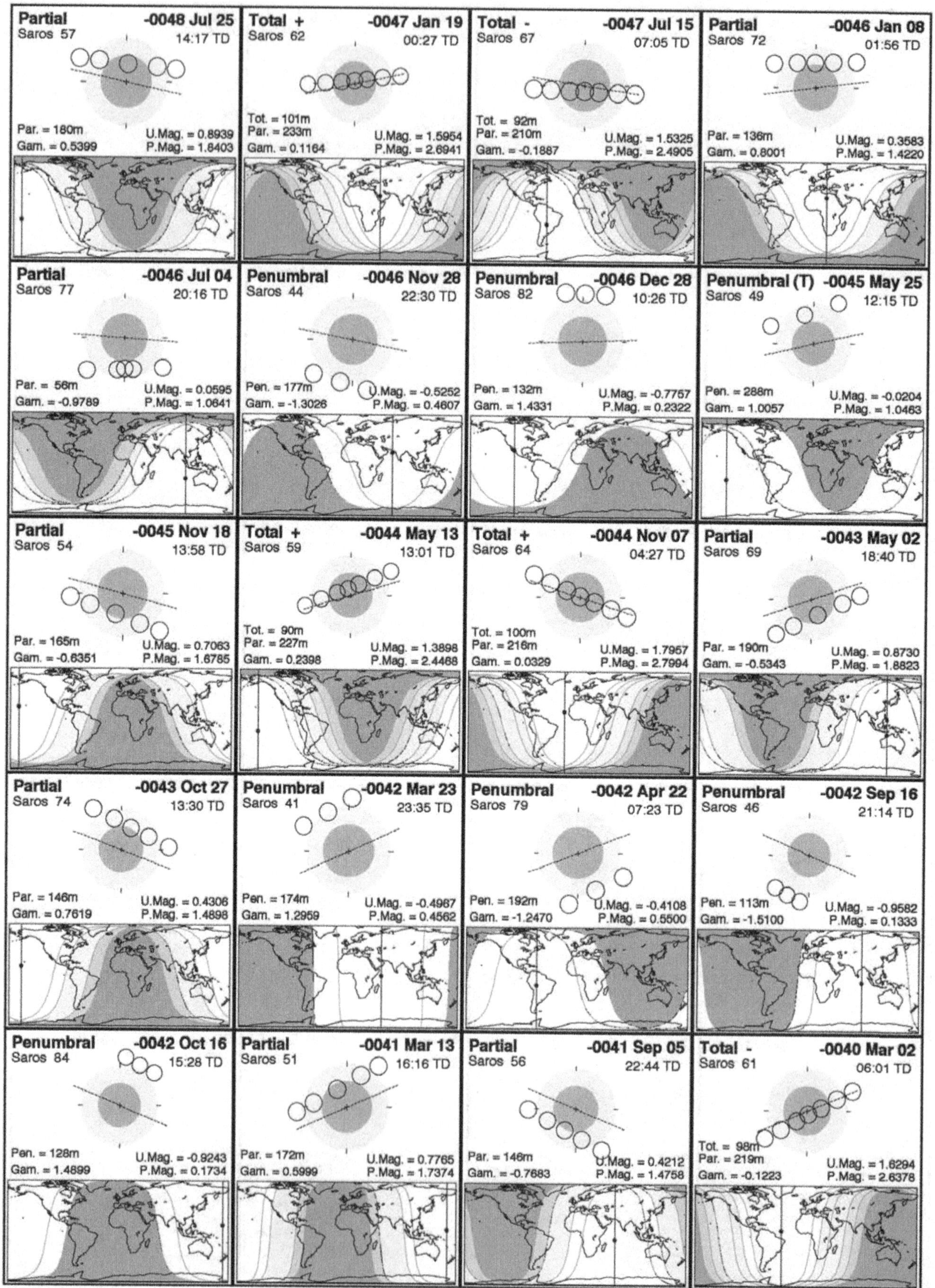

ΔT= 11021 s [= 03h04m] std.err. = ± 279 s [= ± 1.2°] Plate 236

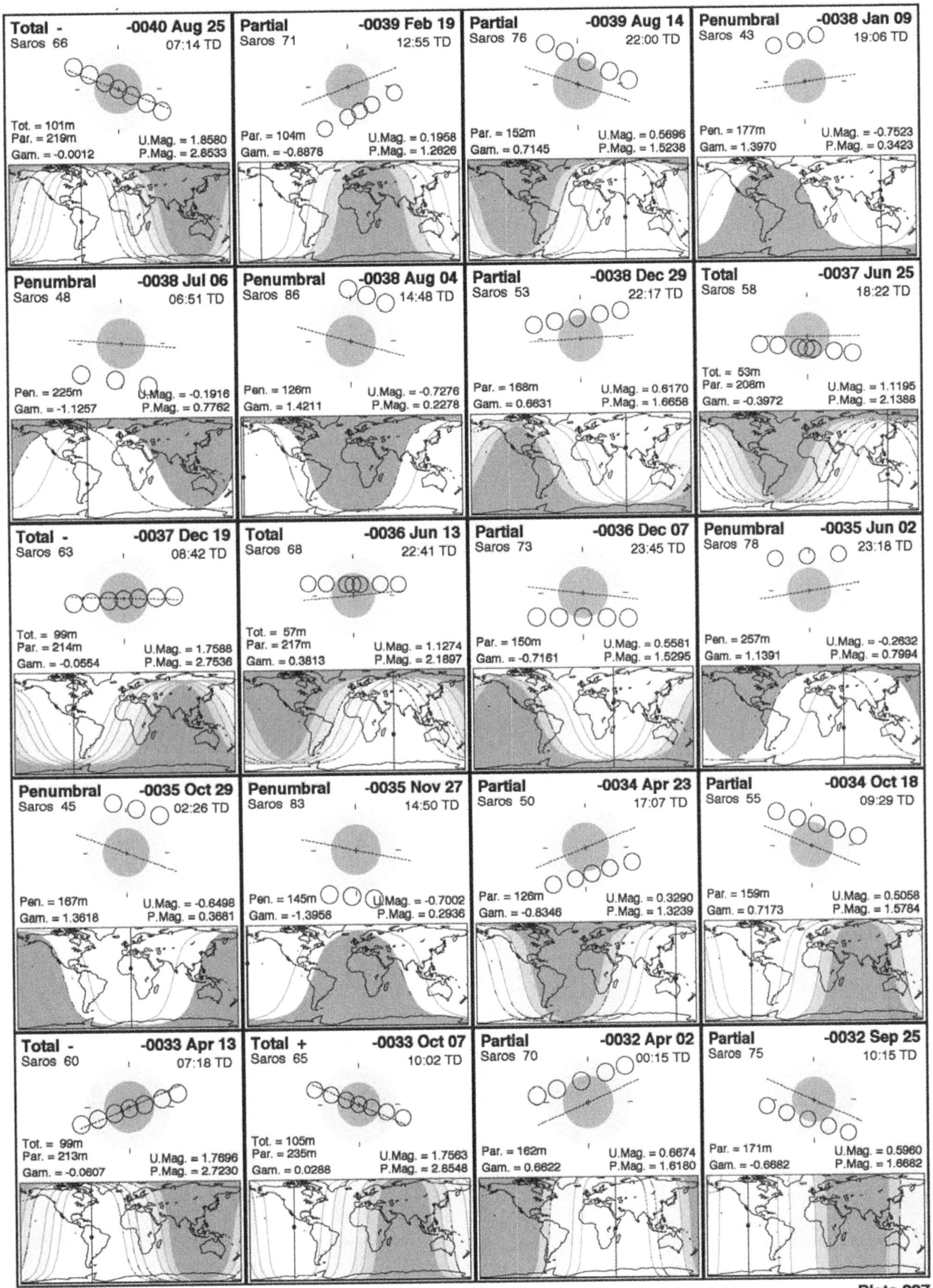

ΔT= 10937 s [= 03h02m] std.err. = ± 277 s [= ± 1.2°] Plate 237

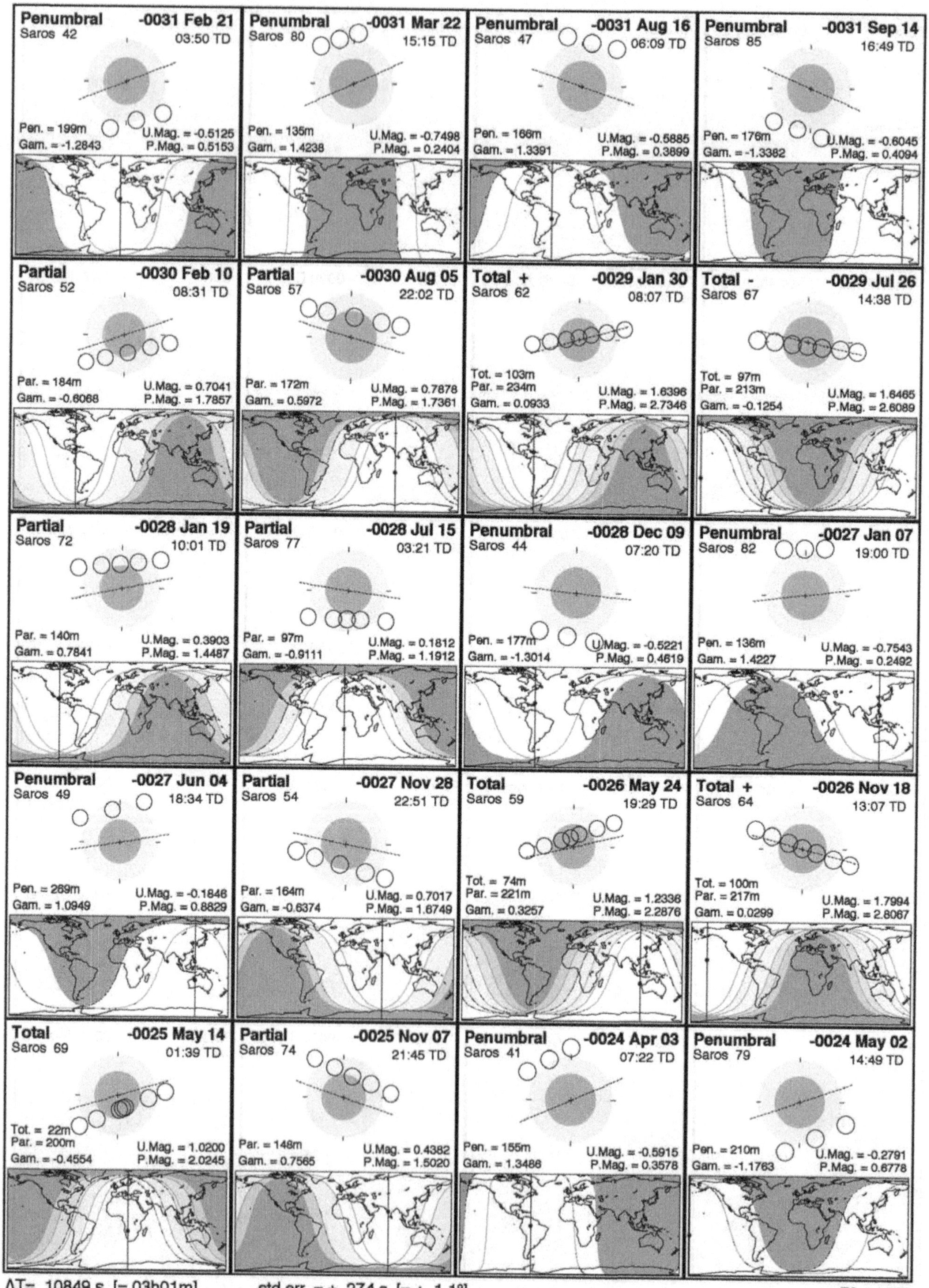

ΔT= 10849 s [= 03h01m] std.err. = ± 274 s [= ± 1.1°] **Plate 238**

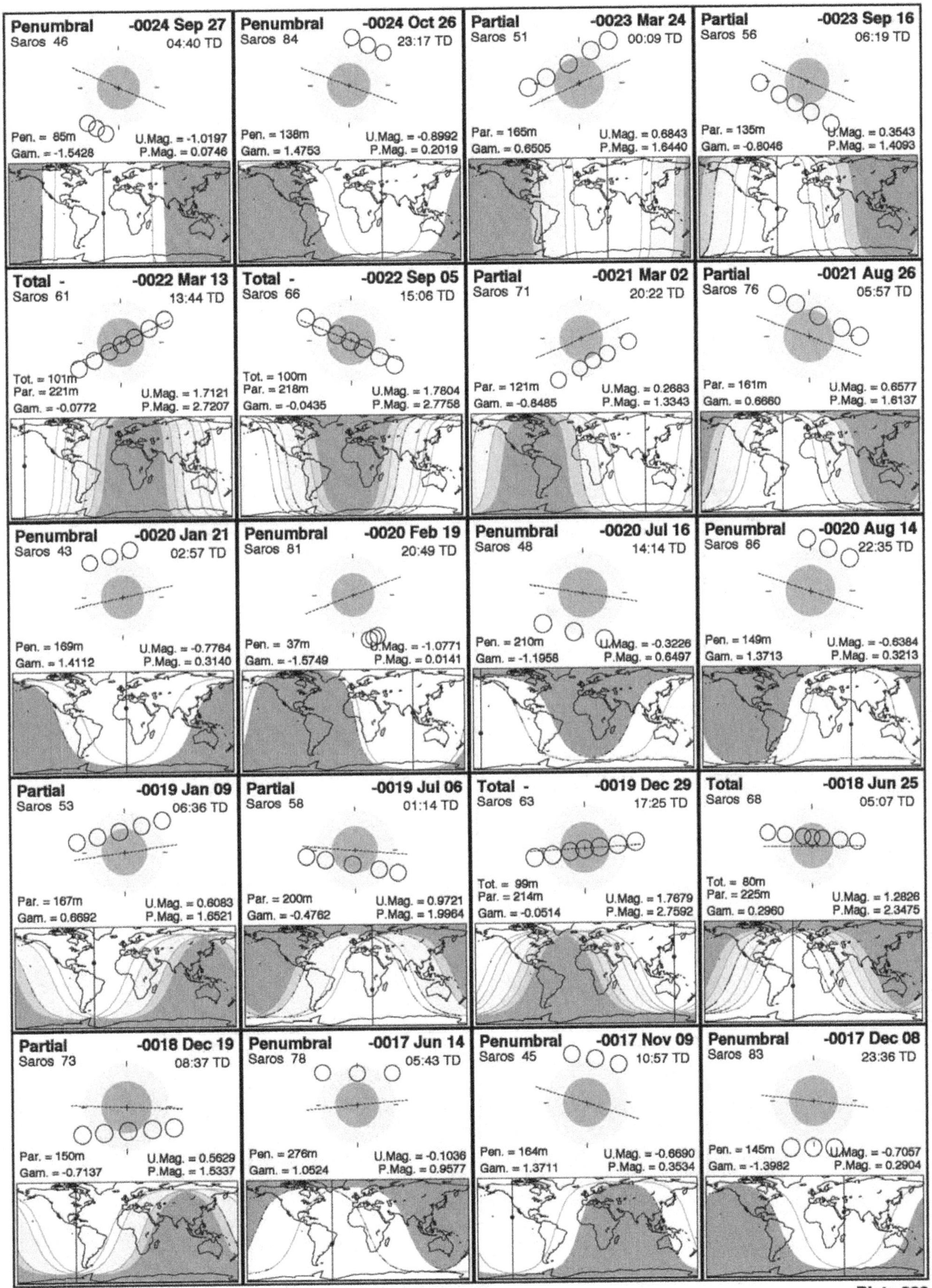

ΔT= 10771 s [= 03h00m] std.err. = ± 272 s [= ± 1.1°] **Plate 239**

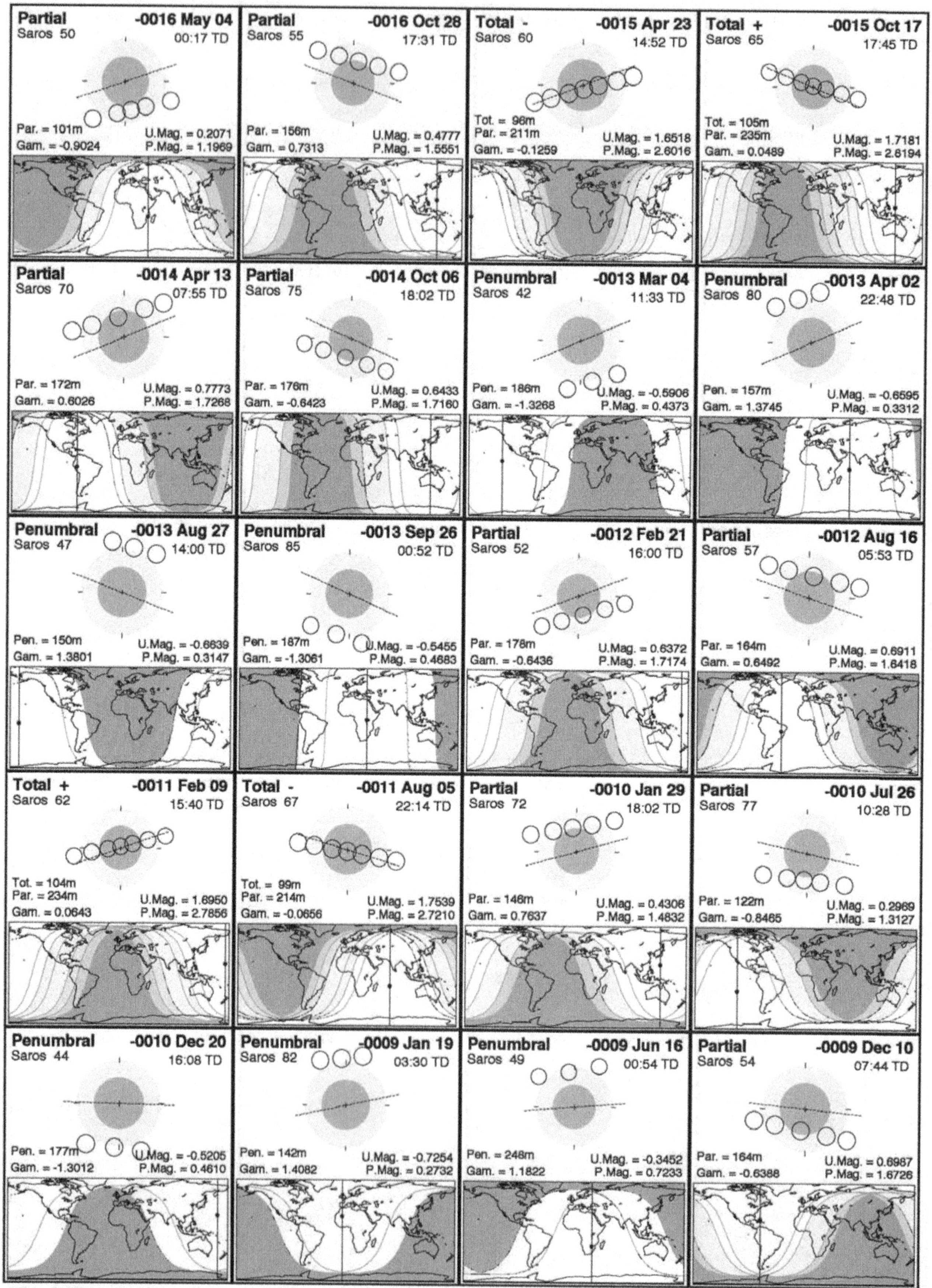

ΔT= 10693 s [= 02h58m] std.err. = ± 270 s [= ± 1.1°] **Plate 240**

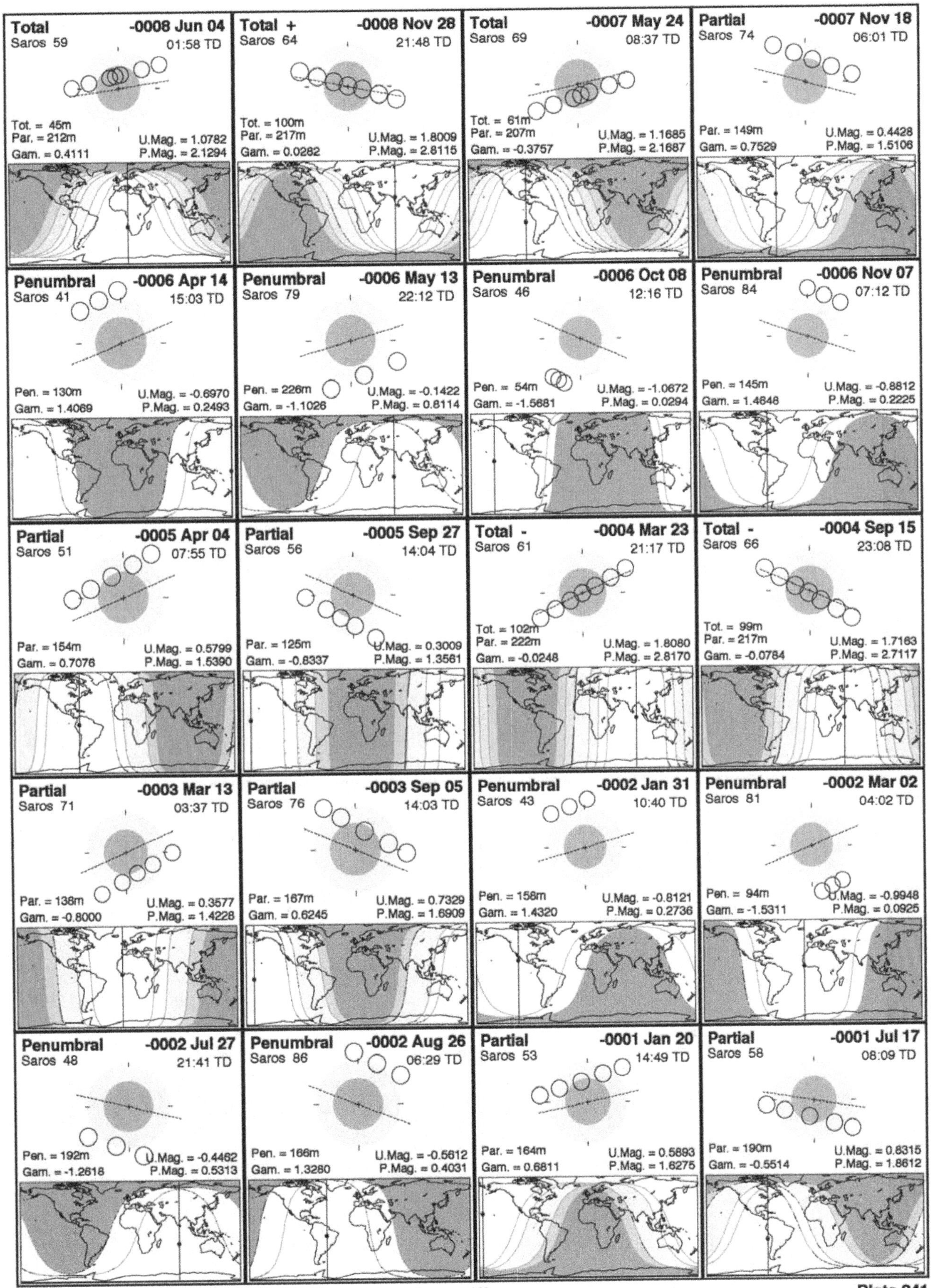

ΔT= 10611 s [= 02h57m] std.err. = ± 267 s [= ± 1.1°] **Plate 241**

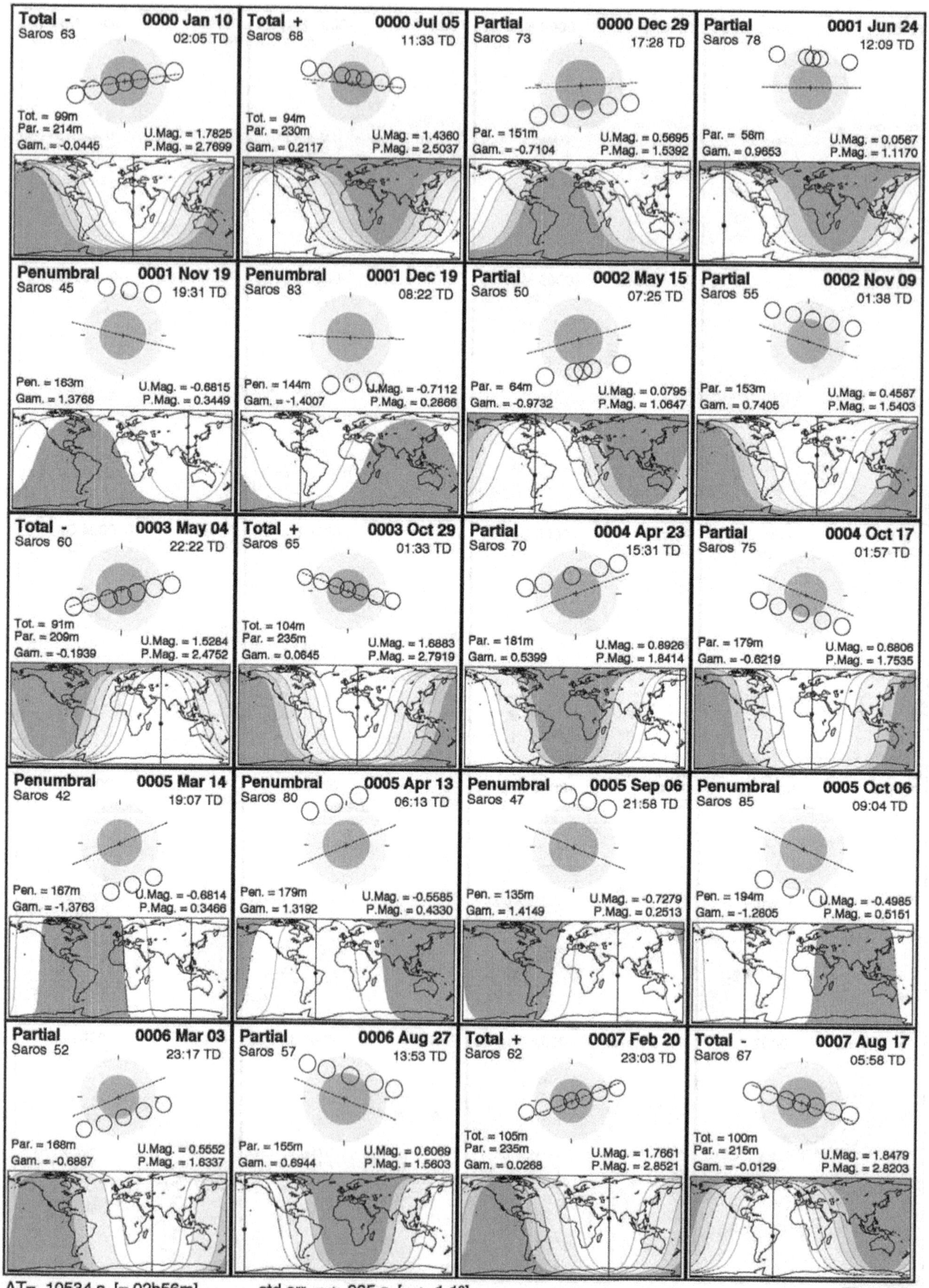

ΔT= 10534 s [= 02h56m] std.err. = ± 265 s [= ± 1.1°] **Plate 242**

Also Available:

Five Millennium Catalog of Lunar Eclipses: -1999 to +3000 Second Edition

The ***Five Millennium Catalog of Lunar Eclipses: -1999 to +3000*** contains a detailed catalog of all 12,064 eclipses of the Moon over 5,000 years. Tabulated data for each eclipse includes the calendar date, time of greatest eclipse, ΔT, Saros number, eclipse type, gamma, eclipse magnitude, and durations of penumbral, partial and total eclipse phases.

The statistics discussed include the lunar eclipse types by month and by century, extremes in eclipse magnitude, maximum durations of each phase, and eclipse duos.

Finally, the periodicity of lunar eclipses is investigated regarding the Saros cycle. Details of every Saros series in progress during the ***Five Millennium Catalog*** are given.

The Catalog is complements the 2-volume ***Five Millennium Canon of Lunar Eclipses*** containing a map of every eclipse.

Number of Pages: 280

For more information visit Astropixels Publishing:
http://astropixels.com/pubs/5MKLE2.html

The ***Five Millennium Catalog of Lunar Eclipses*** *is available from* ***Amazon.com***

Below is a page from the ***Five Millennium Catalog of Lunar Eclipses***

Five Millennium Catalog of Lunar Eclipses –1999 to +3000 (2nd Edition)
Fred Espenak and Jean Meeus

Cat Num	Canon Plate	Calendar Date	TD of Greatest Eclipse	ΔT s	Luna Num	Saros Num	Ecl. Type	QSE	Gamma	Pen. Mag.	Um. Mag.	Eclipse Phase Durations Pen. m	Par. m	Total m	Greatest in Zenith Lat.	Lng.
9721	487	2031 Jun 05	11:45:17	78	388	150	N	a-	1.4731	0.1292	-0.8199	95.6	-	-	21S	176W
9722	487	2031 Oct 30	07:46:45	79	393	117	N	-h	1.1773	0.7161	-0.3204	231.8	-	-	15N	121W
9723	487	2032 Apr 25	15:14:51	79	399	122	T	-a	-0.3558	2.2192	1.1913	342.4	211.2	65.5	14S	131E
9724	487	2032 Oct 18	19:03:40	79	405	127	T	-p	0.4169	2.0830	1.1028	315.4	195.9	47.1	10N	71E
9725	487	2033 Apr 14	19:13:51	80	411	132	T	t-	0.3954	2.1711	1.0944	361.2	215.0	49.2	9S	72E
9726	487	2033 Oct 08	10:56:23	80	417	137	T	p-	-0.2889	2.3057	1.3497	312.6	202.4	78.8	6N	167W
9727	487	2034 Apr 03	19:06:59	80	423	142	N	t-	1.1144	0.8545	-0.2274	265.4	-	-	5S	75E
9728	487	2034 Sep 28	02:47:37	81	429	147	P	a-	-1.0110	0.9911	0.0144	248.7	26.7	-	1N	44W
9729	487	2035 Feb 22	09:06:12	81	434	114	N	-a	-1.0367	0.9652	-0.0535	255.7	-	-	9N	133W
9730	487	2035 Aug 19	01:12:15	81	440	119	P	-t	0.9433	1.1507	0.1037	289.8	76.5	-	12S	17W
9731	487	2036 Feb 11	22:13:06	82	446	124	T	-p	-0.3110	2.2751	1.2995	316.1	201.9	74.5	14N	31E
9732	487	2036 Aug 07	02:52:32	82	452	129	T+	pp	0.2004	2.5266	1.4544	372.1	231.3	95.3	16S	41W
9733	487	2037 Jan 31	14:01:38	82	458	134	T	p-	0.3619	2.1803	1.2074	312.1	197.5	63.7	18N	153E
9734	487	2037 Jul 27	04:09:53	83	464	139	P	t-	-0.5582	1.8584	0.8095	340.8	192.4	-	20S	60W
9735	487	2038 Jan 21	03:49:52	83	470	144	N	a-	1.0710	0.8996	-0.1140	245.8	-	-	21N	54W
9736	487	2038 Jun 17	02:45:02	83	475	111	N	-a	1.3082	0.4422	-0.5275	176.3	-	-	22S	41W
9737	487	2038 Jul 16	11:35:56	84	476	149	N	a-	-1.2837	0.4999	-0.4952	192.4	-	-	23S	172W
9738	487	2038 Dec 11	17:45:00	84	481	116	N	-t	-1.1448	0.8046	-0.2892	258.5	-	-	22N	93E
9739	487	2039 Jun 06	18:54:25	84	487	121	P	-a	0.5460	1.8272	0.8846	296.7	179.3	-	22S	77E
9740	487	2039 Nov 30	16:56:28	85	493	126	P	-t	-0.4721	2.0418	0.9426	360.1	206.0	-	21N	104E
9741	488	2040 May 26	11:46:22	85	499	131	T-	p-	-0.1872	2.4938	1.5348	321.4	210.7	92.2	21S	177W
9742	488	2040 Nov 18	19:04:40	85	505	136	T+	p-	0.2361	2.4525	1.3974	353.6	220.4	87.8	20N	70E
9743	488	2041 May 16	00:43:03	86	511	141	P	t-	-0.9746	1.0747	0.0645	269.7	58.5	-	20S	12W

Astropixels Publications

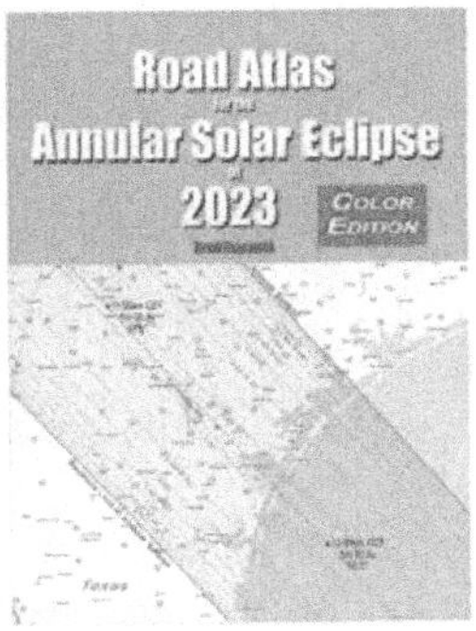

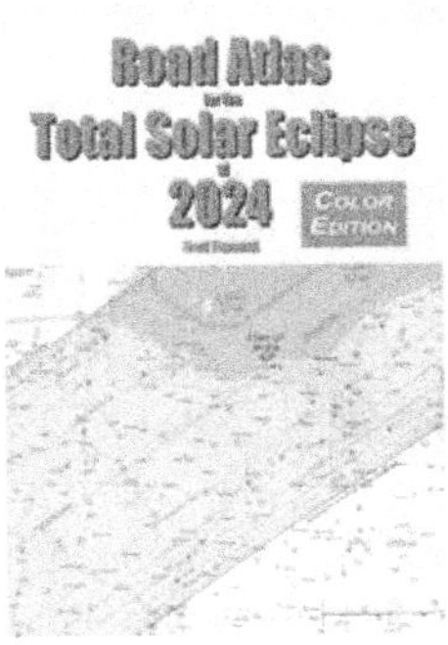

Below is a partial list of some of the books Fred Espenak has written through Astropixels Publishing:

21st Century Canon of Solar Eclipses

The complete guide to every solar eclipse occurring from 2001 tom 2100 (224 eclipses in all). It includes information and maps for all total, annular, hybrid, and partial eclipses. A special world atlas shows detailed full page maps of all central eclipse paths (total, annular and hybrid).

Road Atlas for the Annular Solar Eclipse of 2023

Detailed road maps of the entire eclipse path from the western USA, through Mexico, Central and South America. Information printed on the maps makes it easy to estimate the duration of annularity from any location in the eclipse path. This is the next annular solar eclipse visible from the USA.

Atlas of Central Solar Eclipses in the USA

When was the last total eclipse through the USA and when is the next? How often do they happen? What total eclipse tracks passed across the USA during the 17th, 18th, and 19th centuries, etc., and what states did they include? And how often is a total solar eclipse visible from each of the 50 states? The Atlas of Central Solar Eclipses in the USA answers all of these questions and more with hundreds of maps and tables.

Road Atlas for the Total Solar Eclipse of 2024

This book contains detailed road maps of the entire eclipse path from Mexico, through the USA and Canada. Information printed on the maps makes it easy to estimate the duration of totality from any location in the eclipse path. This is the next total solar eclipse visible from the USA.

21st Century Canon of Lunar Eclipses

The complete guide to every lunar eclipse occurring from 2001 tom 2100 (228 eclipses in all). It includes information and maps for all total, partial, and penumbral eclipses. The predictions use a new model for Earth's elliptical shadows.

For information on these books and more, visit *Astropixels Publishing*: *http://astropixels.com/pubs/index.html*

All Astropixels Publications are available from Amazon.com

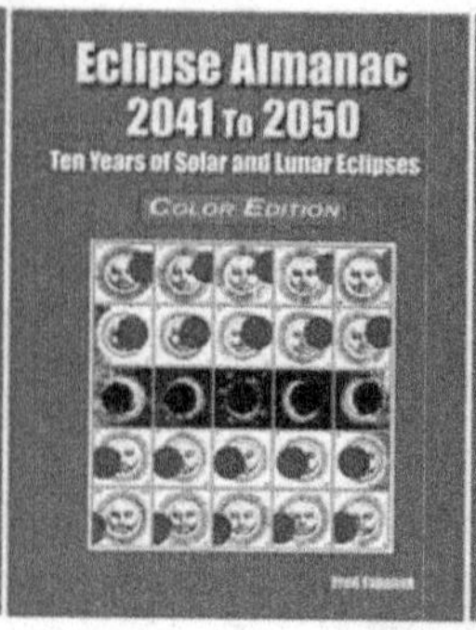

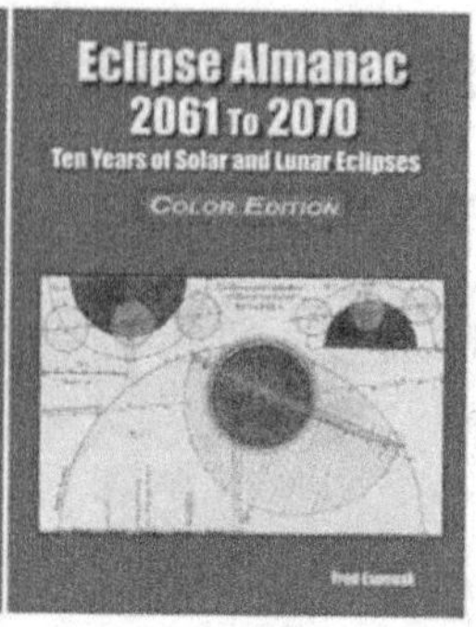

The ***Eclipse Almanac*** contains maps and diagrams of every solar and lunar eclipse over a ten-year period. This permits the reader to look ahead and easily determine when and where each of these events will be seen. Particular details about each eclipse are included as well as a 25-year table looking further into the future.

Section 1 covers solar eclipses, while Section 2 is devoted to lunar eclipses. Explanations are given for each type of eclipse, and the visual appearance of each one is included. Section 3 tabulates the date and time of the Moon's phases over a decade. New Moon and Full Moon phases that coincide with solar and lunar eclipses, are identified.

The ***Eclipse Almanac*** series consists of five volumes, each available in a color, black and white, and Kindle editions. Each volume covers a single decade:

1) Eclipse Almanac 2021 to 2030
2) Eclipse Almanac 2031 to 2040
3) Eclipse Almanac 2041 to 2050
4) Eclipse Almanac 2051 to 2060
5) Eclipse Almanac 2061 to 2070

For information on these books and more, visit *Astropixels Publishing*: *http://astropixels.com/pubs/index.html*

All Astropixels Publications are available from Amazon.com

The **Thousand Year Canon of Solar Eclipses 1501 to 2500** contains maps and data for each of the 2,389 solar eclipses occurring over the ten century period centered on the present era.

The primary content of the Thousand Year Canon is a comprehensive catalog listing the essential characteristics of each eclipse and an atlas of 2,389 maps depicting the geographic regions of visibility of each eclipse.

Available in either black & white or color editions.

The **Thousand Year Canon of Lunar Eclipses 1501 to 2500** contains diagrams, maps and data for each of the 2,424 lunar eclipses occurring over the ten century period centered on the present era.

Appendix A is a comprehensive catalog listing the essential characteristics of each eclipse. Appendix B is an atlas depicting the path of the Moon through Earth's shadows and maps identifying the geographic regions of visibility of each of the 2,424 eclipses.

Available in either black & white or color editions.

All Astropixels Publications are available from Amazon.com

Made in the USA
Monee, IL
08 November 2026

67816258R00177